BASIC PROBABILITY THEORY
AND APPLICATIONS

RAMAKANT KHAZANIE
HUMBOLDT STATE UNIVERSITY

Goodyear Publishing Company, Inc., Pacific Palisades, California

Library of Congress Cataloging in Publication Data

Khazanie, Ramakant
Basic probability theory and applications

(Goodyear mathematics series)
1. Probabilities I. Title.
QA273.K448 519.2 75-11186
ISBN 0-87620-101-X

ISBN 0-87620-101-X

Y-101X-6

Library of Congress Catalog Card Number: 75-11186

Current printing (last digit): 10 9 8 7 6 5 4 3 2 1

Printed in the United States of America

TO MY PARENTS

Contents

1 Building Blocks of the Probability Structure 1

Introduction 1

Section 1. Elements of Set Theory 4
1.1 The Notion of a Set, 1.2 Set Operations, 1.3 Venn Diagrams, 1.4 The Power Set, 1.5 Cartesian Product, 1.6 Limits of Monotone Sequences of Sets

Section 2. Sample Space and Events 15

Section 3. Sigma Fields 24
3.1 The Definition of a Sigma Field, 3.2 The Borel Field of the Real Line

2 Definition of Probability 29

Introduction 29

Section 1. The Axioms of the Theory of Probability 29

Section 2. Finite Sample Spaces 42

Section 3. Sampling and Counting Techniques 45

3 Conditional Probability and Independent Events 67

Introduction 67

Section 1. Conditional Probability 67

Section 2. The Total Probability Theorem and Bayes' Rule 78

Section 3. Independent Events 86

4 Random Variables 107

Introduction 107

Section 1. The Notion of a Random Variable 107

Section 2. The Distribution Function 115

2.1 The Definition of a Distribution Function, 2.2 Properties of a Distribution Function

Section 3. Classification of Random Variables 130

3.1 Discrete Random Variables, 3.2 Absolutely Continuous Random Variables, 3.3 Mixed Distributions, 3.4 Singular Distributions

5 Some Special Distributions 157

Introduction 157

Section 1. Discrete Distributions 157

1.1 Bernoulli Distribution, 1.2 The Binomial Distribution, 1.3 The Hypergeometric Distribution, 1.4 The Geometric Distribution 1.5 The Negative Binomial Distribution, 1.6 The Poisson Distribution

Section 2. Absolutely Continuous Distributions 175

2.1 The Uniform Distribution, 2.2 The Normal Distribution, 2.3 The Gamma Distribution, 2.4 The Cauchy Distribution, 2.5 The Laplace Distribution

6 Functions of a Random Variable 195

Introduction 195

Section 1. The Mathematical Formulation 195

Section 2. The Distribution of a Function of a Random Variable 198

2.1 The Discrete Case, 2.2 The Continuous Case

7 Expectation—A Single Variable 221

Introduction 221

Section 1. Definitions and Basic Results 221

1.1 The Definition of Expectation, 1.2 The Expectation of a Function of a Random Variable, 1.3 Some Properties of Expectation, 1.4 The Variance of a Random Variable, 1.5 Conditional Expectation

Section 2. Expectations of Some Special Distributions 242

8 Joint and Marginal Distributions 261

Introduction 261

Section 1. Joint Distributions 261

1.1 The Notion of a Random Vector, 1.2 The Definition of a Joint Distribution Function, 1.3 Properties of Joint Distribution Functions, 1.4 Classification of Joint Distributions

Section 2. Marginal Distributions 292
2.1 A General Discussion, 2.2 The Discrete Case, 2.3 The Absolutely Continuous Case

9 Conditional Distributions and Independent Random Variables 311

Introduction 311
Section 1. Conditional Distributions 311
1.1 Conditional Distribution Given an Event of Positive Probability, 1.2 Conditional Distribution Given a Specific Value
Section 2. Independent Random Variables 327
Section 3. More Than Two Random Variables 343
3.1 The Joint Distribution Function, 3.2 The Discrete Case, 3.3 The Absolutely Continuous Case

10 Functions of Several Random Variables 351

Introduction 351
Section 1. The Distrete Case 352
Section 2. The Continuous Case 357
2.1 Distribution of the Sum, 2.2 Distribution of the Product, 2.3 Distribution of the Quotient, 2.4 Distribution of the Maximum, 2.5 Distribution of the Minimum
Section 3. Miscellaneous Examples 373

11 Expectation–Several Random Variables 389

Introduction 389
Section 1. Expectation of a Function of Several Random Variables 389
1.1 The Definition, 1.2 Basic Properties of Expectation, 1.3 Covariance and the Correlation Coefficient, 1.4 The Variance of a Linear Combination, 1.5 The Method of Indicator Random Variables, 1.6 Bounds on the Correlation Coefficient
Section 2. Conditional Expectation 422
2.1 The Definition of Conditional Expectation, 2.2 The Expected Value of a Random Variable by Conditioning, 2.3 Probabilities by Conditioning

12 Generating Functions 435

Introduction 435
Section 1. The Moment Generating Function 436
1.1 The Definition, 1.2 How Moments are Generated, 1.3 Some Important Results, 1.4 Reproductive Properties
Section 2. The Factorial Moment Generating Function 452

13 Limit Theorems in Probability 459

Introduction 459

Section 1. Laws of Large Numbers 460
1.1 Chebyshev's Inequality, 1.2 The Laws

Section 2. Convergence in Distribution 472
2.1 The General Notion of Convergence in Distribution, 2.2 Central Limit Theorem

References 491

Solutions to Selected Problems 493

Tables 507

Index 513

Preface

I first conceived the idea of writing this book when I was on the faculty of the University of Vermont. Over the past ten years, I have had the benefit of feedback from a broad spectrum of students. This has included some exceptionally gifted students, as well as students who can gradually work their way to a high degree of proficiency in the subject. It is with this latter class of students foremost in mind that I set out to write this text. The organization of the material, the explanations (which at times may seem too detailed), the comments that follow at the end of some discussions—all prompted in large measure by my experience with such students.

The book is designed for the undergraduate student who is interested in acquiring a sound understanding of introductory probability with an eye to applications. It is hoped that it will be preparatory to many courses such as statistics, operations research, and communication theory, among others, and at the same time meet the needs of a student who intends to pursue serious study of mathematical probability. No previous knowledge of probability theory is assumed. The only prerequisite is a background in calculus which includes a good working knowledge of multiple integration and power series. It is not farfetched to say that any student equipped with this background could work his own way through the book from the first page to the last.

The book might be divided into four general parts. Part 1, consisting of Chapters 1 through 3, is concerned with the basic probability structure, and should be considered an integral part of any curriculum in a modern probability course. In my own teaching, I tend to make short work of the topic on combinatorials, thus allowing more time to discuss the properties of the probability measure and their interpretations. I find this particularly desirable since the student can be made to realize that some problems which seem inaccessible on first appearance can in fact be attempted in a routine way.

Part 2, consisting of Chapters 4 through 7, deals with single random variables, and part 3, consisting of Chapters 8 through 11, treats several random variables. There is a common theme adopted in the development of these two parts. I have found that considerable mileage can be gained if, before embarking on part 3, the student is made aware that the broad approach adopted in part 2 is maintained in part 3. This approach uses the following sequential developments: (1) mathematical description of a function defined on the sample space; (2) introduction of the concept of a distribution function along with its properties; (3) classification of random variables on the basis of the nature of the distribution function; (4) treatment of functions of random variables; and (5) the treatment of expectation. It is also helpful to make the student aware of how, for instance, the definitions of random vector, distribution functions, and so on mimic those in part 2.

Part 4 consists of Chapter 12, treating generating functions, and Chapter 13, which involves the study of limit theorems in probability.

There are a wide variety of illustrative examples throughout the text, and I consider this to be one of its strong points. Thorough explanations are given so that the student can read these on his own, thereby allowing the instructor more time to discuss questions of a more fundamental nature. Some of the examples contain important results which might be invoked as the theory is developed in subsequent chapters. Such examples are indicated by marking them with a solid circle. The reader would be well advised to familiarize himself with their essence. In Section 1 of Chapter 4, some examples are marked with an asterisk. These might be omitted at first reading, especially if the interest of the reader is nonmathematical.

No mathematical book at this level is complete without an adequate number of exercises. I have met this requirement by providing a wealth of exercises which touch on every aspect of the theory discussed in the text. They are given at the end of each section, and—as far as possible—are arranged in the order in which the material is developed in the particular section. No important results which are needed for further development of the subject are relegated to the exercises. The exercises are initiated with simple routine problems which increase in complexity, but none should be considered beyond the prowess of a diligent student. Hints are appended for problems which might call for undue insight.

The extent of coverage in a semester or a quarter will depend largely on the level and background of the students. Even so, it is inconceivable that the entire book would be covered in a one-semester offering. Based on my own experience, a one-semester course can be outlined as follows: most of the topics in Chapters 1 through 9, with varying degrees of emphasis; Section 1 of Chapter 11; a brief touch on the contents of Chapter 12; finally, Chebyshev's inequality and the central limit theorem in Chapter 13.

In a two-quarter course the pace could be more leisurely, allowing more time to discuss topics in Chapters 12 and 13. In this type of offering, the first quarter would consist of the first seven chapters, with the balance of the book to be covered in the second quarter.

There is no denying the fact that I have derived heavily from the existing literature on the subject, and I acknowledge my indebtedness to these sources. Some are mentioned at the end of the text; the interested reader might consult these to broaden his perspective.

On a personal note, during the typing of this manuscript the author lost, in the death of Paul Van Wulven, a good friend and a typist of uncanny genius. The final chapters were typed by Cheryl Richards, who, in spite of no previous experience with mathematical typing, rose to great heights and did a superb job.

Ramakant Khazanie

1 Building Blocks of the Probability Structure

INTRODUCTION

In every walk of life we are accustomed to making statements which are probabilistic in nature and which carry overtones of chance. For example, we might talk about the probability that a bus will arrive on time, or that a child-to-be-born will be a son, or that a player will hit a home run, or that the stock market will go up, and so on. A very conspicuous example is that of the weatherman who diligently reports to us "the probability of precipitation." The weatherman's prediction is likely to be greeted with skepticism. The reasoning might go as follows: "Well, the last time he predicted rain with 'a probability of 90 percent' it did not rain, so he can't be trusted." Are we justified in our criticism of the weatherman? Or did we misread the information that was given? It would seem reasonable to assume that the weatherman does serve some purpose; otherwise he would not have the job for very long.

To argue on behalf of the weatherman, let us consider the following situation: Suppose a box contains fifteen balls which are exactly alike in every respect except that fourteen of the balls are yellow and one is black. If a person who is blindfolded picks one ball from the box, is it possible to say with certainty whether the ball will be black or yellow? Definitely not! However, if one is called upon to predict the color, who wouldn't predict yellow? The weatherman's report, when he predicts rain with high probability, should be interpreted in the same spirit: You cannot be certain that it will rain, but you would be wise to be prepared for it.

Today, probability theory finds applications in diverse disciplines such as biology, economics, operations research, and astronomy, to mention only a few. A biologist might be interested in the distribution of bacteria in a culture, an economist in the economic forecasts, a production engineer in the inventory of a particular commodity, an astronomer in the distribution of stars in different galaxies, and so on. What is the characteristic feature in all the above phenomena? It is that they all lack a deterministic nature. Past information, no matter how voluminous, will not allow us to formulate a rule to determine precisely what will happen when the experiment is repeated. Phenomena of the above type, where the outcome of an experiment is of a fortuitous nature, are called *random phenomena.* The theory of probability involves their study.

A statement like "there is a 90 percent chance that the bus will arrive on time" is not uncommon. When we consider a particular bus, either it arrives on time or it does not. How does "90 percent chance" figure in our statement, and what does it mean? This statement is based on experience over a long period of time. If we were to keep track of all the buses arriving at the bus stop and were to find the ratio of the buses that arrived on time to the total number of buses that stopped at the stop, then the ratio would be approximately 90 percent. This is the empirical idea behind the statement, and it is this kind of rationale that is in general at the root of any probabilistic statement.

As another illustration of the empirical notion of probability, let us consider the experiment of tossing a coin. The results of 40 tosses are presented in Table 0.1. Notice how the entries in column 4 fluctuate in the initial stages (1, 1, 0.667, 0.750, 0.800, etc.) and then stabilize around 0.5 as the number of tosses increases. For example, on toss numbers 37, 38, 39, and 40, the ratios are respectively 0.486, 0.474, 0.487, and 0.475. All these ratios are fairly close to $\frac{1}{2}$. It is this approximation (close to $\frac{1}{2}$) inherent in the situation that lends support to the widely accepted statement that the chance of getting heads on a coin is $\frac{1}{2}$.

In our discussion, the emphasis should not be on the fact that the ratio approaches $\frac{1}{2}$; rather it should be on the fact that it stabilizes. In practice, it is this stable value that we take as the measure of probability. To see this, suppose we toss a coin repeatedly and, as in the above example, compute the ratio of the number of heads to the total number of tosses. Suppose after a stage the ratios fluctuate around $\frac{1}{3}$. In such a situation we would be prone to say, with a high degree of confidence, that the coin is a biased coin, and the chance of getting heads on this particular coin is 1 out of 3.

In general, then, the empirical notion of probability is that of "relative frequency," the ratio of the total number of occurrences of a situation to the total number of times the experiment is repeated. When the number of trials is large, the relative frequency provides a satisfactory measure of the probability associated with a situation of interest.

To bring in the historical perspective, the origin of probability theory goes back to the middle of the seventeenth century. There is general agreement that the early study started with Blaise Pascal (1623-1662). His interest and that of his contemporaries was occasioned by problems in gambling. As interest in natural sciences proliferated, so did demands on developing new laws of probability. Among prominent early contributors in this development may be mentioned the names of the nineteenth-century mathematicians Laplace, De Moivre, Gauss, and Poisson. Their work gave birth to what is known as the *classical theory of probability.*

TABLE 0.1

Toss number (k)	Outcomes	Total no. of heads (i)	Ratio i/k
1	H	1	1.000
2	H	2	1.000
3	T	2	0.667
4	H	3	0.750
5	H	4	0.800
6	T	4	0.667
7	H	5	0.714
8	T	5	0.625
9	T	5	0.556
10	T	5	0.500
11	H	6	0.545
12	T	6	0.500
13	T	6	0.462
14	T	6	0.429
15	H	7	0.467
16	H	8	0.500
17	T	8	0.471
18	T	8	0.444
19	T	8	0.421
20	H	9	0.450
21	T	9	0.429
22	H	10	0.455
23	H	11	0.478
24	H	12	0.500
25	T	12	0.480
26	T	12	0.462
27	T	12	0.444
28	H	13	0.464
29	H	14	0.483
30	T	14	0.467
31	H	15	0.484
32	T	15	0.469
33	T	15	0.455
34	T	15	0.441
35	H	16	0.457
36	H	17	0.472
37	H	18	0.486
38	T	18	0.474
39	H	19	0.487
40	T	19	0.475

H stands for heads and T for tails.

The classical theory was not equipped to handle problems of loaded dice or biased coins. Consideration of problems of this type led to the *axiomatic theory.* A giant step in this direction was taken as a result of pioneering work of A. Kolmogorov, who provided a sound mathematical foundation for the subject of probability.

How does one devise a set of axioms in a mathematical discipline? Of course, it is always possible to propose a system of axioms and derive results from them. The only requirement would be that the axioms be consistent. However, if the axioms are such that they have no connection with reality, then the whole exercise becomes purely academic and of very little practical use. To have any relevance at all, the axioms should be motivated by our experience in real world and should reflect it as closely as possible. In other words, the axioms should serve to provide an idealization of what we observe in nature. Such an axiomatic presentation governing the behavior of chance phenomena was given by A. Kolmogorov (1933) in *The Foundations of Probability Theory.* Our introduction to the subject will be mainly axiomatic. The classical theory will turn out to be a special case.

1. ELEMENTS OF SET THEORY

1.1 The Notion of a Set

Since the concepts of set theory are at the very heart of the treatment of probability, we shall begin by presenting a detailed outline of the basic ideas.

The word *set* is meant to indicate a gathering of objects which we choose to isolate because they have some common characteristic. However, any attempt to define a set is fraught with logical difficulties. For our purpose, we shall adopt the intuitively familiar notion and regard a set as a collection of objects, requiring only that it be possible to determine unambiguously whether or not any given object is a member of the collection.

When a complete list of the members of a set is given, it is customary to write them within braces, separated by commas. For example, a set that contains the four letters a, b, c, and d may be written as $\{a, b, c, d\}$. Since we are talking only about the objects in the set, there is no reason why the members should be written in any particular order. For example, the sets $\{a, b, c, d\}$, $\{d, b, a, c\}$, $\{d, c, a, b\}$ represent the same collection and consequently the same set. Hence *order is irrelevant in listing members of a set.* Also, no purpose is served by repeating the same element, so *only distinct elements are listed in a set.*

We shall denote sets by upper case letters and the elements of these sets by lower case letters. If x is in the set A we shall write

$$x \in A$$

to mean "x is an element of the set A."

If x is not in A, we write

$$x \notin A$$

to mean "x is not an element of the set A."

For example, if $A = \{\text{Tom, Dick, Mary}\}$, then Tom $\in A$ and Jack $\notin A$.

If a set has a large number of elements, it might be tedious or sometimes even impossible to specify the set by a complete list of its elements. A notation that has been devised to describe such sets is the so-called **set-builder notation**. If we represent a typical member of the set by x, then the set of all elements x such that x has some property, say property P, is written as

$$\{x \mid x \text{ has the property } P\}$$

For example, we could write the set of real numbers greater than 4 as

$$\{x \mid x \text{ a real number, } x > 4\}$$

As another example, the set consisting of pairs of real numbers where the first component is twice the second component can be written as

$$\{(u, v) \mid u, v \text{ real numbers, and } u = 2v\}$$

The braces should be read as "the set of all . . . ," and the vertical bar as "such that."

A very important set is the set of all the real numbers. We shall denote it by **R**. Using the set-builder notation

$$\mathbf{R} = \{x \mid x \text{ a real number, } -\infty < x < \infty\}$$

In the sequel we shall also need the following sets: Suppose a and b are real numbers with $a < b$. Then

$$[a, b] = \{x \mid x \in \mathbf{R},\ a \leqslant x \leqslant b\} \quad \text{(closed interval)}$$
$$(a, b) = \{x \mid x \in \mathbf{R},\ a < x < b\} \quad \text{(open interval)}$$
$$[a, b) = \{x \mid x \in \mathbf{R},\ a \leqslant x < b\}$$
$$(a, b] = \{x \mid x \in \mathbf{R},\ a < x \leqslant b\}$$
$$[a, \infty) = \{x \mid x \in \mathbf{R},\ a \leqslant x < \infty\}$$
$$(a, \infty) = \{x \mid x \in \mathbf{R},\ a < x < \infty\}$$
$$(-\infty, a] = \{x \mid x \in \mathbf{R},\ -\infty < x \leqslant a\}$$
$$(-\infty, a) = \{x \mid x \in \mathbf{R},\ -\infty < x < a\}$$

A set which has no elements in it is called the **empty set**, or the **void set**. We shall denote it by the symbol $\emptyset$. The following are examples of empty sets:

(*i*) *The set of all equilateral triangles with one angle equal to 45°*
(*ii*) *The set of all odd integers divisible by 4*
(*iii*) $\{(x, y) \mid x, y \in \mathbf{R},\ |x| + |y| < 0\}$
(*iv*) $\{x \mid x \in \mathbf{R},\ x^2 = -1\}$

A set which is not empty is called **nonempty**.

From now on we shall assume that for any given discussion there is a basic set S relevant to it. We shall refer to S as the **universal set** for the discussion and work strictly within the scope of this set. The universal set will vary from problem to problem, but within a given problem it will be fixed and will act as the basic set from which all other sets are constructed.

Suppose A and B are two sets which consist of members of S. Then B is called a **subset** of A, written $B \subset A$ (or equivalently as $A \supset B$), if every member of B is also in A. Thus we have the following definition

$$B \subset A \qquad \text{means that if } \; x \in B \; \text{ then } \; x \in A$$

B is called a **proper subset** of A if B is a subset of A and there is some member of A which is not in B. We give the following examples:

(*i*) {Tom, Harry} ⊂ {Tom, Dick, Harry}

(*ii*) The set of all the equilateral triangles is a subset of the set of all the isosceles triangles; in fact, a proper subset.

(*iii*) The set consisting of all the spades in a bridge deck of 52 cards is a subset of the set of all the black cards.

(*iv*) $\{(x, y) \mid x, y \in \mathbf{R}, \text{ and } x = y\} \subset \{(x, y) \mid x, y \in \mathbf{R}\}$

We observe in passing that if $B \subset A$, then there is nothing in B which is not in A. In view of this comment, we have

$$\emptyset \subset A \qquad \text{for } \textit{any} \text{ set } A$$

since the empty set has nothing in it that A does not have.

We shall next define what is meant by the **equality of sets.** Two sets A and B are said to be equal, written $A = B$, if A and B represent the same set; that is, they contain the same members. This will be the case if *every element in A is also in B and conversely.* Thus

$$A = B \qquad \text{means } \; A \subset B \; \text{ and } \; B \subset A$$

For example, if A represents the set of triangles with all sides equal, and B the set of triangles with all angles equal, then $A = B$.

If set A is not equal to set B, we write $A \neq B$.

For instance, if $A = \{a, b, c\}$ and $B = \{a, b, e, d\}$, then $A \neq B$.

1.2 Set Operations

We shall now consider how sets can be combined to produce new sets. There are three main operations: set intersection, set union, and set difference.

Intersection of sets

The *intersection* of two sets A and B, written $A \cap B$, is the set of elements that are common to A and B; that is,

$$A \cap B = \{x \mid x \in A \;\; \textit{and} \;\; x \in B\}$$

The notation AB, juxtaposing the two letters A and B, is also used in place of $A \cap B$. We shall find this latter notation more convenient for our purpose.

Consider the following examples:

(*i*) If $A = \{a, b, c, d\}$ and $B = \{a, d, c, f, p\}$ then $AB = \{a, c, d\}$.

(*ii*) If A is the set of people who wear a tie, and B the set of people who wear a jacket, then AB represents the set of people who wear a tie *and* a jacket.

(*iii*) Let

$$A = \{(x, y) \mid x, y \in \mathbf{R},\ x \geqslant 3\}$$
$$B = \{(x, y) \mid x, y \in \mathbf{R},\ y \geqslant -1\}$$

then

$$AB = \{(x, y) \mid x, y \in \mathbf{R},\ x \geqslant 3 \textit{ and } y \geqslant -1\}$$

Two sets A and B are said to be **disjoint** if they have no elements in common; that is, if $AB = \emptyset$. More generally, if $A_1, A_2, \ldots$ is a collection of sets, then we say that the sets are **pairwise disjoint** if, whenever $i \neq j$, A_i and A_j are disjoint.

The concept of intersection carries over to any arbitrary collection of sets and, in particular, to a countable collection of sets. Thus, if $A_1, A_2, \ldots$ represents a countable collection of subsets of S, their intersection, symbolically $\bigcap_{i=1}^{\infty} A_i$, is the set whose elements belong to every set A_i; that is,

$$\bigcap_{i=1}^{\infty} A_i = \{x \mid x \in A_i \quad \text{for } \textit{every } i\}$$

As an example, if $A_i = \left\{x \mid 2 - \frac{1}{i} < x < 5 + \frac{1}{i}\right\}$, $i = 1, 2, \ldots$, then it can be shown that

$$\bigcap_{i=1}^{\infty} \left\{x \mid 2 - \frac{1}{i} < x < 5 + \frac{1}{i}\right\} = \{x \mid 2 \leqslant x \leqslant 5\} = [2, 5]$$

Notice that 5 is in the intersection since $5 \in A_i$ for every i. For the same reason, 2 is also in the intersection.

As another example, we have

$$\bigcap_{i=1}^{\infty} \left(a - \frac{1}{i}, a\right] = \{a\}$$

a set consisting of a singleton.

From the definition of intersection it is obvious that

$$AB = BA \qquad (\textit{commutativity})$$
$$(AB)C = A(BC) \qquad (\textit{associativity})$$

Union of sets

If A and B are two sets, then $A \cup B$, called the *union* of A and B, is the set of elements which are in A or in B (or in both); that is,

$$A \cup B = \{x \mid x \in A \quad or \quad x \in B\}$$

In the sequel, whenever we say "either A or B" it will be used in the sense of "either A or B or both."

Consider the following examples:

(*i*) If $A = \{a, b, c, d\}$ and $B = \{a, d, c, f, p\}$, then $A \cup B = \{a, b, c, d, f, p\}$.

(*ii*) If A is the set of people who wear a tie, and B the set of people who wear a jacket, then $A \cup B$ is the set of people who wear a tie *or* a jacket (or both).

(*iii*) Let

$$A = \{(x, y) \mid x, y \in \mathbf{R},\ x \geqslant 3\}$$
$$B = \{(x, y) \mid x, y \in \mathbf{R},\ y \geqslant -1\}$$

then

$$A \cup B = \{(x, y) \mid x, y \in \mathbf{R},\ x \geqslant 3 \ \ or \ \ y \geqslant -1\}$$

Thus $(4, 2), (2, -\frac{1}{2}), (5, 2)$ are all members of $A \cup B$. However, $(2, -3) \notin A \cup B$.

If $A_1, A_2, A_3, \ldots$ is a countable collection of sets, then their union, symbolically $\bigcup_{i=1}^{\infty} A_i$, is the set whose elements belong to either A_1, or A_2, or A_3, and so on; that is, it represents the set whose elements belong to at least one of the sets $A_1, A_2, \ldots$. Hence

$$\bigcup_{i=1}^{\infty} A_i = \{x \mid x \in A_i \quad \text{for some } i\}$$

As an example, consider a collection of closed intervals A_i, where $A_i = \left\{x \mid 2 \leqslant x \leqslant 6 - \frac{1}{i}\right\}, i = 1, 2, \ldots$. It can be shown that

$$\bigcup_{i=1}^{\infty} \left\{x \mid 2 \leqslant x \leqslant 6 - \frac{1}{i}\right\} = \{x \mid 2 \leqslant x < 6\} = [2, 6)$$

Notice that 6 is not in the union, since 6 is not a member of any of the sets A_i, $i = 1, 2, \ldots$.

As another example, if $A_i = \left\{x \mid 2 + \frac{1}{i} \leqslant x \leqslant 6 - \frac{1}{i}\right\}$, then $\bigcup_{i=1}^{\infty} A_i = (2, 6)$, an open interval.

From the definition of union it is obvious that

$$A \cup B = B \cup A \qquad \textit{(commutativity)}$$
$$(A \cup B) \cup C = A \cup (B \cup C) \qquad \textit{(associativity)}$$

Set difference

Suppose A and B are two sets. The *set difference* $A - B$ is the set whose elements are in A but not in B. Thus we have

$$A - B = \{x \mid x \in A \quad \text{and} \quad x \notin B\}$$

Observe that $B - A = \{x \mid x \in B \text{ and } x \notin A\}$. Hence, in general, $A - B \neq B - A$. Consider the following examples:

(*i*) If $A = \{a, b, c, d\}$ and $B = \{a, d, c, f, p\}$, then $A - B = \{b\}$ and $B - A = \{f, p\}$, and, consequently, $A - B \neq B - A$.

(*ii*) If A is the set of people who wear a tie, and B the set of people who wear a jacket, then $A - B$ is the set of people who wear a tie but not a jacket, and $B - A$ is the set of people who wear a jacket but not a tie.

In particular, $S - B$, where S is the universal set, is called the **complement** of B (with respect to S). We shall write it symbolically as B'. Thus

$$B' = \{x \mid x \in S \quad \text{and} \quad x \notin B\}$$

In finding the complement of a set, it is important to be fully aware of the underlying universal set. Thus, for example, if $S = \{a, b, c, d\}$, then $\{a, c\}' = \{b, d\}$; whereas if $S = \{a, b, c, d, e\}$, then $\{a, c\}' = \{b, d, e\}$.

We have the following useful relations involving complements of sets.

1. $(A')' = A$ for any set A.
2. $S' = \emptyset$ and $\emptyset' = S$.
3. $A \cup A' = S$ for any set A.
4. $AA' = \emptyset$ for any set A.

The proofs of these identities follow immediately from the definition of a complement and will be left to the exercises.

1.3 Venn Diagrams

The set operations can be illustrated by means of **Venn diagrams**. The idea is to represent sets as geometric contours. The shaded portions in Figure 1.1(a), (b), (c), and (d) represent respectively $A \cup B$, AB, $A - B$, and B'.

The representation of sets pictorially as Venn diagrams serves as a powerful tool for establishing informally some basic identities in set theory. For example, on the basis of Figure 1.2 we can see that

$$A(B \cup C) = (AB) \cup (AC)$$

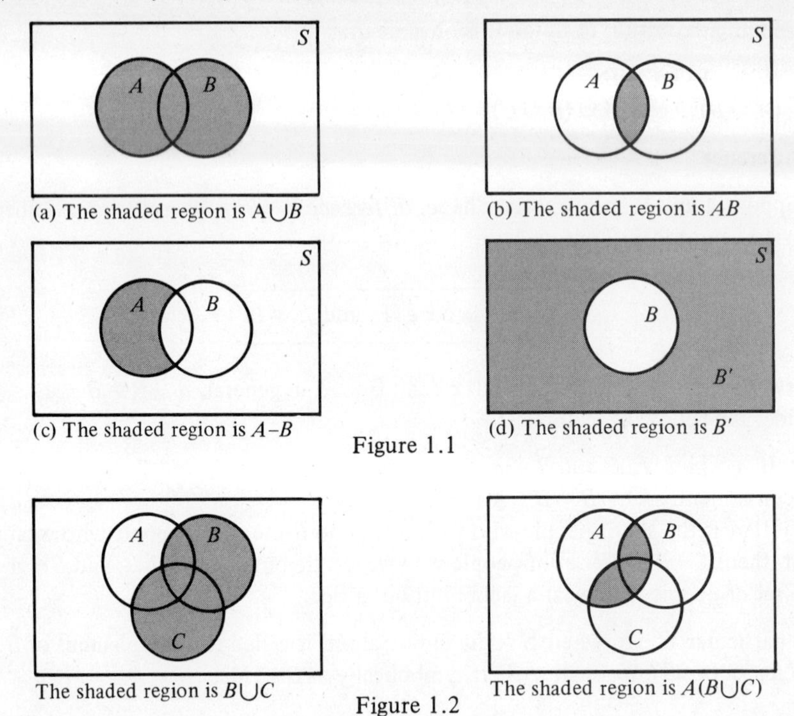

(a) The shaded region is $A \cup B$

(b) The shaded region is AB

(c) The shaded region is $A-B$

(d) The shaded region is B'

Figure 1.1

The shaded region is $B \cup C$

The shaded region is $A(B \cup C)$

Figure 1.2

This is called the **distributive law** (intersection distributes over the union). In fact, the distributive law is more general. We have, for example,

$$A(B_1 \cup B_2 \cup \ldots) = (AB_1) \cup (AB_2) \cup \ldots$$

Also we have

$$A \cup (B_1 \cap B_2 \cap \ldots) = (A \cup B_1) \cap (A \cup B_2) \cap \ldots$$

1.4 The Power Set

The **power set** of a set A is the set whose members are all the subsets of A.

Since $\emptyset$ is a subset of any set A, it is always a member of any power set. Also, since A is a subset of itself, it is also included in the power set. We shall denote the power set of A by $\mathcal{P}(A)$.

As an example, suppose $A = \{x, y\}$. Then $\mathcal{P}(A) = \{\emptyset, \{x\}, \{y\}, \{x, y\}\}$. There are four subsets of A.

The best way to write down the members of $\mathcal{P}(A)$ is to list the empty set first, then the subsets of A consisting of singletons, then the doubletons, and so on, eventually writing down the set A itself.

As another example, if $A = \{\text{Tom, Dick, Harry}\}$ then

$$\mathcal{P}(A) = \{\emptyset, \{\text{Tom}\}, \{\text{Dick}\}, \{\text{Harry}\}, \{\text{Tom, Dick}\}, \{\text{Tom, Harry}\}, \{\text{Dick, Harry}\}, \{\text{Tom, Dick, Harry}\}\}$$

There are eight members in $\mathcal{P}(A)$.

In general, if A has n elements, then $\mathcal{P}(A)$ has 2^n members. (We shall prove this result in Chapter 2.) It is for this reason that the power set of A is often written symbolically as 2^A.

Note: 2^A is merely a symbol and has no numerical value.

1.5 Cartesian Product

A pair of elements a and b, where a is referred to as the first element and b as the second element, is called an **ordered pair** and is written (a, b).

The pair (a, b) should be distinguished from the pair (b, a) where b is the first element. Two ordered pairs (a, b) and (c, d) are said to be equal if and only if their corresponding components are equal, that is, $a = c$ and $b = d$.

For any sets A and B, the **Cartesian* product** of A and B, written $A \times B$, is the set of all ordered pairs of elements where the first entry is from A and the second entry from B. Thus

$$A \times B = \{(a, b) \mid a \in A, \quad b \in B\}$$

For example, if $A = \{a, b, c\}$ and $B = \{p, q\}$, then

$$A \times B = \{(a, p), (a, q), (b, p), (b, q), (c, p), (c, q)\}$$

and

$$B \times A = \{(p, a), (q, a), (p, b), (q, b), (p, c), (q, c)\}$$

Note that in general the Cartesian product of A and B, $A \times B$, is not the same as the Cartesian product of B and A, $B \times A$. Thus, unlike set union and intersection, *the Cartesian product of sets is not commutative.*

What happens if we take the Cartesian product of a set with the empty set? For any set A, we can interpret $A \times \emptyset$ to be the set of ordered pairs whose first entry is from A but which has no second entry. The only sense we can give this is to take $A \times \emptyset = \emptyset$. Similarly $\emptyset \times A = \emptyset$. Thus

$$A \times \emptyset = \emptyset \times A = \emptyset \qquad \text{for any set } A$$

The Cartesian product of more than two sets is defined analogously. Thus, the Cartesian product of $A_1, A_2, \ldots, A_n$ is

$$A_1 \times A_2 \times \ldots \times A_n = \{(x_1, x_2, \ldots, x_n) \mid x_i \in A_i, \ i = 1, 2, \ldots, n\}$$

If $A_1 = A_2 = \ldots = A_n = A$, we denote the Cartesian product $A \times A \times \ldots \times A$ by A^n.

*Named after René Descartes, a French mathematician who lived from 1596 to 1650.

Example 1.1. Let $A = \{1, 2\}, B = \{a\}, C = \{b, c\}$. Find: (*a*) $A \times (B \times C)$ (*b*) $(A \times B) \times C$ (*c*) $A \times B \times C$

Solution. We have

$$\begin{aligned} (a)\ A \times (B \times C) &= \{1, 2\} \times \{(a, b), (a, c)\} \\ &= \{(1, (a, b)), (1, (a, c)), (2, (a, b)), (2, (a, c))\} \\ (b)\ (A \times B) \times C &= \{(1, a), (2, a)\} \times \{b, c\} \\ &= \{((1, a), b), ((1, a), c), ((2, a), b), ((2, a), c)\} \\ (c)\ A \times B \times C &= \{(1, a, b), (1, a, c), (2, a, b), (2, a, c)\} \end{aligned}$$

■

We see from Example 1.1 that $A \times (B \times C)$, $(A \times B) \times C$ and $A \times B \times C$ represent distinct sets. Nevertheless, notice that they have the same number of elements, namely, four. In general, if $\mathbf{n}(A)$ represents the number of elements in A, then the following is true.

$$\mathbf{n}(A_1 \times A_2 \times \ldots \times A_n) = \mathbf{n}(A_1) \cdot \mathbf{n}(A_2) \cdot \ldots \cdot \mathbf{n}(A_n)$$

We shall establish this result in Chapter 2.

An important fact is that the Cartesian product operation is distributive over the union operation and over the intersection operation. That is,

$$A \times (B \cup C) = (A \times B) \cup (A \times C)$$

and

$$A \times (B \cap C) = (A \times B) \cap (A \times C)$$

We shall not prove these results, but will verify them for specific sets A, B, and C in the following example.

Example 1.2. For the sets $A = \{a, b\}$, $B = \{a, b, c\}$, $C = \{b, d\}$, show that (*a*) $A \times (B \cup C) = (A \times B) \cup (A \times C)$, and (*b*) $A \times (B \cap C) = (A \times B) \cap (A \times C)$.

Solution

(*a*) Since $B \cup C = \{a, b, c, d\}$, we have

$$A \times (B \cup C) = \{(a, a), (a, b), (a, c), (a, d), (b, a), (b, b), (b, c), (b, d)\}$$

Now

$$A \times B = \{(a, a), (a, b), (a, c), (b, a), (b, b), (b, c)\}$$

and

$$A \times C = \{(a, b), (a, d), (b, b), (b, d)\}$$

Therefore,

$$\begin{aligned} (A \times B) \cup (A \times C) &= \{(a, a), (a, b), (a, c), (a, d), (b, a), (b, b), (b, c), (b, d)\} \\ &= A \times (B \cup C) \end{aligned}$$

(*b*) We have already found $A \times B$ and $A \times C$ in (*a*). Hence

$$(A \times B) \cap (A \times C) = \{(a, b), (b, b)\}$$

Next, since $B \cap C = \{b\}$, we get

$$A \times (B \cap C) = \{(a, b), (b, b)\} = (A \times B) \cap (A \times C)$$

1.6 Limits of Monotone Sequences of Sets

An infinite collection of sets $A_1, A_2, \ldots$ is called a sequence of sets and is often written as $\{A_n\}$. If

$$A_1 \subset A_2 \subset A_3 \ldots$$

then the sequence is said to be **monotone nondecreasing**. In a more suggestive way, such sequences are also called **expanding** sequences. On the other hand, if

$$A_1 \supset A_2 \supset A_3 \ldots$$

then the sequence is said to be **monotone nonincreasing**, or **contracting**.

Nondecreasing and nonincreasing sequences are together called *monotone* sequences. As examples of monotone sequences, the sequence $\{A_n\}$, where $A_n = \left[2 + \frac{1}{n}, 5 - \frac{1}{n}\right]$, is a nondecreasing sequence; whereas the sequence $\{A_n\}$ with $A_n = \left(2 - \frac{1}{n}, 5 + \frac{1}{n}\right)$ is a nonincreasing sequence.

If $\{A_n\}$ is a nondecreasing sequence of sets, then the limit of this sequence, $\lim_{n\to\infty} A_n$, is defined to be $\bigcup_{n=1}^{\infty} A_n$. Hence, *for a nondecreasing sequence,*

$$\lim_{n\to\infty} A_n = \bigcup_{n=1}^{\infty} A_n$$

For example, if $A_n = \left[2 + \frac{1}{n}, 5 - \frac{1}{n}\right]$, then

$$\lim_{n\to\infty} A_n = \bigcup_{n=1}^{\infty} A_n = \bigcup_{n=1}^{\infty} \left[2 + \frac{1}{n}, 5 - \frac{1}{n}\right] = (2, 5)$$

If $\{A_n\}$ is a nonincreasing sequence of sets, then the limit of this sequence is defined to be $\bigcap_{n=1}^{\infty} A_n$. Hence, *for a nonincreasing sequence,*

$$\lim_{n\to\infty} A_n = \bigcap_{n=1}^{\infty} A_n$$

For example, if $A_n = \left(2 - \frac{1}{n}, 5 + \frac{1}{n}\right)$, then, since the sequence is nonincreasing,

$$\lim_{n\to\infty} A_n = \bigcap_{n=1}^{\infty} A_n = \bigcap_{n=1}^{\infty} \left(2 - \frac{1}{n}, 5 + \frac{1}{n}\right) = [2, 5]$$

EXERCISES–SECTION 1

1. Suppose $S = \{1, 2, 3, \ldots, 12\}$, $A = \{1, 3, 4, 6, 10, 11\}$, $B = \{2, 3, 5, 6, 9, 10, 12\}$. Find:

   (a) $A \cup B$ (b) AB (c) $A - B$
   (d) $B - A$ (e) A'

2. Suppose I^+ is the set of positive integers and A, B, C are three subsets of it defined as follows:

$$A = \{2k \mid k \in I^+\}$$
$$B = \{2k - 1 \mid k \in I^+\}$$
$$C = \{3k \mid k \in I^+\}$$

Describe the sets in words.

3. Let A, B, C be as described in exercise 2. Describe in words the following sets:

   (a) $A \cup B$ (b) $(A \cup C)B$ (c) $A \cup C$
   (d) $A \cup (BC)$ (e) AB (f) BC

4. Use Venn diagrams to establish the following:

   (a) If $A \subset B$ and $B \subset C$, then $A \subset C$.
   (b) $A(B - A) = \emptyset$.
   (c) If $A \subset B$, then $A = AB$.
   (d) If $AB = \emptyset$ and $C \subset B$, then $AC = \emptyset$.
   (e) If $AB = \emptyset$, then $A'B = B$.
   (f) $(A \cup B) - B = A - AB$.
   (g) $(A - AB) \cup B = A \cup B$.
   (h) $(A \cup B) - AB = (AB') \cup (A'B)$.

5. State whether the following are true or false for all sets A and B.

   (a) If $A \cup B = \emptyset$, then $A = B = \emptyset$.
   (b) If $AB = \emptyset$, then $A = B = \emptyset$.
   (c) If $A = \emptyset$ or $B = \emptyset$, then $AB = \emptyset$.
   (d) If $A \subset B$, then $A \cup B = A$.

6. Prove that $A \subset B$ if and only if $B' \subset A'$.

7. Prove the following identities:

   (a) $(A')' = A$ for any set A.
   (b) $S' = \emptyset$ and $\emptyset' = S$.
   (c) $A \cup A' = S$ for any set A.
   (d) $AA' = \emptyset$ for any set A.

8. If S is the universal set, and if A is a subset of S, find:

   (a) SA (b) $(\emptyset \cup A)'$ (c) $S'A'$
   (d) SA' (e) $(\emptyset A)'$ (f) $(A' \cup A)'$

9. Let $S = \{x \in \mathbf{R} \mid 0 \leqslant x \leqslant 4\}$, and let A and B be subsets of S defined as $A = \{x \mid 1 < x \leqslant 3\}$, $B = \{x \mid x \geqslant 2\}$. Find:

   (a) A' (b) B' (c) $A \cup B$
   (d) $A' \cup B'$ (e) $(A \cup B)'$ (f) $A'B'$
   (g) $(AB)'$

10. Suppose $S = \mathbf{R} \times \mathbf{R}$ where $\mathbf{R}$ denotes the set of real numbers.
Let $A = \{(x, y) \in S \mid x \leqslant u\}$, $B = \{(x, y) \in S \mid y \leqslant v\}$.
   (a) Find AB, $A \cup B$, and $A - B$.
   (b) Using the Cartesian plane and the appropriate regions of the plane, indicate (as Venn diagrams) the sets AB, $A \cup B$, and $A - B$.

11. Let $A = \{a, b, c\}$. Answer true or false:
   (a) $\emptyset \notin A$, but $\emptyset \in \mathcal{P}(A)$.
   (b) $a \in A$, but $a \notin \mathcal{P}(A)$.
   (c) $\{a, b\} \notin A$, but $\{a, b\} \in \mathcal{P}(A)$.
   (d) A is an element of $\mathcal{P}(A)$ but is not a subset of it.
   (e) $\{\{a, b\}, \{c\}\} \subset \mathcal{P}(A)$.

12. Let $A = \{a, b, c\}$, $B = \{6, 7\}$, $C = \{7, 8\}$. Find:
   (a) $A \times (B \cup C)$ (b) $(A \times B) \cup (A \times C)$
   (c) $A \times (BC)$ (d) $(A \times B)(A \times C)$

13. Let $A = \{1, 2, 3\}$ and $B = \{1, 2, 5, 6\}$. List the elements in the following sets:
   (a) $(A - B) \times (A - B)$ (b) $A \times (A - B)$ (c) $(A \times A) - (B \times B)$

14. Consider the following sets which are subsets of $\mathbf{R} \times \mathbf{R}$. Indicate which ones are Cartesian product sets, and write them in product form.
   (a) $\{(x, y) \mid x \leqslant 2\}$
   (b) $\{(x, y) \mid 0 \leqslant x < y < 2\}$
   (c) $\{(x, y) \mid 0 \leqslant x < 1, -2 \leqslant y < 3\}$
   (d) $\{(x, y) \mid y > 2\}$
   (e) $\{(x, y) \mid x = 2y\}$

15. Let $S = \{1, 2, 3, \ldots\}$ and $A_n = \{x \in S \mid x \geqslant n\}$. Explain why $\lim_{n\to\infty} A_n$ exists. Find it.

16. Consider the following sequences $\{A_n\}$ of subsets of $\mathbf{R}$. In each casé state whether the sequence is contracting or expanding and find $\lim_{n\to\infty} A_n$.
   (a) $A_n = \left\{x \mid 1 + \frac{1}{n} \leqslant x \leqslant 4 - \frac{1}{n}\right\}$
   (b) $A_n = \left\{x \mid 1 - \frac{1}{n} < x < 4 + \frac{1}{n}\right\}$
   (c) $A_n = \left\{x \mid x > 2 - \frac{1}{n}\right\}$
   (d) $A_n = \left\{x \mid x \geqslant 2 + \frac{1}{n}\right\}$
   (e) $A_n = \left\{x \mid 4 - \frac{1}{n} < x < 4 + \frac{1}{n}\right\}$

2. SAMPLE SPACE AND EVENTS

From the viewpoint of probability theory, we shall be interested in the set of all the possible outcomes of an experiment. The totality of these outcomes will act as the universal set and will be called the **sample space**. As a mathematical entity the sample space is only a set. Throughout we shall denote this set by S. Every conceivable outcome of the experiment judged pertinent to a discussion is in this set, and there is no member in this set which is not a possible outcome of the experiment. Members of S are called **sample points**. Consider the following examples:

(*i*) When we toss a coin, using the letter H for heads and the letter T for tails, we could represent S as $S = \{H, T\}$.

(*ii*) If we roll a die once, we could write S as $S = \{1, 2, 3, 4, 5, 6\}$.

(*iii*) If we roll a die and toss a coin, writing the outcomes as pairs with the first component representing the outcome on the die and the second the outcome on the coin, we could write

$$S = \{(1, H), (2, H), (3, H), (4, H), (5, H), (6, H), (1, T), (2, T), (3, T), (4, T), (5, T), (6, T)\}$$

(*iv*) If we pick one card from a standard deck of 52 cards, then we have

$$S = \{A_{\text{sp}}, K_{\text{sp}}, Q_{\text{sp}}, \ldots, 2_{\text{sp}}, A_{\text{h}}, K_{\text{h}}, Q_{\text{h}}, \ldots, 2_{\text{h}}, A_{\text{d}}, K_{\text{d}}, Q_{\text{d}}, \ldots, 2_{\text{d}}, A_{\text{cl}}, K_{\text{cl}}, Q_{\text{cl}}, \ldots, 2_{\text{cl}}\}$$

with an obvious notation. For example, Q_{sp} denotes the queen of spades.

(*v*) Suppose a basketball player keeps throwing a ball at the basket until he makes the basket for the first time. Once he scores he quits. Let us agree to write m if he misses and s if he scores the basket. What are the possibilities? He could score on the first attempt, or miss on the first attempt and score on the second, or miss on the first two and score on the third attempt, and so on. Writing these possibilities as *s, ms, mms,* and so on, we have

$$S = \{s, ms, mms, mmms, \ldots\}$$

(*vi*) Suppose a person speculates on the length of time he will have to wait at the bus stop for the bus to come. In this case, we have $S = \{x \mid x \geqslant 0,\ x \in \mathbf{R}\}$.

It should be noted that there is no unique way of describing a sample space and considerable skill and insight is called upon in deciding what outcomes will be relevant. Faulty descriptions of the underlying sample spaces have led to several paradoxes in the theory of probability. Let us consider the following examples:

(*vii*) Suppose a coin is flipped. In this case the sample space can be given as $S = \{\text{head, tail}\}$, assuming that the coin will land either heads or tails. However, if we entertain the possibility that the coin might land on its edge, then $S = \{\text{head, tail, edge}\}$ would be a more appropriate sample space.

(*viii*) Suppose a box contains three letters *a, b, c* and we pick two of these letters, one by one, without returning the first letter to the box before the second one is picked. Here one might give the sample space as $S_1 = \{ab, ba, ac, ca, bc, cb\}$, listing the order in which the letters are picked. For example, *ba* would represent that letter b was picked first and the letter a was picked next. With this description of the sample space we are in a position to answer a question such as "Which outcomes resulted in picking the letter b first?" Now suppose we choose to give the sample space as $S_2 = \{ab, ac, cb\}$, recording all the outcomes on the basis of what two distinct letters are picked, but with no regard to the order. This second description, though satisfactory to answer some questions, is totally inadequate to answer the question "Which outcomes resulted in picking the letter b first?" since S_2 fails to record the outcomes according to the order in which the letters are picked. Of the two sample spaces S_1, S_2, it would seem that S_1 is more appropriate for our experiment.

It is not always possible to decide ahead of time the kinds of questions that one might want answered. To be on the safe side, it is a good rule to provide a description of the sample space as fully as possible, so that it is adequate to answer all the possible questions one might pose.

A sample space which has a finite number of outcomes is called a **finite sample space.** A sample space which is not finite is called an **infinite sample space.** Examples (*i*) to (*iv*) are examples of finite sample spaces; (*v*) and (*vi*) are examples of infinite sample spaces.

When we roll a die we might pose a question of the following type: What is the probability that an even number will show up? We will say that we are interested in the *event* "an even number will show up." This event will occur if and only if 2 shows up, or 4 shows up, or 6 shows up. In other words, we are interested in the members of the set $\{2, 4, 6\}$. Thus any verbal description of an event has a set-theoretic representation to it. On account of this, an **event** is defined as a subset of the sample space. An event consisting of a single outcome is called an **elementary event**, or a **simple event**.

An axiomatic treatment of probability requires a more formal definition of an event; this aspect will be discussed in the next section. From now on, we shall identify a verbal description of an event with the underlying set.

An event E will be said to have occurred if one of the outcomes belonging to the set E takes place.

Understanding this is crucial for a good grasp of the algebra of events, where we combine two or more events to get a new event. The algebra of events, as we shall soon see, is precisely the algebra of sets that we have already discussed.

The event consisting of all members of S is called the **sure event**. The name is quite suggestive because one of the outcomes belonging to S is bound to occur when the experiment is performed. Hence the sure event is the set S.

The event which contains no members of S, namely $\emptyset$, is called the **impossible event**. This makes sense because the event $\emptyset$ will never occur since it has no outcomes in it.

Example 2.1. Suppose we roll two dice. Write as sets the following verbal descriptions of events.

(*i*) E_1 = The sum on the two dice is 7.
(*ii*) E_2 = The two dice show the same number.
(*iii*) E_3 = The sum on the two dice is a prime number.

Solution. The sample space can be written as a set of ordered pairs in the following way:

$$\begin{aligned} S = \; & (1, 1), (1, 2), (1, 3), (1, 4), (1, 5), (1, 6), \\ & (2, 1), (2, 2), (2, 3), (2, 4), (2, 5), (2, 6), \\ & (3, 1), (3, 2), (3, 3), (3, 4), (3, 5), (3, 6), \\ & (4, 1), (4, 2), (4, 3), (4, 4), (4, 5), (4, 6), \\ & (5, 1), (5, 2), (5, 3), (5, 4), (5, 5), (5, 6), \\ & (6, 1), (6, 2), (6, 3), (6, 4), (6, 5), (6, 6) \end{aligned}$$

Here, in each outcome, the first component represents the outcome on one die, and the second component, the outcome on the other die. It is easy to see that:

(*i*) $E_1 = \{(6, 1), (1, 6), (2, 5), (5, 2), (3, 4), (4, 3)\}$. There are six outcomes in the event "The sum on the two dice is 7."

(*ii*) $E_2 = \{(1, 1), (2, 2), (3, 3), (4, 4), (5, 5), (6, 6)\}$.

(*iii*) $E_3 = \{(1, 1), (1, 2), (2, 1), (2, 3), (3, 2), (1, 4), (4, 1), (6, 1), (1, 6), (5, 2), (2, 5), (3, 4), (4, 3), (5, 6), (6, 5)\}$.

Example 2.2. Suppose a box contains five lightbulbs, three of which are good and two of which are defective. Two bulbs are picked one at a time—without returning the first bulb to the box before the second bulb is picked.

(*i*) Write down the sample space.
(*ii*) Write down the sets which define the following events:

(*a*) E_1 = There is exactly one good bulb.
(*b*) E_2 = There is at least one good bulb.
(*c*) E_3 = There is at most one good bulb.
(*d*) E_4 = There is no good bulb.

Solution

(*i*) Let us identify the bulbs as g_1, g_2, g_3, d_1, and d_2, where g_1, g_2, g_3 represent the good bulbs and d_1, d_2 the defective ones. The outcomes of the experiment can be written as pairs, the first component representing the result of the first draw and the second component the result of the second draw. Hence

$$\begin{aligned} S = \{&(g_1, g_2), (g_1, g_3), (g_1, d_1), (g_1, d_2), (g_2, g_1), (g_2, g_3), (g_2, d_1), (g_2, d_2), \\ &(g_3, g_1), (g_3, g_2), (g_3, d_1), (g_3, d_2), (d_1, g_1), (d_1, g_2), (d_1, g_3), (d_1, d_2), \\ &(d_2, g_1), (d_2, g_2), (d_2, g_3), (d_2, d_1)\} \end{aligned}$$

(*ii*) It can be easily seen that:

(*a*) $E_1 = \{(g_1, d_1), (g_1, d_2), (g_2, d_1), (g_2, d_2), (g_3, d_1), (g_3, d_2), (d_1, g_1), (d_1, g_2), (d_1, g_3), (d_2, g_1), (d_2, g_2), (d_2, g_3)\}$.

(*b*) Since E_2 represents the event "at least one good bulb," it consists of all the outcomes in S except the two outcomes (d_1, d_2) and (d_2, d_1). Hence we could write E_2 as $S - \{(d_1, d_2), (d_2, d_1)\}$. (Recall the definition of set difference.)

(*c*) $E_3 = \{(g_1, d_1), (g_1, d_2), (g_2, d_1), (g_2, d_2), (g_3, d_1), (g_3, d_2), (d_1, g_1), (d_1, g_2), (d_1, g_3), (d_1, d_2), (d_2, g_1), (d_2, g_2), (d_2, g_3), (d_2, d_1)\}$.

Alternatively, we see that E_3 excludes only those outcomes of S where both bulbs are good. Hence we could also write E_3 as

$$E_3 = S - \{(g_1, g_2), (g_1, g_3), (g_2, g_1), (g_2, g_3), (g_3, g_1), (g_3, g_2)\}$$

(*d*) "There is no good bulb" means that both components represent defective bulbs. Hence $E_4 = \{(d_1, d_2), (d_2, d_1)\}$.

Example 2.3. A person fires four shots at a target. Write the following events as sets:

(*a*) E_1 = The person scores a hit on every shot.
(*b*) E_2 = The person misses at least one shot.
(*c*) E_3 = The person scores exactly two hits.
(*d*) E_4 = The person scores at most two hits.

Solution. Let us write h if the person hits the target and m if he misses it. We can then write the sample space as

$$S = \{(h, h, h, h), (h, h, h, m), (h, h, m, h), (h, m, h, h), (m, h, h, h), (h, h, m, m), (h, m, h, m), (h, m, m, h), (m, h, m, h), (m, h, h, m), (m, m, h, h), (m, m, m, h), (m, m, h, m), (m, h, m, m), (h, m, m, m), (m, m, m, m)\}$$

In each outcome the first letter denotes the result of the first shot, and so on. A more compact way would be to write

$$S = \{(x_1, x_2, x_3, x_4) \mid x_i \in \{h, m\},\ i = 1, 2, 3, 4\}$$

Also notice that S is the cartesian product $\{h, m\} \times \{h, m\} \times \{h, m\} \times \{h, m\}$.

We get set representations for E_1, E_2, E_3, E_4 as follows:

(*a*) $E_1 = \{(h, h, h, h)\}$, consisting of a single outcome.

(*b*) If the person misses at least one shot, then the set E_2 will include all the quadruplets above that have m as one of the components. We see immediately that there are 15 such outcomes. An alternate way is by noting that E_2 excludes only the outcome where he scores on every shot. Hence $E_2 = S - \{(h, h, h, h)\} = \{(h, h, h, h)\}'$.

(*c*) $E_3 = \{(h, h, m, m), (h, m, h, m), (h, m, m, h), (m, h, m, h), (m, h, h, m), (m, m, h, h)\}$.

(*d*) $E_4 = \{(m, m, m, m), (m, m, m, h), (m, m, h, m), (m, h, m, m), (h, m, m, m), (h, h, m, m), (m, m, h, h), (m, h, h, m), (h, m, h, m), (h, m, m, h), (m, h, m, h)\}$. ■

A clear grasp of the following discussion is very important for effective application of the probability laws that we will develop. Our goal is to show how to convert verbal descriptions of events into equivalent set symbolism.

Given a collection of events, we can associate with them new events. For example, if A and B are two events, then we can combine them to get new events such as: "either A or B will occur," "both A and B will occur," "A will occur but not B," "A will not occur," and so on. How do we represent these new events using set symbolism? The clue is in the definition of the occurrence of an event.

The event A or B will occur if and only if either an outcome in A occurs or an outcome in B occurs (or an outcome in both occurs). In other words, A or B will occur precisely when an outcome in $A \cup B$ occurs. Hence the event A or B is, in set language, the event $A \cup B$.

We also see that if $A_1, A_2, \ldots$ is a collection of events, then at least one of these events will occur if an outcome belonging to at least one of the events takes place. As a result the event "at least one A_i" corresponds to the set $\bigcup_{i=1}^{\infty} A_i$.

Arguing as above, it can be shown that "or," "and," and "not" should be interpreted in set language as $\cup$ (union), $\cap$ (intersection), and $'$ (complement), respectively. Table 2.1 gives verbal descriptions of events and equivalent set symbolism. (*Note*: as mentioned earlier, wherever convenient we shall write AB in place of $A \cap B$ by omitting the symbol $\cap$.)

Table 2.1 Verbal and Set Theoretic Equivalences

Events	Sets	Notation
The sure event	The sample space	S
The impossible event	The empty set	$\emptyset$
Not A	Complement of A	A'
Either A or B	Union of A and B	$A \cup B$
At least one A_i	Union of A_i, i=1,2, . . .	$\bigcup_{i=1}^{\infty} A_i$
Both A and B	Intersection of A and B	$A \cap B$ (or AB)
Every A_i	Intersection of A_i, i=1,2, . . .	$\bigcap_{i=1}^{\infty} A_i$
A and B cannot occur simultaneously	A and B are disjoint	$AB = \emptyset$
No two of the events $A_1, A_2, \ldots$ can occur simultaneously	A_i, i=1,2, . . . are pairwise disjoint	$A_iA_j = \emptyset$ if $i \neq j$
A implies B	A is a subset of B	$A \subset B$

Example 2.4. Let A, B, and C be three events which are subsets of a sample space S. Using set symbols, write expressions for the events that, of A, B, and C

(*a*) at least one occurs
(*b*) A and B occur, but not C
(*c*) exactly two occur
(*d*) only A occurs
(*e*) exactly one occurs
(*f*) none occurs
(*g*) at most two occur
(*h*) at least two occur.

Solution

(*a*) At least one occurs means either A occurs *or* B occurs *or* C occurs. Hence the event is represented by the set $A \cup B \cup C$.

(*b*) A and B occur, but not C, means A occurs *and* B occurs *and* (*not* C) occurs. Therefore, the event corresponds to the set ABC'.

(*c*) Exactly two occur precisely when (A and B occur, but not C) *or* (A and C occur, but not B) *or* (B and C occur, but not A). Consequently, the event "exactly two occur" is represented by $(ABC') \cup (AB'C) \cup (A'BC)$.

(*d*) Only A occurs if and only if A occurs *and* (B does *not* occur) *and* (C does *not* occur). Therefore, the set representing the event is $AB'C'$.

(*e*) Exactly one of the events will occur when and only when (only A occurs) *or* (only B occurs) *or* (only C occurs). On the basis of (*d*), the event is therefore described by the set $(AB'C') \cup (A'BC') \cup (A'B'C)$.

(*f*) None of the events occurs means (A does *not* occur) *and* (B does *not* occur) *and* (C does *not* occur). Consequently, the set representation is $A'B'C'$.

(*g*) At most two events occur means either (none of the events occurs) *or* (exactly one event occurs) *or* (exactly two events occur). From (*c*), (*e*), and (*f*) it therefore follows that we can express the event as $(A'B'C') \cup ((AB'C') \cup (A'BC') \cup (A'B'C)) \cup ((ABC') \cup (AB'C) \cup (A'BC))$. Equivalently, a more efficient way is to regard the event "at most two occur" as negating the event that A, B, and C all occur simultaneously. Hence "at most two of the events A, B, C occur" can be represented as $(ABC)'$.

(*h*) At least two occur means either (exactly two occur) *or* (exactly three occur), and therefore can be represented as $(ABC') \cup (AB'C) \cup (A'BC) \cup (ABC)$. ∎

A very important property of set operations is called **De Morgan's law**. We shall state it after discussing the following example.

Example 2.5. Suppose A, B, and C are three events. Using set symbols write expressions for the following events:

(*a*) None of the events occurs.

(*b*) Not all the events occur simultaneously.

Solution

(*a*) Our purpose is to look at the event in two equivalent ways.

None of the events occurs means it is *not* the case that either A occurs or B occurs or C occurs and, consequently, we can write $(A \cup B \cup C)'$ for the event.

On the other hand, an equivalent way is that (A does not occur) *and* (B does not occur) *and* (C does not occur). Hence the event is $A'B'C'$.

In summary, we see that

$$(A \cup B \cup C)' = A'B'C'$$

(*b*) We immediately see that *not* (all the events occur simultaneously) has the set representation $(ABC)'$.

Equivalently, the event also states that (either A does not occur) *or* (B does not occur) *or* (C does not occur). Hence, an equivalent representation is $A' \cup B' \cup C'$. Combining, we get

$$(ABC)' = A' \cup B' \cup C'$$ ∎

We shall now state De Morgan's law. A formal proof of this law can be given by using standard arguments in set theory, but we won't bother to do so.

De Morgan's law

(*i*) The complement of the union of any collection of sets is equal to the intersection of their complements. Thus, in particular, if $A_1, A_2, \ldots$ is a countable collection of events, then

$$\left(\bigcup_{i=1}^{\infty} A_i\right)' = \bigcap_{i=1}^{\infty} A_i'$$

(*ii*) The complement of the intersection of any collection of sets is equal to the union of their complements. That is,

$$\left(\bigcap_{i=1}^{\infty} A_i\right)' = \left(\bigcup_{i=1}^{\infty} A_i'\right)$$

EXERCISES–SECTION 2

1. A box contains three white balls, w_1, w_2, w_3, and one black ball, b. Suppose one ball is picked from the box and we are interested in its color. Discuss the merits of the following sample spaces S_1 and S_2:

$S_1 = \{w_1, w_2, w_3, b\}$
$S_2 = \{\text{white, black}\}$

2. A coin is tossed repeatedly until a head shows up. Discuss the feasibility and the relative merits of the following two sample spaces, S_1 and S_2, describing the experiment:

$S_1 = \{H, TH, TTH, \ldots\}$
$S_2 = \{(x_1, x_2, \ldots, x_i, \ldots) \mid x_i = H \text{ or } T\}$

(S_2 consists of all infinite sequences where the ith component is H or T depending upon whether the coin lands heads or tails.)

3. A box contains six tubes, of which two are defective. If the tubes are tested one by one until a defective tube is found, set up a sample space describing the experiment.

4. A committee of two is selected from a group consisting of five people: Tom, Dick, Mary, Paul, and Jane. Write down the following events as sets:
   (a) S, the sample space
   (b) A, both members of the committee are male
   (c) B, exactly one member is male

5. Tom and John play a game where they simultaneously exhibit their right hands with one, two, three, or four fingers extended.
   (a) Describe an appropriate sample space.
   (b) Give the outcomes in the event that one person extends twice as many fingers as the other.
   (c) Give the outcomes in the event that they both extend the same number of fingers.

6. Four contestants, say Tom, Jane, Mary and Nancy, enter a contest which has two prizes consisting of a first and a second prize. Suppose a contestant can win at most one prize.
   (a) Set up an appropriate sample space describing the results of the contest.
   (b) List the outcomes describing the events that: (*i*) Tom wins the first prize; (*ii*) Tom wins a prize.

7. A box contains two black chips, say b_1, b_2, and three white chips, say w_1, w_2, w_3. An even number of chips are drawn without returning to the box the chips already drawn and without regard to the order in which they are drawn.
   (a) Write down the outcomes in the sample space.
   (b) List all the outcomes in the event E = The same number of black and white chips are drawn.
8. Two numbers are picked in the interval $[0, 1]$.
   (a) Give a sample space appropriate to describe the above experiment.
   (b) Describe the following events as subsets of the sample space: (*i*) the two numbers do not differ by more than $\frac{1}{3}$; (*ii*) the sum of the two numbers is less than 1.
9. A person goes to bed at any time between 9 p.m. and 11 p.m. and wakes up between 6 a.m. and 7 a.m. Assume that time is measured in hours on the time axis with 9 p.m. as the starting point. Let x denote the time when he goes to bed and y the time when he wakes up.
   (a) Describe an appropriate sample space.
   (b) Describe the following events as sets: (*i*) the person wakes up before 6:30 a.m.; (*ii*) the person sleeps less than $9\frac{1}{2}$ hours.
10. Tom has four keys in his pocket, say k_1, k_2, k_3, k_4. Only the key k_1 can open a particular lock. Tom reaches in his pocket, picks a key and tries to open the lock. He tries his keys one at a time (always making sure that he discards the unsuccessful keys) until he finds the right key, namely, k_1.
   (a) Write down the outcomes in the sample space describing the experiment.
   (b) Give the outcomes in the event "Tom opens the lock on the third attempt."
11. From among the students on a university campus a student is picked. Let the event E_1 denote that the student is a mathematics major, E_2 that he is a history minor, and E_3 that he is a senior. Describe in words the following events:
   (a) $E_1 \cup E_2 \cup E_3$ (b) $E_1E_2E_3$ (c) $(E_1E_2)'$
   (d) $(E_1 \cup E_3)'$ (e) $E_1' \cup E_2' \cup E_3'$ (f) $E_1'E_2'E_3'$
   (g) $E_1'E_2E_3$
12. A person wins a contest in which he gets a free trip from San Diego to New York via Los Angeles and Chicago. He has a choice of two buses from San Diego to Los Angeles. Once in Los Angeles, he has a choice of four plane flights to Chicago, and upon arrival in Chicago he has a choice of three trains to New York. Let the event B_i be that he catches bus i ($i = 1, 2$), P_i that he catches plane i ($i = 1, 2, 3, 4$), and T_i that he catches train i ($i = 1, 2, 3$). Express the following events in terms of B_i, P_i, T_i:
   (a) A = The person gets to New York.
   (b) A = The person does not get to New York.
13. Suppose a person goes fishing, and let A, B, C represent respectively that he catches at least one bass, at least one trout, at least one salmon. Express in words the following events:
   (a) $A \cup B \cup C$ (b) ABC (c) $A'BC$
   (d) $A'(B \cup C)$

14. Four cards are drawn from a standard deck of 52 cards. Suppose A and B are two events where A = The cards are face cards, B = The cards are black cards. Describe in words the following events:

(a) AB (b) $A \cup B$ (c) $A - B$
(d) $A' \cup B$ (e) $A' \cup B'$

15. Suppose A and B are two events such that if A does not occur, then B occurs, and if B does not occur, then A occurs. Is it justified to jump to the conclusion that A and B are complements of each other?

3. SIGMA FIELDS

3.1 The Definition of a Sigma Field

The long and short of the discussion that follows is that we have under consideration a collection of subsets of the sample space S. The members of this collection are called events and have the properties that: if "occurrence of A" is an event, then "nonoccurrence of A" is also an event; if "occurrence of A_1," "occurrence of A_2," and so on, are events, then "occurrence of at least one of the events $A_1, A_2, \ldots$" is also an event. As a natural consequence, these stipulations ensure that "occurrence of S," "occurrence of $\emptyset$," "occurrence of *all* the events A_1, $A_2, \ldots$" are also events.

The term used to describe this type of collection is **sigma field**. Though the mathematical formulation might seem forbidding, the reader need only realize that the sigma field is no more than a collection of events wherein, no matter how he combines these events (by using "and," "or," or "negation"), he will always end up getting an event. We now give a formal definition.

Suppose S is the sample space and we form a nonempty collection $\mathcal{F}$ of subsets of S with the following provisoes:

($\mathcal{F}$1) if $A \in \mathcal{F}$, then $A' \in \mathcal{F}$

($\mathcal{F}$2) if $A_i \in \mathcal{F}$, $i = 1, 2, \ldots,$ then $\bigcup_{i=1}^{\infty} A_i \in \mathcal{F}$

The collection $\mathcal{F}$ is then called a **sigma field** or a **sigma algebra** of subsets of S.

Properties ($\mathcal{F}$1) and ($\mathcal{F}$2) are referred to, respectively, as *closure under complementation* and *closure under countable union.* The members of the collection will constitute what are called **random events.**

Example 3.1. Since the power set $\mathcal{P}(S)$ consists of all the subsets of S, it is always a sigma field. Actually, $\mathcal{P}(S)$ is the largest sigma field of subsets of S.

Example 3.2. If S is any set, then, as can be easily verified, $\{\emptyset, S\}$ is a sigma field. As a matter of fact, it is the smallest sigma field of subsets of S.

Example 3.3. If A is any nonempty subset of S, then $\{\emptyset, A, A', S\}$ is always a sigma field. It is a simple matter to check this.

Example 3.4. Suppose $S = \{a, b, c\}$. List some of the sigma fields of subsets of S.

Solution. The set

$$\{\emptyset, \{a\}, \{b\}, \{c\}, \{a, b\}, \{a, c\}, \{b, c\}, \{a, b, c\}\}$$

is a sigma field. It is, in fact, $\mathcal{P}(S)$ and is the largest sigma field. $\{\emptyset, \{a, b\}, \{c\}, \{a, b, c\}\}$ is also a sigma field, and so is $\{\emptyset, \{a\}, \{b, c\}, \{a, b, c\}\}$. However, note, for example, that $\{\emptyset, \{a\}, \{b\}, \{a, b\}\}$ is *not* a sigma field of subsets of $\{a, b, c\}$. This is so because $\{a, b\}$ is in the collection but its complement, namely, $\{c\}$, is not. ■

We have the following consequences as a result of the definition of a sigma field:

$$S \in \mathcal{F}$$

We can prove this as follows: Suppose $A \in \mathcal{F}$. (This is valid since $\mathcal{F}$ is nonempty.) Then by ($\mathcal{F}$1) $A' \in \mathcal{F}$, and by ($\mathcal{F}$2) $A \cup A' \in \mathcal{F}$. Hence $S \in \mathcal{F}$.

$$\emptyset \in \mathcal{F}$$

To prove this, we observe that $S \in \mathcal{F}$. Now, by ($\mathcal{F}$1), we know that if $S \in \mathcal{F}$, then $S' \in \mathcal{F}$. That is, $\emptyset \in \mathcal{F}$.

$$\text{If } A_i \in \mathcal{F}, \quad i = 1, 2, \ldots, \quad \text{then } \bigcap_{i=1}^{\infty} A_i \in \mathcal{F}$$

In other words, a *sigma field is closed under countable intersections*.

We can justify this as follows: Since $A_i \in \mathcal{F}$, $i = 1, 2, \ldots$, by ($\mathcal{F}$1) we must have $A_i' \in \mathcal{F}$, $i = 1, 2, \ldots$. Hence by ($\mathcal{F}$2) it follows that $\bigcup_{i=1}^{\infty} A_i' \in \mathcal{F}$. Now by ($\mathcal{F}$1) in turn $\left(\bigcup_{i=1}^{\infty} A_i'\right)' \in \mathcal{F}$. However, by De Morgan's law, $\left(\bigcup_{i=1}^{\infty} A_i'\right)' = \bigcap_{i=1}^{\infty} (A_i')' = \bigcap_{i=1}^{\infty} A_i$. Hence the result.

3.2 The Borel Field of the Real Line

A sigma field which will be of special interest to us is called the **Borel field**. It will be denoted, using a special symbol, as $\mathcal{B}$. In the discussion that follows we shall specifically consider the Borel field of the real line. The reader can then investigate for himself the Borel field of the interval $[0, 1]$, or the Borel field of the plane $\mathbf{R}^2$, and so on (see exercises 4 and 5). Before describing the Borel field of the subsets of $\mathbf{R}$ in detail, let us consider some rather elementary situations which will serve to motivate us.

Suppose $S = \{a, b, c, d\}$. Then, as can be easily checked, $\{\emptyset, \{a\}, \{a, b\}\}$ is not a sigma field. For instance, the complement of the set $\{a\}$, namely, $\{b, c, d\}$, is missing. Suppose we introduce just those sets we need to make a sigma field—no more, no less. Thus we will need the sets $\{b, c, d\}$, $\{a, b\}' = \{c, d\}$, $\{a\} \cup \{c, d\} = \{a, c, d\}$, $\{a, c, d\}' = \{b\}$, $\{a\} \cup \{b, c, d\} = \{a, b, c, d\}$, and that's it! The set $\{\emptyset, \{a\}, \{a, b\}, \{b, c, d\}, \{c, d\},$

$\{a, c, d\}, \{b\}, \{a, b, c, d\}\}$ is a sigma field. Because of the way we have constructed it, it is the smallest sigma field containing the set $\{\emptyset, \{a\}, \{a, b\}\}$. It is called the **sigma field generated by** the set $\{\emptyset, \{a\}, \{a, b\}\}$.

As a further illustration, if $S = \{a, b, c, d\}$, then it can be easily verified, that (*i*) the sigma field generated by $\{\emptyset, \{a, b\}\}$ is $\{\emptyset, \{a, b\}, \{c, d\}, \{a, b, c, d\}\}$, (*ii*) that generated by $\{\{a\}, \{b\}\}$ is $\{\emptyset, \{a\}, \{b\}, \{a, b\}, \{c, d\}, \{a, c, d\}, \{b, c, d\}, \{a, b, c, d\}\}$, and (*iii*) that generated by $\{\{a\}, \{b\}, \{c\}\}$ is the power set of $\{a, b, c, d\}$.

We are now in a position to discuss the Borel field of the real line. To construct it, we use the following procedure:

(*i*) First start by including *all* the intervals $(-\infty, a]$ where a is any real number. Thus intervals of the form $(-\infty, 2], (-\infty, \frac{1}{2}], (-\infty, \sqrt{3}]$ form a part of the collection.

(*ii*) For $\mathscr{B}$ to be a sigma field we now require, on account of axiom ($\mathscr{F}1$), that it contain complements of the intervals that we included under (*i*). Since the complement of $(-\infty, a]$ is (a, ∞), the collection $\mathscr{B}$ will contain all the intervals of the type (a, ∞), where a is any real number. For example, intervals of the type $(\frac{1}{2}, \infty), (2, \infty), (\sqrt{3}, \infty)$ will be members of $\mathscr{B}$.

(*iii*) Suppose a and b are any two real numbers with $a < b$. Since by (*i*) $(-\infty, b] \in \mathscr{B}$, and by (*ii*) $(a, \infty) \in \mathscr{B}$, their intersection, $(-\infty, b] \cap (a, \infty) = (a, b]$, is also in $\mathscr{B}$. In other words, all the intervals of the type $(a, b]$, where a and b are real numbers with $a < b$, are in $\mathscr{B}$. For example, $(2, 3], (-2, \sqrt{3}]$, and so on, are in $\mathscr{B}$.

The collection $\mathscr{B}$ is thus a very large collection of subsets of $\mathbf{R}$ which, starting with the sets of the type $(-\infty, a]$, is obtained by repeatedly carrying out the operations of union, intersection and complementation. That is, *the Borel field of the real line is the smallest sigma field containing all sets of the form* $(-\infty, a]$. The members of $\mathscr{B}$ are called **Borel sets** of the real line.

Do singletons belong to $\mathscr{B}$? For example, is it true that $\{5\} \in \mathscr{B}$? To answer this, we note from (*iii*) that sets of the type $\left(5 - \frac{1}{n}, 5\right]$ are in $\mathscr{B}$ for every natural number n. Hence, due to closure of $\mathscr{B}$ under intersection, we must have $\bigcap_{n=1}^{\infty} \left(5 - \frac{1}{n}, 5\right] \in \mathscr{B}$. That is, $\{5\} \in \mathscr{B}$. In general, $\{a\} \in \mathscr{B}$ for any real number a. This in turn implies that closed intervals $[a, b]$, open intervals (a, b), and intervals like $[a, b)$ are in $\mathscr{B}$ for any real numbers a, b with $a < b$. (Why?)

In summary, $\mathscr{B}$ is a very large collection of subsets of the real line and it includes "just about" every subset of $\mathbf{R}$ that we can conceive of. However, it should be mentioned that $\mathscr{B}$ is not quite the power set of $\mathbf{R}$. That is, there do exist subsets of $\mathbf{R}$ which are not Borel sets. The proof of this statement requires rather sophisticated tools in mathematics and will not be attempted here.

We have discussed the notion of a sigma field in some detail. It might be a moot point to enquire, "Why not take the power set of S as the underlying sigma field and be done with it?" After all, the power set is the largest sigma field and contains every conceivable subset of A. In most cases, where it does not lead to inconsistencies in defining probabilities, this is precisely what we do. But there are situations where inconsistencies crop up. This happens when S has an uncountable number of sample points. We shall take this problem up in Chapter 2.

EXERCISES—SECTION 3

1. If $S = \{a, b, c, d, e\}$, give three distinct sigma fields of subsets of S.
2. If $\mathcal{F}$ is a sigma field, show that for any $A \in \mathcal{F}$, $B \in \mathcal{F}$, it follows that $(A - B) \cup (B - A) \in \mathcal{F}$.
3. Suppose A and B are subsets of S, neither of which is S or the empty set. Find the sigma field generated by $\{A, B\}$.
4. Discuss the Borel field of the subsets of the interval $[0, 1]$. *Hint*: Start with the set of intervals of the form $\{x \mid 0 < x \leqslant b\}$, where $0 \leqslant b \leqslant 1$.
5. Discuss the Borel field of the subsets of $S = \mathbf{R} \times \mathbf{R} = \{(x, y) \mid x, y \in \mathbf{R}\}$. *Hint*: Start with the collection of sets of the form $I_{a,b} = \{(x, y) \mid -\infty < x \leqslant a, -\infty < y \leqslant b\}$, where a, b are any real numbers.

2 Definition of Probability

INTRODUCTION

With a probability model as our goal, we have so far defined two entities—the sample space which represents the totality of the outcomes of an experiment, and a collection of events called the sigma field. We shall now state the axioms of probability theory. According to Kolmogorov, the three axioms below suffice to construct a theory of probability.

1. The Axioms of the Theory of Probability

Probability is a function defined on the members of the sigma field. It is therefore referred to as a *set function*. We shall denote it by P. For any event A, $P(A)$ is called the probability of the event A; it represents a real number, and has the following properties:

($P1$) $0 \leqslant P(A) \leqslant 1$ for any event A

($P2$) $P(S) = 1$

($P3$) **(Countable additivity)** If $\{A_i\}$ is a sequence of *mutually exclusive* events (that is, $A_iA_j = \emptyset$ if $i \neq j$), then

$$P\left(\bigcup_{i=1}^{\infty} A_i\right) = \sum_{i=1}^{\infty} P(A_i)$$

The triple $(S, \mathcal{F}, P)$ consisting of the set S, the sigma field $\mathcal{F}$ of subsets of S, and the nonnegative set function P with the above properties $P1$, $P2$, and $P3$ is called a **probability space.** We shall often refer to P as the *probability measure.* The following are immediate consequences of the above axioms.

(C1) The probability of the impossible event is zero. That is,

$$P(\emptyset) = 0$$

To prove this, we have for any set A,

$$A = A \cup \emptyset \cup \emptyset \cup \ldots$$

Since $A \cap \emptyset = \emptyset$ and $\emptyset \cap \emptyset = \emptyset$, we get, using axiom ($P3$),

$$P(A) = P(A) + P(\emptyset) + P(\emptyset) + \ldots$$

Now, using the fact that $P(A)$ is finite, this gives $P(\emptyset) = 0$.

Comment. This result incidentally proves that *P is finitely additive.* That is, if $A_1, A_2, \ldots, A_n$ are n mutually exclusive events, then $P\left(\bigcup_{i=1}^{n} A_i\right) = \sum_{i=1}^{n} P(A_i)$. This follows simply from the fact that

$$P\left(\bigcup_{i=1}^{n} A_i\right) = P\left(\left(\bigcup_{i=1}^{n} A_i\right) \cup \emptyset \cup \emptyset \cup \ldots\right)$$
$$= P(A_1) + \ldots + P(A_n) + P(\emptyset) + P(\emptyset) + \ldots$$

by axiom ($P3$). Hence

$$P\left(\bigcup_{i=1}^{n} A_i\right) = P(A_1) + \ldots + P(A_n)$$

since $P(\emptyset) = 0$.

Note: A collection of events is said to be mutually exclusive if no two of them can occur simultaneously.

(C2) For any event A, $P(A') = 1 - P(A)$

In words, the probability that an event will *not* occur is equal to one minus the probability that it will occur.

We can prove this as follows: From the definition of A', we know that $S = A \cup A'$. Now by axiom ($P2$), $P(S) = 1$. Hence

$$P(A \cup A') = 1$$

Therefore, since A and A' are mutually exclusive,

$$P(A) + P(A') = 1$$

Consequently,

$$P(A') = 1 - P(A)$$

For example:

(*i*) If the probability of precipitation is 0.80, then the probability of no precipitation is 0.20.

(*ii*) If the probability that all of ten children are sons is 0.05, then the probability that at least one is a daughter is 0.95.

(C3) If A and B are any two events, then

$$P(A - B) = P(A) - P(AB)$$

(Recall that $A - B$ represents the event that A occurs but not B; that is, *of the two events A and B, only A will occur.*)

To prove this, we note that we can write the event A as the union of two mutually exclusive events $A - B$ and AB. (See the Venn diagram in Figure 1.1.) Hence

$$P(A) = P(A - B) + P(AB)$$

and consequently

$$P(A - B) = P(A) - P(AB)$$

Comment. We must be careful to take note of the fact that, in general, $P(A - B)$ is *not* equal to $P(A) - P(B)$. However, in the special case when $B \subset A$, the result is true. That is,

$$\text{if} \quad B \subset A \quad \text{then} \quad P(A - B) = P(A) - P(B)$$

(C4) (Monotone Property) If events A and B are such that $B \subset A$, then $P(B) \leqslant P(A)$.

The proof of this relies on the comment at the end of the proof of (C3). We know that if $B \subset A$, then

$$P(A - B) = P(A) - P(B)$$

But $P(A - B) \geqslant 0$, by axiom ($P1$). Therefore, $P(A) - P(B) \geqslant 0$ and consequently

$$P(A) \geqslant P(B)$$

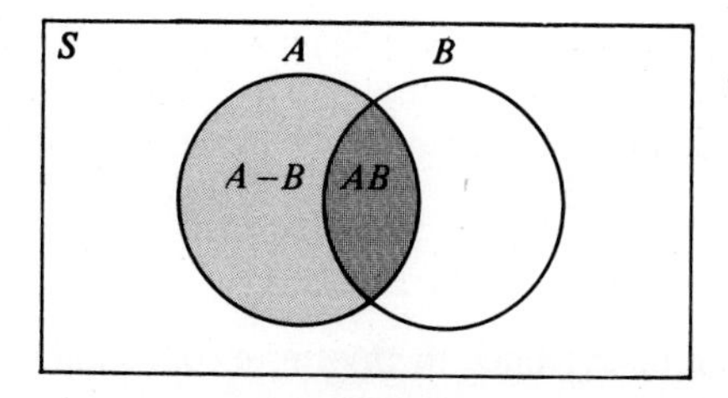

Figure 1.1

For example:

(*i*) The chance of a person being in Iowa can be no greater than his chance of being in the United States–the set of people in Iowa is a subset of the people in the United States.

(*ii*) If we pick an integer, the probability that it is a multiple of 4 can be no greater than the probability that it is even–the set of integers which are multiples of 4 is a subset of the set of even integers.

Note. $A - B$ and AB' represent the same event, and we shall use these representations interchangeably.

(C5) For any two events A and B,

$$P(A \cup B) = P(A) + P(B) - P(AB)$$

(Recall that $A \cup B$ represents at least one of the two events A, B, that is, the event A or B.)

To verify the result, we see that $A \cup B$ can be written as a union of three mutually exclusive events: $A - B$, $B - A$, and AB. See the Venn diagram in Figure 1.2.

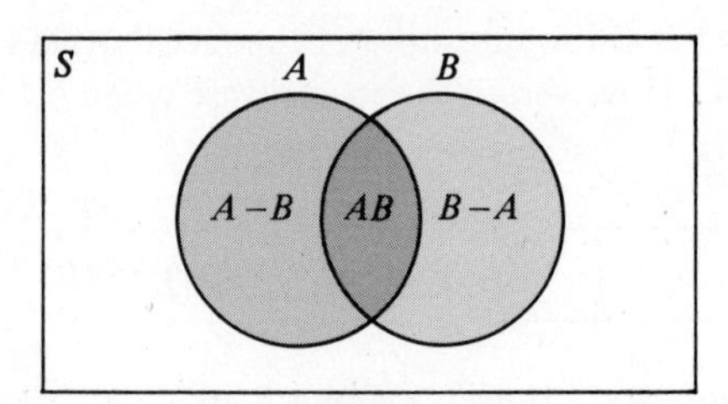

Figure 1.2

Hence

$$P(A \cup B) = P(A - B) + P(B - A) + P(AB)$$

However, by consequence (C3) we know that

$$P(A - B) = P(A) - P(AB) \quad \text{and} \quad P(B - A) = P(B) - P(AB)$$

Upon substitution this gives

$$P(A \cup B) = P(A) + P(B) - P(AB)$$

Note. From the above result and the monotone property, it follows that for any two events A, B,

$$P(AB) \leqslant P(A) \leqslant P(A \cup B) \leqslant P(A) + P(B)$$

The result in (C5) can be extended to any finite collection of events. Let us consider three events A_1, A_2, A_3. Then

$$P(A_1 \cup A_2 \cup A_3) = P(A_1 \cup (A_2 \cup A_3))$$
$$= P(A_1) + P(A_2 \cup A_3) - P(A_1(A_2 \cup A_3))$$

by consequence (C5). Now, by the distributive law, $A_1(A_2 \cup A_3) = (A_1A_2) \cup (A_1A_3)$, and consequently

$$P(A_1(A_2 \cup A_3)) = P((A_1A_2) \cup (A_1A_3))$$
$$= P(A_1A_2) + P(A_1A_3) - P(A_1A_2A_1A_3),$$
$$= P(A_1A_2) + P(A_1A_3) - P(A_1A_2A_3)$$

Therefore,

$$P(A_1 \cup A_2 \cup A_3) = P(A_1) + [P(A_2) + P(A_3) - P(A_2A_3)]$$
$$- [P(A_1A_2) + P(A_1A_3) - P(A_1A_2A_3)]$$
$$= P(A_1) + P(A_2) + P(A_3) - P(A_1A_2)$$
$$- P(A_1A_3) - P(A_2A_3) + P(A_1A_2A_3)$$

We could write the result more compactly as

$$P(A_1 \cup A_2 \cup A_3) = \sum_{i=1}^{3} P(A_i) - \sum_{\substack{i,j=1 \\ i<j}}^{3} P(A_iA_j) + P(A_1A_2A_3)$$

An astute observation might lead one to conjecture that, if $A_1, A_2, \ldots A_n$ are n events, then

$$P\left(\bigcup_{i=1}^{n} A_i\right) = \sum_{i=1}^{n} P(A_i) - \sum_{\substack{i,j=1 \\ i<j}}^{n} P(A_iA_j) + \sum_{\substack{i,j,k=1 \\ i<j<k}}^{n} P(A_iA_jA_k)$$
$$- \ldots + (-1)^{n+1} P(A_1A_2 \ldots A_n).$$

This conjecture is indeed correct and can be proved by induction. The proof of this result will be left to the exercises.

Comment. The preceding formula is useful for finding the probability that at least one of the events $A_1, A_2, \ldots, A_n$ will occur, if we can find the probability of the simultaneous occurrence of any subcollection of the events. Also note that if the events are mutually exclusive, then the probability of the simultaneous occurrence of two or more events is zero, and consequently $P\left(\bigcup_{i=1}^{n} A_i\right)$ will equal $\sum_{i=1}^{n} P(A_i)$, as it should.

Example 1.1. Suppose A and B are two events for which $P(A) = 0.6$, $P(B) = 0.7$, and $P(AB) = 0.4$. Find the following probabilities:

(*a*) $P(A \cup B)$	(*b*) $P(AB')$	(*c*) $P(BA')$
(*d*) $P((AB)')$	(*e*) $P((A \cup B)')$	(*f*) $P(A'B')$

Solution

(*a*) $P(A \cup B) = .06 + .07 - .04 = 0.9.$
(*b*) $P(AB') = P(A) - P(AB) = 0.2.$
(*c*) $P(BA') = P(B) - P(AB) = 0.3.$
(*d*) $P((AB)') = 1 - P(AB) = 0.6.$
(*e*) $P((A \cup B)') = 1 - P(A \cup B) = 0.1.$
(*f*) By De Morgan's law, $A'B' = (A \cup B)'$. Hence $P(A'B') = 0.1$.

Example 1.2. Suppose A, B, and C are three events such that $P(A) = P(B) = P(C) = \frac{1}{4}$, $P(AB) = P(CB) = 0$ and $P(AC) = \frac{1}{8}$. Find the probability that at least one of the events A, B, C will occur.

Solution. We know that

$$P(A \cup B \cup C) = P(A) + P(B) + P(C) - [P(AB) + P(AC) + P(BC)] + P(ABC)$$

We are given every term involved on the right hand side except $P(ABC)$. To find this we note that $ABC \subset AB$ and hence, by the monotone property, $P(ABC) \leqslant P(AB)$; therefore $P(ABC) = 0$ and, as a result, $P(A \cup B \cup C) = \frac{5}{8}$.

Example 1.3. In each of the following, find $P(A \cup (B' \cup C')')$ if A, B, and C are three events with--

(*a*) $P(A) = \frac{1}{2}$, and A, B, C mutually exclusive
(*b*) $P(A) = \frac{1}{2}$, and $P(A) = 2P(BC) = 3P(ABC)$
(*c*) $P(A) = \frac{1}{2}$, $P(BC) = \frac{1}{3}$, and $P(AC) = 0$
(*d*) $P(A'(B' \cup C')) = 0.7$.

Solution. We note first of all, using De Morgan's law, that $A \cup (B' \cup C')' = A \cup (BC)$. Hence

$$P(A \cup (B' \cup C')') = P(A \cup (BC)) = P(A) + P(BC) - P(ABC)$$

(*a*) Since the events A, B, C are mutually exclusive, we have $P(BC) = P(ABC) = 0$. Consequently $P(A \cup (B' \cup C')') = P(A) = \frac{1}{2}$.

(*b*) In this case $P(BC) = \frac{1}{4}$ and $P(ABC) = \frac{1}{6}$. Hence $P(A \cup (B' \cup C')') = \frac{1}{2} + \frac{1}{4} - \frac{1}{6} = \frac{7}{12}$.

(*c*) Since $ABC \subset AC$ and $P(AC) = 0$, by the monotone property it follows that $P(ABC) = 0$. As a result, $P(A \cup (B' \cup C')') = \frac{1}{2} + \frac{1}{3} = \frac{5}{6}$.

(*d*) Here we observe that $(A \cup (B' \cup C')') = (A'(B' \cup C'))'$, and consequently $P(A \cup (B' \cup C')') = 1 - P(A'(B' \cup C')) = 0.3$.

• *Example 1.4.* (*Exactly one of the two events*) Suppose A and B are two events. Show that the probability that *exactly one* of the events occurs is equal to $P(A) + P(B) - 2P(AB)$.

Solution. The event that exactly one of A, B occurs can be written as $(A - B) \cup (B - A)$. Now

$$P((A - B) \cup (B - A)) = P(A - B) + P(B - A)$$

since the events $A - B$ and $B - A$ are mutually exclusive. However,

$$P(A - B) = P(A) - P(AB) \quad \text{and} \quad P(B - A) = P(B) - P(AB)$$

Therefore, the probability that exactly one of the two events occurs is $P(A) + P(B) - 2P(AB)$. ■

It is possible to obtain a general formula giving the probability that *exactly k* of the n events $A_1, A_2, \ldots, A_n$ will occur ($k \leqslant n$). The formula is rather complicated. We shall therefore discuss (in the following example) only the case of exactly k events out of three events A, B, and C ($k = 0, 1, 2, 3$).

• *Example 1.5.* (*Probability of exactly k of the events A, B, C*) If A, B, and C are three events, find the probability that exactly k of the events occur ($k = 0, 1, 2, 3$) in terms of the probabilities of the individual events, probabilities of simultaneous occurrence of pairs of events, and the probability of the simultaneous occurrence of all three events.

Solution.

(*a*) $k = 0$.

If $k = 0$ we want to find $P(A'B'C')$.

$$\begin{aligned} P(A'B'C') &= P(A \cup B \cup C)', && \text{using De Morgan's law} \\ &= 1 - P(A \cup B \cup C), && \text{by (C2)} \end{aligned}$$

Hence

$$\begin{aligned} P(\text{exactly 0 events occur}) = 1 - P(A) - P(B) - P(C) + P(AB) \\ + P(AC) + P(BC) - P(ABC) \end{aligned}$$

(*b*) $k = 1$.

$$\begin{aligned} P(\text{exactly one event occurs}) &= P((AB'C') \cup (A'BC') \cup (A'B'C)) \\ &= P(AB'C') + P(A'BC') + P(A'B'C) \end{aligned}$$

since the events are mutually exclusive.

Let us first find $P(AB'C')$. We have

$$\begin{aligned} P(AB'C') &= P(A(B'C')) \\ &= P(A(B \cup C)'), && \text{using De Morgan's law} \\ &= P(A) - P(A(B \cup C)), && \text{by (C3)} \\ &= P(A) - P(AB \cup AC), && \text{using the distributive law} \\ &= P(A) - [P(AB) + P(AC) - P(AB \cap AC)], && \text{by (C5)} \end{aligned}$$

Hence

$$P(AB'C') = P(A) - P(AB) - P(AC) + P(ABC)$$

Similarly,

$$P(A'BC') = P(B) - P(AB) - P(BC) + P(ABC)$$

and

$$P(A'B'C) = P(C) - P(AC) - P(BC) + P(ABC)$$

Therefore, substituting, we get

$$P(\text{exactly 1 of the events } A, B, C \text{ occurs}) = P(A) + P(B) + P(C) - 2[P(AB) + P(AC) + P(BC)] + 3P(ABC)$$

(*c*) $k = 2$.

$$P(\text{exactly two events occur}) = P((ABC') \cup (AB'C) \cup (A'BC)) = P(ABC') + P(AB'C) + P(A'BC)$$

since ABC', $AB'C$, and $A'BC$ are mutually exclusive events. Now

$$P(ABC') = P(AB) - P(ABC), \qquad \text{by (C3)}$$

and for the same reason

$$P(AB'C) = P(AC) - P(ABC)$$

and

$$P(A'BC) = P(BC) - P(ABC)$$

Therefore,

$$P(\text{exactly two of the events } A, B, C \text{ occur}) = P(AB) + P(AC) + P(BC) - 3P(ABC)$$

(*d*) $k = 3$. If $k = 3$, we simply have $P(ABC)$.

Example 1.6. The probability that a person goes to a concert on Saturday is $\frac{2}{3}$, and the probability that he goes to a baseball game on Sunday is $\frac{4}{9}$. If the probability of going to either program is $\frac{7}{9}$, find the following probabilities:

(*a*) The person goes to both programs.
(*b*) The person goes to the Saturday concert, but not to the baseball game.
(*c*) The person goes to exactly one program.

Solution. Let

A = The person goes to the Saturday concert
B = The person goes to the Sunday game

We are given $P(A) = \frac{2}{3}$, $P(B) = \frac{4}{9}$, and $P(A \cup B) = \frac{7}{9}$.

(*a*) We want to find $P(AB)$, which is equal to $P(A) + P(B) - P(A \cup B)$. Therefore, $P(AB) = \frac{1}{3}$.

(*b*) Here we want to find $P(AB')$. $P(AB') = P(A) - P(AB) = \frac{1}{3}$.

(*c*) The probability that the person goes to exactly one program is equal to $P(A) + P(B) - 2P(AB) = \frac{2}{3} + \frac{4}{9} - \frac{2}{3} = \frac{4}{9}$.

Example 1.7. Suppose a person fires four shots. The probability that he hits the target on the ith shot is $\frac{1}{i+2}$, $(i = 1, 2, 3, 4)$; that he hits the target on the ith and the jth shots is $\frac{1}{5ij}$, $(i, j = 1, 2, 3, 4)$; that he hits the target on the ith, jth, and kth shots is $\frac{1}{20ijk}$, $(i, j, k = 1, 2, 3, 4)$; and finally, that he hits the target on all of the shots is $\frac{1}{960}$. Find the probability that he hits the target at least once.

Solution. Let A_i = The person hits the target on the ith shot. Then

$$P(A_1) = \tfrac{1}{3}, \quad P(A_2) = \tfrac{1}{4}, \quad P(A_3) = \tfrac{1}{5}, \quad P(A_4) = \tfrac{1}{6}$$

$$P(A_1A_2) = \frac{1}{5 \cdot 2}, \quad P(A_1A_3) = \frac{1}{5 \cdot 3}$$

and so on.

$$P(A_1A_2A_3) = \frac{1}{20 \cdot 6}, \quad P(A_1A_3A_4) = \frac{1}{20 \cdot 12}$$

and so on, and

$$P(A_1A_2A_3A_4) = \frac{1}{960}$$

We want to find $P\left(\bigcup_{i=1}^{4} A_i\right)$ which is equal to

$$\begin{aligned} &P(A_1) + P(A_2) + P(A_3) + P(A_4) - [P(A_1A_2) + P(A_1A_3) + P(A_1A_4) + P(A_2A_3) \\ &\quad + P(A_2A_4) + P(A_3A_4)] + [P(A_1A_2A_3) + P(A_1A_3A_4) + P(A_2A_3A_4) \\ &\quad + P(A_1A_2A_4)] - P(A_1A_2A_3A_4) \\ &= \tfrac{1}{3} + \tfrac{1}{4} + \tfrac{1}{5} + \tfrac{1}{6} - \tfrac{1}{5}\left[\tfrac{1}{2} + \tfrac{1}{3} + \tfrac{1}{4} + \tfrac{1}{6} + \tfrac{1}{8} + \tfrac{1}{12}\right] + \tfrac{1}{20}\left[\tfrac{1}{6} + \tfrac{1}{12} + \tfrac{1}{24} + \tfrac{1}{8}\right] - \tfrac{1}{960} \\ &= \tfrac{651}{960} \end{aligned}$$

Example 1.8. Current flows from P to Q through three switches a, b, and c as shown in Figure 1.3.

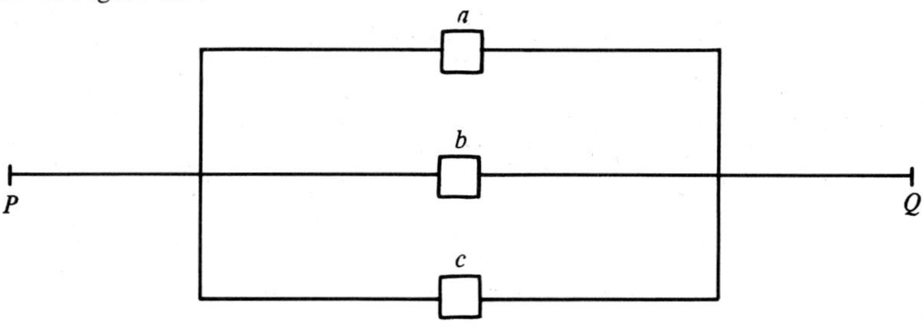

Figure 1.3

The probability that switch a is closed is 0.6, that b is closed is 0.8, and that c is closed is 0.9. Also the probability that a and b are closed is 0.48, that a and c are closed is 0.54, and that b and c are closed is 0.72. Finally, the probability that all the switches are closed is 0.432. Find the probability that–

(*a*) current passes from P to Q
(*b*) current passes from P to Q through precisely one switch
(*c*) current passes from P to Q through precisely two switches..

Solution. Let A, B, and C represent, respectively, the events that current passes through switch a, through switch b, and through switch c.

(*a*) Here we want to find $P(A \cup B \cup C)$ because current will flow from P to Q if and only if it goes through at least one of the switches. Hence, the probability that the current will flow is equal to

$$\begin{aligned} &P(A) + P(B) + P(C) - P(AB) - P(AC) - P(BC) + P(ABC) \\ &= 0.6 + 0.8 + 0.9 - 0.48 - 0.54 - 0.72 + 0.432 \\ &= 0.992 \end{aligned}$$

(*b*) The probability that current will pass from P to Q precisely through one switch is, by Example 1.5(*b*), equal to

$$\begin{aligned} &P(A) + P(B) + P(C) - 2[P(AB) + P(AC) + P(BC)] + 3P(ABC) \\ &= 0.6 + 0.8 + 0.9 - 2[0.48 + 0.54 + 0.72] + 3(0.432) \\ &= 0.116 \end{aligned}$$

(*c*) By Example 1.5(*c*), this probability is equal to $P(AB) + P(AC) + P(BC) - 3P(ABC) = 0.444$. ■

The following consequence is referred to as the **continuity property of the probability measure.** We shall find it useful in the discussion of what are called distribution functions, to be encountered in Chapter 4.

(C6) If $\{A_n\}$ is a monotone sequence of events, then

$$\underbrace{P(\underbrace{\lim_{n\to\infty} A_n}_{\text{set}})}_{\text{real number}} = \underbrace{\lim_{n\to\infty} \underbrace{P(A_n)}_{\text{real number}}}_{\text{real number}}$$

We shall first prove the result for expanding sequences and then for contracting sequences:

(*i*) Assume that the sequence is expanding; that is, $A_1 \subset A_2 \subset \ldots$. In this case we know that $\lim_{n\to\infty} A_n = \bigcup_{n=1}^{\infty} A_n$. This limit can be written as a countable union of mutually exclusive events as

$$\lim_{n\to\infty} A_n = \bigcup_{n=1}^{\infty} A_n = A_1 \cup (A_2 - A_1) \cup (A_3 - A_2) \cup \ldots$$

(See the Venn diagram in Figure 1.4. Notice that the region between A_n and A_{n-1} is $A_n - A_{n-1}$.)

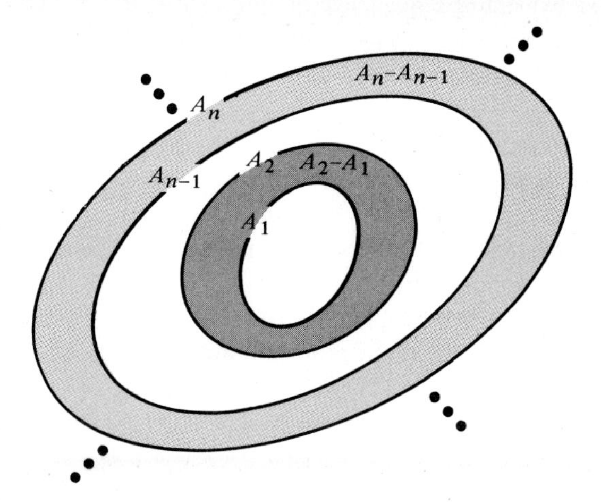

Figure 1.4 An expanding sequence of sets

Since the above represents a disjoint union,

$$P(\lim_{n\to\infty} A_n) = P(A_1) + P(A_2 - A_1) + P(A_3 - A_2) + \ldots$$

Now the partial sum of the first n terms on the right hand side is

$$\begin{aligned} &P(A_1) + P(A_2 - A_1) + \ldots + P(A_n - A_{n-1}) \\ &\quad = P(A_1) + P(A_2) - P(A_1) + \ldots + P(A_n) - P(A_{n-1}), \qquad \text{since } A_{i-1} \subset A_i \\ &\quad = P(A_n) \end{aligned}$$

Hence, in conclusion, $P(\lim_{n\to\infty} A_n) = \lim_{n\to\infty} P(A_n)$.

(*ii*) Now assume that the sequence is contracting, that is, $A_1 \supset A_2 \supset \ldots$. In this case, $\lim_{n\to\infty} A_n = \bigcap_{n=1}^{\infty} A_n$ which, using De Morgan's law, is equal to $\left(\bigcup_{n=1}^{\infty} A_n'\right)'$. Hence

$$P(\lim_{n\to\infty} A_n) = P\left(\left(\bigcup_{n=1}^{\infty} A_n'\right)'\right) = 1 - P\left(\bigcup_{n=1}^{\infty} A_n'\right)$$

Now $\{A_n\}$ is a contracting sequence. Consequently $\{A_n'\}$ is an expanding sequence. Therefore, using the result for expanding sequences,

$$P\left(\bigcup_{n=1}^{\infty} A_n'\right) = \lim_{n\to\infty} P(A_n') = \lim_{n\to\infty} (1 - P(A_n)) = 1 - \lim_{n\to\infty} P(A_n)$$

Thus, substituting, if A_n is a contracting sequence,

$$P(\lim_{n\to\infty} A_n) = 1 - (1 - \lim_{n\to\infty} P(A_n)) = \lim_{n\to\infty} P(A_n)$$

As a simple illustration, if $\{A_n\}$ is a contracting sequence with $\lim_{n\to\infty} A_n = \emptyset$, then

$$\lim_{n\to\infty} P(A_n) = P(\lim_{n\to\infty} A_n) = P(\emptyset) = 0$$

Also, if $\{A_n\}$ is an expanding sequence with $\lim_{n\to\infty} A_n = S$, then

$$\lim_{n\to\infty} P(A_n) = P(\lim_{n\to\infty} A_n) = P(S) = 1$$

Example 1.9. Suppose probabilities are assigned to the subsets of $S = [0, 1]$ in such a way that, if $0 \leqslant a < b \leqslant 1$, then $P((a, b]) = b - a$. (There are some mathematical subtleties that this engenders, but such considerations are extraneous to the thrust of our discussion and will not concern us here. We shall allude to these briefly in the digression that follows.) Find:

(*a*) $P(\{r\})$, for any real number $r \in [0, 1]$
(*b*) $P(\mathbf{Q})$, where $\mathbf{Q}$ is the set of rational numbers
(*c*) $P([a, b])$, where $0 \leqslant a < b \leqslant 1$

Solution

(*a*) For any real number r we can write $\{r\} = \bigcap_{n=1}^{\infty} \left(r - \frac{1}{n}, r\right]$. Since $\left\{\left(r - \frac{1}{n}, r\right]\right\}$ is a contracting sequence of intervals, we get

$$P(\{r\}) = P\left(\bigcap_{n=1}^{\infty} \left(r - \frac{1}{n}, r\right]\right) = \lim_{n\to\infty} P\left(\left(r - \frac{1}{n}, r\right]\right) = \lim_{n\to\infty} \frac{1}{n} = 0$$

Hence singleton sets are assigned zero probability.

(*b*) Since the set of rational numbers is countable, we can enumerate them as $r_1, r_2, \ldots$. Hence

$$P(\mathbf{Q}) = P\left(\bigcup_{n=1}^{\infty} \{r_n\}\right) = \sum_{n=1}^{\infty} P(\{r_n\})$$

by countable additivity. Therefore, $P(\mathbf{Q}) = 0$.

(*c*) We shall find $P([a, b])$ in two ways. One way is simply to note that $P([a, b]) = P(\{a\} \cup (a, b]) = P(\{a\}) + P((a, b]) = b - a$, since $P(\{a\}) = 0$ by part (*a*).

An alternate way is to note that $[a, b] = \bigcap_{n=1}^{\infty} \left(a - \frac{1}{n}, b\right]$, which yields

$$\begin{aligned} P([a, b]) &= P\left(\bigcap_{n=1}^{\infty} \left(a - \frac{1}{n}, b\right]\right) \\ &= \lim_{n\to\infty} P\left(\left(a - \frac{1}{n}, b\right]\right) \qquad \text{(why?)} \\ &= \lim_{n\to\infty} \left(b - a + \frac{1}{n}\right) = b - a \end{aligned}$$

■

Digression. We have dwelled at length on the concept of a sigma field. Is this concept essential? The following discussion will indicate its merits.

Suppose we pick a point at random in the interval $[0, 1]$. Hence, the sample space consists of all real numbers x such that $0 \leqslant x \leqslant 1$; that is, $S = [0, 1]$. If $(a, b]$, where $0 \leqslant a < b \leqslant 1$, is an interval, then it seems reasonable to assume that the probability that the point lies in the interval $(a, b]$ is equal to the length of the interval. Let us therefore define a function P by

$$P((a, b]) = b - a$$

Note that, for any subset A of $[0, 1]$, $0 \leqslant P(A) \leqslant 1$ and $P(S) = 1$. Hence axioms ($P1$) and ($P2$) of probability are satisfied.

Now, using a highly sophisticated argument which relies heavily on what is called the **axiom of choice**, it is possible to express the interval $[0, 1]$ as a countable union of disjoint sets E_i, $i = 1, 2, \ldots,$ with $P(E_i)$ the *same* for each set. Thus, we have

$$[0, 1] = \bigcup_{i=1}^{\infty} E_i$$

where $E_iE_j = \emptyset$ if $i \neq j$, and $P(E_i)$ is the same for each $i = 1, 2, \ldots.$

However, by axiom ($P3$) of probability, we ought to have

$$P([0, 1]) = P\left(\bigcup_{i=1}^{\infty} E_i\right) = \sum_{i=1}^{\infty} P(E_i)$$

This leads to an inconsistency because $P([0, 1]) = 1$, whereas $\sum_{i=1}^{\infty} P(E_i)$ is either zero (when $P(E_i) = 0$) or infinity (when $P(E_i) > 0$).

Hence, if $S = [0, 1]$, there does not exist a set function that coincides with the length on the subintervals and which at the same time satisfies the third axiom ($P3$) in a consistent way. We would not face this situation if we had confined ourselves to the Borel field of subsets of $[0, 1]$. The sets E_i mentioned above are *not* members of the Borel field. However, they are, of course, members of the power set of $[0, 1]$.

EXERCISES—SECTION 1

1. Suppose A and B are mutually exclusive events for which $P(A) = 0.4$ and $P(B) = 0.3$. Find the following probabilities:
   (a) $P(A')$ (b) $P(AB)$ (c) $P(A \cup B)$
   (d) $P(AB')$ (e) $P((AB)')$ (f) $P((A \cup B)')$
2. Suppose A, B, and C are mutually exclusive events for which $A \cup B \cup C = S$. If $P(A) = 2P(B) = 3P(C)$, find:
   (a) $P(A \cup B)$ (b) $P(AB')$ (c) $P(A'B'C')$
   (d) $P(A' \cup B' \cup C')$ (e) $P(A'(B \cup C))$ (f) $P(A(B' \cup C'))$
3. (a) If $P(ABC) = 0.2$ and $P(A) = 0.8$, find $P(A(B' \cup C'))$.
   (b) If $P(A) = 0.6$, $P(AB) = P(AC) = 0.35$, and $P(ABC) = 0.2$, find $P(AB'C')$.
4. Show that $P(A) = P(B)$ if and only if $P(AB') = P(A'B)$.
5. If A, B, C are three events, show that $P(ABC) = P(AC) + P(BC) - P((A \cup B)C)$.
6. If A and B are any two events, show that $|P(A) - P(B)| \leqslant P((AB') \cup (A'B))$.
7. The probability that a person is a lawyer is 0.64, the probability that he is a liar is 0.75, and the probability that he is a liar but not a lawyer is 0.25. Find the probability that—
   (a) he is a lawyer and a liar
   (b) he is a lawyer or a liar
   (c) he is neither a lawyer nor a liar.

8. A student is taking two courses, History and English. If the probability that he will pass either of the courses is 0.7, that he will pass both the courses is 0.2, and that he will fail in History is 0.6, find the probability that–
   (a) he will pass History
   (b) he will pass English
   (c) he will pass exactly one course.
9. Suppose A, B, C, D are four events. Derive an expression for the probability that exactly k of the events occur ($k = 1, 2, 3, 4$) in terms of the probabilities of their intersections.
10. Ann, Betty, Cathy, and Dorothy are invited to attend a party. Let $A, B, C,$ and D represent respectively the events that Ann, Betty, Cathy, and Dorothy attend the party. If $P(A) = P(B) = P(C) = P(D) = 0.6$, $P(AB) = P(AC) = P(AD) = P(BC) = P(BD) = P(CD) = 0.36$, $P(ABC) = P(ABD) = P(ACD) = P(BCD) = 0.216$, and $P(ABCD) = 0.1296$, find the probability that exactly k girls attend the party, $k = 0, 1, 2, 3, 4$.
11. Suppose $S = \{1, 2, \ldots\}$, and $P(\{i\}) = k/3^i$ for all $i \in S$, where k is a constant.
   (a) Determine k.
   (b) Find the probability of (i) the set of even numbers, (ii) the set of odd numbers.
12. Prove by induction that

$$P\left(\bigcup_{i=1}^{n} A_i\right) = \sum_{i=1}^{n} P(A_i) - \sum_{\substack{i,j=1\\ i<j}}^{n} P(A_iA_j) + \sum_{\substack{i,j,k=1\\ i<j<k}}^{n} P(A_iA_jA_k) - \ldots + (-1)^{n+1} P(A_1A_2 \ldots A_n)$$

13. Establish the following inequalities:
   (a) $P(AB) \geqslant 1 - P(A') - P(B')$
   (b) $P\left(\bigcup_{i=1}^{\infty} A_i\right) \leqslant \sum_{i=1}^{\infty} P(A_i)$
   (c) $P\left(\bigcap_{i=1}^{\infty} A_i\right) \geqslant 1 - \sum_{i=1}^{\infty} P(A_i')$
14. Suppose $\{A_i\}$ is a sequence of events where $P(A_i) \geqslant 1 - [(0.1)/3^i]$, $i = 1, 2, \ldots$. Determine a lower bound for $P\left(\bigcap_{i=1}^{\infty} A_i\right)$.

2. FINITE SAMPLE SPACES

We shall devote the rest of this chapter to the discussion of finite sample spaces; that is, sample spaces which have only a finite number of outcomes. Consider a sample space S with N outcomes so that we can write S as $S = \{s_1, s_2, \ldots, s_N\}$. Recall that the power set of S has 2^N members. In other words, there are 2^N possible events. If we define a function P which assigns numerical values to these events in a way which is consistent with the three probability axioms (($P1$), ($P2$), and ($P3$)), then the function P is a probability measure. Thus, in order to define a probability function on a sample space with N outcomes, we need specify at most 2^N values. In practice, this is accomplished by assigning probabilities to the

N elementary events. This is sufficient. Probabilities are then assigned in a natural way to all the events as follows:

Suppose $A = \{s_{i_1}, s_{i_2}, \ldots, s_{i_k}\}$ with k outcomes. Then A can be expressed as the union of k mutually exclusive elementary events as

$$A = \{s_{i_1}\} \cup \{s_{i_2}\} \cup \ldots \cup \{s_{i_k}\}$$

Using axiom ($P3$), we therefore get

$$P(A) = P(\{s_{i_1}\}) + \ldots + P(\{s_{i_k}\})$$

Thus, for a finite sample space, the probability of an event A is equal to the sum of the probabilities assigned to each of the outcomes that make up the event A.

The classical definition of the probability of an event is based on two fundamental assumptions. One of these is to assume that the performance of an experiment results in a *finite* number of outcomes. The other is to assume that all the elementary events have the same probability; that is, the outcomes are **equally likely** or **equiprobable**.

In what follows let us assume that the outcomes are equally likely; that is,

$$P(\{s_1\}) = P(\{s_2\}) = \ldots = P(\{s_N\}) = p$$

Then, since $\sum_{i=1}^{N} P(\{s_i\}) = 1$, we get $Np = 1$, so that $p = \frac{1}{N}$.

Hence, if the outcomes of a sample space S with N outcomes are equally likely, then the probability of each elementary event is $1/N$, the reciprocal of the number of outcomes in S.

Next, suppose A is an event with k outcomes, $s_{i_1}, s_{i_2}, \ldots, s_{i_k}$. Then, as we have already seen,

$$P(A) = P(\{s_{i_1}\}) + P(\{s_{i_2}\}) + \ldots + P(\{s_{i_k}\}) = \frac{k}{N}$$

Thus we have a very fundamental formula of classical probability:

If the sample space is finite and the outcomes are equally likely, then the probability of an event A is equal to the ratio of the number of outcomes in A to the number of outcomes in the sample space.

Comment. We shall have occasion to use phrases like "an *unbiased* coin is tossed," "a *fair* die is rolled," "an object is picked *at random,*" and so on. They are all meant to suggest that the outcomes in the sample space are equally likely.

Example 2.1. Let the sample space $S = \{s_1, s_2, s_3, s_4, s_5, s_6\}$ be given. Probabilities are assigned to some events as follows:

$$P(\{s_1, s_2, s_3\}) = P(\{s_2, s_4\}) = P(\{s_4, s_5, s_6\})$$

and

$$P(\{s_2\}) = 2P(\{s_4\})$$

Find: (*a*) $P(\{s_1, s_3\})$ (*b*) $P(\{s_1, s_2, s_3, s_4\})$

Solution. Let $P(\{s_4\}) = x$. Then $P(\{s_2, s_4\}) = P(\{s_2\}) + P(\{s_4\}) = 3x$, and consequently $P(\{s_1, s_2, s_3\}) = 3x$ and $P(\{s_4, s_5, s_6\}) = 3x$. Now

$$P(\{s_1, s_2, s_3, s_4, s_5, s_6\}) = P(\{s_1, s_2, s_3\}) + P(\{s_4, s_5, s_6\})$$

Hence,

$$1 = 3x + 3x \quad \text{and therefore} \quad x = \frac{1}{6}$$

As a result, $P(\{s_2\}) = \frac{1}{3}$.

$$(a)\ P(\{s_1, s_3\}) = P(\{s_1, s_2, s_3\}) - P(\{s_2\}) = \frac{1}{2} - \frac{1}{3} = \frac{1}{6}$$

$$(b)\ P(\{s_1, s_2, s_3, s_4\}) = P(\{s_1, s_2, s_3\}) + P(\{s_4\}) = \frac{1}{2} + \frac{1}{6} = \frac{2}{3}$$

Example 2.2. One integer is chosen at random from the numbers 1, 2, 3, . . . , 1000. What is the probability that the chosen number is (*a*) divisible by 6 or 8; (*b*) divisible by exactly two of the numbers 6, 8, 10.

Solution

(*a*) Let A be the event that the number is divisible by 6 and B the event that it is divisible by 8. Since we are picking one number,

$$S = \{1, 2, 3, \ldots, 1000\}, \quad \text{with 1000 outcomes}$$
$$A = \{6, 12, 18, \ldots, 990, 996\}, \quad \text{with 166 outcomes}$$
$$B = \{8, 16, 24, \ldots, 992, 1000\}, \quad \text{with 125 outcomes}$$

Now AB represents the event that the number is divisible by 6 *and* 8; that is, divisible by 24. Hence

$$AB = \{24, 48, 72, \ldots, 984\}, \quad \text{with 41 outcomes}$$

Hence

$$P(A) = \frac{166}{1000}, \quad P(B) = \frac{125}{1000}, \quad \text{and} \quad P(AB) = \frac{41}{1000}$$

The probability that the number is divisible by 6 or 8 is equal to $P(A \cup B) = P(A) + P(B) - P(AB) = \frac{1}{4}$.

(*b*) Let C be the event that the number is divisible by 10. Essentially, what we want to find is the probability that exactly two of the events A, B, C occur. This, by Example 1.5, is equal to $P(AB) + P(BC) + P(AC) - 3P(ABC)$. Now, arguing as in part (*a*), we get

$$P(AC) = \frac{33}{1000}, \quad P(BC) = \frac{25}{1000}, \quad \text{and} \quad P(ABC) = \frac{8}{1000}$$

Substituting, the desired probability is equal to

$$\frac{41}{1000}+\frac{25}{1000}+\frac{33}{1000}-3\cdot\frac{8}{1000}=\frac{3}{40}$$

EXERCISES–SECTION 2

1. Suppose a sample space with four outcomes is given as $S=\{s_1, s_2, s_3, s_4\}$. Probabilities are assigned to the elementary events so that $P(\{s_1\})=P(\{s_2\})$, and $P(\{s_3\})=P(\{s_4\})=2P(\{s_2\})$. Find:
   (a) $P(\{s_1, s_3\})$ (b) $P(\{s_1, s_2\}\cup\{s_1, s_3\})$
2. Four hundred people attending a party are each given a number, 1 to 400. One number is called out at random. What is the probability that the number picked–
   (a) is 123?
   (b) has the same three digits?
   (c) ends in 9?
   (d) is divisible by exactly two of the integers 6, 8, 10?
3. Four cards–a spade, a heart, a diamond, and a club–are laid on the table with their faces down. A card is picked. If the probability of picking a spade is twice that of picking a heart, the probability of picking a heart is three times that of picking a diamond, and the probability of picking a diamond is four times that of picking a club, find the probability of the following events:
   (a) The card is red.
   (b) The card is a spade or a diamond.
   (c) The card is black or a diamond.
4. Tom and John play a game in which they simultaneously exhibit their right hands with one, two, three, or four fingers extended. Find the probability that Tom and John extend the same number of fingers.
5. A letter is chosen at random from the letters of the English alphabet. What is the probability of each of the following events?
   (a) The letter is *j*.
   (b) The letter is one of the letters in the word *board*.
   (c) The letter is not in the word *board*.
   (d) The letter is in the word *card* or the word *board*.
   (e) The letter is in exactly one of the words *board* and *card*.

3. SAMPLING AND COUNTING TECHNIQUES

In dealing with problems in applied probability, one often comes across situations where there is a collection of distinct objects, say M in number, and one picks from this collection a certain number, say n. The procedure of picking objects is called **sampling**. For brevity, we shall say that we have drawn a sample of size n.

First of all, the mechanics of the sampling procedure are quite important. It is possible that an object is returned to the lot before the next one is picked. We refer to this as **sampling with replacement**. In this case, since the same object can be picked more than once, there is no limit to the size of n and it could be any positive integer. On the other hand, the object may not be returned to the lot after it is picked. This is called **sampling without replacement**. Obviously, with this type of sampling there is an upper limit on n; it can be at most equal to M.

In each case, whether the sampling is carried out with or without replacement, we may or may not be interested in the order in which the objects are picked. As a result, we have the following four situations:

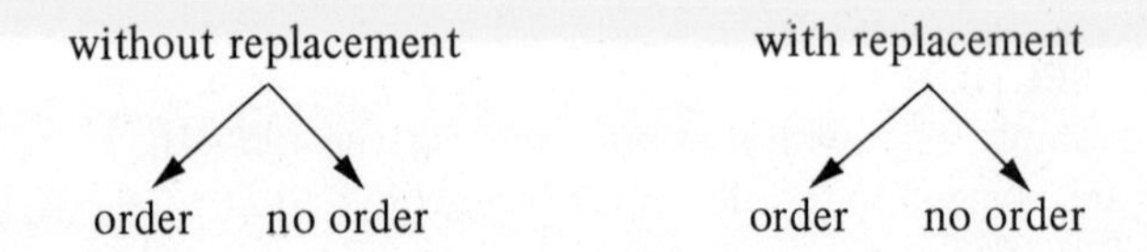

Let us consider the following illustration which brings out the essential ingredients of our discussion: Suppose there are four distinct objects represented by the letters *a, b, c, d,* and two of these letters are picked. The following four cases are possible:

	a	*b*	*c*	*d*
a		*ab*	*ac*	*ad*
b	*ba*		*bc*	*bd*
c	*ca*	*cb*		*cd*
d	*da*	*db*	*dc*	

Case 1. Without replacement, with order

	a	*b*	*c*	*d*
a		*ab*	*ac*	*ad*
b			*bc*	*bd*
c				*cd*
d				

Case 2. Without replacement, without order

	a	*b*	*c*	*d*
a	*aa*	*ab*	*ac*	*ad*
b	*ba*	*bb*	*bc*	*bd*
c	*ca*	*cb*	*cc*	*cd*
d	*da*	*db*	*dc*	*dd*

Case 3. With replacement, with order

	a	*b*	*c*	*d*
a	*aa*	*ab*	*ac*	*ad*
b		*bb*	*bc*	*bd*
c			*cc*	*cd*
d				*dd*

Case 4. With replacement, without order

In cases 1 and 2, the sampling is carried out without replacement, and consequently there are no possibilities like *aa, bb, cc, dd*. This explains why there are no entries along the diagonal in these cases. In case 2, moreover, we are not interested in order so that, for example, *ab* is listed, but not *ba*.

In case 3, we list all sixteen possibilities. In case 4, since the sampling is with replacement, we certainly have outcomes like *aa, bb, cc, dd*. However, since order is not relevant, *ab* is the same as *ba*, and so on. Hence there are no entries below the diagonal.

Comment. When order matters each possibility is called an **arrangement**, or a **permutation.** If order does *not* matter, it is called a **combination.**

Comment. When n objects are picked and the order is important, it is convenient to write the sample points as ordered n-tuples $(x_1, x_2, \ldots, x_n)$ where the ith component x_i represents the ith object picked. Thus x_1 represents the result of the *first* draw, x_2 of the *second* draw, and so on.

We shall now provide a general formula in each of the above four cases. Towards this, we state the following basic rule of counting techniques.

The Basic Counting Principle If a certain experiment can be performed in r ways and, corresponding to each of these ways, another experiment can be performed in k ways, then the combined experiment can be performed in rk ways.

To understand this principle, suppose the outcomes of the first experiment are written as $A = \{a_1, a_2, \ldots, a_r\}$ and those of the second experiment as $B = \{b_1, b_2, \ldots, b_k\}$. Then the outcomes of the combined experiment can be represented in a rectangular array as ordered pairs (a_i, b_j):

	b_1	b_2	$\ldots$	b_j	$\ldots$	b_k
a_1	(a_1, b_1)	(a_1, b_2)	$\ldots$	(a_1, b_j)	$\ldots$	(a_1, b_k)
a_2	(a_2, b_1)	(a_2, b_2)	$\ldots$	(a_2, b_j)	$\ldots$	(a_2, b_k)
$\vdots$	$\vdots$					
a_i	(a_i, b_1)	(a_i, b_2)	$\ldots$	(a_i, b_j)	$\ldots$	(a_i, b_k)
$\vdots$	$\vdots$					
a_r	(a_r, b_1)	(a_r, b_2)	$\ldots$	(a_r, b_j)	$\ldots$	(a_r, b_k)

In other words, the outcomes of the combined experiment can be represented as the Cartesian product $A \times B$. Clearly, there are rk pairs. Indeed, this shows that $\mathbf{n}(A \times B) = \mathbf{n}(A) \times \mathbf{n}(B)$.

Another way of illustrating the above principle is by a **tree diagram**, as shown in Figure 3.1. First we list all the outcomes of one experiment, and then, corresponding to each of these, those of the other experiment. The total number of branches, namely rk, gives all the combined possibilities.

The basic counting principle can be extended to any number of experiments in an obvious way. We shall now give some examples.

(*i*) If a die is tossed twice, then there are $6 \times 6 = 36$ possible outcomes.

(*ii*) If a person has 3 different shirts, 6 different ties, and 5 different jackets, then he can get dressed for an occasion in $8 \times 6 \times 5 = 240$ ways.

(*iii*) If the purchaser of an automobile has a choice of 3 makes, 5 body styles, and 6 colors, then he can choose from $3 \times 5 \times 6 = 90$ different models.

(*iv*) Suppose license plates are formed with three distinct letters followed by three distinct digits. Then there are 26 choices for the first letter, 25 for the second, and 24 for the third. Also, there are 10 choices for the first digit, 9 for the second, and 8 for the third. Therefore, there are $26 \times 25 \times 24 \times 10 \times 9 \times 8 = 11{,}232{,}000$ different license plates.

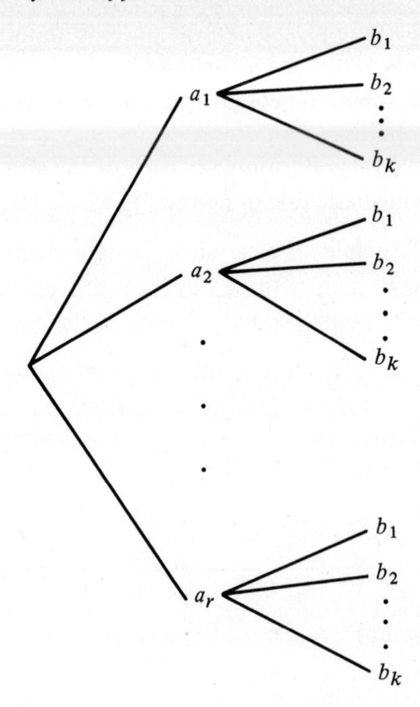

Figure 3.1 A tree diagram

Case 1: **Without replacement, with order (permutations)**

We shall begin with an example. Suppose the initial collection of objects is the set $\{a, b, c, d\}$ and we pick three objects. If we list all the possibilities we get the following arrangements:

abc	*abd*	*acd*	*bcd*
acb	*adb*	*adc*	*bdc*
bca	*bad*	*cad*	*cbd*
bac	*bda*	*cda*	*cdb*
cab	*dab*	*dac*	*dbc*
cba	*dba*	*dca*	*dcb*

The order in which the letters are written is important. For example, *abc* and *cab* are different arrangements, even though both use the same letters *a, b, c*. Any particular arrangement is called a permutation. In the above listing there are 24 permutations in all. The reason is simple. We are picking three letters out of four. The first letter can be picked in 4 ways; once this is done, the second letter can be picked in 3 ways; and after drawing the first two letters, the third can be chosen in 2 ways. Therefore, by the basic counting rule, the total number of ways is $4 \cdot 3 \cdot 2 = 24$.

Turning to the general case, suppose there are M distinct objects and n of these are picked without replacement. Any particular arrangement of n objects is called a **permutation.**

Let us denote the total number of permutations by $(M)_n$. There are M ways to draw the first object, $M - 1$ ways to draw the second object, $M - 2$ ways to draw the third object, and so on. Finally, there are $M-(n-1)$ ways to draw the nth object. Thus a direct application of the basic rule gives:

The number of permutations (or arrangements) of n objects selected from a collection of M distinct objects is

$$(M)_n = M(M-1)(M-2)\ldots(M-n+1)$$

For example, the number of permutations when 4 objects are picked from 8 distinct objects is $8 \times 7 \times 6 \times 5 = 1680$.

In particular, $(M)_M$, the number of permutations (or arrangements) of M objects taken all together, is $M(M-1)(M-2)\ldots 3 \cdot 2 \cdot 1$. Such a product of consecutive integers is denoted by $M!$ and is read M **factorial.**

Using the factorial notation, we can write $(M)_n$ as

$$(M)_n = \frac{M!}{(M-n)!}$$

We know that $(M)_M = M!$. Hence, for the above formula to be meaningful when $n = M$, it is convenient to define

$$0! = 1$$

We give the following examples:

(*i*) If there are 100 entries in a contest, the number of ways in which three different prizes—say a first, second, and third prize—can be awarded if no contestant can win more than one prize is $100 \times 99 \times 98 = 970{,}200$. (Keep in mind that, say, Henry winning the first prize, June the second prize, and Raquel the third prize is different from, say, Raquel winning the first prize, Henry the second prize, and June the third prize).

(*ii*) The number of ways of arranging the letters in the word *number* is 6 factorial, that is, 720.

(*iii*) Suppose there are 10 flags of different colors. Using 4 different flags in a row one can give $10 \cdot 9 \cdot 8 \cdot 7 = 5040$ different signals. (Notice, for example, that a signal red, blue, green, yellow, in that order, is different from red, yellow, green, blue.)

(*iv*) The number of ways in which 4 people can be seated in a row of 7 chairs is $7 \times 6 \times 5 \times 4 = 840$.

(*v*) The number of three-digit numbers that can be formed from 2, 3, 6, 7, 8, 9, if no digit is to be repeated, is $6 \times 5 \times 4 = 120$.

Case 2: **Without replacement, without order (combinations)**

We shall discuss this case in conjunction with case 1. We have seen that if we pick three letters out of *a, b, c,* and *d,* and if order is important, then we get 24 permutations. In the present case, however, we are not interested in order, and as such there are just 4 possibilities, namely, *abc, abd, acd,* and *bcd*. Each of these possibilities is called a combination. Among the 24 permutations of case 1 the first column consists of the permutations of the letters *a, b, c* and, as we know, there are 3! of these. This is why there are 3! = 6 arrangements in column 1. The same is true of columns 2, 3, and 4. Consequently, we get from our example, that the number of combinations, *multiplied* by 3!, is the number of permutations.

Let us now take up the general case where we pick n objects without replacement from M distinct objects, where order is not important. Symbolically, we shall denote the number of ways of doing this by $\binom{M}{n}$ and call it the number of combinations of n objects from a set of M. Our objective is to derive an expression for $\binom{M}{n}$.

Towards this, we see that if a combination has n elements, then there are $n!$ possible arrangements of its elements. *Each* combination gives rise to $n!$ arrangements, thereby giving rise to all the permutations, namely, $M(M-1)\ldots(M-n+1)$. Hence we have

$$\binom{M}{n}\cdot n! = M(M-1)\ldots(M-n+1) = \frac{M!}{(M-n)!}$$

Therefore,

$$\boxed{\binom{M}{n} = \frac{M!}{n!(M-n)!}}$$

is the number of unordered samples of size n that can be drawn without replacement from M distinct objects.

For example:

(*i*) The number of ways of choosing a set of 3 books to read from a set of 8 books is $\binom{8}{3} = \frac{8!}{3!5!} = 56$. (Note that we are not interested in the order in which the books are read.)

(*ii*) The number of ways in which a five-card poker hand can be dealt from a deck of 52 cards is

$$\binom{52}{5} = \frac{52!}{5!47!} = 2{,}598{,}960$$

(*iii*) From a group of 8 seniors, 6 juniors, and 4 sophomores, there are $\binom{18}{5}$ ways of picking a five-member committee.

For convenience, the following conventions are adopted:

$$\binom{M}{0} = 1, \quad \text{and} \quad \binom{M}{n} = 0 \quad \text{if } n < 0 \quad \text{or} \quad n > M$$

Comments. (1) Picking n objects out of M to form a group is tantamount to picking $M - n$ objects out of M *not* to belong to the group. Thus, for example, the number of ways of choosing 3 books to read from a set of 8 books is the same as the number of ways of picking 5 books not to read from the 8. Therefore, we always have

$$\binom{M}{n} = \binom{M}{M-n}$$

This can also be seen by observing that $\binom{M}{n}$ and $\binom{M}{M-n}$ are both equal to $\frac{M!}{n!(M-n)!}$.

(2) For *any* two real numbers x and y the expansion of $(x + y)^M$ can be written as

$$(x + y)^M = \sum_{n=0}^{M} \binom{M}{n} x^n y^{M-n}$$

This is called the binomial expansion. Since $\binom{M}{n}$ occurs as the coefficient of $x^n y^{M-n}$ in the binomial expansion, $\binom{M}{n}$, $n = 0, 1, \ldots, M$, are called the **binomial coefficients.**

(3) If a set has M objects, then *the number of different subsets of size n is* $\binom{M}{n}$. This is because, as we know, order is not important in listing the members of a set.

(4) We have mentioned above the following identity, which holds for any real numbers x, y:

$$(x + y)^M = \binom{M}{0} y^M + \binom{M}{1} y^{M-1} x + \binom{M}{2} y^{M-2} x^2 + \ldots + \binom{M}{M} x^M$$

In particular, if we set $x = y = 1$ we get

$$2^M = \binom{M}{0} + \binom{M}{1} + \ldots + \binom{M}{M}$$

This shows that *the total number of subsets that can be formed from a set with M elements is* 2^M. (Recall that we mentioned in Chapter 1 that the power set of a set with n elements has 2^n members.)

Case 3: **With replacement, with order**

The number of ways of picking n objects from M distinct objects is M^n when the objects are picked with replacement and when order is important. This is easy to see because at every draw there are M different choices.

For example:

(*i*) With the eight digits 1, 2, 3, 4, 5, 7, 8, 9, one can form 8^3 distinct three-digit numbers.

(*ii*) If there are M cells, then n objects can be placed in them in M^n ways. (We are assuming that a cell can have more than one object.) Placing an object in a cell amounts to picking one of the M cells, and allowing a cell to have more than one object amounts to sampling with replacement.

(*iii*) If 10 people are in a train which stops at 6 stations, then there are 6^{10} possible ways that the 10 can get off the train. Notice that a person can get off at any one of the 6 stations so that he has 6 choices. This is true of each of the 10 people. Also, if one person gets off at a station, it does not preclude other persons from getting off at that same station.

Case 4: **With replacement, without order**

The derivation of a general formula in this case is rather tricky and we shall not pursue the matter here. For our purpose it will suffice to know that *the number of unordered samples of size n when objects are picked with replacement from M distinct objects is* $\binom{M+n-1}{n}$.

For example, the number of ways of placing n *nondistinguishable* balls into M cells is $\binom{M+n-1}{n}$. (Try to see the analogy between the indistinguishable balls and the irrelevance of order.)

We shall now define a rather important term used in statistics. A sample is said to be a **random sample** if the sampling procedure assigns the same probability to each sample point. Thus a sample is random if its probability of being drawn is–

(1) $\dfrac{1}{(M)_n}$ if sampling is ordered, without replacement

(2) $\dfrac{1}{\binom{M}{n}}$ if sampling is unordered, without replacement

(3) $\dfrac{1}{M^n}$ if sampling is ordered, with replacement

(4) $\dfrac{1}{\binom{M+n-1}{n}}$ if sampling is unordered, with replacement.

Example 3.1. Suppose 5 cards are picked without replacement from a standard bridge deck of 52 cards. What is the probability that there are 3 black cards and 2 red cards? We shall consider the following two cases:

(*a*) the cards are seen one by one;

(*b*) the cards are seen all at once.

Solution

(*a*) In this case we are interested in the order. Since we are picking 5 cards without replacement and the order is relevant, there are $(52)_5 = 52 \times 51 \times 50 \times 49 \times 48$ possible outcomes in the sample space. For example, three of the outcomes in this sample space can be written as $(K_{sp}, J_h, 3_h, 7_d, 8_{cl})$, $(J_h, 7_d, 3_h, 8_{cl}, K_{sp})$, $(A_h, Q_h, J_d, 8_{cl}, 6_d)$. How many of these $(52)_5$ outcomes are favorable to the event that there are 3 black cards and 2 red cards? Let us call this event the event A. First of all we observe that there are 5 locations, of which 3 are to be assigned to the black cards and 2 to the red cards. This can be done in $\binom{5}{3} = 10$ ways. Consider just one of these, and say we have black cards in the first, third, and fourth locations, and red cards in the second and fifth. There are $26 \times 25 \times 24$ ways of filling the first, third, and fourth locations with the black cards, and corresponding to any of these there are 26×25 ways to fill locations two and five with red cards. Hence, by the basic rule of counting, there are $\binom{5}{3} \times 26 \times 25 \times 24 \times 26 \times 25$ outcomes favorable to A. Hence

$$P(A) = \frac{\binom{5}{3}(26)_3(26)_2}{(52)_5} = \frac{\binom{26}{3}\binom{26}{2}}{\binom{52}{5}}$$

(*b*) In this case order is not of interest. Hence there are $\binom{52}{5}$ possible samples of size 5. Now there are $\binom{26}{3}$ ways of picking 3 black cards out of the 26 black cards, and corresponding to each of these ways there are $\binom{26}{2}$ ways of picking the 2 red cards. Therefore, by the basic counting rule, there are $\binom{26}{3}\binom{26}{2}$ possible samples of size 5 where each sample has exactly 3 black cards and 2 red cards. Consequently,

$$P(A) = \frac{\binom{26}{3}\binom{26}{2}}{\binom{52}{5}}$$

■

Comment. In Example 3.1 we see that $P(A)$ is the same in both cases so that, for finding $P(A)$, it does not make any difference whether we observe the cards all at once or one by one. It should be borne in mind that the event A specifies that there are so many cards of one kind (black) and so many of the other kind (red).

Example 3.2. If a person is dealt 13 cards from a standard deck of cards, what is the probability that he is dealt:

(*a*) the complete suit of spades;

(*b*) a complete suit?

Solution

(*a*) There are $\binom{52}{13}$ possible hands of 13 cards each. Of these, there is only one hand with all spades. (This also follows from the fact that there are $\binom{13}{13}\binom{39}{0} = 1$ hand with 13 spades.) Hence, the probability of a complete suit of spades is $1 \div \binom{52}{13}$.

(*b*) A complete suit is dealt if a person gets a spade suit *or* a heart suit *or* a diamond suit *or* a club suit. Since the events are mutually exclusive, and since each has a probability of $1 \div \binom{52}{13}$, the desired probability is $4 \div \binom{52}{13}$.

Example 3.3. Suppose 8 cards are dealt from a standard deck of 52 cards. Find the probability that the hand will contain:

(*a*) all face cards in spades (that is, king, queen, and jack of spades);
(*b*) all face cards in at least one suit.

Solution. Since 8 cards are dealt, there are $\binom{52}{8}$ possible hands.

(*a*) How many hands are there with the king, queen, and jack of spades? Since the hands contain 8 cards and they are supposed to contain the king, queen, and jack of spades, this is equal to $\binom{1}{1}\binom{1}{1}\binom{1}{1}\binom{49}{5}$, or, arguing that 3 spade face cards are picked out of 3, $\binom{3}{3}\binom{49}{5}$. In either case, there are $\binom{49}{5}$ hands. Hence the probability is equal to $\binom{49}{5} \div \binom{52}{8}$.

(*b*) Let A represent the event that the hand contains K, Q, J of spades, B that the hand contains K, Q, J of hearts, C that it contains K, Q, J of diamonds, and D that it contains K, Q, J of clubs.

We are interested in finding $P(A \cup B \cup C \cup D)$, which, as we know, is equal to $P(A) + P(B) + P(C) + P(D) - [P(AB) + P(AC) + P(BC) + P(AD) + P(BD) + P(CD)] + [P(ABC) + P(ABD) + P(ACD) + P(BCD)] - P(ABCD)$.

Notice that A and B are not mutually exclusive because a hand could contain all face cards in spades and all face cards in hearts. A similar comment holds for any two events taken at a time.

However, a hand of 8 cards cannot contain all face cards in three different suits, or four different suits. Therefore,

$$ABC = ABD = ACD = BCD = ABCD = \emptyset$$

and hence their probabilities are zero.

Now, from (*a*),

$$P(A) = \frac{\binom{49}{5}}{\binom{52}{8}} = P(B) = P(C) = P(D)$$

Also, the probability of the intersection of *any pair* of events is the same, and, for example,

$$P(AB) = \frac{\binom{46}{2}}{\binom{52}{8}} \qquad \text{(why?)}$$

Hence

$$P(A \cup B \cup C \cup D) = \frac{4\binom{49}{5}}{\binom{52}{8}} - \frac{6\binom{46}{2}}{\binom{52}{8}}$$

Example 3.4. An urn contains the following numbers: 2, 3, 7, 8, 12, 15, 17, 21, 28. Six numbers are picked without replacement. What is the probability that the third largest number is 15?

Solution. There are $\binom{9}{6}$ ways of picking 6 numbers from the 9 numbers.

If the third largest number is 15, then 2 of the numbers picked must be larger than 15. These will have to come from 17, 21, 28. There are $\binom{3}{2}$ ways of picking 2 numbers out of 3.

Three of the numbers picked will be less than 15. These will have to be picked from 2, 3, 7, 8, 12. The number of ways of doing this is $\binom{5}{3}$.

Combining these results, by the basic counting rule, there are $\binom{3}{2}\cdot\binom{5}{3}$ outcomes in our event.

Hence the probability that the third largest number is 15 is

$$\frac{\binom{3}{2}\binom{5}{3}}{\binom{9}{6}} = \frac{5}{14}$$

■

Hypergeometric probabilities Some of the problems that we have encountered so far bring out a fairly common and important feature of some sampling problems. The setup is as follows: There are a certain number of objects in a collection and they represent a mixture of two kinds of objects. A sample of size n is picked *without replacement,* and the interest is in finding the probability that among the objects picked there are so many of one kind, with the remaining objects being of the other kind. These probabilities are called *hypergeometric probabilities*.

From Example 3.1, we see that these probabilities are the same whether the objects are seen one by one or all at once. We shall now derive a general expression.

Suppose the lot consists of M objects of which D are defective and $M - D$ are nondefective. If n of these objects are picked *without replacement,* we want to find the probability that there are exactly k defective objects. Naturally, $0 \leq n \leq M$; also $k \leq D$ and $k \leq n$. The situation is depicted in Figure 3.2.

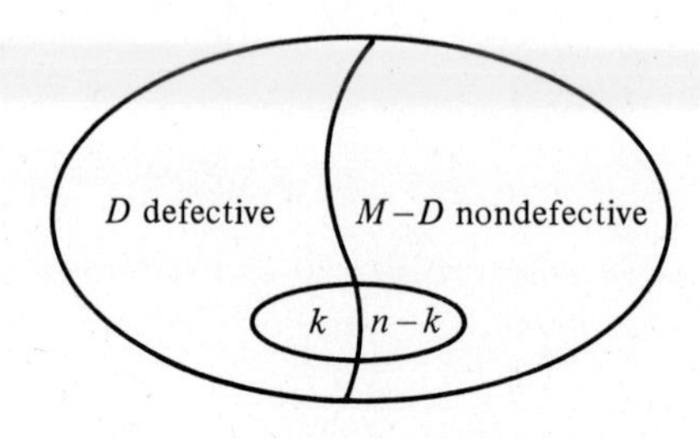

Figure 3.2

Let us denote the event of interest by A_k. Since we are picking n objects out of M, there are $\binom{M}{n}$ possible samples. There are $\binom{D}{k}$ ways of picking k of the D defective objects. Corresponding to each of these, there are $\binom{M-D}{n-k}$ ways of picking $n-k$ nondefective objects. Therefore, by the basic counting rule, there are $\binom{D}{k}\binom{M-D}{n-k}$ possible samples with k defective and $n-k$ nondefective objects. Hence

$$P(A_k) = \frac{\binom{D}{k}\binom{M-D}{n-k}}{\binom{M}{n}}, \qquad k = 0, 1, \ldots, n$$

Comments. (1) $P(A_k) = 0$ if $k > D$, since we cannot pick more defective items than there are in the lot. Also $P(A_k) = 0$ if $k > n$, since in a sample of size n we cannot have more than n defective items. These situations are taken care of in the above formula for $P(A_k)$ by the fact that $\binom{a}{b} = 0$ if $b < 0$ or $b > a$.

(2) Take note of how the indices on top in the numerator, namely D, $M-D$, add up to the index on top in the denominator, and the indices at the bottom in the numerator, namely k, $n-k$, add up to the index at the bottom in the denominator.

Example 3.5. Suppose a box contains 100 ball bearings. The following inspection plan has been formulated: Twenty randomly drawn ball bearings are to be tested. If the number of defective ball bearings in the sample does not exceed 2, then the box is accepted as good and shipped. What is the probability that the box is shipped if in fact it has 8 defective ball bearings? (We shall interpret this to mean *exactly* 8 defective ball bearings, as opposed to *at least* 8. *In the sequel, "at least" will always be explicitly indicated.*)

Solution. Let A_i, $i = 0, 1, 2$, be the event that there are i defective ball bearings among the 20 that are randomly picked. The box will be shipped if the number of defectives is less than or equal to 2. Since the events A_0, A_1, and A_2 are mutually exclusive, the desired probability is equal to $P(A_0) + P(A_1) + P(A_2)$. Now, if indeed there are 8 defective ball bearings,

$$P(A_i) = \frac{\binom{8}{i}\binom{92}{20-i}}{\binom{100}{20}}, \qquad i = 0, 1, 2, \ldots, 8$$

Hence the probability that the box will be shipped is equal to

$$\frac{\binom{8}{0}\binom{92}{20}}{\binom{100}{20}} + \frac{\binom{8}{1}\binom{92}{19}}{\binom{100}{20}} + \frac{\binom{8}{2}\binom{92}{18}}{\binom{100}{20}}$$

Example 3.6. A box contains 20 numbers, of which 8 are negative and 12 are positive. If 6 numbers are picked without replacement, find the probability that their product will be positive.

Solution. Six numbers are picked from 20. This can be done in $\binom{20}{6}$ ways.

The product of the 6 numbers will be positive if all the numbers are positive, *or* 4 are positive and 2 are negative, *or* 2 are positive and 4 are negative, or all are negative numbers. Letting A_i represent the event that among the 6 numbers there are i positive numbers (and consequently $6 - i$ negative numbers), what we want is $P(A_6 \cup A_4 \cup A_2 \cup A_0)$. Since the events are mutually exclusive, this probability is equal to $P(A_6) + P(A_4) + P(A_2) + P(A_0)$.

Now

$$P(A_i) = \frac{\binom{12}{i}\binom{8}{6-i}}{\binom{20}{6}}, \qquad i = 0, 1, 2, \ldots, 6$$

Hence the probability that the product is positive is

$$\frac{\binom{12}{6}\binom{8}{0} + \binom{12}{4}\binom{8}{2} + \binom{12}{2}\binom{8}{4} + \binom{12}{0}\binom{8}{6}}{\binom{20}{6}}$$

■

The hypergeometric probabilities have an obvious generalization. For example, suppose there are M objects of which D_1 are white, D_2 are black, D_3 are red, and the remaining $M - D_1 - D_2 - D_3$ are green. If n objects are picked from such a lot without replacement, what is the probability that there are k_1 white, k_2 black, k_3 red, and, naturally, $n - k_1 - k_2 - k_3$ green objects?

Letting A be the event of interest and carrying out an argument similar to the one adopted earlier, the sample space has $\binom{M}{n}$ outcomes in it and the event A has

$\binom{D_1}{k_1}\binom{D_2}{k_2}\binom{D_3}{k_3}\binom{M-D_1-D_2-D_3}{n-k_1-k_2-k_3}$ outcomes favorable to it. Hence

$$P(A) = \frac{\binom{D_1}{k_1}\binom{D_2}{k_2}\binom{D_3}{k_3}\binom{M-D_1-D_2-D_3}{n-k_1-k_2-k_3}}{\binom{M}{n}}$$

Example 3.7. If 10 cards are dealt from a standard deck of 52 cards, what is the probability of getting 3 spades, 2 hearts, 4 diamonds, and 1 club?

Solution. Since there are 13 cards in each suit, the probability is equal to $\binom{13}{3}\binom{13}{2}\binom{13}{4}\binom{13}{1} \div \binom{52}{10}$.

Example 3.8. Consider a lottery that sells 100 tickets and offers five cars and eight motorcycles as prizes. If a person buys 4 tickets, find the probabilities of the following events:

(*a*) A = He wins one car and two motorcycles.
(*b*) B = He wins no prize.
(*c*) C = He wins at least one prize.
(*d*) D = He wins exactly one prize.
(*e*) E = He wins with every ticket.

Solution. There are 13 tickets that carry prizes and, consequently, 87 that do not.

(*a*) If a person wins one car and two motorcycles, then he has to pick one nonwinning ticket. Hence

$$P(A) = \frac{\binom{5}{1}\binom{8}{2}\binom{87}{1}}{\binom{100}{4}}$$

(*b*) The event B specifies that the person picks 0 tickets from among those that offer a car, 0 from those that offer a motorcycle, and 4 from those that offer no prize. Hence

$$P(B) = \frac{\binom{5}{0}\binom{8}{0}\binom{87}{4}}{\binom{100}{4}} = \frac{\binom{87}{4}}{\binom{100}{4}}$$

(*c*) The event C is simply the complement of the event B above. Therefore,

$$P(C) = 1 - \frac{\binom{87}{4}}{\binom{100}{4}}$$

(*d*) Let A_1 = He wins one car and no other prize, and A_2 = He wins one motorcycle and no other prize. Then A_1 and A_2 are mutually exclusive and $D = A_1 \cup A_2$. Also, it can be easily seen that

$$P(A_1) = \frac{\binom{5}{1}\binom{87}{3}\binom{8}{0}}{\binom{100}{4}} = \frac{\binom{5}{1}\binom{87}{3}}{\binom{100}{4}}$$

$$P(A_2) = \frac{\binom{8}{1}\binom{87}{3}\binom{5}{0}}{\binom{100}{4}} = \frac{\binom{8}{1}\binom{87}{3}}{\binom{100}{4}}$$

Consequently,

$$P(D) = \frac{\binom{5}{1}\binom{87}{3}}{\binom{100}{4}} + \frac{\binom{8}{1}\binom{87}{3}}{\binom{100}{4}} = \frac{\binom{13}{1}\binom{87}{3}}{\binom{100}{4}}$$

It should be observed that we could have found $P(D)$ alternatively by simply noting that there are 13 prize-winning tickets, of which the person has to pick 1, and 87 non-prize-winning tickets, of which he has to pick 3.

(*e*) The event E signifies that the person picks 4 winning tickets from among the 13 tickets and 0 nonwinning tickets from the remaining 87. Hence

$$P(E) = \frac{\binom{13}{4}\binom{87}{0}}{\binom{100}{4}} = \frac{\binom{13}{4}}{\binom{100}{4}}$$

■

In the following, we present a slight variation of the hypergeometric probabilities, in that the sampling is carried out with replacement. To start with, consider the following example:

Example 3.9. Suppose 5 cards are picked from a standard deck of 52 cards *with replacement*. What is the probability that there are 3 black cards and 2 red cards?

Solution. Since we are picking 5 cards with replacement, there are 52^5 ways of doing this. For example, $(K_{\text{sp}}, J_{\text{h}}, Q_{\text{cl}}, J_{\text{h}}, K_{\text{sp}})$ is a possible outcome. We shall next find the number of outcomes favorable to the event A representing 3 black cards and 2 red cards. To find this number, we note that there are 5 locations, of which 3 are to be assigned to the black cards and 2 to the red cards. There are $\binom{5}{3} = 10$ ways of doing this. Consider one of these ways, and suppose we have black cards in the first, third, and fourth locations, and red cards in the second and fifth. There are $26 \times 26 \times 26$ ways of filling the first, third, and fourth locations with the black cards, and corresponding to any of these there are 26×26 ways of filling locations two and five with red cards. Applying the basic counting rule, there are $\binom{5}{3} \cdot 26^3 \cdot 26^2$ outcomes favorable to A. Hence

$$P(A) = \frac{\binom{5}{3}(26)^3(26)^2}{52^5}$$

■

As a general example, suppose there is a box containing M objects of which D are defective and $M - D$ are nondefective. *If n of these objects are picked with replacement,* then, arguing as in Example 3.9, it can be easily seen that *the probability of getting exactly k defective objects is*

$$\frac{\binom{n}{k} D^k (M-D)^{n-k}}{M^n}$$

This probability can be rewritten as

$$\binom{n}{k}\left(\frac{D}{M}\right)^k \left(1 - \frac{D}{M}\right)^{n-k}$$

We shall say more on this when we discuss the binomial probabilities in Chapter 3.

Example 3.10. A mathematics department which has ten graduate students is offering three courses, one in probability theory and two in statistics. A student has to take precisely one of these courses. If each student selects a course at random, what is the probability that 3 students will take probability theory and 7 will take statistics?

Solution. The problem boils down to the following: There are 3 courses and we are picking 10 times with replacement. Notice that there is only 1 course in probability for the 3 students to pick from and there are 2 courses in statistics for the 7 students to pick from. The probability of the desired event is therefore equal to

$$\frac{\binom{10}{3} 1^3 \cdot 2^7}{3^{10}} = \binom{10}{3}\left(\frac{1}{3}\right)^3\left(\frac{2}{3}\right)^7$$

■

In the rest of this section we shall consider miscellaneous examples which unify the different ideas developed thus far.

Example 3.11. Find the probability that in a bridge game North, East, South, and West get, respectively, i, j, k, and l spades ($i + j + k + l = 13$).

Solution. The number of ways of dealing 13 cards to one player is $\binom{52}{13}$. There are 39 cards left from which the second player can receive 13 cards in $\binom{39}{13}$ ways. Continuing the argument, the third player can be dealt 13 cards in $\binom{26}{13}$ ways, and, finally, the fourth player can be dealt the remaining cards in $\binom{13}{13}$ ways. By the basic counting rule, there are $\binom{52}{13}\binom{39}{13}\binom{26}{13}\binom{13}{13} = \binom{52}{13}\binom{39}{13}\binom{26}{13}$ ways to deal four bridge hands.

The number of outcomes favorable to the event is

$$\binom{13}{i}\binom{39}{13-i}\binom{13-i}{j}\binom{26+i}{13-j}\binom{13-i-j}{k}\binom{13+i+j}{13-k}$$

(Why?) Hence the probability is

$$\frac{\binom{13}{i}\binom{39}{13-i}\binom{13-i}{j}\binom{26+i}{13-j}\binom{13-i-j}{k}\binom{13+i+j}{13-k}}{\binom{52}{13}\binom{39}{13}\binom{26}{13}}$$

Example 3.12. In a bridge game, find the probability that North gets exactly k aces, $k = 0, 1, 2, 3, 4$.

Solution. From Example 3.11, we know that there are $\binom{52}{13}\binom{39}{13}\binom{26}{13}$ ways of dealing cards to the four players.

Now it can be easily seen that the number of hands where North gets exactly k aces is

$$\binom{4}{k}\binom{48}{13-k}\binom{39}{13}\binom{26}{13}$$

Hence the probability that North gets exactly k aces is equal to

$$\frac{\binom{4}{k}\binom{48}{13-k}\binom{39}{13}\binom{26}{13}}{\binom{52}{13}\binom{39}{13}\binom{26}{13}} = \frac{\binom{4}{k}\binom{48}{13-k}}{\binom{52}{13}}$$

We observe that this probability is the same as the probability that an arbitrary hand of 13 cards contains exactly k aces.

Example 3.13. Find the probability that eight players on a team will all have their birthdays on–

(*a*) Monday or Tuesday (but not all on one day)

(*b*) exactly two days of the week.

Solution. There are 7 days of the week on which each of the players could be born. Hence there are 7^8 possibilities.

(*a*) If each person is born on Monday or Tuesday, then each person has two choices of days, and as a result there are 2^8 possible ways this can happen. However, the men cannot all have birthdays on Monday, nor all on Tuesday. Therefore, there are $2^8 - 2$ outcomes favorable to the event, and consequently the desired probability is equal to $(2^8 - 2)/7^8$

(*b*) There are $\binom{7}{2}$ ways of picking 2 days out of 7. Hence the probability of having all of the birthdays on exactly 2 days of the week is $\left[\binom{7}{2}(2^8 - 2)\right]/7^8$.

- *Example 3.14.* (*The matching problem*) Suppose n people attend a party. Before the party starts, each person deposits his coat in the cloakroom, and, at the end of the party, picks one coat at random. Find the probability that at least one person picks his own coat.

Solution. Let us denote the n people as $\pi_1, \pi_2, \ldots, \pi_n$ and suppose they are arranged in this order. Then picking coats for these people amounts to arranging the coats in all possible ways. Hence there are $n!$ possibilities in all.

Now let A_k be the event that the person π_k picks his own coat. What we are interested in finding is $P\left(\bigcup_{k=1}^{n} A_k\right)$, which, as we know, is given by

$$P\left(\bigcup_{k=1}^{n} A_k\right) = \sum_{i=1}^{n} P(A_i) - \sum_{i<j}^{n} P(A_iA_j) + \sum_{i<j<k}^{n} P(A_iA_jA_k) - \ldots + (-1)^{n+1} P(A_1A_2 \ldots A_n)$$

To find $P(A_i)$, we note that if person π_i gets his coat, the remaining $n-1$ coats can be in any order. Hence there are $(n-1)!$ possibilities where person π_i gets his coat. Therefore,

$$P(A_i) = \frac{(n-1)!}{n!} = \frac{1}{n}$$

(Observe that $P(A_i) = 1/n$ no matter what i is.)

Next, fo find $P(A_iA_j)$ we see that if persons π_i and π_j get their respective coats, the remaining $n-2$ coats can be in any order, and, therefore, there are $(n-2)!$ possibilities favorable to A_iA_j. Hence

$$P(A_iA_j) = \frac{(n-2)!}{n!} = \frac{1}{n(n-1)}$$

for any combination i, j. In general,

$$P(A_{i_1}A_{i_2} \ldots A_{i_r}) = \frac{(n-r)!}{n!}$$

Finally, since there are $\binom{n}{1}$ terms in the sum $\sum_{i=1}^{n} P(A_i)$, $\binom{n}{2}$ terms in the sum $\sum_{i<j}^{n} P(A_iA_j)$, and, in general, $\binom{n}{r}$ terms in the sum $\sum_{i_1<i_2<\ldots<i_r}^{n} P(A_{i_1}A_{i_2}\ldots A_{i_r})$, we get

$$P\left(\bigcup_{k=1}^{n} A_k\right) = \binom{n}{1}\frac{(n-1)!}{n!} - \binom{n}{2}\frac{(n-2)!}{n!} + \ldots + (-1)^{n+1}\frac{1}{n!}$$

or, simplifying,

$$P\left(\bigcup_{k=1}^{n} A_k\right) = 1 - \frac{1}{2!} + \frac{1}{3!} - \ldots + (-1)^{n+1}\frac{1}{n!}$$

Thus if n is large, $P\left(\bigcup_{k=1}^{n} A_k\right)$ is approximately equal to $1 - e^{-1}$. ∎

The following example, called the *birthday problem*, indicates a context in which one might be interested in finding the probability that there is no repetition when sampling is carried out with replacement.

Example 3.15. If there are n people in a room, what is the probability that no two of them will have the same birthday. (Assume $n \leqslant 365$.)

Solution. We shall ignore the fact that there are leap years and assume 365 days to a year. When we consider the birthdays of n people, in essence we pick n days with replacement from the 365 days. (Remember, if Tom is born on January 21, it does not preclude Jane from being born on that date. Also, order is relevant because Tom being born on February 3 and Jane on October 9 is different from Tom being born on October 9 and Jane on February 3.) Hence there are 365^n possibilities in the sample space.

If no two people have the same birthday, then the first person has 365 choices for his birthday, the second 364 days, and so on, and, eventually, the last person has $(365 - n + 1)$ choices. (Here first, second, and so on are used in the sense of writing the outcomes as n-tuples.) Hence there are $365 \times 364 \times \ldots \times (365 - n + 1) = (365)_n$ outcomes favorable to the event. Therefore, the probability that no two people have the same birthday is

$$\frac{(365)_n}{365^n} = \left(1 - \frac{1}{365}\right)\left(1 - \frac{2}{365}\right) \cdots \left(1 - \frac{n-1}{365}\right)$$ ∎

In general, prompted by Example 3.15, *if we pick n objects with replacement from M distinct objects (with $n \leqslant M$), then the probability of no repetition is $(M)_n/M^n$.*

EXERCISES–SECTION 3

1. Find n if–

   (a) $\binom{n}{10} = \binom{n}{16}$

   (b) $\binom{18}{n} = \binom{18}{n-6}$

2. Show that

$$\binom{m}{n} = \binom{m-1}{n-1} + \binom{m-1}{n}, \quad 0 \leqslant n \leqslant m$$

Use this result to complete the following pattern for $m = 5, 6, 7$.

$m = 0$					1				
$m = 1$				1		1			
$m = 2$			1		2		1		
$m = 3$		1		3		3		1	
$m = 4$	1		4		6		4		1

(Notice that each number in the mth row is obtained by adding appropriate members of the $(m - 1)$th row. The array is called *Pascal's triangle.*)

3. In the following, n is a nonnegative integer. Use the fact that

$$(1+x)^n = \sum_{i=0}^{n} \binom{n}{i} x^i \qquad \text{for any } x$$

to establish the following identities:

(a) $\sum_{i=0}^{n} (-1)^i \binom{n}{i} = 0$

(b) $\sum_{i=0}^{n} i \binom{n}{i} = n \cdot 2^{n-1}$ *Hint:* Differentiate and set $x = 1$.

(c) $\sum_{i=0}^{n} (-1)^i i \binom{n}{i} = 0$

(d) $\sum_{i=0}^{n} \binom{n}{i}^2 = \binom{2n}{n}$ *Hint:* Consider $(1+x)^{2n} = (1+x)^n(1+x)^n$

4. Find the different ways a person can dress for an occasion if he has 8 different shirts, 6 different ties, and 5 different jackets.
5. A die is tossed six times and a coin is tossed four times. How many outcomes are there in the sample space?
6. On a test a student has a choice of answering eight out of ten questions.
   (a) How many ways are there to answer the test?
   (b) How many ways are there if questions 1 and 2 are obligatory?
7. Find how many positive, integral divisors 3600 has.
8. Find the number of ways of arranging the letters of the word SUCCESSION.
9. If the letters of the word VOLUMES are arranged in all possible ways, find the probability that–
   (a) the word ends with a vowel
   (b) the word starts with a consonant and ends with a vowel.
10. Three digits are picked at random, without replacement, from the digits 1 through 9. Find the probability that the digits are consecutive digits.
11. Two numbers are picked without replacement from the following numbers: 2, 6, 8, 9, 11, 12, 17, 18. Find the probability that they are relatively prime.
12. A box contains eight balls marked 1, 2, 3, . . . , 8. If four balls are picked at random, find the probability that the balls marked 1 and 5 are among the four selected balls.
13. A laundry bag contains four black and eight white gloves. If two gloves are picked one by one at random, what is the probability that–
   (a) they are both black
   (b) they are of the same color.
14. A box contains ten items, of which four are defective. If three items are picked without replacement, find the probability that–
   (a) all are defective
   (b) exactly two are defective
   (c) at most two are defective
   (d) at least two are defective.

15. From an ordinary deck of 52 cards, 4 cards are picked at random. Find the probability that–

(a) exactly one is an ace
(b) exactly one is a face card
(c) all are black cards
(d) each is from a different suit
(e) at least two are aces.

16. Find the probability that a hand of five cards selected from a standard deck has–

(a) an ace, king, queen, jack, and ten of spades
(b) an ace, king, queen, jack, and ten of the same suit
(c) an ace, king, queen, jack, and ten.

17. Two cards are drawn, with replacement, from an ordinary deck of 52 cards. Find the probability that both cards belong to the same suit.

18. Work exercise 17, this time assuming that the cards are drawn without replacement.

19. In a five-card poker hand, find the probability that–

(a) there are three kings and two aces
(b) there are exactly three kings
(c) there are exactly three kings and one ace
(d) there is one ace and at least three kings
(e) there are at least three kings.

20. From a group of 10 lawyers, 8 doctors, 6 businessmen, and 9 professors, a committee of six is selected at random. Find the probability that–

(a) the committee consists of 2 lawyers, 2 doctors, one businessman, and one professor
(b) the committee contains no lawyers
(c) the committee contains at least one lawyer.

21. From a group of 5 lawyers, 7 accountants, and 9 doctors, a committee of three is selected at random. What is the probability that the committee has more lawyers than doctors?

22. Suppose eight books are arrayed on a shelf in a random order. What is the probability that three particular books will be next to each other?

23. The numbers $1, 2, \ldots, n$ are arranged in all possible ways. Assuming that all the arrangements are equally likely, find the probability that–

(a) 1, 2, 3, and 4 appear next to each other in the order indicated
(b) 1, 2, 3, and 4 appear next to each other.

24. Suppose 4 letters are placed at random in 4 addressed envelopes. What is the probability that exactly k of the letters are in their correct envelope, $k = 0, 1, 2, 3, 4$.

25. Eight cards are dealt from a standard deck of 52 cards. Find the probability of obtaining either 3 aces or 3 kings or 3 queens. (This does *not* preclude, for instance, getting 3 aces *and* 3 kings.)

26. Suppose 13 cards are dealt from a standard deck of 52 cards. Find the probability that the hand will contain all the face cards in at least one suit.

27. On a ten-question true-false test, a student guesses on every question. Find the probability that the student answers–
    (a) no question correctly
    (b) at least one question correctly
    (c) exactly r questions correctly, $r = 0, 1, 2, \ldots, 10$.

28. An elevator stops at ten floors. If there are six people in the elevator, find the probability that–
    (a) no two of them will get off on the same floor.
    (b) all of them will get off on the same floor.

29. A box contains six pencils of lengths 1, 3, 4, 5, 7, 8. If three pencils are picked at random, find the probability that a triangle can be formed with them. *Hint:* The sum of the lengths of any two of the pencils must be greater than the length of the third pencil.

30. A box contains six pencils of length 1 foot, four of length 2 feet and three of length 5 feet. If four pencils are picked at random (without replacement)–
    (a) what is the probability of being able to form a rectangle with them?
    (b) what is the probability of being able to form a square?

31. A party is attended by twenty males and twenty females. If the party is divided at random into two equal groups, find the probability that there are an equal number of males and females in each group.

32. A party consists of n males and n females. If these persons are seated at random in a row, find the probability that no two members of the same sex will be seated next to each other.

33. A box contains four books on mathematics and twelve books on history. If the books are distributed equally at random among four students, find the probability that each student will get a book on mathematics.

3 Conditional Probability and Independent Events

INTRODUCTION

The groundwork for an understanding of basic probability was laid down in the previous two chapters. In this chapter we shall consider principally two topics which come under the purview of probability theory. The first of these topics will cover conditional probability, and the second, independent events.

1. CONDITIONAL PROBABILITY

To discuss conditional probability, suppose we pick a person at random and pose the following three questions:

(*i*) What is the probability that the person is in the United States? Assuming that there are 200 million people in the United States and 3 billion people in the world, the answer would be $\frac{2}{30}$.

(*ii*) Given that the person is in Australia, what is the probability that he is in the United States? The answer would be, obviously, 0.

(*iii*) Given that the person is in Iowa, what is the probability that he is in the United States? Here the probability is, of course, 1.

We see from the above three situations that prior knowledge of the person's location influences the probability of his being found in the United States. Thus it often happens that partial information is available about the outcome of the underlying experiment and this, in turn, leads to appropriate adjustment of the probabilities of the associated events. In summary, the notion of conditional probability involves the probability of an event, say A, given the information that an event B has occurred.

It will help us to consider the following example, which will give us the intuitive idea behind the concept of conditional probability:

Suppose in a club there are five males, two of whom wear glasses and six females, four of whom wear glasses. We pick one person at random. Given that the person is a male, what is the probability that he wears glasses?

Since we are picking one person, the sample space has eleven outcomes which we might write as $S = \{M1G, M2G, M3, M4, M5, F1G, F2G, F3G, F4G, F5, F6\}$. (M1G, M2G represent males who use glasses and M3, M4, M5 males who do not. Similar notation is used for females.)

Let A = Person wears glasses, and B = Person is a male. Then $A = \{M1G, M2G, F1G, F2G, F3G, F4G\}$, $B = \{M1G, M2G, M3, M4, M5\}$, and $AB = \{M1G, M2G\}$. Therefore, $P(A) = \frac{6}{11}$, $P(B) = \frac{5}{11}$, and $P(AB) = \frac{2}{11}$.

We shall now answer the question that we posed earlier. Given the person is a male, what is the probability that he wears glasses? We will introduce the notation $P(A|B)$ to denote this probability. (The vertical line is to be read "given." Thus, $P(A|B)$ is read as "the probability of A given B.")

With the information at hand, we need concentrate only on the set {M1G, M2G, M3, M4, M5} to answer the question. There are 5 males and 2 of them wear glasses. Therefore, we can say

$$P(A|B) = \frac{2}{5}$$

We can rewrite this to get

$$P(A|B) = \frac{2}{5} = \frac{\frac{2}{11}}{\frac{5}{11}} = \frac{P(AB)}{P(B)}$$

The above example provides a motivation for the definition of $P(A|B)$ in the general case.

Suppose $(S, \mathcal{F}, P)$ is a probability space and A and B are two events. Then the **conditional probability** of the event A given the event B, denoted $P(A|B)$, is defined by:

$$P(A|B) = \frac{P(AB)}{P(B)}, \quad \text{provided } P(B) > 0$$

Comment. We require that $P(B)$ be greater than zero not just for the mathematical reason that we cannot divide by zero. We require this for common sense reasons also. For, suppose the club we are considering consists of ten females, six of whom wear glasses, and no males. In this case there is little sense in contemplating the question "Given the persion is a male, what is the probability that he wears glasses?" *$P(A|B)$ is not defined if $P(B) = 0$.*

According to the definition, $P(A|B)$ is the ratio of $P(AB)$ to $P(B)$. Therefore, for a fixed event B, $P(A|B)$ will be large or small depending upon whether $P(AB)$ is large or small, and this in turn will depend upon the extent of overlap between A and B.

The formula for conditional probability can be rewritten to give

$$P(AB) = P(B) \cdot P(A|B)$$

This is called the **general multiplication rule** of probabilities and is often useful for determining the probability that two events A and B will occur simultaneously. It is also referred to as the **theorem of compound probabilities.**

The multiplication rule can be extended to find the probability of the simultaneous occurrence of three events A, B, C as follows:

$$P(ABC) = P(A) \cdot P(B|A) \cdot P(C|AB)$$

The verification of this result follows immediately by regarding AB as one event and C as the other event, and is left to the reader.

Generalized to n events $A_1, A_2, \ldots, A_n$, the multiplication rule gives the following:

$$P\left(\bigcap_{i=1}^{n} A_i\right) = P(A_1) \cdot P(A_2|A_1) \cdot P(A_3|A_1A_2) \ldots P\left(A_n \middle| \bigcap_{i=1}^{n-1} A_i\right)$$

Example 1.1. Suppose the sample space S is given as $S = \{s_1, s_2, s_3, s_4, s_5\}$. If probabilities of the simple events are defined by $P(\{s_1\}) = \frac{1}{10}$, $P(\{s_2\}) = \frac{1}{10}$, $P(\{s_3\}) = \frac{3}{20}$, $P(\{s_4\}) = \frac{3}{20}$, and $P(\{s_5\}) = \frac{1}{2}$, find the conditional probability of–

(*a*) $\{s_1, s_2, s_3\}$ given $\{s_2, s_5\}$

(*b*) $\{s_2, s_5\}$ given $\{s_1, s_2, s_3\}$

Solution

(*a*) We have

$$P(\{s_1, s_2, s_3\}|\{s_2, s_5\}) = \frac{P(\{s_1, s_2, s_3\} \cap \{s_2, s_5\})}{P(\{s_2, s_5\})} = \frac{P(\{s_2\})}{P(\{s_2, s_5\})}$$

$$= \frac{\frac{1}{10}}{\frac{1}{10} + \frac{1}{2}} = \frac{1}{6}$$

(*b*) Proceeding as in (*a*), it follows that

$$P(\{s_2, s_5\}|\{s_1, s_2, s_3\}) = \frac{P(\{s_2\})}{P(\{s_1, s_2, s_3\})} = \frac{\frac{1}{10}}{\frac{1}{10} + \frac{1}{10} + \frac{3}{20}} = \frac{2}{7}$$

Example 1.2. If A and B are two events for which $P(A) = 0.5$, $P(B) = 0.7$, and $P(A|B) = 0.4$, find:

(*a*) $P(AB)$ (*b*) $P(A \cup B)$ (*c*) $P(A - B)$

Solution. We have:

(*a*) $P(AB) = P(B) \cdot P(A|B) = 0.7 \times 0.4 = 0.28$

(*b*) $P(A \cup B) = P(A) + P(B) - P(AB) = 0.5 + 0.7 - 0.28 = 0.92$

(*c*) $P(A - B) = P(A) - P(AB) = 0.5 - 0.28 = 0.22$

Example 1.3. Two fair dice are rolled. Given that the sum on the dice is greater than 7, find the probability that the two dice show the same number.

Solution. Let A = The sum on the dice is greater than 7, and B = The two dice show the same number. Then

$$A = \{(2,6),(6,2),(3,5),(5,3),(4,4),(3,6),(6,3),(4,5),(5,4),(5,5),(6,4),(4,6),(5,6),(6,5),(6,6)\}$$

$$B = \{(1,1),(2,2),(3,3),(4,4),(5,5),(6,6)\}$$

and

$$AB = \{(4,4),(5,5),(6,6)\}$$

Therefore

$$P(B|A) = \frac{P(BA)}{P(A)} = \frac{\frac{3}{36}}{\frac{15}{36}} = \frac{1}{5}$$

Example 1.4. A committee of four people is formed from a group consisting of eight lawyers and seven doctors. Given that there is at least one lawyer on the committee, what is the probability that there are exactly three lawyers?

Solution. Let A = There is at least one lawyer, and B = There are exactly three lawyers. We want to find $P(B|A)$. This is given by

$$P(B|A) = \frac{P(BA)}{P(A)} = \frac{P(B)}{P(A)}, \qquad \text{since } B \subset A$$

Now

$$P(A) = 1 - P(\text{no lawyer on the committee})$$

$$= 1 - \frac{\binom{8}{0}\binom{7}{4}}{\binom{15}{4}} = \frac{\binom{15}{4} - \binom{7}{4}}{\binom{15}{4}}$$

and

$$P(B) = \frac{\binom{8}{3}\binom{7}{1}}{\binom{15}{4}}$$

Hence

$$P(B|A) = \frac{\left[\binom{8}{3}\binom{7}{1}\right] \div \binom{15}{4}}{\left[\binom{15}{4} - \binom{7}{4}\right] \div \binom{15}{4}} = \frac{\binom{8}{3}\binom{7}{1}}{\binom{15}{4} - \binom{7}{4}}$$

Example 1.5. The probability that the market goes up on Monday is 0.6; given that it went up on Monday, the probability that it goes up on Tuesday is 0.3; and, finally, given that it went up on Monday and Tuesday, the probability that it goes up on Wednesday is 0.4. Find the probability that (*a*) the market goes up on all three days, (*b*) the market goes up on Monday and Tuesday, but not on Wednesday.

Solution. Let A = The market goes up on Monday, B = The market goes up on Tuesday, and C = The market goes up on Wednesday.

(*a*) We know that $P(A) = 0.6, P(B|A) = 0.3$, and $P(C|AB) = 0.4$, and we want to find $P(ABC)$. Using the formula $P(ABC) = P(A) \cdot P(B|A) \cdot P(C|AB)$, it follows that

$$P(ABC) = 0.6 \times 0.3 \times 0.4 = 0.072$$

(*b*) In this case we want to find $P(ABC')$ which, by consequence (C3) of probability (see Section 1 of Chapter 2), is equal to $P(AB) - P(ABC)$. Now $P(AB) = P(A) \cdot P(B|A) = 0.6 \times 0.3 = 0.18$, and, by part (*a*), $P(ABC) = 0.072$.

Hence

$$P(ABC') = 0.18 - 0.072 = 0.108$$

An alternate approach is to use the relation

$$P(ABC') = P(A) \cdot P(B|A) \cdot P(C'|AB) = 0.6 \times 0.3 \times 0.6 = 0.108$$

Example 1.6. From the first ten integers, $1, 2, \ldots, 10$, a number is picked at random. If this number is i, then another number is picked at random from $1, 2, \ldots, i$; if this number is j, a third number is picked from $1, 2, \ldots, j$. What is the probability that the three numbers are distinct prime numbers. (*Note:* 1 is not considered prime.)

Solution. We are interested in finding the probability of getting a sequence (7, 5, 3) or (7, 5, 2) or (7, 3, 2) or (5, 3, 2). We shall first find the probability of getting the sequence (7, 5, 3). The probabilities for other sequences are found analogously.

Let A represent the event that 7 is picked on the first draw, B that 5 is picked on the second draw, and C that 3 is picked on the third draw. Then:

$P(A) = \frac{1}{10}$, since the first number is picked from the first ten integers

$P(B|A) = \frac{1}{7}$, since, given the first number drawn is 7, the second number is picked from $1, 2, \ldots, 7$

and $P(C|AB) = \frac{1}{5}$, since the third number is picked from $1, 2, \ldots, 5$

Consequently, $P(ABC) = \frac{1}{10} \cdot \frac{1}{7} \cdot \frac{1}{5} = \frac{1}{350}$.

Similarly, the probability of getting the sequence (7, 5, 2) is $\frac{1}{350}$, that of sequence (7, 3, 2) is $\frac{1}{10} \cdot \frac{1}{7} \cdot \frac{1}{3} = \frac{1}{210}$, and that of (5, 3, 2) is $\frac{1}{10} \cdot \frac{1}{5} \cdot \frac{1}{3} = \frac{1}{150}$. Hence, in conclusion, the probability of three distinct primes is

$$\frac{1}{350} + \frac{1}{350} + \frac{1}{210} + \frac{1}{150} = \frac{3}{175}$$

Example 1.7. Three balls are picked at random one by one and without replacement from a box containing four white and eight black balls. Let

A = The first ball is white
B = The second ball is white
C = The third ball is white

Find: (*a*) $P(A)$ (*b*) $P(B|A)$ (*c*) $P(C|AB)$

Solution

(*a*) Since the three balls are picked *one by one,* there are $12 \times 11 \times 10$ outcomes in the sample space, and $4 \times 11 \times 10$ outcomes favorable to the event A (why?). Hence $P(A) = \frac{4}{12}$.

We could also expect this answer by arguing that, as far as the first draw is concerned, there are 12 choices and 4 of these result in picking a white ball.

(*b*) We have $P(B|A) = P(BA)/P(A)$. Now to find $P(BA)$ we observe that there are $4 \times 3 \times 10$ outcomes in the event BA. Hence $P(BA) = (4 \times 3 \times 10)/(12 \times 11 \times 10)$, and consequently $P(B|A) = \frac{3}{11}$.

This makes good sense because given that the first ball is white, there are 11 balls left when the second ball is picked; of these 3 are white. Actually in many situations this will be the nature of our argument in finding conditional probabilities.

(*c*) To find $P(C|AB)$ we observe that $P(C|AB) = P(CAB)/P(AB)$. Now there are $4 \times 3 \times 2$ outcomes favorable to the event CAB so that $P(CAB) = (4 \times 3 \times 2)/(12 \times 11 \times 10)$. Hence

$$P(C|AB) = \left(\frac{4 \times 3 \times 2}{12 \times 11 \times 10}\right) \div \left(\frac{4 \times 3 \times 10}{12 \times 11 \times 10}\right) = \frac{2}{10}$$

Once again, given that the first 2 balls are white, there are 10 balls left when the third ball is picked, and 2 of these are white. Wherever convenient, the reader should adopt this type of argument. It is assumed that he is aware of the background leading to such an argument.

Example 1.8. Three balls are picked at random one by one from a box containing four white and eight black balls. Find the following probabilities:

(*a*) The first and third ball are white.

(*b*) The third ball is white given that the first ball is white.

Solution

(*a*) Let A = The first ball is white, B = The second ball is white, and C = The third ball is white. Then

$$\begin{aligned} P(AC) &= P((ABC) \cup (AB'C)) \qquad \text{(why?)} \\ &= P(ABC) + P(AB'C) \qquad \text{(why?)} \\ &= P(A) \cdot P(B|A) \cdot P(C|AB) + P(A) \cdot P(B'|A) \cdot P(C|AB') \\ &= \frac{4}{12} \cdot \frac{3}{11} \cdot \frac{2}{10} + \frac{4}{12} \cdot \frac{8}{11} \cdot \frac{3}{10} = \frac{1}{11} \end{aligned}$$

(*b*) Here we want to find $P(C|A)$. We have

$$P(C|A) = \frac{P(CA)}{P(A)} = \frac{\frac{1}{11}}{\frac{4}{12}} = \frac{3}{11}$$

■

Suppose $(S, \mathcal{F}, P)$ is a probability space and B is a *fixed* event with $P(B) > 0$. We have just defined the concept of conditional probability. In essence we have introduced a function $P(\cdot|B)$, which, for any event A (that is, $A \in \mathcal{F}$), is given by $P(A|B) = P(AB)/P(B)$. To make the following result more appealing, we write P_B in place of $P(\cdot|B)$.

The function P_B has the following properties:

(*i*) $0 \leqslant P_B(A) \leqslant 1$ for any event A

(*ii*) $P_B(B) = 1$

(*iii*) If $\{A_i\}$ is a sequence of mutually exclusive events, then

$$P_B\left(\bigcup_{i=1}^{\infty} A_i\right) = \sum_{i=1}^{\infty} P_B(A_i)$$

The properties (*i*) and (*ii*) are obvious and we shall prove only (*iii*). We have

$$P_B\left(\bigcup_{i=1}^{\infty} A_i\right) = P\left(\bigcup_{i=1}^{\infty} A_i|B\right) = \left[P\left(\left(\bigcup_{i=1}^{\infty} A_i\right)B\right)\right] \div P(B), \qquad \text{by definition}$$

$$= \left[P\left(\bigcup_{i=1}^{\infty} (A_iB)\right)\right] \div P(B), \qquad \text{by the distributive rule}$$

Now since the events $\{A_i\}$ are mutually exclusive, so are the events $\{A_iB\}$. Hence

$$P_B\left(\bigcup_{i=1}^{\infty} A_i\right) = \left(\sum_{i=1}^{\infty} P(A_iB)\right) \div P(B) = \sum_{i=1}^{\infty} (P(A_iB)/P(B)) = \sum_{i=1}^{\infty} P(A_i|B) = \sum_{i=1}^{\infty} P_B(A_i)$$

It follows from properties (*i*), (*ii*), (*iii*) that the function P_B, that is, $P(\cdot|B)$, satisfies all the axioms of a probability function, and hence is a probability measure.

Comment. For completeness, we should add the following–keeping in mind that our treatment of the subject will not be impaired if such fine details are relegated to obscurity: Conditioning on the event B amounts to choosing B as the new sample space. As such, the appropriate sigma field is a sigma field of subsets of B. Denoting this sigma field by $\mathcal{F}_B$, it is given by

$$\mathcal{F}_B = \{AB|A \in \mathcal{F}\}$$

where $\mathcal{F}$ is the original sigma field.

Since, for a given event B, $P(\cdot|B)$ is a probability function, all the consequences appropriate to a probability function follow. For example, if A_1 and A_2 are any two events, then

$$P(A_1 - A_2|B) = P(A_1|B) - P(A_1A_2|B)$$
$$P(A_1 \cup A_2|B) = P(A_1|B) + P(A_2|B) - P(A_1A_2|B)$$
$$P(A_1'|B) = 1 - P(A_1|B)$$

and so on. The reader might wish to prove these directly.
Comment. We are *not* claiming that $P(A|B') = 1 - P(A|B)$. As a matter of fact, this is *not* true in general.

Example 1.9. Suppose A and B are two events with $P(AB') = 0.4$ and $P(B) = 0.2$. Find: (*a*) $P(A|B')$ (*b*) $P(A'|B')$ (*c*) $P(A'B')$ (*d*) $P(A \cup B)$

Solution. We have:

(*a*) $P(A|B') = \dfrac{P(AB')}{P(B')} = \dfrac{0.4}{0.8} = \dfrac{1}{2}$

(*b*) $P(A'|B') = 1 - P(A|B') = 1 - \dfrac{1}{2} = \dfrac{1}{2}$

(*c*) $P(A'B') = P(B') \cdot P(A'|B') = 0.8 \times \dfrac{1}{2} = 0.4$

(*d*) $P(A \cup B) = 1 - P(A \cup B)' = 1 - P(A'B') = 1 - 0.4 = 0.6$

We could have found $P(A \cup B)$ in an alternate way by noting that $P(A \cup B) = P(AB') + P(B)$. (Draw a Venn diagram to see this.)

Example 1.10. A number is picked at random from the integers 1, 2, 3, . . . , 1000. If the number is known to be divisible by four, what is the probability that–

(*a*) it is divisible by six or eight?
(*b*) it is divisible by six, but not by eight?
(*c*) it is divisible by exactly one of the integers six, eight?

Solution. Let B represent that the number is a multiple of 4, A_1 that it is a multiple of 6, and A_2 that it is a multiple of 8.

(*a*) We want to find $P(A_1 \cup A_2|B)$ which, as we know, is equal to $P(A_1|B) + P(A_2|B) - P(A_1A_2|B)$. Now

$$P(A_1|B) = \frac{P(A_1B)}{P(B)} = \frac{\frac{83}{1000}}{\frac{250}{1000}} = \frac{83}{250}$$

$$P(A_2|B) = \frac{P(A_2B)}{P(B)} = \frac{\frac{125}{1000}}{\frac{250}{1000}} = \frac{125}{250}$$

$$P(A_1A_2|B) = \frac{P(A_1A_2B)}{P(B)} = \frac{\frac{41}{1000}}{\frac{250}{1000}} = \frac{41}{250}$$

Hence

$$P(A_1 \cup A_2|B) = \frac{83}{250} + \frac{125}{250} - \frac{41}{250} = \frac{167}{250}$$

(*b*) In this case we want to find $P(A_1 - A_2|B)$; this is equal to $P(A_1|B) - P(A_1A_2|B)$. Hence

$$P(A_1 - A_2|B) = \frac{83}{250} - \frac{41}{250} = \frac{21}{125}$$

(*c*) Given B, the probability that the number is divisible by exactly one of the integers six, eight is equal to $P(A_1|B) + P(A_2|B) - 2P(A_1A_2|B) = \frac{63}{125}$.

Example 1.11. The probability that a person passes a proficiency test on the first attempt is 0.5, that he passes on the second attempt is 0.7 (of course, given that he failed on the first attempt), and the probability that he passes on the third attempt is 0.8 (given that he failed on the first two attempts). If the person is allowed three attempts, what is the probability that he will pass the test?

Solution. Let A_i represent the event that the person passes the test on the ith attempt, $i = 1, 2, 3$. Then $A_1 \cup (A_1'A_2) \cup (A_1'A_2'A_3)$ is the event that the person passes the test. The probability of this event is equal to $P(A_1) + P(A_1'A_2) + P(A_1'A_2'A_3)$ (why?). Next,

$$P(A_1) = 0.5, \quad P(A_1'A_2) = P(A_1')P(A_2|A_1') = 0.5 \times 0.7 = 0.35$$

and

$$P(A_1'A_2'A_3) = P(A_1')P(A_2'|A_1')P(A_3|A_1'A_2') = 0.5 \times 0.3 \times 0.8 = 0.12$$

Hence the probability of passing the test is equal to

$$0.5 + 0.35 + 0.12 = 0.97$$

EXERCISES–SECTION 1

1. Suppose probabilities are assigned to the simple events of $S = \{s_1, s_2, s_3, s_4, s_5, s_6\}$ as follows:

$$P(\{s_1\}) = 2P(\{s_2\}) = 3P(\{s_3\}) = 4P(\{s_4\}) = 5P(\{s_5\}) = 6P(\{s_6\})$$

Find the conditional probability of:
   (a) $\{s_1, s_3, s_4\}$ given $\{s_2, s_3\}$
   (b) $\{s_2, s_3\}$ given $\{s_1, s_3, s_4\}$
   (c) $\{s_1, s_2, s_3, s_4\}$ given $\{s_2, s_4, s_5, s_6\}$
   (d) $\{s_1, s_4\}$ given $\{s_1, s_2, s_3, s_4\}$.

2. Suppose A and B are two events with $P(A|B) = 0.3$, $P(A'|B') = 0.4$, and $P(B) = 0.7$. Find:
   (a) $P(A|B')$ (b) $P(A)$ (c) $P(B|A)$

3. Suppose $P(A_1|B) = 0.7$, $P(A_2|B) = 0.4$, and $P(A_1A_2|B) = 0.3$. Given that B has occurred, find the probability that:
   (a) at least one of the events A_1, A_2 occurs
   (b) exactly one of the events A_1, A_2 occurs
   (c) only A_1 occurs.

4. Suppose $P(A_1|B) = 0.7$, $P(A_2|B) = 0.4$, $P(A_3|B) = 0.5$, $P(A_1A_3|B) = 0.3$, $P(A_2A_1|B) = 0.35$, and $P(A_2A_3|B) = 0$. Given the event B, find the probability that, of the events A_1, A_2, A_3:

(a) at least one occurs
(b) exactly two occur.

5. (a) If $P(E) = \frac{1}{4}, P(F|E) = \frac{1}{2}$, and $P(E|F) = \frac{1}{3}$, determine $P(F)$.
(b) If $P(E) = x, P(F|E) = y$, show that $P(E|F) \geqslant xy$.

6. If A and B are two events with $P(B) > 0$, state whether the following statements are true or false. Justify your answer in each case.
(a) $P(A|B) + P(A|B') = 1$
(b) $P(A|B) + P(A'|B) = 1$
(c) $P(A|B) + P(A'|B') = 1$
(d) $P(A|B') + P(A'|B') = 1$

7. If $P(A|B) = 1$, show that, for any event C–
(a) $P(C|AB) = P(C|B)$
(b) $P(AC|B) = P(C|B)$

8. Suppose $0 < P(A) < 1$ and $0 < P(B) < 1$. Show that A and B are not mutually exclusive if $P(A|B) = P(A|B')$.

9. If A and B are two events for which $P(A|B) \geqslant P(A)$, then show that $P(B|A) \geqslant P(B)$.

10. Suppose A and B are two events.
(a) If $P(A) = P(B) = \frac{4}{5}$, show that $P(A|B) \geqslant \frac{3}{4}$.
(b) If $P(A) = x$ and $P(B) = y$, show that $P(A|B) \geqslant (x + y - 1)/y$.
Hint: $P(AB) \geqslant P(A) + P(B) - 1$.

11. A fair coin is tossed three times. Find the conditional probability of getting three heads, given that there are at least two heads.

12. A fair die is tossed twice. Find the conditional probability that–
(a) the sum is 9, given that the sum is greater than 6
(b) the sum is 7, given that the sum is odd
(c) the sum is even, given that at least one of the tosses shows an odd number
(d) the sum is 8, given that the two tosses show the same number
(e) the two tosses show the same number, given that the sum is 8.

13. Tom and Mary agree to meet at some place at a certain time. The probability that Tom keeps his appointment is 0.7, and that either of them keeps the appointment is 0.9. Given that Tom did not keep the appointment, what is the probability that Mary didn't either.

14. Suppose current flows through two switches A and B to a radio and back to the battery as shown in Figure 1.1. If the probability that switch A is closed is 0.8, and,

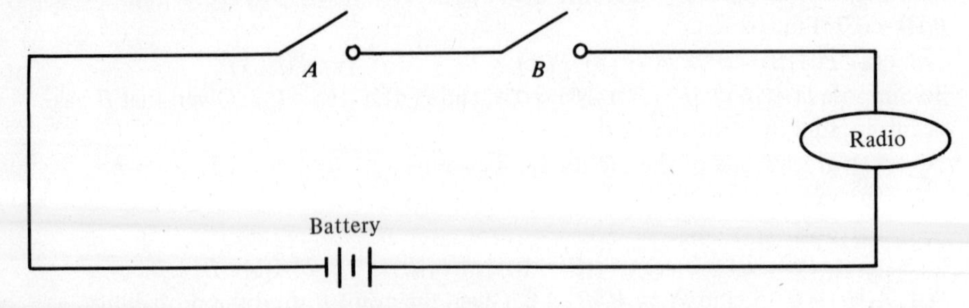

Figure 1.1

given that switch A is closed, the probability that switch B is closed is 0.7, find the probability that the radio is playing.

15. Suppose current flows through two switches A and B to a radio and back to the battery as shown in Figure 1.2. Suppose the probability that switch A is closed is

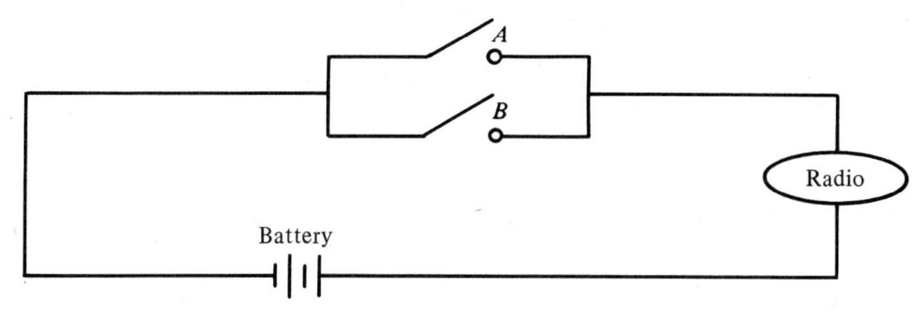

Figure 1.2

0.8, the probability that switch B is closed is 0.6, and, given that switch A is closed, the probability that switch B is closed is 0.7. Find the probability that the radio is playing.

16. An athlete participates in three events—the 100 meter, 400 meter, and 800 meter sprints. From past experience, it is known that the probability that he wins the 100 meter dash is 0.40, the probability that he wins the 400 meter run given that he won the 100 meter dash is 0.60, and the probability that he wins the 800 meter run given that he won the other two events is 0.80. What is the probability he wins all three events?

17. Thirteen cards are dealt from a standard deck of bridge cards.

(a) Find the probability that there are i aces and j kings, $i, j = 0, 1, 2, 3, 4$.

(b) If the hand contains i aces, find the probability that it contains j kings.

18. Three bulbs are picked one by one without replacement from a box containing twelve bulbs, of which five are defective.

(a) What is the conditional probability that the first bulb is defective, given that there are two defective bulbs in the sample?

(b) What is the conditional probability that the first and the third bulb are defective, given that there are two defective bulbs in the sample?

19. A box contains M objects of which W are white. Given that a sample of size n contains k ($k < W$) white objects, find the probability that they were picked on the first k draws. Also, see if you can justify the answer intuitively.

20. A bowl contains M chips of which W are white ($W < M$). If the chips are picked without replacement, find the probability that a white chip is picked for the first time on the ith draw, $i = 1, 2, \ldots, M - W + 1$.

21. Give a probabilistic argument to justify why

$$\sum_{i=1}^{M-W+1} \frac{(M-W)_{(i-1)}}{(M)_{(i-1)}} \cdot \frac{W}{M-i+1} = 1$$

Hint: Use exercise 20 and the fact that it is certain that a white chip will be picked either on the first draw or on the second draw or . . . on the $(M - W + 1)$th draw.

2. THE TOTAL PROBABILITY THEOREM AND BAYES' RULE

We shall initiate this topic with an example. Suppose a desk has two drawers. One of the drawers has three green balls, say D_1G_1, D_1G_2, D_1G_3, and two white balls, say D_1W_1, D_1W_2. The second drawer has four green balls, which we might represent by $D_2G_1, D_2G_2, D_2G_3, D_2G_4$, and three white balls, say D_2W_1, D_2W_2, D_2W_3. A drawer is picked at random and a ball drawn from it. We are interested in finding the probability that this ball is green.

Let us denote the event "the ball is green" by A. We shall find $P(A)$ by two methods. The first of these consists of writing the sample space explicitly.

Method 1. The sample space has twelve outcomes and can be written as

$$S = \{D_1G_1, D_1G_2, D_1G_3, D_1W_1, D_1W_2, D_2G_1, D_2G_2, D_2G_3, D_2G_4, D_2W_1, D_2W_2, D_2W_3\}$$

These outcomes are not equally likely. As a matter of fact, the first five outcomes in S, namely, $D_1G_1, D_1G_2, D_1G_3, D_1W_1, D_1W_2$ are *each* assigned a probability of $\frac{1}{10}$, whereas the rest of the outcomes in S are each assigned a probability of $\frac{1}{14}$. (Why?)

Now

$$A = \{D_1G_1, D_1G_2, D_1G_3, D_2G_1, D_2G_2, D_2G_3, D_2G_4\}$$

Hence

$$P(A) = \frac{1}{10} + \frac{1}{10} + \frac{1}{10} + \frac{1}{14} + \frac{1}{14} + \frac{1}{14} + \frac{1}{14} = \frac{41}{70}$$

Method 2. There is a more elegant way of finding $P(A)$. The method presented here does not require us to write down S. We have

$$\text{we pick a green ball} \Leftrightarrow \begin{pmatrix}\text{we pick drawer 1}\\ \text{and}\\ \text{pick green ball}\end{pmatrix} \text{ or } \begin{pmatrix}\text{pick drawer 2}\\ \text{and}\\ \text{pick green ball}\end{pmatrix}$$

(*Note:* In the sequel we shall use the double arrow to indicate that the two statements imply each other, that is, are equivalent.)

From this, letting B_i = Drawer i is picked, $i = 1, 2$, we can write

$$A = (B_1A) \cup (B_2A)$$

Therefore,

$$\begin{aligned} P(A) &= P(B_1A) + P(B_2A), \qquad \text{since } B_1A \text{ and } B_2A \text{ are mutually exclusive} \\ &= P(B_1) \cdot P(A|B_1) + P(B_2) \cdot P(A|B_2) \end{aligned}$$

Now $P(A|B_1) = \frac{3}{5}, P(A|B_2) = \frac{4}{7}$, and $P(B_1) = P(B_2) = \frac{1}{2}$. Hence,

$$P(A) = \frac{1}{2} \cdot \frac{3}{5} + \frac{1}{2} \cdot \frac{4}{7} = \frac{41}{70}$$

We shall now consider the above problem in greater generality. First of all, we shall give a definition: A collection of events is said to be **exhaustive** if their union gives the sample space S. For example, if a coin is tossed ten times and B_i represents the event that there are i heads, $i = 0, 1, 2, \ldots, 10$, then the collection of events $B_0, B_1, B_2, \ldots, B_{10}$ is exhaustive, since $\bigcup_{i=0}^{10} B_i = S$.

The Theorem of Total Probability Suppose $\{B_i\}$ is a countable set of mutually exclusive and exhaustive events such that $P(B_i) > 0$, $i = 1, 2, \ldots$. If A is any event, then

$$P(A) = \sum_{i=1}^{\infty} P(B_i)P(A|B_i)$$

(*Note:* With reference to the illustration we have discussed, $B_1 = \{D_1G_1, D_1G_2, D_1G_3, D_1W_1, D_1W_2\}$, and $B_2 = \{D_2G_1, D_2G_2, D_2G_3, D_2G_4, D_2W_1, D_2W_2, D_2W_3\}$.)

Pictorially, the situation is represented in Figure 2.1. The event A is represented by the region enclosed in the dotted contour.

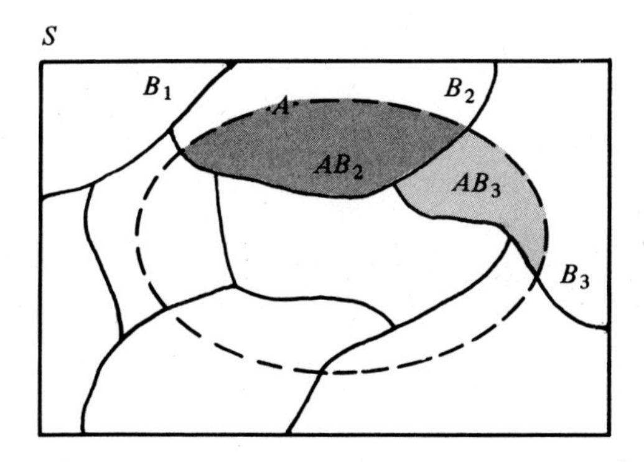

Figure 2.1

To prove the theorem, since $A = AS$, we have

$$\begin{aligned} P(A) &= P(AS) \\ &= P\left(A\left(\bigcup_{i=1}^{\infty} B_i\right)\right), \quad \text{since } B_i \text{ are exhaustive} \\ &= P\left(\bigcup_{i=1}^{\infty} (AB_i)\right), \quad \text{by the distributive law} \end{aligned}$$

Now, since $B_i \cap B_j = \emptyset$ if $i \neq j$, we have $(AB_i) \cap (AB_j) = \emptyset$ if $i \neq j$. Therefore,

$$P(A) = \sum_{i=1}^{\infty} P(AB_i) = \sum_{i=1}^{\infty} P(B_i)P(A|B_i)$$

Hence the result.

Comments. (1) The theorem of total probability is useful for finding $P(A)$ if $P(B_i)$ and $P(A|B_i)$ can be computed from the information available.

(2) In the statement of the theorem, we stipulated that the events B_i be exhaustive. Actually, we can relax this. It is sufficient that $P\left(\bigcup_{i=1}^{\infty} B_i\right) = 1$.

Example 2.1. Suppose two cards are picked from a standard deck of 52 cards, one at a time, without returning the first card to the deck. What is the probability that the second card is a spade.

Solution. Let A be the event that the second card is a spade. We will have to consider the occurrence of this event in conjunction with what happened on the first draw. Let B_1 represent that the first card was a spade and B_2 that the first card is not a spade. Then

$$P(A) = P(A|B_1) \cdot P(B_1) + P(A|B_2) \cdot P(B_2)$$
$$= \frac{12}{51} \cdot \frac{13}{52} + \frac{13}{51} \cdot \frac{39}{52} = \frac{1}{4}$$ ■

Comment. In Example 2.1, we see that the probability of picking a spade on the second draw is the same as the probability of picking a spade on the first draw. This result is true in general. For example, if we picked ten cards as above, then the probability of picking a spade on the seventh draw, say, would also be $\frac{1}{4}$.

The following result is referred to as **Bayes' rule** or the **rule of a posteriori probabilities.**

Bayes' Rule Let $\{B_i\}$ be a countable set of exhaustive and mutually exclusive events such that $P(B_i) > 0$. Let A be an arbitrary event with $P(A) > 0$. Then

$$P(B_i|A) = \frac{P(A|B_i)P(B_i)}{\sum_{i=1}^{\infty} P(A|B_i)P(B_i)} \qquad \text{for all } i$$

The proof follows immediately from the theorem of total probability. We have, since $P(A) > 0$,

$$P(B_i|A) = \frac{P(B_iA)}{P(A)}$$

But $P(B_iA) = P(A|B_i)P(B_i)$ and by the theorem of total probability

$$P(A) = \sum_{i=1}^{\infty} P(A|B_i)P(B_i)$$

Hence, substituting, the result follows.

Bayes' rule is often called *Bayes' rule for the probability of causes.* The reason for this is the following: The event A can occur in conjunction with the events B_i, $i = 1, 2, \ldots$, and as such we can regard the events B_i as causes for A. In trying to find $P(B_i|A)$, we are interested in the probability of B_i given that A has occurred. In other words, given that A has occurred, we want the probability that it was caused by B_i.

The probabilities $P(B_i)$, $i = 1, 2, \ldots$, are called the *a priori*, or prior, probabilities, and $P(B_i|A)$ the *a posteriori*, or posterior, probabilities. Bayes' rule can be used to compute the posterior probabilities if we know the a priori probabilities $P(B_i)$, $i = 1, 2, \ldots$, and the conditional probabilities $P(A|B_i)$, $i = 1, 2, \ldots$.

Example 2.2. A box contains three coins. One of the coins is a two-headed coin, the second is a fair coin, and the third is a biased coin with $P(\text{head}) = p$. One of the coins is picked at random and flipped.

(*a*) Find the probability that the coin shows heads.

(*b*) If the coin shows heads, find the probability that it is the fair coin.

Solution

(*a*) Let

A = Coin shows heads
B_1 = Coin is two-headed
B_2 = Coin is fair
B_3 = Coin is biased with $P(\text{head})$ equal to p

Then

$$\begin{aligned} P(A) &= P(A|B_1)P(B_1) + P(A|B_2)P(B_2) + P(A|B_3)P(B_3) \\ &= 1 \cdot \frac{1}{3} + \frac{1}{2} \cdot \frac{1}{3} + p \cdot \frac{1}{3} = \frac{1}{3}\left(\frac{3}{2} + p\right) \end{aligned}$$

Note that the maximum value of $P(A)$ is $\frac{5}{6}$.

(*b*) We want to find $P(B_2|A)$. This is given by

$$P(B_2|A) = \frac{P(A|B_2)P(B_2)}{P(A)} = \frac{\frac{1}{2} \cdot \frac{1}{3}}{\frac{1}{3}(\frac{3}{2} + p)} = \frac{1}{3 + 2p}$$

The maximum value of $P(B_2|A)$ occurs when $p = 0$ and is equal to $\frac{1}{3}$. The minimum value occurs when $p = 1$ and is equal to $\frac{1}{5}$.

Example 2.3. The probability that an executive is promoted to a higher position is $\frac{5}{8}$. In case he is promoted, he will go on vacation with a probability of $\frac{5}{6}$; however, if he is not promoted, there is a probability of $\frac{1}{3}$ that he will take a vacation.

(*a*) Find the probability that the executive will go on a vacation.

(*b*) Given that he does go on a vacation, find the probability that it was due to his having been promoted.

Solution

(*a*) Let A represent the event that the executive takes a vacation, and B the event that he is promoted. Then

$$\begin{aligned} P(A) &= P(A|B) \cdot P(B) + P(A|B') \cdot P(B') \\ &= \frac{5}{6} \cdot \frac{5}{8} + \frac{1}{3} \cdot \frac{3}{8} = \frac{31}{48} \end{aligned}$$

(*b*) Using Bayes' rule,

$$P(B|A) = \frac{P(A|B) \cdot P(B)}{P(A)} = \frac{\frac{5}{6} \cdot \frac{5}{8}}{\frac{31}{48}} = \frac{25}{31}$$

Example 2.4. A box contains b black balls and r red balls. One ball is drawn at random. The ball is returned to the box together with c additional balls of the same color as the drawn ball. If we now pick one ball from the new composition of balls, find the probability that the first ball was red given that the second ball is red.

Solution. Let us identify the events as follows:

A = The second ball is red
B_1 = The first ball is red
B_2 = The first ball is black

Then, by Bayes' rule,

$$P(B_1|A) = \frac{P(A|B_1)P(B_1)}{P(A|B_1)P(B_1) + P(A|B_2)P(B_2)}.$$

Now

$$P(A|B_1) = \frac{r+c}{b+r+c}, \qquad P(A|B_2) = \frac{r}{b+r+c}$$

$$P(B_1) = \frac{r}{b+r} \quad \text{and} \quad P(B_2) = \frac{b}{b+r}$$

Hence, substituting and simplifying, we get

$$P(B_1|A) = \frac{r+c}{b+r+c}$$

Example 2.5. Four machines are in operation. Machine 1 produces five percent defective items, machine 2 three percent, machine 3 four percent, and machine 4 eight percent. A machine is selected at random and an item picked.

(*a*) Given that the item is defective, find the probability that it came from machine 2.

(*b*) Given that the item is defective, find the probability that it did *not* come from machine 2.

(*c*) Given that the item is *not* defective, what is the probability that it is from machine 2?

Solution

(*a*) Let A be the event that the item is defective, and B_i, $i = 1, 2, 3, 4$, the event that it is from machine i. Then

$$P(B_2|A) = \frac{P(A|B_2)P(B_2)}{\sum_{i=1}^{4} P(A|B_i)P(B_i)}$$

$$= \frac{(0.03)\frac{1}{4}}{(0.05)\frac{1}{4} + (0.03)\frac{1}{4} + (0.04)\frac{1}{4} + (0.08)\frac{1}{4}} = \frac{3}{20}$$

(*b*) In this case we want to find $P(B_2'|A)$ which is equal to $1 - P(B_2|A) = \frac{17}{20}$.
(*c*) We are interested in $P(B_2|A')$. Using Bayes' rule appropriately, we get

$$P(B_2|A') = \frac{P(A'|B_2)P(B_2)}{\sum_{i=1}^{4} P(A'|B_i)P(B_i)}$$

$$= \frac{(0.97)\frac{1}{4}}{(0.95)\frac{1}{4} + (0.97)\frac{1}{4} + (0.96)\frac{1}{4} + (0.92)\frac{1}{4}} = \frac{97}{380}$$

Example 2.6. Sixteen people participate on a quiz show. Of these, four are on team 1 and can answer a question correctly with probability 0.8; six are on team 2 and can answer correctly with probability 0.6; four are on team 3 and can answer correctly with probability 0.7; and two are on team 4 and can answer correctly with probability 0.4. A randomly selected person answers the question and he is correct. Which team is he most likely to be on?

Solution. Let A be the event that the question is answered correctly and B_i the event that he is on team i, $i = 1, 2, 3, 4$. Then

$$P(B_i|A) = \frac{P(A|B_i)P(B_i)}{P(A)}, \qquad i = 1, 2, 3, 4$$

Since the denominator is fixed, we see that $P(B_i|A)$ is directly proportional to $P(A|B_i)P(B_i)$. Now

$$P(A|B_1) \cdot P(B_1) = 0.8 \times \frac{4}{16} = \frac{3.2}{16}$$

$$P(A|B_2) \cdot P(B_2) = 0.6 \times \frac{6}{16} = \frac{3.6}{16}$$

$$P(A|B_3) \cdot P(B_3) = 0.7 \times \frac{4}{16} = \frac{2.8}{16}$$

$$P(A|B_4) \cdot P(B_4) = 0.4 \times \frac{2}{16} = \frac{0.8}{16}$$

Hence the person is most likely to be on the second team.

Example 2.7. A bag contains 9 white balls and another bag contains 18 balls, of which 8 are white. A bag is picked at random and 6 balls are picked without replacement. They all turn out to be white. What is the probability that the bag with all white balls was chosen?

Solution. Let

A = All the six balls are white
B_1 = The bag with all white balls was chosen
B_2 = The other bag was chosen

We have

$$P(B_1|A) = \frac{P(A|B_1)P(B_1)}{P(A|B_1)P(B_1) + P(A|B_2)P(B_2)}$$

Now $P(B_1) = P(B_2) = \frac{1}{2}, P(A|B_1) = 1$, and

$$P(A|B_2) = \frac{\binom{8}{6}}{\binom{18}{6}}$$

Therefore,

$$P(B_1|A) = \frac{\frac{1}{2}}{\frac{1}{2} + \frac{\binom{8}{6}}{\binom{18}{6}} \cdot \frac{1}{2}} = \frac{\binom{18}{6}}{\binom{18}{6} + \binom{8}{6}}$$

Example 2.8. Suppose there are two batches of students in a school. In the first batch, all students have been inoculated, whereas in the second batch two-thirds of the students have been inoculated. A batch is picked at random, and a student picked from it at random turns out to have been inoculated. If this student is returned to his batch, what is the probability that a second student picked from *this* batch will not have been inoculated?

Solution. Let

A = The first student has been inoculated
C = The second student has not been inoculated
and B_i = Batch i is picked, $i = 1, 2$

We want to find $P(C|A)$. Now

$$P(C|A) = \frac{P(AC)}{P(A)} = \frac{P(AC|B_1)P(B_1) + P(AC|B_2)P(B_2)}{P(A|B_1)P(B_1) + P(A|B_2)P(B_2)}$$

by applying the theorem of total probability to the events AC and A. We are given that

$$P(B_1) = P(B_2) = \frac{1}{2}, \quad P(A|B_1) = 1, \quad P(A|B_2) = \frac{2}{3}$$

Also, it can be easily seen that $P(AC|B_1) = 0$, and $P(AC|B_2) = P(A|B_2) \cdot P(C|AB_2)$ $= \frac{2}{3} \cdot \frac{1}{3} = \frac{2}{9}$. ($P(C|AB_2) = \frac{1}{3}$, since the first student is returned to the batch.)

Thus, finally,

$$P(C|A) = \frac{\frac{2}{9} \cdot \frac{1}{2}}{1 \cdot \frac{1}{2} + \frac{2}{3} \cdot \frac{1}{2}} = \frac{2}{15}$$

■

EXERCISES–SECTION 2

1. Three boxes contain white, red, and black balls in the numbers given below.

Box	White	Black	Red
1	8	2	6
2	4	5	3
3	2	7	3

A box is picked at random, and a ball is drawn from it at random.

(a) Find the probability that the ball is red.

(b) Given that the ball is red, what is the probability that it came from box i ($i = 1, 2, 3$).

2. Suppose 3 cards are picked without replacement from a standard deck of 52 cards. Find the probability that the third card is a spade.

3. The probability that a student studies for a test is 0.7. Given that he studies, the probability is 0.8 that he will pass the test. Given that he does not study, the probability is 0.3 that he will pass the test. What is the probability that the student will pass the test?

4. On a true-false test, the probability that a student knows the answer to a question is equal to 0.7. If he knows the answer, he checks the correct answer; otherwise, he answers the question by flipping a fair coin.

(a) What is the probability that he answers a question correctly?

(b) Given that he answers the question correctly, what is the probability that he knew the answer?

5. A bowl contains n white chips and n black chips. A number is picked at random from the even integers $2, 4, 6, \ldots, 2n$ and that many chips are drawn at random from the bowl, without replacement.

(a) Find the probability that the same number of chips of each color are picked.

(b) Given that the same number of chips of each color are picked, what is the probability that $2k$ chips were picked, where $1 < 2k \leqslant 2n$?

6. A box contains ten white and eight black objects. A fair die is rolled. If the number is even, then as many white objects are added to the box, and if it is odd, then as many black objects are added. From the new composition of the box three objects are picked at random without replacement.

(a) Find the probability of picking two white and one black object.

(b) Given that two white and one black object were picked, what is the probability that (*i*) a 4 showed up on the die, (*ii*) an even number showed up on the die.

7. In a high school the sophomore class has 6 girls and 6 boys, the junior class has 8 girls and 10 boys, and the senior class has 3 girls and 9 boys. Suppose a student is picked at random from each class. Given that the sample contains exactly 2 boys, find the probability that the student picked from the sophomore class is a boy.

8. A party is divided into two groups, one of which has 6 males and 4 females, and the other, 9 males and 11 females. Given that a person is a male, the probability

that he smokes is 0.4, and given that the person is a female, the probability that she smokes is 0.2. A group is picked at random and a person selected from it at random. Find the probability that the person selected smokes.

9. The morning section of a class has 9 male and 7 female students, and the afternoon section has 6 male and 3 female students. Five students picked at random from the morning section are transferred to the afternoon section. Find the probability of picking a male student from the afternoon section after making the transfer.

10. There are n people in a room, all of different ages. Two people are picked from this group with the understanding that if the first person picked is the youngest in the group he is kept; otherwise he is returned. What is the probability that the second person picked is the next youngest?

11. Messages are sent on a wireless by using the signals dot and dash to code the messages. The probability that a signal is a dash is 0.6. From past experience, it is found that a signal which originates as a dash has a probability of 0.2 of being received as a dot and a signal which originates as a dot has a probability of 0.1 of being received as a dash.

 (a) What is the probability that a signal is received as a dash?
 (b) Given that a signal is received as a dash, what is the probability that it originated as a dash?

12. A lot contains nine items, of which four are defective and five are nondefective. Two items are picked at random and discarded without examination. What is the probability that a third item picked from the remaining lot is defective?

13. (Generalization of problem 12.) A lot contains M items of which D are defective. Suppose r items are picked without replacement and discarded without examination. Find the probability that the $(r + 1)$th item picked is defective.

14. A box contains 18 beads, of which 10 are white and 8 are black. Three beads are picked and discarded and replaced by three new black beads. If three beads are now picked from the new composition, find the probability that they are all white.

3. INDEPENDENT EVENTS

The notion of conditional probability was prompted by the fact that the knowledge that an event B has occurred might lead us to reassess the probability of another event A. It can, of course, turn out that the occurrence of one event has no influence on the occurrence of the other. For example, given the sex of the first child it seems reasonable to assume that it would not influence the sex of the second child. This concept leads to what are called *independent events*. We shall begin the discussion with an example.

Suppose we toss a die and a coin. Consider two events A and B where A = Multiple of 3 on the die, and B = Heads on the coin. The physical nature of the experiment tells us that what happens on the coin should not influence the outcome on the die, and conversely. In that sense, using our everyday interpretation of the word "independent," we might say that the outcome on one is independent of the outcome on the other. How does this carry over in the context of probability theory?

Writing the events as sets we get

$$A = \{(H, 3), (H, 6), (T, 3), (T, 6)\}$$
$$B = \{(H, 1), (H, 2), (H, 3), (H, 4), (H, 5), (H, 6)\}$$
and $AB = \{(H, 3), (H, 6)\}$

Since there are 12 outcomes in S, this gives

$$P(A) = \frac{4}{12}, \quad P(B) = \frac{6}{12} \quad \text{and} \quad P(AB) = \frac{2}{12}$$

Now

$$P(A|B) = \frac{P(AB)}{P(B)}, \qquad \text{by definition}$$
$$= \frac{\frac{2}{12}}{\frac{6}{12}} = \frac{1}{3}$$

But $\frac{1}{3}$ is precisely the probability of A. Thus we find that $P(A|B) = P(A)$. In other words, the conditional probability of A given B is equal to the unconditional probability of A. Of course, if $P(A|B) = P(A)$, then we have

$$P(AB) = P(B) \cdot P(A|B) = P(A) \cdot P(B)$$

We are thus led to the following definition of independent events:

Two events A and B are said to be **stochastically independent**, or simply **independent**, if

$$P(AB) = P(A) \cdot P(B)$$

Two events that are not independent are said to be **dependent events.**

Comments. (1) The above definition applies even in the cases where $P(A) = 0$ or $P(B) = 0$, though the conditional probabilities are not defined.

(2) If the probability of one of the events is 0, then A and B are independent events. This follows because in this case $P(AB) = 0$ (why?), giving $P(AB) = P(A) \cdot P(B)$ since both sides are equal to 0.

(3) The impossible event $\emptyset$ and any event B are always independent in view of comment (2) above.

(4) The sure event S and any event A are always independent since $P(AS) = P(A) = 1 \cdot P(A) = P(S) \cdot P(A)$.

Example 3.1. Suppose a fair coin is tossed twice. Let A = More than one head, and B = Exactly one head. Are A and B independent events?

Solution. Since

$$S = \{HH, HT, TH, TT\}, \quad A = \{HH\}, \quad B = \{HT, TH\} \quad \text{and } AB = \emptyset,$$

we get

$$P(A) = \frac{1}{4}, \quad P(B) = \frac{1}{2}, \quad \text{and} \quad P(AB) = 0$$

Hence $P(AB) \neq P(A) \cdot P(B)$, so that A and B are not independent events.

Example 3.2. Suppose a coin is tossed twice, with the following assignment of probabilities to the simple events:

$$P(\{HH\}) = 0.16, \quad P(\{HT\}) = 0.24, \quad P(\{TH\}) = 0.24 \quad \text{and} \quad P(\{TT\}) = 0.36$$

Show that the two events A = Heads on the first toss, and B = Tail on the second toss, are independent.

Solution. It follows that

$$\begin{aligned} & P(A) = P(\{HH, HT\}) = 0.16 + 0.24 = 0.40 \\ & P(B) = P(\{HT, TT\}) = 0.24 + 0.36 = 0.60 \\ \text{and} \quad & P(AB) = P(\{HT\}) = 0.24 \end{aligned}$$

Hence $P(AB) = P(A) \cdot P(B)$, and, consequently, A and B are independent.

Example 3.3. It is known from past experience that the probability that a randomly picked person has cancer is 0.2, and the probability that he has heart disease is 0.1. Assuming the two events are independent, what is the probability that a person has (*a*) at least one ailment? (*b*) precisely one ailment?

Solution. Let A = The person has cancer, and B = The person has heart disease.

(*a*) We want to find $P(A \cup B)$. Now

$$\begin{aligned} P(A \cup B) &= P(A) + P(B) - P(AB) \\ &= P(A) + P(B) - P(A) \cdot P(B), \qquad \text{since the events are independent} \\ &= 0.2 + 0.1 - 0.2 \times 0.1 = 0.28 \end{aligned}$$

(*b*) The probability of precisely one ailment is equal to $P(A) + P(B) - 2P(A) \cdot P(B) = 0.26$. ■

Comments. The reader should be aware of the distinction between independent events and mutually exclusive events. The two concepts are often confused.

Two events are mutually exclusive when they are not compatible, that is, they cannot occur together. "Mutually exclusive" is a property of sets. In this case, $AB = \emptyset$ so that $P(AB) = 0$.

Two events are independent when the occurrence of one event does not influence the occurrence of the other. Thus no inference can be drawn regarding the occurrence of one event on the basis of the knowledge of the occurrence of the other. *Independence is a property of the probability measure.* In this case $P(AB) = P(A) \cdot P(B)$.

As a matter of fact, the following result shows how divergent the two concepts really are:

If A and B are two events with $P(A) > 0$ and $P(B) > 0$, then, as can be seen immediately,

(*i*) if A and B are independent, they cannot be mutually exclusive; and
(*ii*) if A and B are mutually exclusive, they cannot be independent.

Example 3.4. Suppose we draw a card from a standard bridge deck of cards. Give two events B and D which are–

(*a*) mutually exclusive and independent
(*b*) mutually exclusive, but not independent
(*c*) not mutually exclusive, but are independent
(*d*) not mutually exclusive and not independent.

Solution. In each case we shall give the events B and D. The reader is asked to provide the justifications.

(*a*) Note that one of the events has to have zero probability since, by previous discussion, mutually exclusive events can be independent only if one of the events has zero probability.

Let

D = The card is an ace
B = The card is a black card of hearts

(*b*) Let

D = The card is an ace of spades
B = The card is a 4 of a red suit

(*c*) Let

D = The card is an ace
B = The card is a spade

(*d*) Let

D = The card is a king
B = The card is a face card ■

Two events are independent if the occurrence of one does not influence the occurrence of the other. If this is the case, then it seems intuitively obvious that the nonoccurrence of one should not influence the occurrence of the other. Nor should the nonoccurrence of one influence the nonoccurrence of the other. The following result spells this out.

If A and B are independent events, then

(*i*) A and B' are independent
(*ii*) A' and B are independent
(*iii*) A' and B' are independent

We shall prove (*i*) and leave the other cases to the reader. We want to show that $P(AB') = P(A) \cdot P(B')$. Now

$$
\begin{aligned}
P(AB') &= P(A) - P(AB) \\
&= P(A) - P(A) \cdot P(B), \qquad \text{since } A \text{ and } B \text{ are independent} \\
&= P(A)[1 - P(B)] = P(A) \cdot P(B')
\end{aligned}
$$

Hence, by definition, A and B' are independent.

We shall now extend the notion of independence to more than two events. As a first step in this direction, let us consider three events A, B, and C. For these events to be independent–granted that the conditional probabilities are defined–we might require, intuitively, that

$$
\begin{aligned}
P(A|BC) &= P(A|B) = P(A|C) = P(A) \\
P(B|AC) &= P(B|A) = P(B|C) = P(B) \\
P(C|AB) &= P(C|A) = P(C|B) = P(C)
\end{aligned}
$$

These requirements would in turn imply that we must have $P(AB) = P(A) \cdot P(B)$, $P(AC) = P(A) \cdot P(C)$, $P(BC) = P(B) \cdot P(C)$, and $P(ABC) = P(A) \cdot P(B) \cdot P(C)$.

We are thus led to the following definition of the independence of three events.

Three events A, B, C are said to be **mutually independent**, or simply **independent**, if

$$
\left.
\begin{aligned}
&1.\ P(AB) = P(A) \cdot P(B) \\
&2.\ P(AC) = P(A) \cdot P(C) \\
&3.\ P(BC) = P(B) \cdot P(C)
\end{aligned}
\right\} \quad \text{pairwise independence}
$$

$$4.\ P(ABC) = P(A) \cdot P(B) \cdot P(C)$$

Comment. *All* four conditions have to be satisfied. There are situations where the events are pairwise independent, but not independent, because $P(ABC) \neq P(A) \cdot P(B) \cdot P(C)$. Also, there are situations where $P(ABC) = P(A) \cdot P(B) \cdot P(C)$, but the events are not independent because the events are not pairwise independent.

In Examples 3.5 and 3.6 below we give three events which are pairwise independent but are not independent.

Example 3.5. Suppose the sample space S is given by

$$S = \{(1, 0, 0), (0, 1, 0), (0, 0, 1), (1, 1, 1)\}$$

with four equally likely outcomes. Let

A = The first coordinate is 1
B = The second coordinate is 1
C = The third coordinate is 1

Show that the events are pairwise independent but $P(ABC) \neq P(A) \cdot P(B) \cdot P(C)$ and, consequently, the events are not independent.

Solution. Writing the events as sets,

$$
\begin{aligned}
&A = \{(1, 0, 0), (1, 1, 1)\}, \quad B = \{(0, 1, 0), (1, 1, 1)\} \\
&C = \{(0, 0, 1), (1, 1, 1)\}, \quad AB = \{(1, 1, 1)\} \\
&AC = \{(1, 1, 1)\}, \quad BC = \{(1, 1, 1)\}, \quad ABC = \{(1, 1, 1)\}
\end{aligned}
$$

Hence

$$P(A) = P(B) = P(C) = \frac{1}{2}$$

$$P(AB) = P(AC) = P(BC) = \frac{1}{4}$$

and $P(ABC) = \dfrac{1}{4}$

We see immediately that the events are pairwise independent, but $P(ABC) \neq P(A) \cdot P(B) \cdot P(C)$.

Example 3.6. Consider the following three events in the experiment of tossing two fair dice.

A = The first die shows an even number
B = The second die shows an odd number
C = The sum on the two dice is even

Show that the three events are pairwise independent, but not independent.

Solution. It can be seen that $P(A) = P(B) = P(C) = \frac{1}{2}$. Now, $AB = \{(2, 1), (2, 3), (2, 5), (4, 1), (4, 3), (4, 5), (6, 1), (6, 3), (6, 5)\}$. Consequently, $P(AB) = \frac{1}{4}$. Similarly, $P(AC) = \frac{1}{4}$ and $P(BC) = \frac{1}{4}$. Hence, $P(AB) = P(A) \cdot P(B)$, $P(BC) = P(B) \cdot P(C)$, and $P(AC) = P(A) \cdot P(C)$. Therefore, the events A, B, and C are pairwise independent.

However, since $ABC = \emptyset$, we have $P(ABC) = 0$, and therefore

$$P(ABC) \neq P(A) \cdot P(B) \cdot P(C)$$

proving that A, B, and C are not independent. ■

The following example gives three events for which $P(ABC) = P(A) \cdot P(B) \cdot P(C)$, but the events are not independent because they are not pairwise independent.

Example 3.7. Suppose two dice are rolled. Define three events A, B, C as follows:

A = The first die shows an even number
B = The sum on the two dice is 4
C = The outcomes on the two dice differ at most by 2

That is,

$$C = \{(x, y) \mid |x - y| \leqslant 2\}$$

Verify that $P(ABC) = P(A) \cdot P(B) \cdot P(C)$, but that the events are not pairwise independent.

Solution. The outcomes of the sample space are listed below in Figure 3.1, and the outcomes in the shaded region belong to the event C.

It can be easily seen that $P(A) = \frac{1}{2}$, $P(B) = \frac{1}{12}$, and $P(C) = \frac{2}{3}$. Also, since $ABC = \{(2, 2)\}$, $P(ABC) = \frac{1}{36}$. Hence, $P(ABC) = P(A) \cdot P(B) \cdot P(C)$.

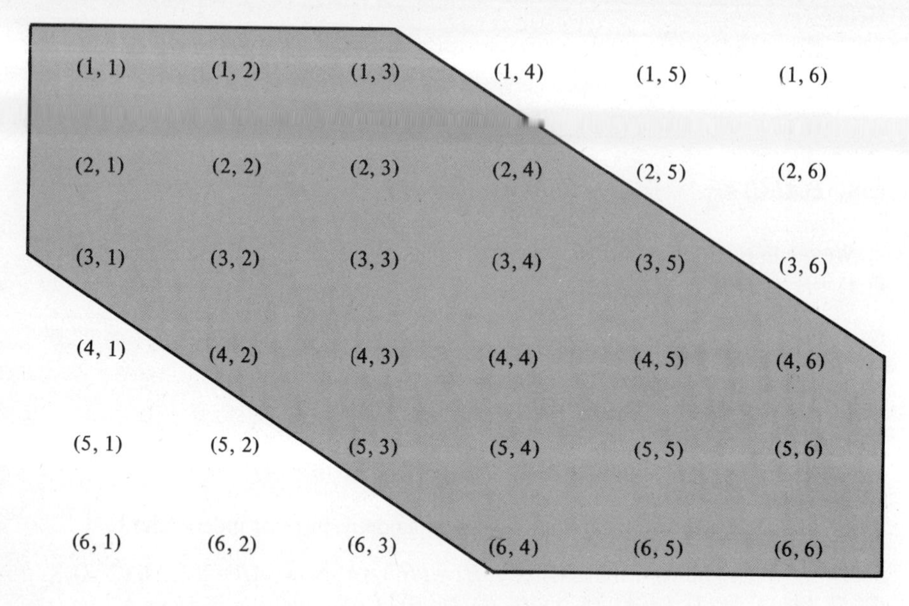

Figure 3.1

However, note, for example, that $P(AB) = \frac{1}{36}$ and consequently $P(AB) \neq P(A) \cdot P(B)$. The events A, B, and C are not mutually independent. ■

If there are n events $A_1, A_2, \ldots, A_n$, they are said to be **mutually independent**, or simply **independent**, if

$$\binom{n}{2} \text{ conditions} \rightarrow P(A_iA_j) = P(A_i)P(A_j), \qquad 1 \leqslant i < j \leqslant n$$

$$\binom{n}{3} \text{ conditions} \rightarrow P(A_iA_jA_k) = P(A_i)P(A_j)P(A_k), \qquad 1 \leqslant i < j < k \leqslant n$$

$$\vdots$$

$$\binom{n}{n} \text{ conditions} \rightarrow P(A_1A_2 \ldots A_n) = P(A_1)P(A_2) \ldots P(A_n)$$

There are

$$\binom{n}{2} + \binom{n}{3} + \ldots + \binom{n}{n} = 2^n - n - 1$$

conditions that must hold. In short, $A_1, A_2, \ldots, A_n$ *are mutually independent if the probability of the intersection of* **any** *subcollection of these events is equal to the product of the probabilities of the events in the subcollection.*

Intuitively, it seems reasonable to expect that if $A_1, A_2, \ldots, A_n$ are mutually independent events, then the occurrence of some of these events should not influence the nonoccurrence of the others. Specifically, if we take any k of the above events ($k \leqslant n$), and the complements of the remaining events, then the n events so obtained should be mutually independent. This is indeed true. For instance, we have the following result:

If A, B, and C are mutually independent events, so are A', B', C'; so are A, B', C', and so on.

Recall that we have proved a similar result for two events A, B. We shall prove here that if A, B, C are mutually independent, so are A', B', C'. The reader might enjoy proving the other variations.

Since A, B, C are mutually independent, they must be pairwise independent. Consequently, A', B', C' must be pairwise independent.

Next, we shall show that $P(A'B'C') = P(A')P(B')P(C')$. This follows because

$$\begin{aligned}
P(A'B'C') &= P((A \cup B \cup C)'), \qquad \text{by De Morgan's law} \\
&= 1 - P(A \cup B \cup C) \\
&= 1 - [P(A) + P(B) + P(C) - P(AB) - P(AC) - P(BC) + P(ABC)] \\
&= 1 - P(A) - P(B) - P(C) + P(A)P(B) + P(A)P(C) + P(B)P(C) \\
&\qquad - P(A)P(B)P(C), \qquad \text{since } A, B, C \text{ are independent} \\
&= (1 - P(A))(1 - P(B))(1 - P(C)) \\
&= P(A')P(B')P(C')
\end{aligned}$$

Hence A', B', C' are independent.

Example 3.8. Suppose A, B, C are three events where A and C are independent and B and C are independent. Show that $A \cup B$ and C are independent if AB and C are independent, and conversely.

Solution

(*i*) Assume AB and C are independent. We shall show that $A \cup B$ and C are independent. We have

$$\begin{aligned}
P((A \cup B)C) &= P(AC \cup BC) \\
&= P(AC) + P(BC) - P(ABC) \\
&= P(A)P(C) + P(B)P(C) - P(AB)P(C)
\end{aligned}$$

since A and C are independent, B and C are independent, and AB and C are independent. Hence

$$\begin{aligned}
P((A \cup B)C) &= [P(A) + P(B) - P(AB)]P(C) \\
&= P(A \cup B) \cdot P(C)
\end{aligned}$$

As a result, $A \cup B$ and C are independent.

(*ii*) To prove the converse, assume $A \cup B$ and C are independent. We shall establish that $P((AB)C) = P(AB) \cdot P(C)$. We have

$$\begin{aligned}
P((AB)C) &= P(AC) + P(BC) - P((A \cup B)C) \qquad \text{(why?)} \\
&= P(A) \cdot P(C) + P(B) \cdot P(C) - P(A \cup B) \cdot P(C) \qquad \text{(why?)} \\
&= [P(A) + P(B) - P(A \cup B)]P(C) \\
&= P(AB) \cdot P(C)
\end{aligned}$$

Therefore, AB and C are independent.

Example 3.9. If A, B, and C are mutually independent events with $P(A) = 0.5$, $P(B) = 0.6$, and $P(C) = 0.3$, find:

(a) $P((AB') \cup C)$
(b) $P(A \cup B' \cup C')$
(c) $P[(C'(A \cup B))']$

Solution. We have:

$$(a)\quad P((AB') \cup C) = P(AB') + P(C) - P(AB'C) = P(A) \cdot P(B') + P(C) - P(A) \cdot P(B') \cdot P(C) = 0.44$$

$$(b)\quad P(A \cup B' \cup C') = P(A) + P(B') + P(C') - P(AB') - P(AC') - P(B'C') + P(AB'C')$$
$$= P(A) + P(B') + P(C') - P(A)P(B') - P(A)P(C') - P(B')P(C') + P(A)P(B')P(C') = 0.91$$

Alternatively,

$$P(A \cup B' \cup C') = 1 - P(A'BC) \qquad \text{(why?)}$$
$$= 1 - P(A')P(B)P(C) = 0.91$$

$$(c)\quad P[(C'(A \cup B))'] = 1 - P(C'(A \cup B))$$
$$= 1 - P((C'A) \cup (C'B))$$
$$= 1 - P(C'A) - P(C'B) + P(ABC')$$
$$= 1 - P(C')P(A) - P(C')P(B) + P(A)P(B)P(C') = 0.44$$

Example 3.10. Current flows through a circuit through a series of relays as shown in the circuit diagram in Figure 3.2.

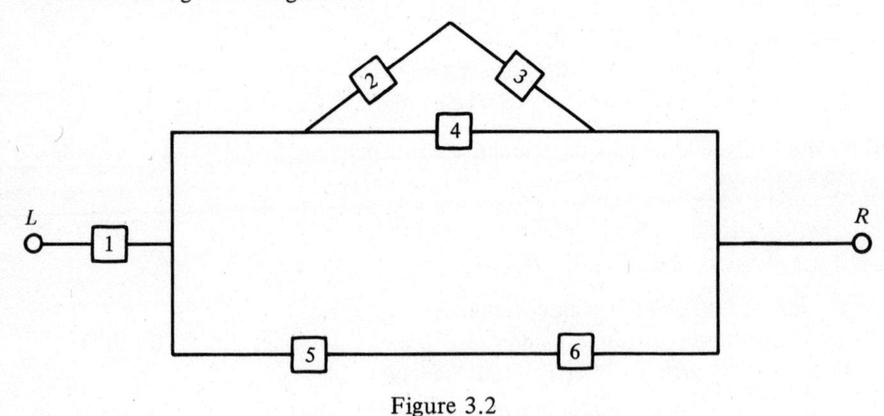

Figure 3.2

The probability that a relay is closed is p. Assuming that the relays function independently, find the probability that current flows from L to R.

Solution. Let

A_i = Relay i is closed, $\quad i = 1, 2, 3, 4, 5, 6$

and $\quad B$ = Current flows from L to R

Then

$$B = (A_1A_2A_3) \cup (A_1A_4) \cup (A_1A_5A_6)$$

Therefore

$$\begin{aligned}
P(B) &= P((A_1A_2A_3) \cup (A_1A_4) \cup (A_1A_5A_6)) \\
&= P(A_1A_2A_3) + P(A_1A_4) + P(A_1A_5A_6) - [P(A_1A_2A_3A_4) \\
&\quad + P(A_1A_2A_3A_5A_6) + P(A_1A_4A_5A_6)] + P(A_1A_2A_3A_4A_5A_6) \quad \text{(why?)} \\
&= p^3 + p^2 + p^3 - [p^4 + p^5 + p^4] + p^6 \\
&= p^2 + 2p^3 - 2p^4 - p^5 + p^6
\end{aligned}$$

since, for example, $P(A_1A_5A_6) = P(A_1)P(A_5)P(A_6) = p^3$, and so on.

Independent trials

Up to this point, our discussion has involved independent events. We shall now describe what we mean by **independent trials.** The intuitive notion is that the basic experiment is repeated under identical conditions and the outcome of one trial does not influence the outcome of any other trial. The precise mathematical formulation of this concept is in terms of independent events.

Consider an experiment which consists of tossing a coin twice. We shall refer to the experiment of tossing the coin once as the basic experiment and call it a *trial*. The experiment consisting of tossing the coin twice, looked at as a whole, is called the *composite experiment*. The basic experiment has the sample space $S = \{H, T\}$ with, let us say, the probability of heads equal to p. The composite experiment leads to four outcomes which can be written as $(H, H), (H, T), (T, H), (T, T)$. Indeed, the sample space of the combined experiment can be regarded as the Cartesian product $\{H, T\} \times \{H, T\} = S^2$.

The set $A = \{(H, H), (H, T)\} = \{H\} \times S$ represents the event that the first toss falls heads, and hence refers only to the first trial. We shall say that the event A is *determined by the first trial.*

Similarly, the event $B = \{(H, H), (T, H)\} = S \times \{H\}$ is said to be *determined by the second trial.*

The independence of the two tosses or trials is meant to suggest that the events A and B are independent. Also, we would require that the probability of A be equal to p because, intuitively, the probability that the first toss falls heads should be the same whether we consider this event in the context of the composite experiment or in the context of the basic experiment of the first toss of the coin. Thus the question that faces us is the following: How should the probabilities be assigned to the outcomes of the composite experiment so that the probability of A will be equal to p (the probability of heads on a single toss) and the events A and B will be independent?

Suppose we make the following assignment of probabilities to the outcomes of the composite experiment:

$$P(\{(H, H)\}) = p^2 + \epsilon, \qquad P(\{(H, T)\}) = p(1-p) - \epsilon$$
$$P(\{(T, H)\}) = p(1-p) - \epsilon, \qquad P(\{(T, T)\}) = (1-p)^2 + \epsilon$$

where ϵ is any real number satisfying $0 < \epsilon \leqslant p(1-p)$. (Thus there is an infinite number of ways of making the above assignment.)

With this assignment we see that

$$P(A) = p^2 + \epsilon + p(1-p) - \epsilon = p$$

and, similarly,

$$P(B) = p$$

This assignment of probabilities to the outcomes of the composite experiment is acceptable to the extent that it results in a probability of heads on any toss equal to p, as we would intuitively demand. However, we see that the events A and B are not independent, since $P(AB) = P\{(H, H)\} = p^2 + \epsilon \neq P(A) \cdot P(B)$, and consequently the trials are not independent. In that sense, the above assignment of probabilities is not satisfactory for the purpose of having the trials independent.

Now consider the assignment

$$P(\{(H, H)\}) = p^2, \qquad P(\{(H, T)\}) = p(1-p)$$
$$P(\{(T, H)\}) = p(1-p), \qquad P(\{(T, T)\}) = (1-p)^2$$

It is easily verified that the probability of heads on any toss is equal to p and also that the events A and B are independent. As a matter of fact, it can be seen that any event determined only by the first toss and any event determined only by the second toss are independent. *In order for the trials to be independent, this is the only way to assign probabilities to the outcomes of the composite experiment.*

To generalize from the above discussion, consider an experiment consisting of n identical trials, each trial defined by the sample space S with a finite number of outcomes. Let P_1 be the probability measure of the events of S. The sample space appropriate for the composite experiment consisting of n trials is the Cartesian product S^n where

$$S^n = \{(s_1, s_2, \ldots, s_n) \mid s_i \text{ is the outcome of the } i\text{th trial}, \ i = 1, 2, \ldots, n\}$$

An event B (that is, a subset of S^n) is said to be determined by the ith trial if

$$B = S \times S \times \ldots \times S \times \underset{\substack{\uparrow \\ i\text{th trial}}}{C} \times S \times \ldots \times S$$

where C is some subset of S.

We define the n trials to be **independent** if every set of events $B_1, B_2, \ldots, B_n$, where B_i is determined by the ith trial, is a set of mutually independent events.

To define a probability measure on S^n, it suffices to define it for each sample point in S^n. Let us denote this probability measure by P. *If the trials are to be independent this can be accomplished in one and only one way:*

$$\boxed{P(\{(s_1, s_2, \ldots, s_n)\}) = P_1(\{s_1\}) \cdot P_1(\{s_2\}) \cdot \ldots \cdot P_1(\{s_n\})}$$

This assignment or probabilities is acceptable, since it can be easily verified that the probabilities add to unity.

Furthermore, with this assignment, it can be shown that if B_i depends only on the ith trial $(i = 1, 2, \ldots, n)$, that is, say

$$B_1 = C_1 \times S \times S \times \ldots \times S$$
$$B_2 = S \times C_2 \times S \times \ldots \times S$$
$$\vdots$$
$$B_n = S \times S \times \ldots \times S \times C_n$$

then

$$P(B_1) = P_1(C_1), \quad P(B_2) = P_1(C_2), \quad \ldots \quad P(B_n) = P_1(C_n)$$

and

$$P(B_1 \cap B_2 \cap \ldots \cap B_n) = P_1(C_1) \cdot P_1(C_2) \cdot \ldots \cdot P_1(C_n)$$

The verification of these results is omitted.

Comment. What is the benefit of all this discussion of independent trials? *The important fact is that if the trials are independent, then we can compute the probabilities of the events in the composite experiment on the basis of the probabilities of the events in the basic experiment.* For instance, if we want to find the probability that, in rolling a fair die three times, we get an even number on the first toss, a 5 on the second toss, and a multiple of 3 on the third toss, we do not have to consider the set of triplets $\{(x, 5, z) \mid x \text{ an even number}, z \text{ a multiple of } 3\}$ from among the 6^3 outcomes in the composite experiment. Instead, we can argue as follows: the probability of getting an even number on a roll of a die is $\frac{1}{2}$, of getting a 5 is $\frac{1}{6}$, and of getting a multiple of 3 is $\frac{1}{3}$, and, consequently, the probability of the desired event is $\frac{1}{2} \cdot \frac{1}{6} \cdot \frac{1}{3} = \frac{1}{36}$.

At least one, and exactly k of n independent events

We open this discussion with the following example:

Example 3.11. Suppose A, B, C are mutually independent events with $P(A) = P(B) = P(C) = p$. Find the probability that (a) exactly k $(k = 0, 1, 2, 3)$ of the events occur, (b) at least one of the events occurs.

Solution

(a) We shall calculate only the case $k = 2$. We see that

$$\begin{aligned} P\begin{pmatrix}\text{exactly two of the} \\ \text{events } A, B, C\end{pmatrix} &= P((ABC') \cup (AB'C) \cup (A'BC)) \\ &= P(ABC') + P(AB'C) + P(A'BC) \\ &= P(A)P(B)P(C') + P(A)P(B')P(C) + P(A')P(B)P(C), \\ &\qquad \text{since the events are independent} \\ &= p^2(1-p) + p^2(1-p) + p^2(1-p) = 3p^2(1-p) \end{aligned}$$

There is an alternate approach. Recall that

$$\begin{aligned} P\begin{pmatrix}\text{exactly two of the} \\ \text{events } A, B, C\end{pmatrix} &= P(AB) + P(AC) + P(BC) - 3P(ABC) \\ &= P(A)P(B) + P(A)P(C) + P(B)P(C) - 3P(A)P(B)P(C) \\ &= 3p^2 - 3p^3 = 3p^2(1-p) \end{aligned}$$

(*b*) We have

$$P\begin{pmatrix}\text{at least one of the}\\ \text{events } A, B, C\end{pmatrix} = P(A \cup B \cup C)$$
$$= P(A) + P(B) + P(C) - [P(AB) + P(AC) + P(BC)] + P(ABC)$$
$$= 3p - 3p^2 + p^3, \qquad \text{due to independence}$$

An alternate method is as follows:

$$P(A \cup B \cup C) = 1 - P((A \cup B \cup C)') = 1 - P(A'B'C')$$
$$= 1 - P(A')P(B')P(C') = 1 - (1-p)^3 = 3p - 3p^2 + p^3$$ ■

We shall now extend the results of Example 3.11 to the case of n independent events to find the probabilities of "at least one of the n events" and "exactly k ($k = 0, 1, \ldots, n$) of the events."

At least one of *n* independent events In Chapter 2, we developed a general expression for the probability of at least one of the events $A_1, A_2, \ldots, A_n$:

$$P\left(\bigcup_{i=1}^{n} A_i\right) = \sum_{i=1}^{n} P(A_i) - \sum_{i<j}^{n} P(A_iA_j) + \ldots + (-1)^{n+1} P(A_1A_2 \ldots A_n)$$

Now if the events are mutually independent, then we know that for any subcollection of events $A_{i_1}, A_{i_2}, \ldots, A_{i_k}$,

$$P(A_{i_1}A_{i_2} \ldots A_{i_k}) = P(A_{i_1})P(A_{i_2}) \ldots P(A_{i_k})$$

and consequently we have the following:

If $A_1, A_2, \ldots, A_n$ are independent, then

$$P\left(\bigcup_{i=1}^{n} A_i\right) = \sum_{i=1}^{n} P(A_i) - \sum_{i<j}^{n} P(A_i)P(A_j) + \ldots + (-1)^{n+1} P(A_1) \ldots P(A_n)$$

There is an alternate expression for $P\left(\bigcup_{i=1}^{n} A_i\right)$ if the events are mutually independent:

$$P\left(\bigcup_{i=1}^{n} A_i\right) = 1 - P\left(\left(\bigcup_{i=1}^{n} A_i\right)'\right) = 1 - P\left(\bigcap_{i=1}^{n} A_i'\right), \qquad \text{by De Morgan's law}$$
$$= 1 - P(A_1')P(A_2') \ldots P(A_n'), \qquad \text{due to independence}$$

Hence we get the following:

If $A_1, A_2, \ldots, A_n$ are independent, then

$$P\left(\bigcup_{i=1}^{n} A_i\right) = 1 - (1 - P(A_1))(1 - P(A_2)) \ldots (1 - P(A_n))$$

The two expressions for $P\left(\bigcup_{i=1}^{n} A_i\right)$ are equivalent, as a simple algebraic expansion will show.

Exactly k of n independent events Suppose $A_1, A_2, \ldots, A_n$ are n independent events with $P(A_i) = p,\ i = 1, 2, \ldots, n$. We shall prove the following:

(*a*) The probability that k specified events $A_{i_1}, A_{i_2}, \ldots, A_{i_k}$ occur and none of the other events occurs is $p^k(1-p)^{n-k},\ k = 0, 1, \ldots, n$.

(*b*) The probability that exactly k of the events occur is $\binom{n}{k}p^k(1-p)^{n-k}$, $k = 0, 1, \ldots, n$.

To prove (*a*), we shall find the probability that $A_1, A_2, \ldots, A_k$ occur and $A_{k+1}, A_{k+2}, \ldots, A_n$ do not occur. The argument is analogous in the general case. We have

$$\begin{aligned} P(A_1A_2\ldots A_kA'_{k+1}A'_{k+2}\ldots A'_n) &= P(A_1)\ldots P(A_k)P(A'_{k+1})\ldots P(A'_n) \\ &= p^k(1-p)^{n-k} \end{aligned}$$

since $P(A_i) = p$ and $P(A'_i) = 1 - p$

To prove (*b*), we note that there are $\binom{n}{k}$ ways of picking k events out of n so that these k events will occur and the remaining $n - k$ events will not. Consider just one such situation where k specified events $A_{i_1}, A_{i_2}, \ldots, A_{i_k}$ occur and the others do not. From (*a*), we know that the probability of this sequence is $p^k(1-p)^{n-k}$. Hence, since there are $\binom{n}{k}$ such sequences, the probability that exactly k events occur is $\binom{n}{k}p^k(1-p)^{n-k}$.

The expressions $\binom{n}{k}p^k(1-p)^{n-k},\ k = 0, 1, 2, \ldots, n$, are called the **binomial probabilities** because $\binom{n}{k}p^k(1-p)^{n-k}$ is the kth term in the binomial expansion $[p + (1-p)]^n$, since $[p + (1-p)]^n = \sum_{k=0}^{n}\binom{n}{k}p^k(1-p)^{n-k}$.

Example 3.12. A person fires ten shots at a target and each shot has a probability of $\frac{1}{3}$ of hitting the target. What is the probability that he hits the target at least once?

Solution. The probability of hitting at least once is

$$1 - (1 - \tfrac{1}{3})^{10} = 1 - (\tfrac{2}{3})^{10}$$

(*Note:* We are assuming that the outcome of one shot does not influence the outcome of another shot.)

Example 3.13. The probability that a test rocket is fired successfully is 0.6. How many rockets should be manufactured to ensure that the probability of at least one rocket being fired successfully is at least 0.99?

Solution. Suppose n rockets are needed. Then the probability that at least one rocket is fired successfully is equal to $1 - (1 - 0.6)^n$. Hence we must pick n such that

$$1 - (0.4)^n \geqslant 0.99, \quad \text{i.e.,} \quad 0.01 \geqslant (0.4)^n$$

Hence

$$n \geqslant \frac{-2}{\log_{10} 0.4} = 5, \quad \text{approximately}$$

Therefore, 5 rockets should be manufactured.

Example 3.14. A pair of fair dice are rolled ten times. What is the probability of getting a total of 7 six times?

Solution. Let A_i, $i = 1, 2, \ldots, 10$, represent a total of 7 on the ith roll. Then $P(A_i) = \frac{1}{6}$, $i = 1, 2, \ldots, 10$. Hence, by the formula for binomial probabilities, the probability of getting a total of 7 six times is equal to

$$\binom{10}{6}\left(\frac{1}{6}\right)^6\left(1 - \frac{1}{6}\right)^{10-6} = \binom{10}{6}\frac{5^4}{6^{10}} \qquad \blacksquare$$

(*Note:* For applying the formula, $n = 10$, $k = 6$, and $p = \frac{1}{6}$.)

Comment. The reader should be able to get the solution in Example 3.14 above purely by using a combinatorial argument as was done in Chapter 2 in Example 3.10.

Example 3.15. Suppose there are four coins, say C_1, C_2, C_3, C_4, in a bag. For coin C_i the probability of getting a head on a toss is $\dfrac{i}{i+2}$. A coin is picked at random and tossed five times.

(*a*) What is the probability of getting three heads?

(*b*) Given that there are exactly three heads in the five tosses, what is the probability that it was the coin C_3 that was picked?

Solution. Let B_i = Coin C_i is picked, $i = 1, 2, 3, 4$, and A = Three heads show on five tosses.

(*a*) By the theorem of total probability,

$$P(A) = \sum_{i=1}^{4} P(A \mid B_i)P(B_i)$$

Now $P(B_i) = \frac{1}{4}$. Also, given that B_i has occurred, it means that coin C_i was picked so that the probability of heads on any toss is $\dfrac{i}{i+2}$. Hence, applying the formula for binomial probabilities, we get

$$\begin{aligned} P(A \mid B_i) &= \binom{5}{3}\left(\frac{i}{i+2}\right)^3\left(1 - \frac{i}{i+2}\right)^{5-3} \\ &= \binom{5}{3}\frac{2^2 i^3}{(i+2)^5}, \qquad i = 1, 2, 3, 4 \end{aligned}$$

Therefore

$$P(A) = \sum_{i=1}^{4} \binom{5}{3} \frac{4 \cdot i^3}{(i+2)^5} \cdot \frac{1}{4} = 10 \sum_{i=1}^{4} \frac{i^3}{(i+2)^5}$$

(*b*) We want to find $P(B_3|A)$. Application of Bayes' rule yields

$$P(B_3|A) = \frac{P(A|B_3)P(B_3)}{P(A)}$$

$$= \frac{\binom{5}{3} \frac{2^2 \cdot 3^3}{(3+2)^5} \cdot \frac{1}{4}}{10 \sum_{i=1}^{4} \frac{i^3}{(i+2)^5}} = \frac{27}{5^5 \sum_{i=1}^{4} \frac{i^3}{(i+2)^5}}$$

Example 3.16. A player plays fifteen games. The probability that he wins a game is p and the outcome of one game does not influence the outcome of another. What is the probability that he wins six games, the sixth game being won on the fifteenth play?

Solution. We have the following:

$$\left\{ \begin{array}{l} \text{The player wins} \\ \text{the 6th game on} \\ \text{the 15th play} \end{array} \right\} \iff \left\{ \begin{array}{l} \text{The player wins 5 games} \\ \text{in the first 14 plays, and} \\ \text{he wins the 15th play} \end{array} \right\}$$

Let

A = The player wins 5 games in the 14 plays
B = The player wins the 15th play

We are interested in finding $P(AB)$ which, due to the independence of A and B, is equal to $P(A)P(B)$.

Now, applying the formula for binomial probabilities with $n = 14$ and $k = 5$, we have

$$P(A) = \binom{14}{5} p^5 (1-p)^{14-5}$$

Of course, $P(B) = p$. Hence, the probability that the player wins the 6th game on the 15th play is equal to

$$\binom{14}{5} p^5 (1-p)^9 \cdot p = \binom{14}{5} p^6 (1-p)^9$$

Example 3.17. Tom and Jane agree to play a game of Ping-Pong. According to the rules of the game, whoever wins 21 points first wins the game. The probability that Tom wins a point is p. (Consequently, the probability for Jane to win a point is $1-p$.) Find the probability that Tom wins the game.

Solution. Tom can win the game in the following mutually exclusive ways:

(1) He wins all of the first 21 points; that is, of the first 20 points, he wins 20 points, and then he wins the 21st.

or (2) Of the first 21 points, he wins 20 points, and then he wins the 22nd.

or (3) Of the first 22 points, he wins 20 points, and then he wins the 23rd.

⋮

or (21) Of the first 40 points, he wins 20 points, and then he wins the 41st.

Now the probability that in the first i points $(i \geqslant 20)$ Tom wins 20 points and then wins the $(i+1)$th point is

$$\binom{i}{20} p^{20}(1-p)^{i-20} \cdot p = \binom{i}{20} p^{21}(1-p)^{i-20}$$

(See Example 3.16). Hence the probability that Tom wins the game is

$$\sum_{i=20}^{40} \binom{i}{20} p^{21}(1-p)^{i-20} = p^{21} \sum_{i=20}^{40} \binom{i}{20} (1-p)^{i-20}$$

EXERCISES–SECTION 3

1. A and B are two independent events such that $P(A) > 0$ and $A \subset B$. Find $P(B)$.
2. Suppose A and B are two events with $0 < P(A) < 1$ and $0 < P(B) < 1$. Show that A and B are independent if $P(A|B) = P(A|B')$. Also prove the converse.
3. If A, B, C are independent events, show that so are (a) A, B', C', and (b) A', B, C.
4. Suppose A, B, and C are mutually independent. Show that:
   (a) $P(A|BC) = P(A)$ (b) $P(A \cup B|C) = P(A \cup B)$
5. If A, B, and C are independent events with $P(A) = 0.2$, $P(B) = 0.4$, and $P(C) = 0.5$, find:
   (a) $P(A \cup (BC'))$ (b) $P((AB') \cup (BC'))$ (c) $P([A \cup (B'C)']')$
6. Show that it is not possible to have two independent events A and B for which $A \cup B = S$, $0 < P(A) < 1$, and $0 < P(B) < 1$.
7. If $A_1, A_2, \ldots, A_n$ are mutually independent events with $0 < P(A_i) < 1$, show that we cannot have $P\left(\bigcup_{i=1}^{n} A_i\right) = 1$.
8. Suppose A and B are two independent events and F is an event with $P(F) > 0$. Is it true that $P(AB|F) = P(A|F) \cdot P(B|F)$?
9. Let $S = \{1, 2, \ldots, 16\}$, where each outcome is equally likely. Determine whether the three events $A = \{1, 2, 3, 4\}$, $B = \{1, 6, 7, 9\}$, and $C = \{3, 6, 8, 14\}$ are independent.
10. Tom, Dick, and Jane are of different ages, but their relative age standing is not known, and a fourth person makes a guess. Let E be the event that Tom is younger than Jane and F the event that Dick is younger than Jane. Determine whether the events E and F are independent.

11. Suppose $S = \{1, 2, 3, 4\}$, and let $A = \{1, 2\}$, $B = \{1, 3\}$, and $C = \{1, 4\}$. Define two probability measures, say P_1 and P_2, on subsets of S such that–

(a) the events A, B, C are pairwise independent according to P_1

(b) the events are not pairwise independent according to P_2

12. A pair of fair dice are thrown. Find the probability of obtaining a 2 or a 3 on one die and an even number or a number greater than 4 on the other.

13. Suppose events A, B, C are independent with $P(A) = \frac{1}{2}$, $P(B) = \frac{1}{3}$, and $P(C) = \frac{1}{5}$. Find the probability that–

(a) exactly one of the events occurs

(b) only B occurs

(c) only B and C occur

(d) at most one of the events occurs.

14. A die is thrown until a 3 appears twice in a row. Find the probability that this will happen within the first four attempts.

15. A store has three kinds of meat: beef, lamb, and pork. From past experience, it is known that a prospective customer buys beef, lamb, and pork with respective probabilities 0.5, 0.2, and 0.3. Find the probability that, of three customers–

(a) all will buy the same kind of meat

(b) all will buy different kinds of meat

(c) two of the customers will buy the same kind of meat.

(Assume that the customers decide independently.)

16. The probabilities that a marksman hits a target on the first, second, third, fourth, and fifth shot are respectively 0.5, 0.6, 0.7, 0.8, and 0.9. If the outcomes of the attempts are independent, find the probability that he scores–

(a) exactly four hits

(b) at least one hit.

17. Persons are classified according to their blood types as O, A, B, AB, and it is known that the probability that a randomly picked person is of type O is 0.4, of type A is 0.3, of type B is 0.2, and of type AB is 0.1. If three people are picked at random, find the probability that–

(a) none is of type O

(b) two are of type A and one is of type AB

(c) two are of type A

(d) at least one is of type AB

(e) they are all of different blood types.

18. In the circuit diagrams given in Figure 3.3, assume that the probability that each relay is closed is p, and that each relay is open or closed independently of any other relay. Find the probability that current flows from L to R for circuits (a) and (b).

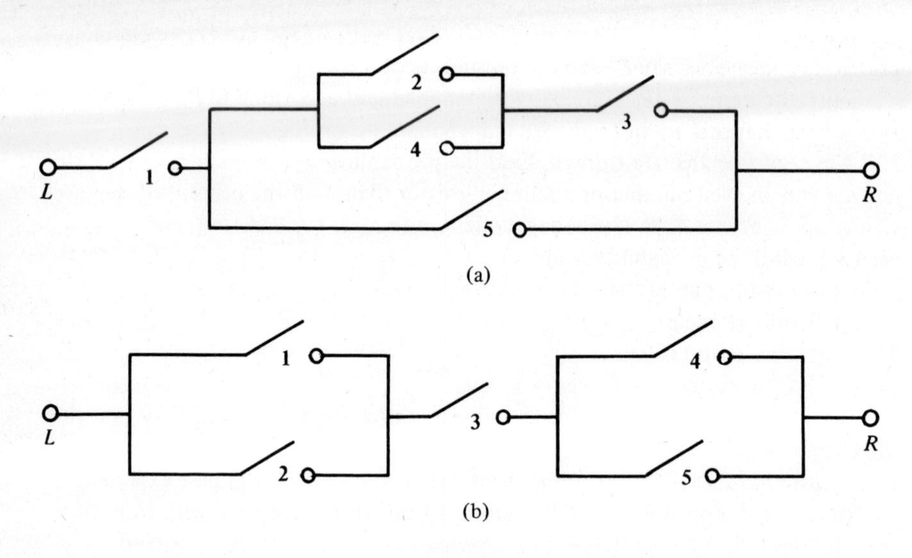

Figure 3.3

19. A coin with $P(\text{heads}) = 0.3$ is tossed until both heads and tails appear at least once. What is the probability that this will require–

 (a) two tosses
 (b) four tosses
 (c) r tosses, where $r \geqslant 2$.

20. Assume that the probability that a child is a son is the same on any birth, and that there are eight children in a family. Given that five of the children are sons, find the probability that they were all born on consecutive births.

21. A group of women consists of r_1 librarians and s_1 teachers, while a group of men consists of r_2 librarians and s_2 teachers. A person is picked at random from each group independently and from the two persons so chosen one person is picked at random. Find the probability that the person is a librarian.

22. Tom, Jane, and Dick fire at a target simultaneously and independently, and the respective probabilities that they score a hit are 0.6, 0.4, and 0.8. If there are exactly two hits, find the probability that it was Jane who missed.

23. A panel of six judges is shown a painting which is as likely to be the work of a master artist as it is to be the work of an impostor. The judges evaluate the painting independently of each other, and the probability for each of them to judge correctly is $\frac{4}{5}$. If four of the judges declare that the painting is a fraud, what is the probability that it was actually the work of a master artist?

24. Suppose a and b are real numbers such that $0 < a < 1$ and $0 < b < 1$. A positive integer is picked with $P(\{i\}) = (1-a)a^{i-1}$, $i = 1, 2, \ldots$. If the number is i, a coin with the probability of heads equal to b^i is tossed N times. Find the probability of getting heads on every toss.

25. If a pair of fair dice is rolled repeatedly, find the probability that a sum of 5 will appear before a sum of 4.

26. Suppose Tom and Jane each toss an unbiased coin n times. Find the probability they get the same number of heads.

27. A box contains M balls, of which W are white. If n balls are picked at random, let B_j, $j = 1, 2, \ldots, n$, represent the event that a white ball is picked on the jth draw and A_k, $k = 1, 2, \ldots, n$, represent the event that the sample contains exactly k white balls. Show that $P(B_j|A_k) = k/n$. Is this answer to be anticipated intuitively? (Consider the two cases, with and without replacement.)

4 Random Variables

INTRODUCTION

So far we have concentrated on building the basic probability structure. We have the sample space S which consists of all the conceivable outcomes of an experiment; we have the sigma field $\mathcal{F}$ whose members we have called the events; and, finally, we have assigned probabilities to these events. Together, we have called $(S, \mathcal{F}, P)$ the probability space.

It often happens that when we perform an experiment our interest is not so much in the outcomes of the experiment as it is in the numerical values assigned to these outcomes. For instance, a gambler who stands to win 5 dollars if heads show up and lose 4 dollars if tails show up will think in terms of the numbers 5 and -4 rather than in terms of heads and tails. Formally, he is interested in his winnings and is assigning the value 5 to heads and -4 to tails. This leads us to the concept of what is called a random variable.

1. The Notion of a Random Variable

As a first step towards introducing the concept, a **random variable** (abbreviated r.v.) *is a function which assigns numerical values to the outcomes of S, the sample space.* If we denote the function by X, then for any outcome s in S the random variable X takes on the numerical value $X(s)$. This is often written compactly as

$$X : S \to \mathbf{R}$$

where $\mathbf{R}$, as usual, represents the set of real numbers.

It is customary to denote random variables by capital letters U, V, W, X, Y, Z and their typical values by the corresponding small letters u, v, w, x, y, z. This is merely a matter of convenience and one should not feel constrained to follow this convention. It should be borne in mind that a random variable is neither random nor is it a variable. It is simply a function defined on the sample space.

Example 1.1. (*i*) Suppose three satellites are fired into space. Let X denote the number of satellites that go into orbit. Writing h if the satellite attains an orbit successfully, and m if it does not, we have the situation given in the following table, where the first row lists the outcomes in S and the second row the corresponding value of X.

s	hhh	hhm	hmh	mhh	mmh	mhm	hmm	mmm
$X(s)$	3	2	2	2	1	1	1	0

(*ii*) A bowl contains five beads, of which three are white and two are black. Three beads are picked without replacement. Let X represent the number of white beads in the sample. Then the sample points together with the corresponding values of X are as given below. (We identify the white beads as w_1, w_2, w_3 and the black beads as b_1, b_2.)

s	$w_1w_2w_3$	$w_1w_2b_1$	$w_1w_2b_2$	$w_1w_3b_1$	$w_1w_3b_2$	$w_2w_3b_1$	$w_2w_3b_2$	$w_1b_1b_2$	$w_2b_1b_2$	$w_3b_1b_2$
$X(s)$	3	2	2	2	2	2	2	1	1	1

(*iii*) A person shoots at a target on the wall. Let X denote the amount by which he misses the target. In this case the sample space S consists of all the points on the wall. For any point s on the wall, $X(s)$ represents the distance of s from the target. Hence for any s, $0 \leqslant X(s) < \infty$ (assuming a wall of infinite size). ■

We have defined a random variable as a real-valued function. It turns out that we cannot allow the function to be totally arbitrary. Certain considerations force us to impose some restrictions on it. Consider the following situation:

Suppose we roll two dice. The sample space can be given as

$$S = \{(i, j) \mid i, j = 1, 2, 3, 4, 5, 6\}$$

and has 36 outcomes. Let X represent "the sum on the two dice." Then

$$X((1, 1)) = 2, \qquad X((1, 2)) = X((2, 1)) = 3$$

and so on. As a matter of fact, the range of X is $\{2, 3, 4, \ldots, 12\}$. Now let us pose the following question:

Question: What is the probability that the sum on the two dice is less than 4?

To find this probability, we naturally look for the points in S for which the sum is less than 4. In other words we consider the subset of S namely $\{(i, j) \mid X((i, j)) < 4\}$. Since this set is $\{(1, 1), (1, 2), (2, 1)\}$, the probability is $\frac{3}{36}$, assuming that the outcomes are equally likely.

Instead of making the long-winded statement "What is the probability . . . less than 4?" we could ask in an equivalent way the following question:

Question: What is the probability of the interval $(-\infty, 4)$?

Notice that in this question the event for which we want to find the probability is a subset of **R** and, consequently, we are basically considering a new probability measure different from P. (Recall that P is defined on the subsets of S.) If we denote the new probability measure by P^X, using the superscript X, what we want is $P^X((-\infty, 4))$. Naturally, we will make the following assignment

$$P^X((-\infty, 4)) = P(\{(i, j) \mid X((i, j)) < 4\}) = \frac{3}{36}$$

What class of subsets of **R** shall we take as the collection of events on which P^X is defined? It is conceivable that for any real number a we would be interested in finding "the probability that the sum on the two dice is less than or equal to a," or, equivalently, referring to the real line, "the probability of the interval $(-\infty, a]$." It is possible that we might want to find "the probability that the sum on the two dice is between a and b, both inclusive," or, equivalently, "the probability of the interval $[a, b]$." Indeed, we are led to consider the Borel sets of **R** as the collection of events on which the probability measure P^X is defined.

How are P and P^X related to each other? We have indicated above briefly the nature of the relation. In a general setting, suppose B is a Borel set of the real line. In order to find $P^X(B)$ it seems natural to refer back to the sample space S and see precisely what points of S went to the set B under X. In other words, we would look for the set

$$\{s \in S \mid X(s) \in B\}$$

This is indicated in Figure 1.1.

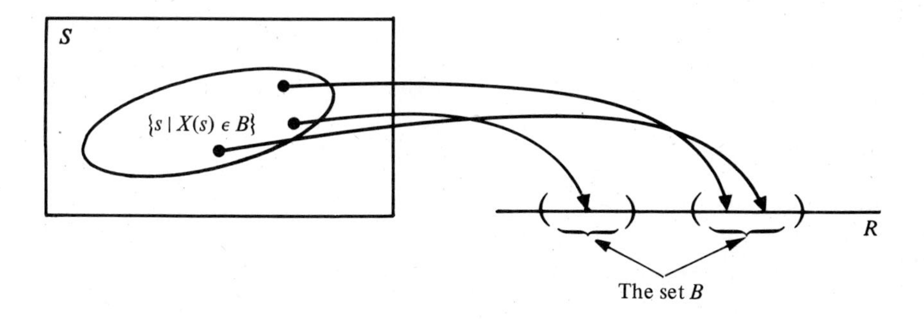

Figure 1.1

We now make the following assignment:

$$\boxed{P^X(B) = P(\{s \in S \mid X(s) \in B\})}$$

For example,

$$P^X((a, b]) = P(\{s \mid X(s) \in (a, b]\})$$

that is,

$$P^X((a, b]) = P(\{s \mid a < X(s) \leqslant b\})$$

(*Note:* It will be left to the exercise set to show that P^X as defined above is indeed a probability measure on Borel sets of **R**. In what follows we shall accept this fact.)

At this point, recall that the probability measure P is defined only on the members of $\mathcal{F}$. Therefore, in order to be able to define $P^X(B)$ as equal to $P(\{s \in S \mid X(s) \in B\})$, the set $\{s \in S \mid X(s) \in B\}$ had better belong to $\mathcal{F}$. Hence it seems appropriate that we put the restriction on X that $\{s \in S \mid X(s) \in B\}$ be a member of $\mathcal{F}$ for every Borel set B.

We are now in a position to give the formal mathematical definition of a random variable.

Suppose $(S, \mathcal{F}, P)$ is a probability space. A **random variable** X is a real-valued function defined on S ($X: S \to \mathbf{R}$), and has the following property: For any Borel set B of the real numbers, the set $\{s \mid X(s) \in B\}$ is a member of $\mathcal{F}$.

It can be shown that the above definition of a random variable is equivalent to the following definition:

Suppose X is a function from S to **R**. Then X is called a random variable if, for each $u \in \mathbf{R}$, $\{s \mid X(s) \leqslant u\}$ is a member of $\mathcal{F}$.

It is this latter definition which is most helpful in proving whether a given real-valued function defined on S is a random variable.

Observe that the probability measure P has not figured in the actual definition of the random variable. The definition clearly implies that *not every real-valued function defined on S is a random variable.* Whether a given real-valued function is a random variable or not will depend on the underlying sigma field $\mathcal{F}$. To see what we are driving at, consider the following examples:

*Example 1.2.** Let S be a sample space and let $\mathcal{F}$ be the *power set* of the set S. Show that *any* real-valued function defined on S is a random variable.

Solution. Suppose $X: S \to \mathbf{R}$, and let $B \in \mathcal{B}$. Then $\{s \mid X(s) \in B\}$ is some subset of S. Since $\mathcal{F}$ is the power set of S, naturally

$$\{s \mid X(s) \in B\} \in \mathcal{F}$$

Hence, by the definition, X is a random variable.

*Asterisked examples in this section are rather specialized in nature and may be omitted at first reading without impairing continuity. However, the reader should familiarize himself with the definitions of indicator random variable and degenerate random variable.

Comment. For finite or countable sample spaces, the underlying sigma field $\mathscr{F}$ will usually be taken as the power set, and hence in such cases any real-valued function will always turn out to be a random variable.

*Example 1.3.** Suppose we have a finite sample space $S = \{a, b, c, d\}$ and the sigma field $\mathscr{F}$ is $\{\emptyset, \{a, b\}, \{c, d\}, S\}$. Define–

(*a*) a real-valued function X on S where X is an r.v.

(*b*) a real-valued function Y on S where Y is not an r.v.

Solution. We see that $\mathscr{F}$ is a sigma field but not the power set of S.

(*a*) Define the function X as follows:

$$X(a) = X(b) = 0 \quad \text{and} \quad X(c) = X(d) = 2$$

Then,

$$\begin{array}{ll} \text{if } u < 0, & \{s \mid X(S) \leqslant u\} = \emptyset \in \mathscr{F} \\ \text{if } 0 \leqslant u < 2, & \{s \mid X(s) \leqslant u\} = \{a, b\} \in \mathscr{F} \end{array}$$

(For example, if $u = 1.5$, $\{s \mid X(s) \leqslant 1.5\} = \{a, b\}$.)

$$\text{if } u \geqslant 2, \qquad \{s \mid X(s) \leqslant u\} = \{a, b, c, d\} \in \mathscr{F}$$

Hence, for any real number u, $\{s \mid X(s) \leqslant u\}$ is a member of $\mathscr{F}$. Consequently, X is a random variable.

(*b*) Suppose Y is a function for which $Y(a) = 0$, $Y(b) = 2$, $Y(c) = 4$, and $Y(d) = 5$. Consider the Borel set $(-\infty, 4]$. We see that $\{s \mid Y(s) \in (-\infty, 4]\} = \{a, b, c\}$. However, $\{a, b, c\} \notin \mathscr{F}$. Hence Y is not a random variable.

*Example 1.4.** Let the sample space S consist of the set of real numbers; that is, $S = \mathbf{R}$. Suppose

$$\mathscr{F} = \{A \mid A \text{ is countable or } A' \text{ is countable}\}$$

(*a*) Let $X: S \to \mathbf{R}$ such that $X(s) = s$ for any real number s. Show that X is not a random variable.

(*b*) Let $Y: S \to \mathbf{R}$, such that

$$Y(s) = \begin{cases} 1 & \text{if } s \text{ is an irrational number} \\ 0 & \text{if } s \text{ is a rational number} \end{cases}$$

Show that Y is a random variable.

Solution. It can be verified that $\mathscr{F}$ is indeed a sigma field.

(*a*) That X is not a random variable follows immediately because, for example,

$$\{s \mid X(s) \leqslant 2\} = (-\infty, 2]$$

Now, neither is $(-\infty, 2]$ countable, nor is its complement $(2, \infty)$ countable. Hence $(-\infty, 2] \notin \mathscr{F}$, and, consequently, X is not a random variable.

(*b*) Now let us consider the random variable Y. Suppose $\mathbf{Q}$ represents the set of rational numbers. Now

$$\begin{array}{ll} \text{if } u < 0, & \{s \mid Y(s) \leqslant u\} = \emptyset \in \mathscr{F} \\ \text{if } 0 \leqslant u < 1, & \{s \mid Y(s) \leqslant u\} = \mathbf{Q} \in \mathscr{F} \end{array}$$

($\mathbf{Q} \in \mathcal{F}$ because, $\mathbf{Q}$ is countable)

$$\text{if } u \geqslant 1, \quad \{s \mid Y(s) \leqslant u\} = S \in \mathcal{F}$$

($S \in \mathcal{F}$ because its complement, $\emptyset$, is countable.) Hence, for every real number u, $\{s \mid Y(s) \leqslant u\}$ is a member of $\mathcal{F}$, and, consequently, Y is a random variable.

*Example 1.5.**

(*a*) If X is a random variable, show that $|X|$ is also a random variable.

(*b*) Give an example to show that the converse is false.

Solution

(*a*) Note that the function $|X|$ is defined by $|X|(s) = |X(s)|$ for every $s \in S$. Assume that X is an r.v. Then, by definition, $\{s \mid X(s) \in B\}$ is a member of $\mathcal{F}$ for any Borel set B.

Now let $a \in \mathbf{R}$. Then

$$\begin{aligned}\{s \mid |X|(s) \leqslant a\} &= \{s \mid |X(s)| \leqslant a\} \\ &= \{s \mid X(s) \in [-a, a]\}\end{aligned}$$

Since $[-a, a]$ is a Borel set and X is an r.v., $\{s \mid X(s) \in [-a, a]\}$ is a member of $\mathcal{F}$. Hence $\{s \mid |X|(s) \leqslant a\} \in \mathcal{F}$. Therefore, $|X|$ is an r.v.

(*b*) Let $S = \mathbf{R}$ and let $\mathcal{F}$ be as given in Example 1.4; that is, $\mathcal{F}$ contains every subset A of S where either A or A' is countable.

Define a function $X: S \to \mathbf{R}$ by

$$X(s) = \begin{cases} -1 & \text{if } s \text{ is a } \textit{negative} \text{ irrational number} \\ 0 & \text{if } s \text{ is a rational number} \\ 1 & \text{if } s \text{ is a } \textit{positive} \text{ irrational number} \end{cases}$$

Now consider, for example, the set $\{s \mid X(s) \leqslant -\frac{1}{2}\}$. This is simply the set of *negative* irrational numbers. But the set of negative irrational numbers is not a countable set; nor is its complement, which consists of the union of the set of rational numbers and the set of positive irrational numbers, a countable set. Hence $\{s \mid X(s) \leqslant -\frac{1}{2}\} \notin \mathcal{F}$ and, as a result, X is not a random variable. However, note that $|X|$ is a random variable, since $|X|$ is the function Y of Example 1.4(b).

*Example 1.6.** Suppose $(S, \mathcal{F}, P)$ is a probability space and $A \in \mathcal{F}$. Define a function $I_A, I_A: S \to \mathbf{R}$, as follows:

$$I_A(s) = \begin{cases} 1 & \text{if } s \in A \\ 0 & \text{if } s \notin A \end{cases}$$

Show that I_A is a random variable.

Solution. The nature of the mapping is indicated in Figure 1.2.

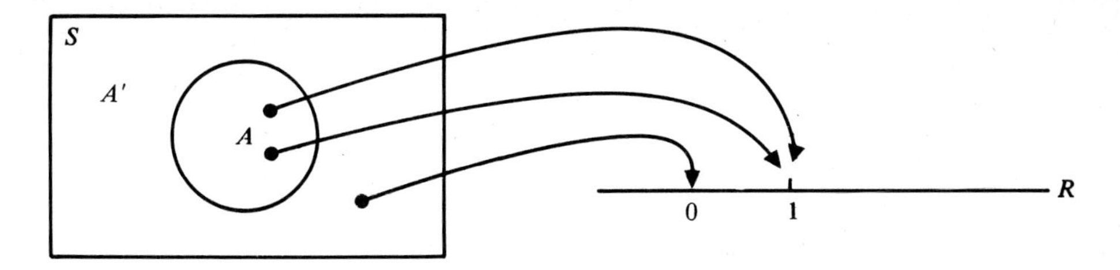

Figure 1.2

We see immediately that

$$\begin{array}{ll} \text{if } u<0, & \{s \mid I_A(s) \leq u\} = \emptyset \in \mathcal{F} \\ \text{if } 0 \leq u < 1, & \{s \mid I_A(s) \leq u\} = A' \in \mathcal{F} \\ \text{if } u \geq 1, & \{s \mid I_A(s) \leq u\} = S \in \mathcal{F} \end{array}$$

Hence, for every real number u, $\{s \mid I_A(s) \leq u\}$ is a member of $\mathcal{F}$. Consequently, I_A is an r.v. ■

Suppose A is an event. The random variable I_A defined by

$$I_A(s) = \begin{cases} 1 & \text{if } s \in A \\ 0 & \text{if } s \notin A \end{cases}$$

is called the **indicator random variable** for the event A.

*Example 1.7.** Suppose a function X is defined on S by $X(s) = a$ for each $s \in S$, where a is some (fixed) real number. Show that X is a random variable.

Solution. We make an observation that $\{s \mid X(s) = a\} = S$. To show that X is a random variable:

(*i*) Suppose B is a Borel set with $a \in B$; then $\{s \mid X(s) \in B\} = S \in \mathcal{F}$.

(*ii*) Suppose B is a Borel set with $a \notin B$; then $\{s \mid X(s) \in B\} = \emptyset \in \mathcal{F}$.

Hence, no matter what Borel set B we pick, $\{s \mid X(s) \in B\}$ is a member of $\mathcal{F}$. Consequently X is a random variable. ■

The function X defined by $X(s) = a$ for every $s \in S$ is a constant function and the random variable X is said to be **degenerate.**

In what follows, we shall adopt the convention of denoting $\{s \mid X(s) \in B\}$ by $\{X \in B\}$ and reading it as the event "X takes on a value in B." The probability of this event will be denoted by $P(X \in B)$. For example, with a and b two given numbers, $\{a \leq X \leq b\}$ is the set $\{s \mid a \leq X(s) \leq b\}$. It represents the event that X assumes a value between a and b, both inclusive; the probability of this event is $P(a \leq X \leq b)$.

The long and short of our discussion thus far is the following:

Suppose $(S, \mathcal{F}, P)$ is a probability space. *Then a random variable X is a real-valued function* (with a special property) *defined on S, and it induces a new probability space* $(\mathbf{R}, \mathcal{B}, P^X)$, *where the class of Borel sets* $\mathcal{B}$ *becomes the new collection of events and*

$$P^X(B) = P(\{s \mid X(s) \in B\})$$

for any Borel set B.

EXERCISES–SECTION 1

1. If a die is thrown twice, give the values that the random variable in each of the following cases can assume.
   (a) X = The absolute value of the difference on the two throws.
   (b) X = The larger of the values on the two throws.
   (c) X = The smaller of the values on the two throws.
2. Suppose $(S, \mathcal{F}, P)$ is a probability space where $S = \{s_1, s_2, s_3, s_4, s_5, s_6\}$ and $P(\{s_1\}) = P(\{s_6\}) = \frac{1}{12}$, $P(\{s_2\}) = \frac{1}{3}$, $P(\{s_3\}) = P(\{s_4\}) = P(\{s_5\}) = \frac{1}{6}$. The assignment that X makes to the outcomes in S is shown in Figure 1.3.

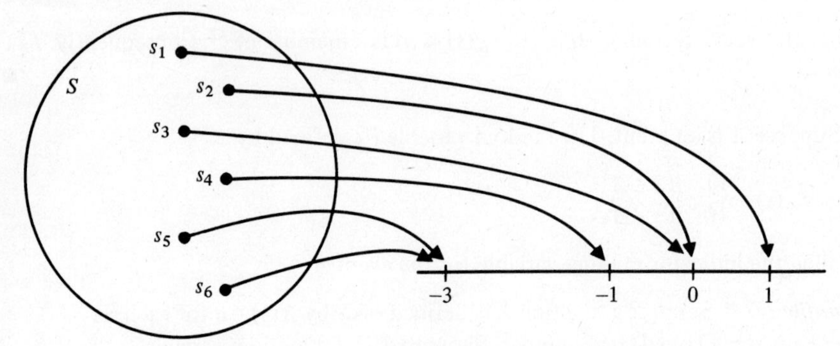

Figure 1.3

Find: (a) $P^X(\{0\})$ (b) $P^X((-2, 1.5])$
(c) $P^X((0, 3])$ (d) $P^X([-1, 3.5])$
(e) $P^X([-1, \infty))$ (f) $P^X((-\infty, -1))$
(g) $P^X((-\infty, -1) \cup [0.5, \infty))$

3. A die is weighted so that the probability of 1, 2, 3, 4, 5, 6 showing up is proportional to the number. Describe the random variable in each of the following cases and find $P^X((-\infty, 3])$, $P^X((1, 3))$, $P^X([2, \infty))$.
   (a) A person wins as many dollars as the number that shows up and X represents the person's winnings.
   (b) The same as in part (a), but this time the person pays one dollar for the privilege of playing and X represents the net winnings.
   (c) A person wins four dollars if an even number shows up, and loses three dollars if an odd number shows up, and X represents the person's winnings.

4. Suppose there are three children in a family, all of different ages. Describe the random variable in each of the following cases and find $P^X(\{1\}), P^X([1, \infty))$, $P^X((-\infty, 0.5])$.

(a) X = The number of sons among three children.

(b) X = The number of sons minus the number of daughters.

(c) X = The number of sons on the second birth.

(Assume that all sex distributions are equally likely.)

5. Suppose $S = \{s \mid 0 < s \leqslant 1\}$, and if $0 < a < b \leqslant 1$, let $P((a, b]) = b - a$. If the random variable X is defined by $X(s) = 2s$, find

(a) $P^X((0, \frac{1}{2}])$ (b) $P^X((\frac{1}{4}, \frac{1}{2}])$

(c) $P^X((\frac{1}{4}, \infty))$ (d) $P^X((-\infty, \frac{1}{2}])$

6. Suppose A and B are two independent events for which $P(A) = \frac{1}{3}, P(B) = \frac{1}{4}$. Define a random variable X on S as follows:

$$X(s) = \begin{cases} 0 & \text{if } s \in (A \cup B)' \\ 2 & \text{if } s \in AB \\ 3 & \text{if } s \in A - B \\ 5 & \text{if } s \in B - A \end{cases}$$

Find: (a) $P^X((1.5, 3.5))$ (b) $P^X([3.5, \infty))$

(c) $P^X(\{4\})$ (d) $P^X((-\infty, 1))$

7. Suppose $(S, \mathcal{F}, P)$ is a probability space. Show that the set function P^X defined for any Borel set $B \subset \mathbf{R}$ by

$$P^X(B) = P\{s \in S \mid X(s) \in B\}$$

satisfies the following relations:

(a) $0 \leqslant P^X(B) \leqslant 1$

(b) $P^X(\mathbf{R}) = 1$

(c) If $\{B_i\}$ is a sequence of mutually exclusive Borel sets of $\mathbf{R}$, then

$$P^X\left(\bigcup_{i=1}^{\infty} B_i\right) = \sum_{i=1}^{\infty} P^X(B_i).$$

What can you conclude from this?

2. THE DISTRIBUTION FUNCTION

2.1 The Definition of a Distribution Function

For any real number u, we have defined

$$P^X((-\infty, u]) = P\{s \mid -\infty < X(s) \leqslant u\}$$

or, using the convention,

$$P^X((-\infty, u]) = P(X \leqslant u)$$

Now $P(X \leqslant u)$—being a probability—is a real number, and its value depends on u. In other words, it is a real-valued function of u. This function is called the **cumulative distribution function** of X, or simply the **distribution function** of X, and it plays a very prominent role in the theory of probability, because knowledge of it completely determines the probability distribution of a random variable.

A real-valued function F defined for any real number u by

$$F(u) = P^X((-\infty, u]) = P(X \leqslant u)$$

is called the **distribution function** (abbreviated D.F.) of the random variable X.

It is sometimes convenient to use X as a subscript and denote the particular distribution function by F_X. This might be necessary when more than one r.v. is involved in the discussion. When the context is clear, we shall omit the subscript.

Example 2.1. Suppose that in tossing a coin a person stands to win \$2.00 if he rolls heads, and to lose \$1.50 if he rolls tails. Let X represent the winnings of the person on a toss. Find the distribution function of X, assuming that the probability of heads is p, where $0 \leqslant p \leqslant 1$.

Solution. The random variable makes the following assignment to each outcome of S

$$X(H) = 2 \quad \text{and} \quad X(T) = -1.5$$

To find the D.F., we observe the following:

(*i*) From Figure 2.1(a),

$$\{s \mid X(s) \leqslant u\} = \emptyset \quad \text{if } u < -1.5$$

Therefore, $F(u) = P(X \leqslant u) = P(\emptyset) = 0$ if $u < -1.5$.

(*ii*) From Figure 2.1(b),

$$\{s \mid X(s) \leqslant u\} = \{T\} \quad \text{if } -1.5 \leqslant u < 2$$

Therefore, $F(u) = P(\{T\}) = 1 - p$ if $-1.5 \leqslant u < 2$.

(*iii*) From Figure 2.1(c),

$$\{s \mid X(s) \leqslant u\} = \{H, T\} \quad \text{if } 2 \leqslant u < \infty$$

Therefore, $F(u) = P(\{H, T\}) = 1$, if $u \geqslant 2$.

Combining (*i*), (*ii*), and (*iii*), we get the D.F. of X as

$$F(u) = \begin{cases} 0 & \text{if } u < -1.5 \\ 1 - p & \text{if } -1.5 \leqslant u < 2 \\ 1 & \text{if } u \geqslant 2 \end{cases}$$

The graph of F is given in Figure 2.2.

Note: The dotted vertical lines in the graph are not part of the graph. They are drawn merely to indicate that at -1.5 and 2 the values of F are respectively $1 - p$ and 1.

Comment. Briefly, the procedure adopted simply consists of taking different values of u in $\mathbf{R}$ and then, as u increases from $-\infty$ to ∞, finding $P(X \leqslant u)$ by referring to the original sample space S.

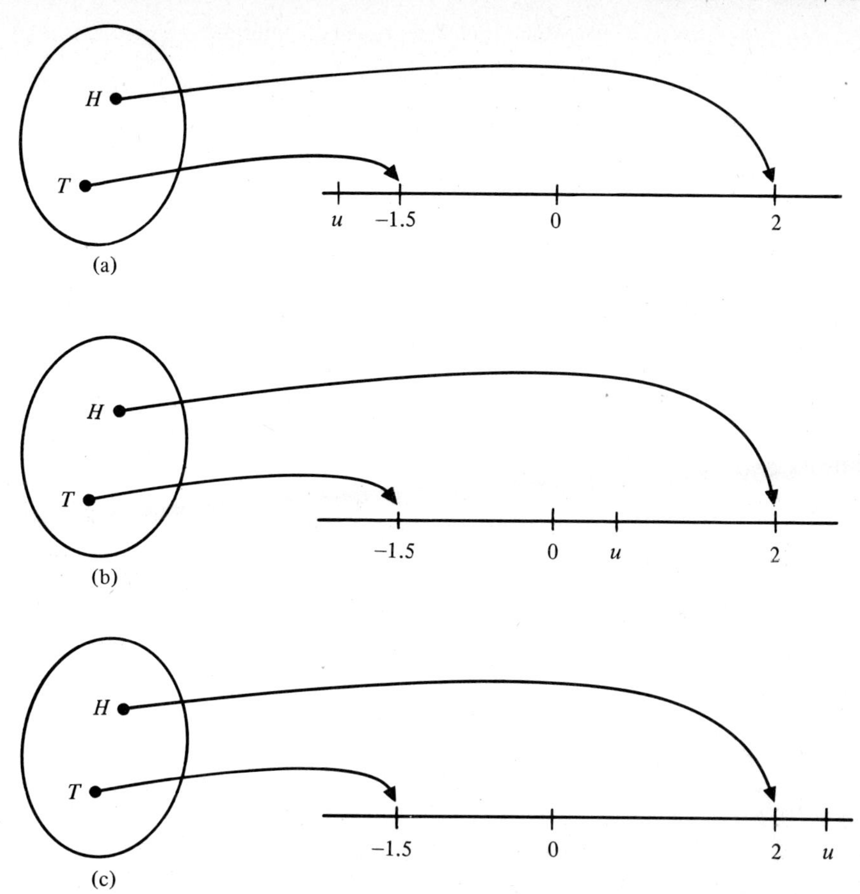

Figure 2.1

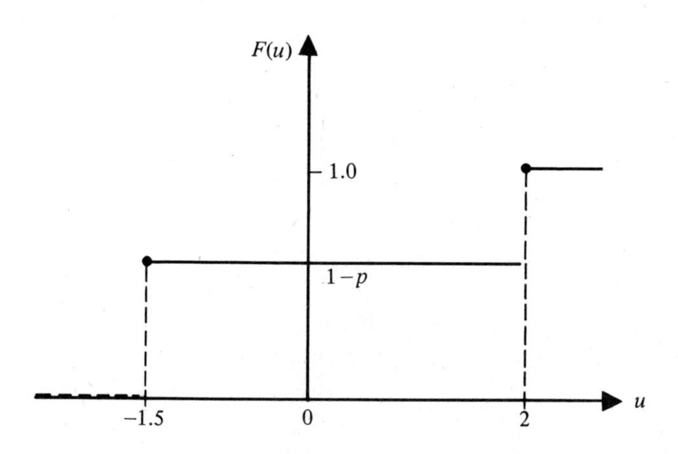

Figure 2.2

Example 2.2. Toss an unbiased die. Let X represent the number on the die. Find the D.F. of X.

Solution. The D.F. of X is given by

$$F(u) = \begin{cases} 0, & u < 1 \\ \frac{1}{6}, & 1 \leqslant u < 2 \\ \frac{2}{6}, & 2 \leqslant u < 3 \\ \frac{3}{6}, & 3 \leqslant u < 4 \\ \frac{4}{6}, & 4 \leqslant u < 5 \\ \frac{5}{6}, & 5 \leqslant u < 6 \\ 1, & u \geqslant 6 \end{cases}$$

To see this, consider, for example, the case $2 \leqslant u < 3$, say $u = 2.5$; then $\{s \mid X(s) \leqslant 2.5\} = \{1, 2\}$. As a matter of fact, $\{s \mid X(s) \leqslant u\} = \{1, 2\}$ as long as u satisfies $2 \leqslant u < 3$. Therefore, $F(u) = P(\{1, 2\}) = \frac{2}{6}$, if $2 \leqslant u < 3$.

The graph of F is given in Figure 2.3.

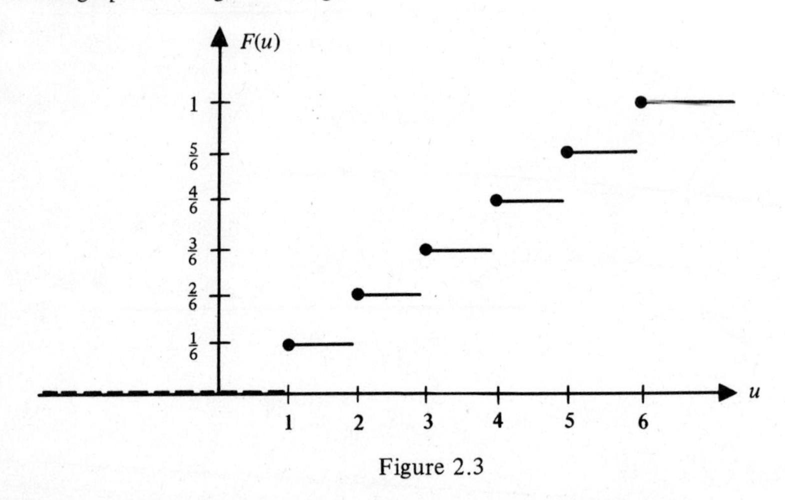

Figure 2.3

Example 2.3. Three beads are picked at random without replacement from a bowl containing three white beads and two black. If X represents the number of white beads in the sample, find the distribution function of X.

Solution. This random variable has been described in Example 1.1(*ii*) of Section 1. To find the D.F., we see that

if $u < 1$, $\{s \mid X(s) \leqslant u\} = \emptyset$; hence $F(u) = 0$

if $1 \leqslant u < 2$, $\{s \mid X(s) \leqslant u\} = \{w_1 b_1 b_2, w_2 b_1 b_2, w_3 b_1 b_2\}$ and consequently $F(u) = \frac{3}{10}$

if $2 \leqslant u < 3$, $\{s \mid X(s) \leqslant u\} = \{w_1 w_2 b_1, w_1 w_2 b_2, w_1 w_3 b_1, w_1 w_3 b_2, w_2 w_3 b_1, w_2 w_3 b_2, w_1 b_1 b_2, w_2 b_1 b_2, w_3 b_1 b_2\}$ and therefore $F(u) = \frac{9}{10}$

if $u \geqslant 3$, $\{s \mid X(s) \leqslant u\} = S$, and hence $F(u) = 1$

In summary:

$$F(u) = \begin{cases} 0 & \text{if } u < 1 \\ \frac{3}{10} & \text{if } 1 \leqslant u < 2 \\ \frac{9}{10} & \text{if } 2 \leqslant u < 3 \\ 1 & \text{if } u \geqslant 3 \end{cases}$$

The graph of the D.F. is given in Figure 2.4.

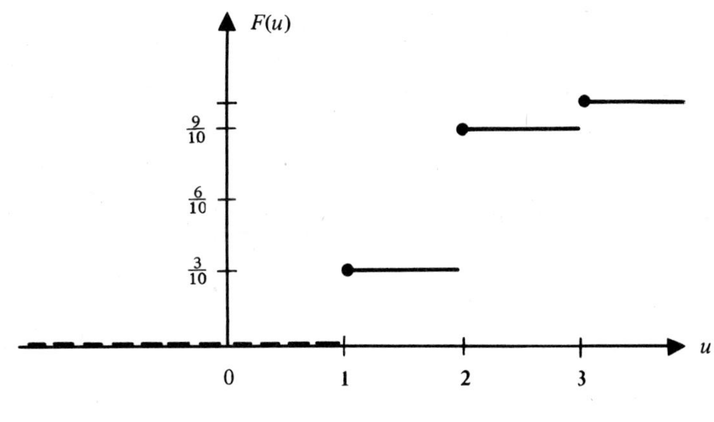

Figure 2.4

2.2 Properties of a Distribution Function

We shall now discuss the properties of distribution functions. Before we do that, let us give the following definition:

If h is a function of a real variable defined for $-\infty < x < \infty$, then we write

$$h(\infty) = \lim_{x\to\infty} h(x) \quad \text{and} \quad h(-\infty) = \lim_{x\to-\infty} h(x)$$

if these limits exist. For example, if $h(x) = 1/x$, then $h(\infty) = 0$; and if $h(x) = \cos x$, then neither $h(\infty)$ nor $h(-\infty)$ exists.

We have defined a distribution function equivalently in terms of the probability measure P^X and the probability measure P. Hence we are free to prove the following properties either by using P^X or P. We shall prove the results by using the probability measure P^X defined on the subsets of **R**. The reader will find it informative to argue by using the measure P and referring to the basic sample space S.

(1) For any real number u,

$$\boxed{0 \leqslant F(u) \leqslant 1}$$

This follows immediately because $F(u) = P^X((-\infty, u])$ and represents the probability of an event.

The result states that if we plot $F(u)$ against u, then the graph of F is between the two horizontal lines $y = 0$ and $y = 1$. In other words, the lines $y = 0$ and $y = 1$ are the bounds for the function F.

(2) F is nondecreasing. That is,

$$\text{if} \quad a < b, \quad \text{then} \quad F(a) \leqslant F(b)$$

To prove this notice that if $a < b$, then $(-\infty, a] \subset (-\infty, b]$. Therefore, by the monotone property of P^X,

$$P^X((-\infty, a]) \leqslant P^X((-\infty, b])$$

Hence

$$F(a) \leqslant F(b)$$

Notice in Figure 2.5 that the height at a cannot exceed the height at b, which in turn cannot exceed the height at c.

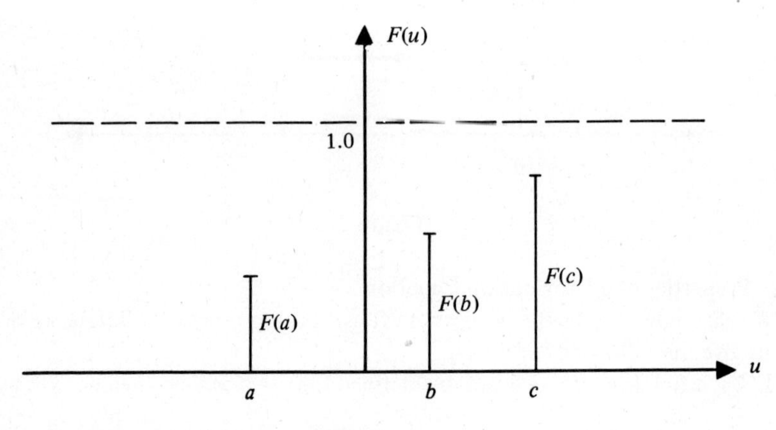

Figure 2.5

(3) If a and b are real numbers with $a < b$, then

$$P(a < X \leqslant b) = F(b) - F(a)$$

Since $a < b$, $(-\infty, b] = (-\infty, a] \cup (a, b]$. Therefore,

$$P^X((-\infty, b]) = P^X((-\infty, a]) + P^X((a, b]) \qquad \text{(why?)}$$

Consequently,

$$F(b) = F(a) + P^X((a, b])$$

That is,

$$P^X((a, b]) = F(b) - F(a)$$

Hence

$$P(a < X \leqslant b) = F(b) - F(a)$$

This is shown in Figure 2.6.

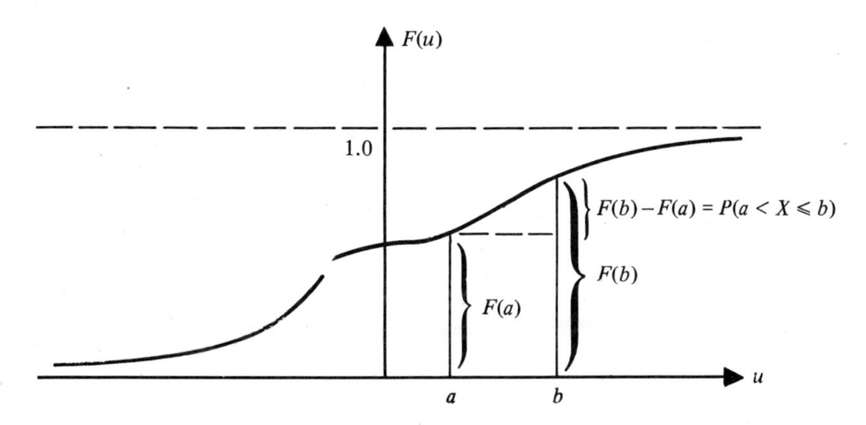

Figure 2.6

(4)
$$\lim_{u \to \infty} F(u) = 1; \quad \text{that is,} \quad F(\infty) = 1$$

(*Note:* In the proofs that follow we shall invoke the continuity property of the probability measure discussed in Section 1 of Chapter 2. The reader might wish to familiarize himself with these concepts once again.)

To prove (4), let $\{u_n\}$ be an increasing sequence of real numbers such that $\lim_{n \to \infty} u_n = \infty$. Then $\{(-\infty, u_n]\}$ is an increasing sequence of intervals and $\lim_{n \to \infty} (-\infty, u_n] = \bigcup_{n=1}^{\infty} (-\infty, u_n] = (-\infty, \infty)$. Hence

$$\begin{aligned}
\lim_{n \to \infty} F(u_n) &= \lim_{n \to \infty} P^X((-\infty, u_n]) \\
&= P^X\left(\bigcup_{n=1}^{\infty} (-\infty, u_n]\right), \qquad \text{by continuity property} \\
&= P^X((-\infty, \infty)) = 1
\end{aligned}$$

(5)
$$\lim_{u \to -\infty} F(u) = 0; \quad \text{that is,} \quad F(-\infty) = 0$$

Consider a decreasing sequence $\{u_n\}$ of real numbers such that $\lim_{n \to \infty} u_n = -\infty$. Then $\{(-\infty, u_n]\}$ is a decreasing sequence of intervals, and therefore $\lim_{n \to \infty} (-\infty, u_n] = \bigcap_{n=1}^{\infty} (-\infty, u_n] = \emptyset$. Hence

$$\lim_{n\to\infty} F(u_n) = \lim_{n\to\infty} P^X((-\infty, u_n])$$

$$= P^X\left(\bigcap_{n=1}^{\infty} (-\infty, u_n]\right), \qquad \text{by continuity property}$$

$$= P^X(\emptyset) = 0$$

Properties **(4)** and **(5)** suggest that the graph of F must eventually approach the line $y = 0$ at one end and the line $y = 1$ at the other end as shown in Figure 2.7. In other words, the lines $y = 0$ and $y = 1$ are the asymptotes.

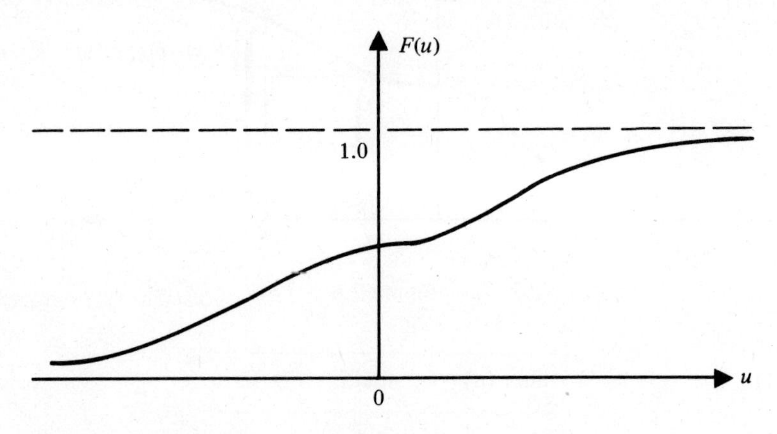

Figure 2.7

(6) A distribution function is continuous from the right; that is, $\lim_{x\to a^+} F(x) = F(a)$ *for* every real number a.

Remark. $\lim_{x\to a^+} F(x)$ is defined as the limit of $F(x)$ as x tends to a through values greater than a. Since F is monotone and bounded, this limit always exists. It is sometimes denoted by $F(a^+)$.

Let us now prove the result. Suppose $\{u_n\}$ is a decreasing sequence of real numbers such that $\lim_{n\to\infty} u_n = a$. Then $\{(-\infty, u_n]\}$ is a decreasing sequence of intervals, and therefore $\lim_{n\to\infty} (-\infty, u_n] = \bigcap_{n=1}^{\infty} (-\infty, u_n] = (-\infty, a]$. Consequently,

$$\lim_{n\to\infty} F(u_n) = \lim_{n\to\infty} P^X((-\infty, u_n])$$

$$= P^X\left(\bigcap_{n=1}^{\infty} (-\infty, u_n]\right), \qquad \text{by continuity property}$$

$$= P^X((-\infty, a]) = F(a)$$

(See Figure 2.8.)

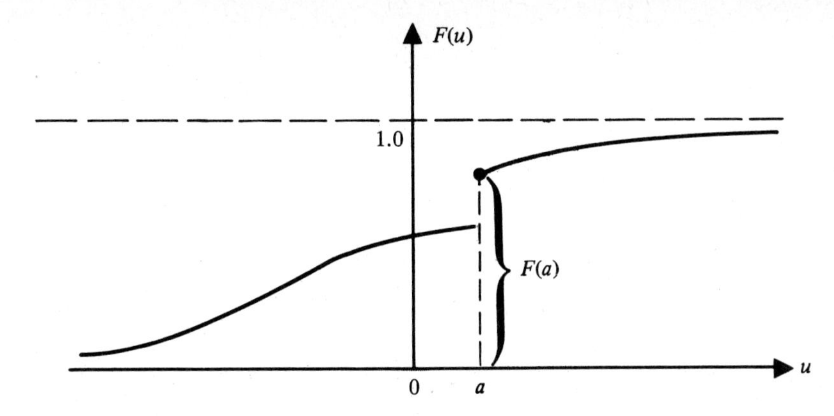

Figure 2.8

(7) $$\lim_{x \to a^-} F(x) = P(X < a)$$

Remark. $\lim_{x \to a^-} F(x)$ is defined as the limit of $F(x)$ as x approaches a through values less than a; it is sometimes denoted as $F(a^-)$.

To prove (7), suppose $\{u_n\}$ is an increasing sequence of real numbers such that $\lim_{n\to\infty} u_n = a$. Then $\{(-\infty, u_n]\}$ is an increasing sequence of intervals, and therefore $\lim_{n\to\infty} (-\infty, u_n] = \bigcup_{n=1}^{\infty} (-\infty, u_n] = (-\infty, a)$. (Note that the limit interval is open at a. Why?) Hence

$$\begin{aligned}
\lim_{n\to\infty} F(u_n) &= \lim_{n\to\infty} P^X((-\infty, u_n]) \\
&= P^X\left(\bigcup_{n=1}^{\infty} (-\infty, u_n]\right), \qquad \text{by continuity property} \\
&= P^X((-\infty, a)) \\
&= P(X < a)
\end{aligned}$$

Hence $F(a^-) = P(X < a)$. (See Figure 2.9.)

(8) $$P(X = a) = F(a) - F(a^-) \quad \text{for any real number } a$$

To prove this, we note that

$$\begin{aligned}
F(a) = P^X((-\infty, a]) &= P^X((-\infty, a) \cup \{a\}) \\
&= P^X((-\infty, a)) + P^X(\{a\})
\end{aligned}$$

Therefore,

$$F(a) = F(a^-) + P(X = a)$$

and the result follows. (See Figure 2.9.)

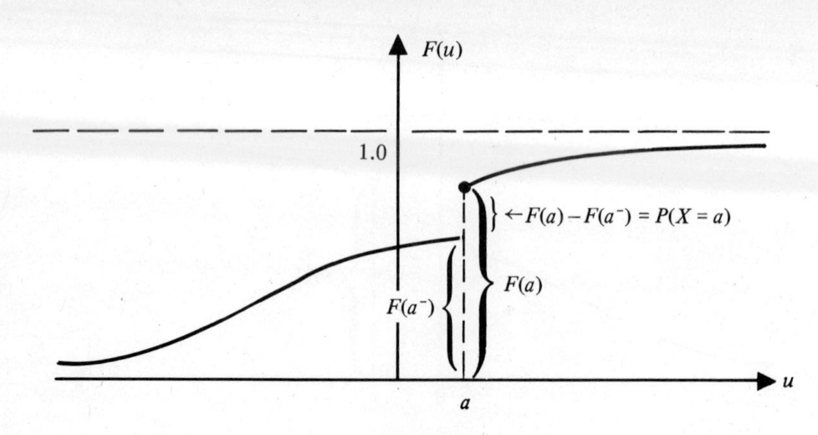

Figure 2.9

Comment. $P(X = a) = F(a) - F(a^-)$, *and hence is the amount of jump in F at a.* If the D.F. is continuous at a, then certainly the amount of jump in F at a is zero, and consequently $P(X = a) = 0$. Thus, it is important to know that *if the D.F. is continuous, then the probability that the underlying r.v. takes on any particular value is zero.*

Summarizing, we have the following properties of a D.F.:

1. $0 \leqslant F(x) \leqslant 1$ for every real number x
2. F is nondecreasing. That is, if $a < b$, then $F(a) \leqslant F(b)$
3. $\lim_{x \to \infty} F(x) = 1$ and $\lim_{x \to -\infty} F(x) = 0$
4. F is continuous from the right, that is, $F(a^+) = F(a)$

We already know that $P(a < X \leqslant b) = F(b) - F(a)$. Further, it can be easily shown that

$$P(a \leqslant X \leqslant b) = F(b) - F(a) + P(X = a)$$
$$P(a \leqslant X < b) = F(b) - F(a) + P(X = a) - P(X = b)$$
$$P(a < X < b) = F(b) - F(a) - P(X = b)$$

Comment. If the D.F. is continuous at the extremities a and b, then, of course, $P(X = a) = P(X = b) = 0$, and consequently $P(a < X \leqslant b) = P(a \leqslant X \leqslant b)$ $= P(a \leqslant X < b) = P(a < X < b) = F(b) - F(a)$. Otherwise, one must be careful to take into account the probabilities at the end points of the intervals.

Example 2.4. Give a reason why each of the graphs of F given in Figure 2.10 does not represent a distribution function.

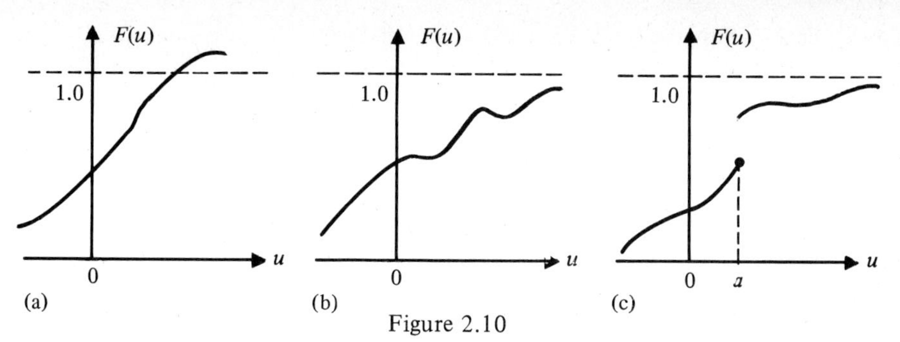

Figure 2.10

Solution. In graph (a), $F(u) > 1$ for some u; in graph (b), F is not nondecreasing; and in (c), F is not right continuous at a.

Example 2.5. Suppose a function F is defined by

$$F(x) = \begin{cases} 0, & x \leqslant 1 \\ \frac{1}{2}, & 1 < x < 3 \\ 1, & x \geqslant 3 \end{cases}$$

Show why F does not represent a D.F.

Solution. We observe that $0 \leqslant F(u) \leqslant 1$ for every u; $F(\infty) = 1$; $F(-\infty) = 0$; and F is nondecreasing. However, $\lim_{x \to 1^+} F(x) = \frac{1}{2}$, whereas $F(1) = 0$. Hence $F(1^+) \neq F(1)$. Consequently, F is not right continuous at 1 and, therefore, F cannot represent a D.F.

Note that if the function F is defined slightly differently as

$$F(x) = \begin{cases} 0, & x < 1 \\ \frac{1}{2}, & 1 \leqslant x < 3 \\ 1, & x \geqslant 3 \end{cases}$$

then F does represent a D.F.

Example 2.6. Show that the function F defined by

$$F(x) = \begin{cases} 0, & x < 0 \\ c \ln x, & 0 \leqslant x < 2 \\ 1, & x \geqslant 2 \end{cases}$$

cannot represent a D.F. for any nonzero constant c.

Solution. We know that a D.F. has to be nonnegative. Now note that $\ln x < 0$ if $0 < x < 1$. This would necessitate that c be negative. On the other hand, $\ln x > 0$ if $1 < x < 2$, and this would require that c be positive. Hence there is no constant c which would make $c \ln x \geqslant 0$ for all x in the interval $(0, 2)$. Therefore, F as defined above cannot represent a D.F. for any nonzero constant c. (Of course, if we let c be equal to zero, then

$$F(x) = \begin{cases} 0, & x < 2 \\ 1, & x \geqslant 2 \end{cases}$$

does represent a D.F.)

Example 2.7. Suppose the D.F. of an r.v. is given by

$$F(x) = \begin{cases} \frac{1}{2}e^x, & x < 0 \\ 1 - \frac{1}{4}e^{-x}, & x \geqslant 0 \end{cases}$$

Find the following probabilities:

(*a*) $P(X = 0)$ (*b*) $P(0 < X \leqslant 2)$
(*c*) $P(0 \leqslant X \leqslant 2)$ (*d*) $P(-3 < X < 0)$
(*e*) $P(-3 \leqslant X \leqslant 0)$

Solution. The graph of the D.F. is indicated in Figure 2.11.

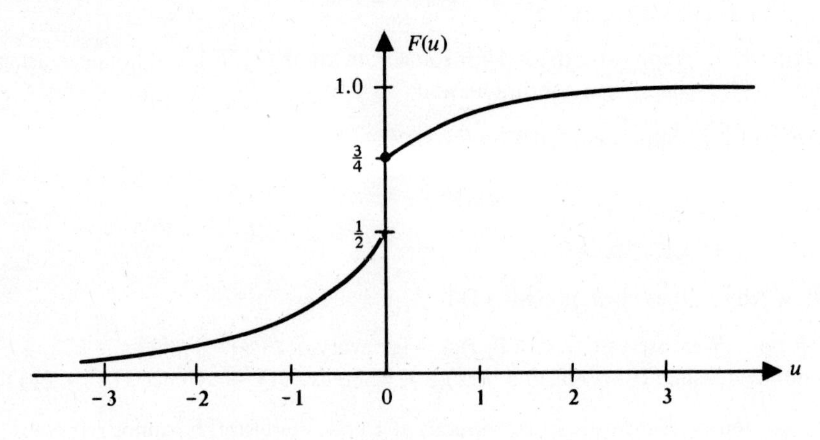

Figure 2.11

(*a*) We see that $F(0^-) = \lim_{x \to 0^-} F(x) = \lim_{x \to 0} \frac{1}{2}e^x = \frac{1}{2}$. Therefore,

$$P(X = 0) = F(0) - F(0^-) = (1 - \tfrac{1}{4}e^{-0}) - \tfrac{1}{2} = \tfrac{1}{4}$$

(*b*) $P(0 < X \leqslant 2) = F(2) - F(0) = (1 - \frac{1}{4}e^{-2}) - (1 - \frac{1}{4}e^{-0}) = \frac{1}{4}(1 - e^{-2})$

(*c*) $P(0 \leqslant X \leqslant 2) = F(2) - F(0) + P(X = 0) = \frac{1}{4}(1 - e^{-2}) + \frac{1}{4} = \frac{1}{2} - \frac{1}{4}e^{-2}$

(*d*) $P(-3 < X < 0) = F(0) - F(-3) - P(X = 0) = \frac{3}{4} - (\frac{1}{2}e^{-3}) - (\frac{1}{4}) = \frac{1}{2} - \frac{1}{2}e^{-3}$

(*e*) $$\begin{aligned} P(-3 \leqslant X \leqslant 0) &= F(0) - F(-3) + P(X = -3) \\ &= \tfrac{3}{4} - \tfrac{1}{2}e^{-3} + 0, \qquad \text{since } P(X = -3) = 0 \\ &= \tfrac{3}{4} - \tfrac{1}{2}e^{-3} \end{aligned}$$

Example 2.8. Let X denote the number of hours a student studies during a randomly selected day. Suppose the probability law is specified by the D.F. given by

$$F(x) = \begin{cases} 0, & x < 0 \\ \frac{1}{8}x + \frac{1}{8}, & 0 \leqslant x < 1 \\ \frac{1}{2}, & 1 \leqslant x < 2 \\ \frac{1}{8}x + \frac{1}{2}, & 2 \leqslant x < 4 \\ 1, & x \geqslant 4 \end{cases}$$

(*a*) Plot the D.F.
(*b*) Find the probability that the student:

(*i*) studies exactly 2 hours; that is, $P(X = 2)$
(*ii*) studies exactly 3 hours; that is, $P(X = 3)$
(*iii*) studies; that is, $P(X > 0)$
(*iv*) studies more than 2 hours; that is, $P(X > 2)$
(*v*) studies less than 2 hours; that is, $P(X < 2)$
(*vi*) studies between 1 and 3 hours; that is, $P(1 < X < 3)$
(*vii*) studies more than 2 hours given that he does study; that is, $P(X > 2 | X > 0)$
(*viii*) studies less than 3 hours given that he studies at least 1 hour; that is, $P(X < 3 | X > 1)$.

Solution
(*a*) The graph of F is given in Figure 2.12.

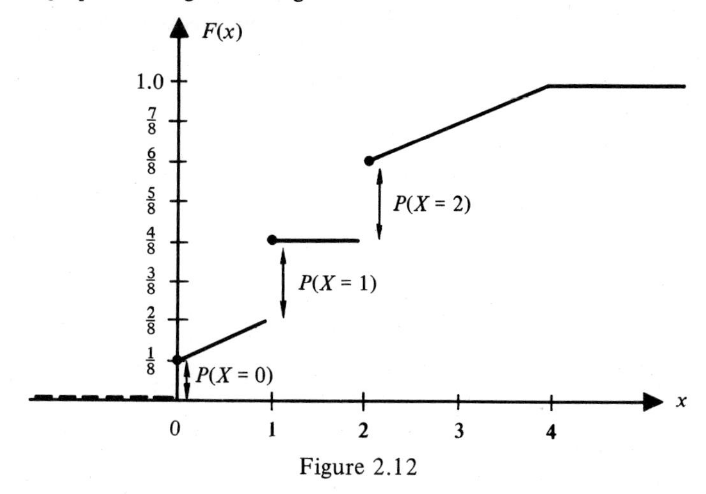

Figure 2.12

(*b*) (*i*) $P(X = 2)$ is the amount of jump in F at $x = 2$. Therefore, $P(X = 2) = \frac{1}{4}$.
(*ii*) The amount of jump in F at $x = 3$ is 0. Hence $P(X = 3) = 0$.
(*iii*) $P(X > 0) = 1 - P(X \leqslant 0) = 1 - F(0) = 1 - [\frac{1}{8} \cdot 0 + \frac{1}{8}] = \frac{7}{8}$.
(*iv*) $P(X > 2) = 1 - P(X \leqslant 2) = 1 - F(2) = 1 - [\frac{1}{8} \cdot 2 + \frac{1}{2}] = \frac{1}{4}$.
(*v*) $P(X < 2) = F(2^-) = \frac{1}{2}$.
(*vi*) $P(1 < X < 3) = F(3) - F(1) - P(X = 3) = [\frac{1}{8} \cdot 3 + \frac{1}{2}] - \frac{1}{2} + 0 = \frac{3}{8}$.

$$(vii)\quad P(X > 2 | X > 0) = \frac{P(\{X > 2\} \cap \{X > 0\})}{P(X > 0)}$$

(*Note:* We are using the formula $P(A|B) = P(AB)/P(B)$.) Now $\{X > 2\} \cap \{X > 0\} = \{X > 2\}$. Therefore,

$$P(X > 2 | X > 0) = \frac{P(X > 2)}{P(X > 0)} = \frac{\frac{1}{4}}{\frac{7}{8}}, \qquad \text{by } (iii) \text{ and } (iv)$$

$$= \frac{2}{7}$$

$$(viii)\quad P(X<3 \mid X>1) = \frac{P(\{X<3\}\cap\{X>1\})}{P(X>1)}$$
$$= \frac{P(1<X<3)}{P(X>1)}$$
$$= \frac{\frac{3}{8}}{1-F(1)}, \qquad \text{since } P(1<X<3)=\tfrac{3}{8} \text{ by } (vi)$$
$$= \frac{\frac{3}{8}}{1-\frac{1}{2}} = \frac{3}{4}$$

EXERCISES–SECTION 2

1. A fair die in the shape of a tetrahedron with faces marked 1, 2, 3, 4 is rolled. A person stands to win four dollars if 1 shows up and three dollars if an even number shows up, but he stands to lose two dollars if 3 shows up. If X represents the winnings, find the distribution function and plot it.

2. Two random variables X and Y are said to be equal, written $X = Y$, if $X(s) = Y(s)$ for every $s \in S$. On the other hand, two random variables are said to be *identically distributed* if $P(X \in B) = P(Y \in B)$ for every Borel set B, or equivalently, $F_X(u) = F_Y(u)$ for every real number u.

Let $S = \{s_1, s_2, s_3, s_4\}$. Give an example of two random variables X and Y defined on S which are identically distributed but such that $X \neq Y$.

3. Indicate which of the functions given below represent genuine distribution functions.

(a)
$$F(x) = \begin{cases} 0, & x \leqslant 0 \\ \frac{1}{2} + \frac{1}{4}x, & 0 < x \leqslant 1 \\ 1, & x > 1 \end{cases}$$

(b)
$$F(x) = \begin{cases} -\frac{1}{x} - \frac{1}{2x^2}, & -\infty < x < -1 \\ \frac{1}{2}, & -1 \leqslant x < 1 \\ 1 - \frac{1}{x} + \frac{1}{2x^2}, & x \geqslant 1 \end{cases}$$

(c)
$$F(u) = \begin{cases} 0, & u \leqslant 0 \\ 1 - e^{-u} + e^{-1/u}, & 0 < u < \frac{1}{2} \\ 1 - e^{-u} + e^{-2}, & \frac{1}{2} \leqslant u < 2 \\ 1, & u \geqslant 2 \end{cases}$$

(d)
$$F(x) = \begin{cases} 0, & x < -1 \\ x, & -1 \leqslant x < 0 \\ 1, & x \geqslant 0 \end{cases}$$

(e)
$$F(x) = \begin{cases} 0, & x < 0 \\ x, & 0 \leqslant x < 1 \\ 1, & x \geqslant 1 \end{cases}$$

4. Suppose the lifetime T (measured in hours) of an electronic tube has the following D.F.:

$$F(x) = \begin{cases} 0, & x < 10 \\ 1 - \left(\frac{10}{x}\right)^2, & x \geqslant 10 \end{cases}$$

Find the probability that–

(a) the tube will last beyond 20 hours
(b) the tube will burn out before 15 hours
(c) the tube will last beyond 20 hours, given that it lasts at least 15 hours
(d) the tube will last between 18 and 20 hours, given that it lasts at least 15 hours.

5. The distribution function of a random variable is given as follows:

$$F(x) = \begin{cases} 0, & x < -2 \\ \frac{1}{12}, & -2 \leqslant x < 2 \\ \frac{1}{3}, & 2 \leqslant x < 5 \\ \frac{2}{3}, & 5 \leqslant x < 6 \\ 1, & x \geqslant 6 \end{cases}$$

Plot the distribution function. Find:

(a) $P(-1 < X < 5.5)$ (b) $P(X = 2)$
(c) $P(-2 \leqslant X \leqslant 5)$ (d) $P(-2 < X < 5)$
(e) $P(-2 \leqslant X < 5)$ (f) $P(-2 < X \leqslant 5)$
(g) $P(X \geqslant 2)$ (h) $P(X > 2)$
(i) $P(X < 2)$ (j) $P(X \leqslant 2)$

6. Suppose the distribution of X is given by the following D.F.:

$$F(x) = \begin{cases} 0, & x < -5 \\ \frac{1}{20}x + \frac{1}{3}, & -5 \leqslant x < 0 \\ \frac{1}{16}x^2 + \frac{1}{2}, & 0 \leqslant x < 2 \\ 1, & x \geqslant 2 \end{cases}$$

Find:

(a) $P(X = -5)$ (b) $P(X = 0)$ (c) $P(X = 1)$
(d) $P(X = 2)$ (e) $P(0 \leqslant X < 2)$ (f) $P(0 \leqslant X \leqslant 1)$
(g) $P(X \geqslant 2)$ (h) $P(0.5 < X \leqslant 1.5)$

7. Suppose a random variable X has a distribution function F. The real number m which satisfies $F(m) \geqslant \frac{1}{2}$ and $F(m^-) \leqslant \frac{1}{2}$ is called a *median* for the distribution of X. Show that this condition is equivalent to the following: $P(X \leqslant m) \geqslant \frac{1}{2}$ and $P(X \geqslant m) \geqslant \frac{1}{2}$.

8. Show that a median always exists but is not unique.

9. Find a median of the distribution of X where the D.F. is given by–

(a)
$$F(x) = \begin{cases} 0, & x < 0 \\ 1 - \frac{1}{1 + x}, & x \geqslant 0 \end{cases}$$

(b)
$$F(x) = \begin{cases} 0, & x < -3 \\ \frac{1}{3}, & -3 \leqslant x < -1 \\ \frac{1}{2}, & -1 \leqslant x < 0 \\ 1, & x \geqslant 0 \end{cases}$$

(c)
$$F(x) = \begin{cases} 0, & x < -3 \\ \frac{1}{4}, & -3 \leqslant x < 4 \\ \frac{1}{3}x - 1, & 4 \leqslant x < 6 \\ 1, & x \geqslant 6 \end{cases}$$

10. A random variable X has a distribution described by the following relation:

$$P(X > u) = \begin{cases} 1, & u \leqslant 0 \\ e^{-u^2}, & u > 0 \end{cases}$$

Find:

(a) $P(1 < X < 2)$
(b) $P(X > 1.5 \mid 1 < X < 2)$

11. Suppose F is a D.F.

(a) Show that the function G defined by

$$G(x) = 1 - F(-x^-)$$

for every real number x is also a D.F.

(b) Show that if F is the D.F. of X, then G is the D.F. of $-X$.

12. *Convex combination of two D.F.'s* Suppose F_1 and F_2 are two D.F.'s and λ is any real number such that $0 \leqslant \lambda \leqslant 1$. If $\lambda F_1 + (1 - \lambda)F_2$ is defined for any real number x by

$$(\lambda F_1 + (1 - \lambda)F_2)(x) = \lambda F_1(x) + (1 - \lambda)F_2(x)$$

show that $\lambda F_1 + (1 - \lambda)F_2$ is a D.F. *Hint:* Prove the following:

(*i*) $0 \leqslant (\lambda F_1 + (1 - \lambda)F_2)(x) \leqslant 1$ for any x.

(*ii*) If $a < b$, then $(\lambda F_1 + (1 - \lambda)F_2)(a) \leqslant (\lambda F_1 + (1 - \lambda)F_2)(b)$.

(*iii*) $\lim_{x \to -\infty} (\lambda F_1 + (1 - \lambda)F_2)(x) = 0$

$\lim_{x \to \infty} (\lambda F_1 + (1 - \lambda)F_2)(x) = 1$

(*iv*) $(\lambda F_1 + (1 - \lambda)F_2)(a^+) = (\lambda F_1 + (1 - \lambda)F_2)(a)$ for any real number a.

3. CLASSIFICATION OF RANDOM VARIABLES

In Section 2 we were concerned with the properties of distribution functions. Specifically, we saw that, if a D.F. is continuous at a point x_0, then $P(X = x_0) = 0$, whereas, if the D.F. is not continuous at x_0, then $P(X = x_0)$ is positive and is equal to the amount of jump in the D.F. at x_0. We shall now classify random variables on the basis of this.

3.1 Discrete Random Variables

It is known from analysis in real variables that a bounded monotone function can be discontinuous at most at a countably infinite number of points. Therefore, *a D.F., being a bounded monotone function, can have at most a countably infinite number of discontinuities.*

If the distribution function of a random variable is discontinuous and is of the staircase form shown in Figure 3.1, then the r.v. is said to be a **discrete random variable**.

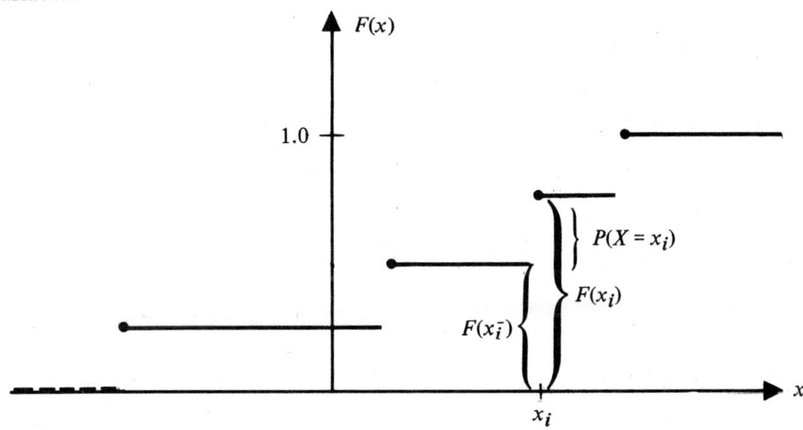

Figure 3.1

Suppose F is discontinuous at most at $x_1, x_2, \ldots$. (Because the set of discontinuities is at most countably infinite we can indeed label them as $x_1, x_2, \ldots$, using the subscripts $1, 2, \ldots$.) Then $P(X = x)$ is positive at most for $x = x_1, x_2, \ldots$ and zero for all other values of x. If we let $p(x) = P(X = x)$, then p is a nonnegative function which vanishes everywhere except perhaps at $x_1, x_2, \ldots$. We therefore have an alternate definition of a discrete r.v.:

A random variable is called a **discrete random variable** if there exists a nonnegative function p which vanishes everywhere except at a finite or countably infinite number of points. In this case, the D.F. of X can be expressed in terms of p by

$$F(u) = \sum_{x_i \leqslant u} p(x_i)$$

for any real number u.

The notation $\sum_{x_i \leqslant u}$ states that the summation is to be carried over all the x values which are less than or equal to u. Specifically, if $x_1, x_2, \ldots, x_k$ are all less than or equal to u, then $F(u) = p(x_1) + p(x_2) + \ldots + p(x_k)$.

The function p defined by $p(x) = P(X = x)$ is called the **probability function** of X. It is also referred to as a **probability mass function**.

For a discrete random variable, the values that it assumes with positive probabilities are called its possible values. To describe the probability function of a discrete random variable, we shall give all the possible values of the r.v., together with the corresponding probabilities.

There is a property of the function p that we have not mentioned so far. We already know that $F(u) = \sum_{x_i \leq u} p(x_i)$ for any real number u. Also, we know that $\lim_{u \to \infty} F(u) = 1$. This, in view of

$$\lim_{u\to\infty} F(u) = \lim_{u\to\infty} \sum_{x_i \leq u} p(x_i) = \sum_{x_i < \infty} p(x_i)$$

leads to

$$\sum_{x_i < \infty} p(x_i) = 1, \quad \text{that is,} \quad \sum_{i=1}^{\infty} p(x_i) = 1.$$

To summarize, p has the following properties:

$p(x) \geq 0$	for any real number x
$\sum_{i=1}^{\infty} p(x_i) = 1,$	where the x_i's represent possible values of X

Example 3.1. For what value of the constant C does the function given by

$$P(X = k) = C \cdot \frac{1}{k!}, \qquad k = 0, 1, 2, \ldots$$

represent a probability function?

Solution. C has to be positive and such that $\sum_{k=0}^{\infty} p(k) = 1$. That is, C must be such that

$$\sum_{k=0}^{\infty} C \cdot \frac{1}{k!} = 1$$

Now

$$\sum_{k=0}^{\infty} C \cdot \frac{1}{k!} = C \cdot \sum_{k=0}^{\infty} \frac{1}{k!} = C \cdot e, \quad \text{since} \quad \sum_{k=0}^{\infty} \frac{1}{k!} = e$$

Hence $C \cdot e = 1$, and consequently $C = e^{-1}$ (incidentally, note that $C > 0$). ■

Given the D.F. of an r.v. X, we know that

$$P(a < X \leq b) = F(b) - F(a)$$

But since $F(u) = \sum_{x_i \leq u} p(x_i)$, we get

$$P(a < X \leq b) = \sum_{x_i \leq b} p(x_i) - \sum_{x_i \leq a} p(x_i)$$

$$= \sum_{a < x_i \leq b} p(x_i)$$

We therefore have

$$P(a < X \leq b) = \sum_{a < x_i \leq b} p(x_i)$$

Example 3.2. Suppose a probability function p is given by

$$p(x) = \begin{cases} \frac{1}{6} & \text{if } x = -4, -1, \text{ or } 2 \\ \frac{1}{3} & \text{if } x = -2.5 \\ \frac{1}{12} & \text{if } x = 1.5 \\ \frac{1}{12} & \text{if } x = 3 \end{cases}$$

(*a*) Plot the probability function.
(*b*) Find:
(*i*) $P(-2 < X \leqslant 2.2)$ (*ii*) $P(-1 < X \leqslant 1.5)$
(*iii*) $P(-1 \leqslant X \leqslant 1.5)$ (*iv*) $P(-1 \leqslant X < 1.5)$

Solution

(*a*) The probability function of X is plotted in Figure 3.2.

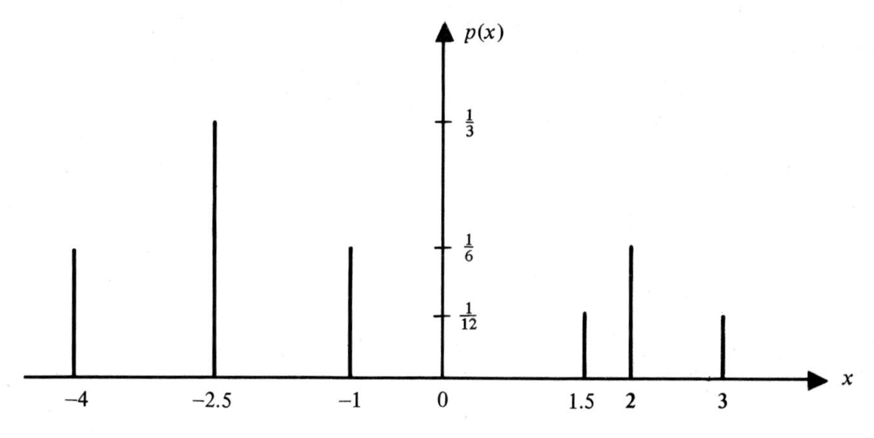

Figure 3.2

(*b*) (*i*) $\{-2 < X \leqslant 2.2\} = \{X = -1\} \cup \{X = 1.5\} \cup \{X = 2\}$. Therefore, since the events on the right-hand side are mutually exclusive,

$$P(-2 < X \leqslant 2.2) = p(-1) + p(1.5) + p(2) = \frac{5}{12}$$

Arguing as in (*i*), it follows that–
(*ii*) $P(-1 < X \leqslant 1.5) = p(1.5) = \frac{1}{12}$
(*iii*) $P(-1 \leqslant X \leqslant 1.5) = p(-1) + p(1.5) = \frac{1}{4}$
(*iv*) $P(-1 \leqslant X < 1.5) = p(-1) = \frac{1}{6}$

Example 3.3. Let X denote the number of dollars a laborer earns during a randomly selected hour. Suppose the probability function of X is given by

$$P(X = x) = \begin{cases} 0.3 & \text{if } x = 0 \\ Kx & \text{if } x = 2 \text{ or } 3 \\ K(x-2) & \text{if } x = 4 \\ 0 & \text{elsewhere} \end{cases}$$

where K is a constant.

(*a*) Find the constant K.

(*b*) Find the probability that the laborer has positive earnings for the hour.

(*c*) Find the conditional probability that he earns at least three dollars, given that he has positive earnings for the hour.

Solution

(*a*) Since $\sum_{x_i} p(x_i) = 1$, we must have

$$0.3 + 2K + 3K + K(4-2) = 1$$

Therefore,

$$K = 0.1$$

Consequently, $p(0) = 0.3, p(2) = 2K = 0.2, p(3) = 3K = 0.3$, and $p(4) = 2K = 0.2$.

(*b*) The probability that the laborer has positive earnings is

$$P(X > 0) = p(2) + p(3) + p(4) = 0.7$$

Alternatively, notice that

$$P(X > 0) = 1 - P(X \leqslant 0) = 1 - P(X = 0) = 0.7$$

(*c*) Here we want to find $P(X \geqslant 3 \mid X > 0)$. This is equal to

$$\frac{P(\{X \geqslant 3\} \cap \{X > 0\})}{P(X > 0)} = \frac{P(X \geqslant 3)}{P(X > 0)} = \frac{p(3) + p(4)}{0.7} = \frac{5}{7} \qquad \blacksquare$$

The equations that relate a distribution function and the corresponding probability function are

$$\text{(i)} \quad F(u) = \sum_{x_i \leqslant u} p(x_i)$$

$$\text{and} \quad \text{(ii)} \quad p(x) = F(x) - F(x^-)$$

From (*i*), we see that *if we are given the probability function, we can obtain the distribution function as follows: simply add all the p-values to the left of u for any u.*

Conversely, from (*ii*), *given the D.F. we can always obtain the probability function as follows: see where the discontinuities occur; these are the possible values of X. If x_0 is a possible value, then $p(x_0)$ is the amount of jump in F at x_0.*

Example 3.4. Suppose the D.F. of an r.v. is given by

$$F(x) = \begin{cases} 0, & x < -1 \\ \frac{1}{6}, & -1 \leqslant x < 0 \\ \frac{1}{2}, & 0 \leqslant x < 1.5 \\ \frac{5}{6}, & 1.5 \leqslant x < 2.5 \\ 1, & x \geqslant 2.5 \end{cases}$$

(*a*) Draw the D.F.
(*b*) Find the probability function of X and draw it.

Solution

(*a*) The graph of the D.F. is given in Figure 3.3.

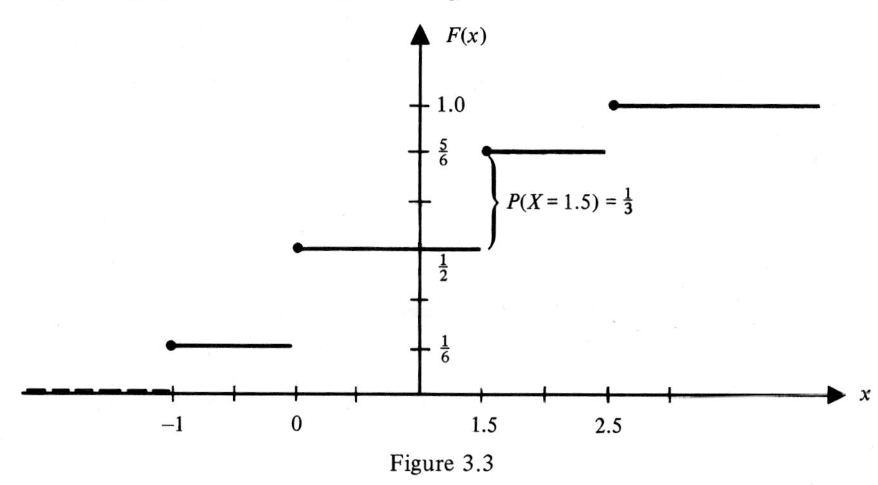

Figure 3.3

Observe that, since the D.F. is of the staircase type, the r.v. is discrete.

(*b*) There are jumps in the D.F. at $-1, 0, 1.5$, and 2.5, and the magnitudes of these jumps are respectively $\frac{1}{6}, \frac{1}{3}, \frac{1}{3}$, and $\frac{1}{6}$. Hence, the probability function p is given by $p(-1) = \frac{1}{6}, p(0) = \frac{1}{3}, p(1.5) = \frac{1}{3}, p(2.5) = \frac{1}{6}$, and $p(x) = 0$ for all other x. (Notice that $p(x)$ is nonnegative, and $p(-1) + p(0) + p(1.5) + p(2.5) = 1$.) The probability function of X is plotted in Figure 3.4.

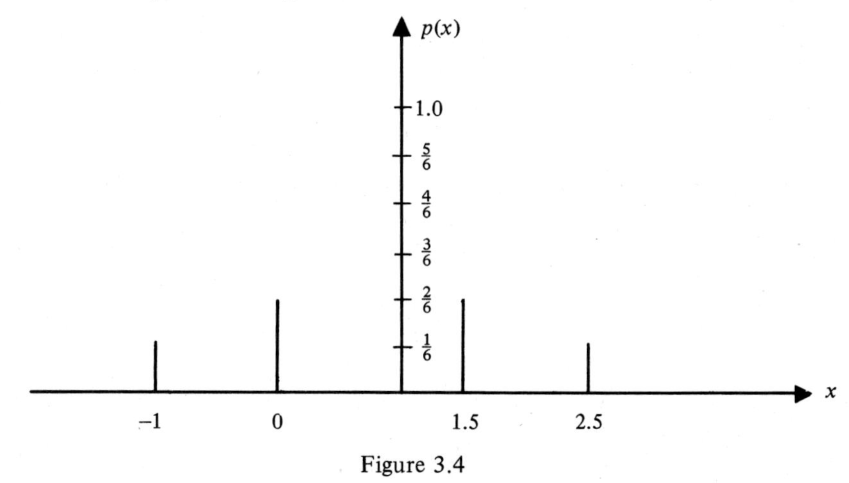

Figure 3.4

Example 3.5. Let $[x]$ denote the largest integer less than or equal to x. (For example, $[1.5] = 1, [-0.5] = -1$, and so on.) Suppose the D.F. of an r.v. X is given by

$$F(x) = \begin{cases} 0, & x < 0 \\ 1 - (\frac{1}{3})^{[x]}, & x \geqslant 0 \end{cases}$$

Find the probability function of X.

Solution. We see that

$$\begin{aligned} &\text{if } 0 \leqslant x < 1, \text{ then } [x] = 0, \text{ and therefore } F(x) = 0 \\ &\text{if } 1 \leqslant x < 2, \text{ then } [x] = 1, \text{ and therefore } F(x) = 1 - (\tfrac{1}{3}) \\ &\text{if } 2 \leqslant x < 3, \text{ then } [x] = 2, \text{ and therefore } F(x) = 1 - (\tfrac{1}{3})^2 \end{aligned}$$

In general, for any positive integer i,

$$\text{if } i \leqslant x < i+1, \text{ then } [x] = i \text{ and } F(x) = 1 - (\tfrac{1}{3})^i$$

To find the probability function of X, we note that

$$\begin{aligned} P(X = i) &= F(i) - F(i^-) \\ &= [1 - (\tfrac{1}{3})^i] - [1 - (\tfrac{1}{3})^{i-1}] \end{aligned}$$

Thus,

$$P(X = i) = 2 \cdot (\tfrac{1}{3})^i, \quad i = 1, 2, \ldots$$

(*Note:* Since $\sum_{k=0}^{\infty} x^k = \frac{1}{1-x}$ if $|x| < 1$, we see that

$$\sum_{i=1}^{\infty} P(X = i) = 2 \sum_{i=1}^{\infty} (\tfrac{1}{3})^i = 2 \cdot \left[\frac{1}{1 - \frac{1}{3}} - 1\right] = 1$$

so that $P(X = i) = 2(\frac{1}{3})^i$, $i = 1, 2, \ldots$, is a genuine probability function.)

Example 3.6. Suppose the probability function p of a random variable is given by

$$p(-2) = \tfrac{1}{6}, \quad p(2) = \tfrac{1}{6}, \quad p(4) = \tfrac{1}{3}, \quad p(5) = \tfrac{1}{6}, \text{ and } p(8) = \tfrac{1}{6}$$

Find the D.F. of the r.v.

Solution. The graph of the probability function is given in Figure 3.5.

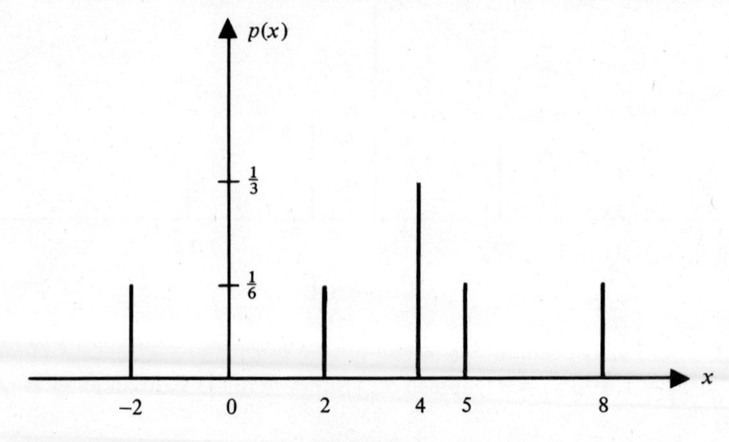

Figure 3.5

Recall that to find $F(u)$ for any u we add up all the probabilities corresponding to the points that are to the left of u. Thus, for instance, $F(4.5) = p(-2) + p(2) + p(4) = \frac{2}{3}$. As a matter of fact, $F(u) = \frac{2}{3}$ for any u for which $4 \leqslant u < 5$. Adopting the above argument, we get the D.F. as

$$F(u) = \begin{cases} 0 & \text{if } u < -2 \\ \frac{1}{6} & \text{if } -2 \leqslant u < 2 \\ \frac{1}{3} & \text{if } 2 \leqslant u < 4 \\ \frac{2}{3} & \text{if } 4 \leqslant u < 5 \\ \frac{5}{6} & \text{if } 5 \leqslant u < 8 \\ 1 & \text{if } u \geqslant 8 \end{cases}$$

Example 3.7. Suppose the probability function of an r.v. X is given by

$$p(k) = a(1-a)^k, \qquad k = 0, 1, 2, \ldots$$

where a is a fixed number, $0 < a < 1$.

(*a*) Show that p is indeed a probability function.

(*b*) Find the D.F. of X.

Solution

(*a*) Since $0 < a < 1$, $a(1-a)^k \geqslant 0$; therefore, $p(k) \geqslant 0$. Also,

$$\begin{aligned} \sum_{k=0}^{\infty} p(k) &= a \sum_{k=0}^{\infty} (1-a)^k \\ &= \frac{a}{1-(1-a)}, \qquad \text{since } |1-a| < 1 \\ &= 1 \end{aligned}$$

Therefore, p is a probability function.

(*b*) To find the D.F. of X, we note that

$$F(u) = \begin{cases} 0, & u < 0 \\ a, & 0 \leqslant u < 1 \\ a + a(1-a), & 1 \leqslant u < 2 \\ a + a(1-a) + a(1-a)^2, & 2 \leqslant u < 3 \\ \text{and so on} \end{cases}$$

In fact, after some reflection we see that, for any $u \geqslant 0$,

$$F(u) = a + a(1-a) + \ldots + a(1-a)^{[u]}$$

where $[u]$ represents the largest integer less than or equal to u. Hence

$$\begin{aligned} F(u) &= a[1 + (1-a) + \ldots + (1-a)^{[u]}] \\ &= a\left(\frac{1-(1-a)^{[u]+1}}{1-(1-a)}\right), \qquad \text{adding the geometric progression} \\ &= 1 - (1-a)^{[u]+1} \end{aligned}$$

Hence the D.F. is given by

$$F(u) = \begin{cases} 0, & u < 0 \\ 1 - (1-a)^{[u]+1}, & u \geqslant 0 \end{cases}$$

3.2 Absolutely Continuous Random Variables

We have seen that for discrete r.v.'s the entire probability mass is concentrated at a number of points (finite or countably infinite). We shall now discuss another class of random variables called the absolutely continuous random variables. It is a common practice to omit the word absolutely and refer to these r.v.'s as continuous random variables.

A random variable X is said to be **absolutely continuous** if there is a nonnegative function f defined on the real line such that

$$F(x) = \int_{-\infty}^{x} f(u)\, du$$

for all real numbers x.

Here the integral is taken as the improper Riemann integral. Thus what we are saying is that the D.F. is obtainable by integrating a nonnegative function.

Example 3.8. Consider a D.F. F given by

$$F(x) = \begin{cases} 0, & x < 0 \\ 1 - e^{-x}, & x \geqslant 0 \end{cases}$$

Show that the underlying r.v. is absolutely continuous.

Solution. (We can check in the usual way that F is indeed a D.F.) Consider the function f given by

$$f(u) = \begin{cases} 0, & u < 0 \\ e^{-u}, & u \geqslant 0 \end{cases}$$

This function is obviously nonnegative. Also,

$$\text{if } x < 0, \text{ then } \int_{-\infty}^{x} f(u)\, du = \int_{-\infty}^{x} 0\, du = 0$$

$$\text{and if } x \geqslant 0, \text{ then } \int_{-\infty}^{x} f(u)\, du = \int_{0}^{x} e^{-u}\, du = 1 - e^{-x}$$

Thus, if we take

$$f(u) = \begin{cases} 0, & u < 0 \\ e^{-u}, & u \geqslant 0 \end{cases} \quad \text{then} \quad F(x) = \begin{cases} 0, & x < 0 \\ 1 - e^{-x}, & x \geqslant 0 \end{cases}$$

can be expressed as

$$F(x) = \int_{-\infty}^{x} f(u)\, du$$

for any real number x. Hence the random variable is absolutely continuous. ∎

The reader might wonder how we happened to pick the function f as given above. We shall address ourselves to this question shortly.

The nonnegative function f for which

$$F(x) = \int_{-\infty}^{x} f(u)\, du$$

is called the **probability density function**, or pdf, for short.

As mentioned, it is customary to refer to an absolutely continuous r.v. simply as a continuous r.v. It can be shown that, if an r.v. is absolutely continuous, then the D.F. is continuous. The reader should *not* assume, however, that a continuous r.v. is one whose D.F. is just continuous. Not only is it continuous, but—also—it can be obtained by integrating a nonnegative function as mentioned above. Later we shall give an example of a continuous D.F. where the underlying r.v. is *not* (absolutely) continuous. Such distributions are called **singular distributions**. The notion of "absolute continuity" is deep and refers to the condition that necessitates the existence of the nonnegative function f, the probability density function, from which the D.F. can be recovered by integrating appropriately.

Since $F(x) = \int_{-\infty}^{x} f(u)\, du$, we have the following consequences:

(*i*)

$$\int_{-\infty}^{\infty} f(u)\, du = 1$$

This follows because

$$\int_{-\infty}^{\infty} f(u)\, du = \lim_{x\to\infty} \int_{-\infty}^{x} f(u)\, du = \lim_{x\to\infty} F(x) = 1$$

Geometrically, this result states that the total area under the graph of the pdf f is one.

(*ii*)

$$P(a < X \leqslant b) = \int_{a}^{b} f(u)\, du$$

This follows because

$$\begin{aligned} P(a < X \leqslant b) &= F(b) - F(a) \\ &= \int_{-\infty}^{b} f(u)\, du - \int_{-\infty}^{a} f(u)\, du = \int_{a}^{b} f(u)\, du \end{aligned}$$

The result merely states that $P(a < X \leqslant b)$ is the area under the curve f, between a and b, as shown in Figure 3.6. In fact, it can be shown that, for any Borel set B,

$$P(X \in B) = \int_{B} f(x)\, dx$$

In short, all the probabilities associated with the r.v. X are determined by the function f.

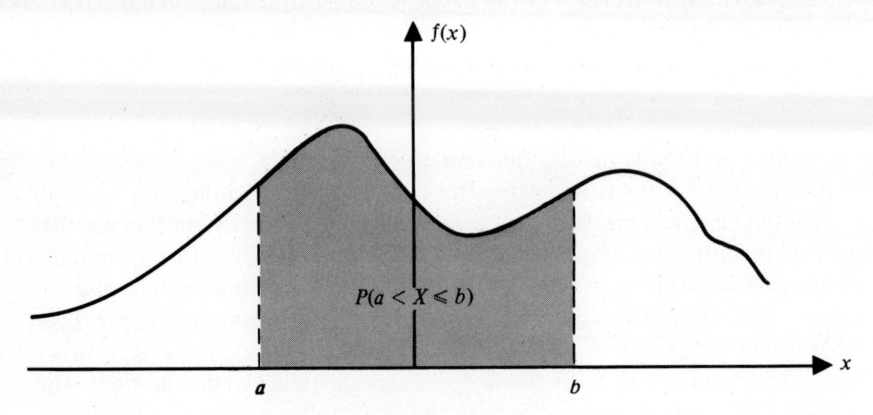

Figure 3.6

In particular, for any real number a,

$$P(X = a) = P(a \leqslant X \leqslant a) = \int_a^a f(x)\, dx = 0$$

Comment. The reader might have anticipated that $P(X = a)$ ought to be zero in view of the remark we made earlier that absolutely continuous random variables have continuous distribution functions. Of course, as we now know, if the D.F. is continuous at a, then $P(X = a) = 0$.

(iii)	$f(u) = \frac{d}{du} F(u)$	at every point where the pdf is continuous

This follows immediately by the fundamental theorem of calculus.

Comment. Recall from calculus that

$$\frac{d}{du} F(u) = \lim_{h \to 0} \frac{F(u+h) - F(u)}{h} = \lim_{h \to 0} \frac{P(u < X \leqslant u + h)}{h}$$

The pdf f therefore represents the limit of the ratio of the probability over an interval of length h to the length of the interval as $h \to 0$. It is this rationale that is behind calling f the probability density function.

To bring out the essence of *(iii)*, consider the D.F. F given by

$$F(x) = \begin{cases} 0 & \text{if } x < 0 \\ \frac{1}{2}x & \text{if } 0 \leqslant x < 1 \\ \frac{1}{6}x + \frac{1}{3} & \text{if } 1 \leqslant x < 4 \\ 1 & \text{if } x \geqslant 4 \end{cases}$$

This D.F. is plotted in Figure 3.7.

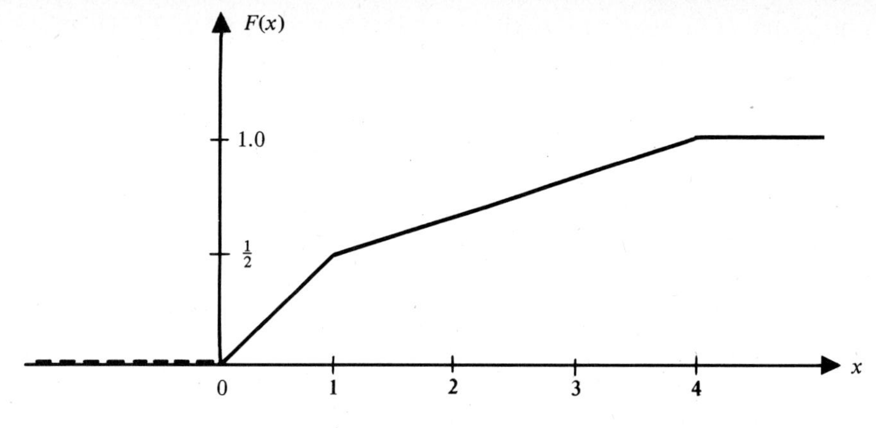

Figure 3.7

At $x = 0$, $x = 1$, and $x = 4$ the function F is not differentiable. (At these points the slope of F changes abruptly.)

$$\begin{aligned}
&\text{For } x < 0, && \frac{d}{dx}F(x) = 0, && \text{that is,} \quad f(x) = 0 \\
&\text{for } 0 < x < 1, && \frac{d}{dx}F(x) = \tfrac{1}{2}, && \text{that is,} \quad f(x) = \tfrac{1}{2} \\
&\text{for } 1 < x < 4, && \frac{d}{dx}F(x) = \tfrac{1}{6}, && \text{that is,} \quad f(x) = \tfrac{1}{6} \\
\text{and } &\text{for } x > 4, && \frac{d}{dx}F(x) = 0, && \text{that is,} \quad f(x) = 0
\end{aligned}$$

What about the values of f at $x = 0, 1$, and 4, where the derivatives of F do not exist? At these points we are free to assign to f any value we please. This will not affect the basic probability law associated with the given random variable since, after all, $\int_a^a f(x)\,dx = 0$ for any real number a. Thus, in the above case we could have given the pdf as

$$f(x) = \begin{cases} 0 & \text{if } x < 0 \\ \frac{1}{2} & \text{if } 0 < x < 1 \\ \frac{1}{6} & \text{if } 1 < x < 4 \\ 0 & \text{if } x > 4 \\ 200 & \text{if } x = 0 \\ 300 & \text{if } x = 1 \\ 10^{60} & \text{if } x = 4 \end{cases}$$

However, it might be more convenient to give the pdf in the above case as

$$f(x) = \begin{cases} 0 & \text{if } x \leqslant 0 \\ \frac{1}{2} & \text{if } 0 < x \leqslant 1 \\ \frac{1}{6} & \text{if } 1 < x \leqslant 4 \\ 0 & \text{if } x > 4 \end{cases}$$

In general, for an absolutely continuous r.v. it is possible that the D.F. is not differentiable at some points. These are the points where the pdf will not be continuous. (Geometrically, these are the points where the graph of the D.F. has sharp corners.) At such points, we are free to assign to f any value we please, without disturbing the basic probability law associated with the r.v. Usually we shall do this in the most convenient way possible.

In view of the preceding discussion, the relations that connect the D.F. F and the corresponding pdf f are

$$F(x) = \int_{-\infty}^{x} f(u)\, du \qquad \text{for any real number } x$$

$$\text{and} \quad f(x) = \frac{d}{dx} F(x) \qquad \text{for any real number } x$$

Example 3.9. If the D.F. F of an r.v. X is given by

$$F(u) = \begin{cases} 0, & u < 0 \\ u^2, & 0 \leqslant u < \frac{1}{2} \\ 1 - \frac{3}{25}(3-u)^2, & \frac{1}{2} \leqslant u < 3 \\ 1, & u \geqslant 3 \end{cases}$$

find the pdf of X.

Solution. The D.F. is not differentiable at the points $0, \frac{1}{2}$, and 3. Except at these values, the D.F. is differentiable everywhere. Differentiating, the pdf is given by

$$f(u) = \begin{cases} 2u, & 0 \leqslant u < \frac{1}{2} \\ \frac{6}{25}(3-u), & \frac{1}{2} \leqslant u < 3 \\ 0, & \text{elsewhere} \end{cases}$$

(*Note:* At $0, \frac{1}{2}$, and 3 we have set the pdf arbitrarily equal to $0, \frac{3}{5}$, and 0, respectively.)

Example 3.10. Find the pdf of an r.v. X whose D.F. is given by

$$F(x) = \begin{cases} 0, & x < 0 \\ 1 - e^{-x^a}, & x \geqslant 0 \end{cases}$$

where a is a positive constant. The pdf corresponding to this D.F. is referred to as the *Weibull density function*. It has applications in reliability theory.

Solution. (*Note:* Since $\lim_{x\to\infty} F(x) = 1$, we must have $\lim_{x\to\infty} e^{-x^a} = 0$. Therefore, $\lim_{x\to\infty} x^a = \infty$, and, consequently, a must be positive.)

Differentiating, the pdf f is given by

$$f(x) = \begin{cases} 0, & x \leqslant 0 \\ a x^{a-1} e^{-x^a}, & x > 0 \end{cases}$$

■

In the following examples, we shall obtain the D.F. when the pdf is given. The reader should realize the advantage of obtaining the D.F. from the pdf. As soon as the D.F. is available, the problem of integration is eliminated once and for all. For example, we would rather compute $P(a < X \leq b)$ from the formula $F(b) - F(a)$ than from $\int_a^b f(x)\,dx$.

Recall that $F(x_0) = \int_{-\infty}^{x_0} f(u)\,du$, and therefore represents the area under the curve f to the left of x_0, as indicated in Figure 3.8.

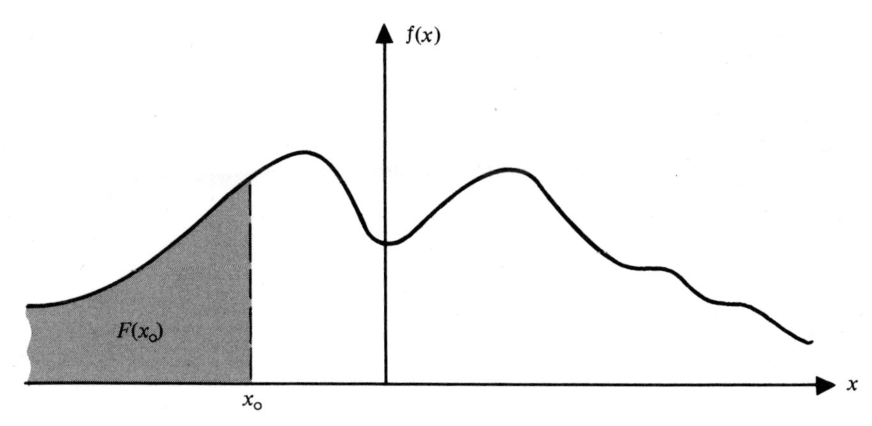

Figure 3.8

Example 3.11. Find the D.F. of an r.v. if its pdf f is given by

$$f(x) = \begin{cases} \frac{3}{8}x^2, & 0 < x < 2 \\ 0, & \text{elsewhere} \end{cases}$$

Solution. The pdf f is sketched in Figure 3.9. We see that: if $t < 0$, then the area to the left of t is zero; if $0 \leq t < 2$, then the area to the left of t is equal to the shaded area as shown in the figure; and if $t \geq 2$, then the area to the left of t is the entire area under the pdf and is equal to 1. In terms of integrals, we have

$$F(t) = \int_{-\infty}^{t} f(x)\,dx$$

$$= \begin{cases} \int_{-\infty}^{t} 0\,dx \quad \text{if } t < 0 & \text{(since the function } f \text{ is 0 to the left of 0)} \\ \int_{-\infty}^{0} 0\,dx + \int_0^t \frac{3}{8}x^2\,dx \quad \text{if } 0 \leq t < 2 & \text{(since } f(x) \text{ is equal to } \frac{3}{8}x^2\text{, if } x \text{ is between 0 and 2)} \\ \int_{-\infty}^{0} 0\,dx + \int_0^2 \frac{3}{8}x^2\,dx + \int_2^t 0\,dx \quad \text{if } t \geq 2 & \text{(since the function } f \text{ is zero for } x \geq 2) \end{cases}$$

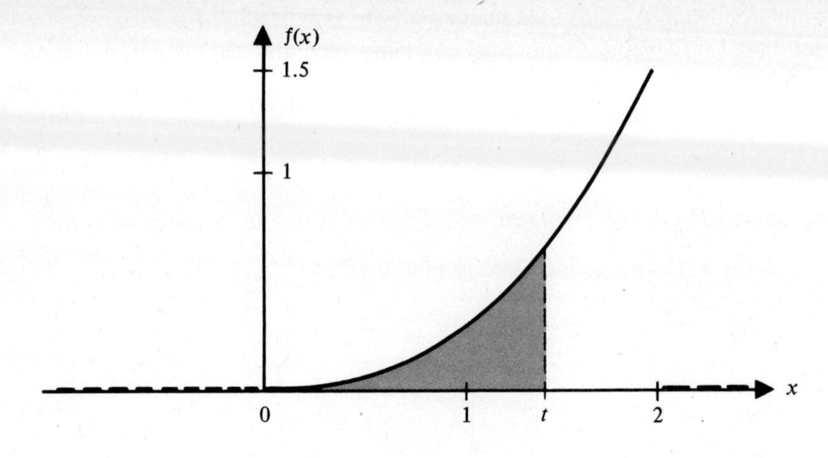

Figure 3.9

Hence

$$F(t) = \begin{cases} 0, & t < 0 \\ \dfrac{t^3}{8}, & 0 \leqslant t < 2 \\ 1, & t \geqslant 2 \end{cases}$$

Important. The D.F. must be defined for every real number. To say that $F(t) = t^3/8$ in the above example would not be sufficient.

Example 3.12. The pdf of an r.v. X is given by

$$f(x) = \begin{cases} \frac{1}{2}, & -1 < x < 1 \\ 0, & \text{elsewhere} \end{cases}$$

Find the D.F. of X.

Solution. The graph of the pdf is given in Figure 3.10. To find the D.F. of X we shall make use of the geometric shape of the graph.

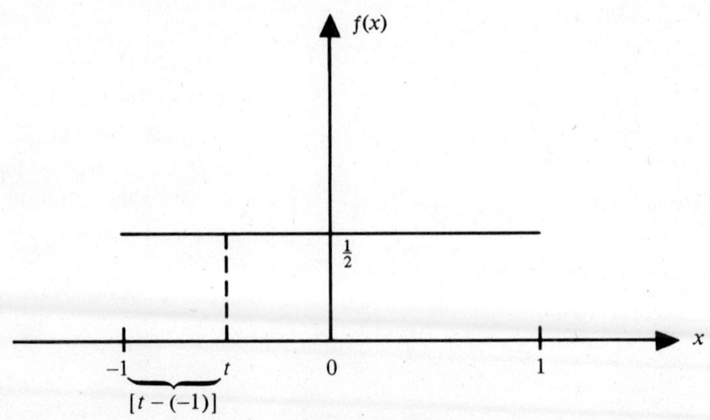

Figure 3.10

If $t < -1$, then the area to the left of t is 0; if $-1 \leqslant t < 1$, then the area to the left of t is $[t - (-1)]\frac{1}{2} = \frac{1}{2}(t + 1)$; and if $t \geqslant 1$, then the area to the left of t is 1.

Therefore,

$$F(t) = \begin{cases} 0, & t < -1 \\ \frac{1}{2}(t+1), & -1 \leqslant t < 1 \\ 1, & t \geqslant 1 \end{cases}$$

The D.F. is plotted in Figure 3.11.

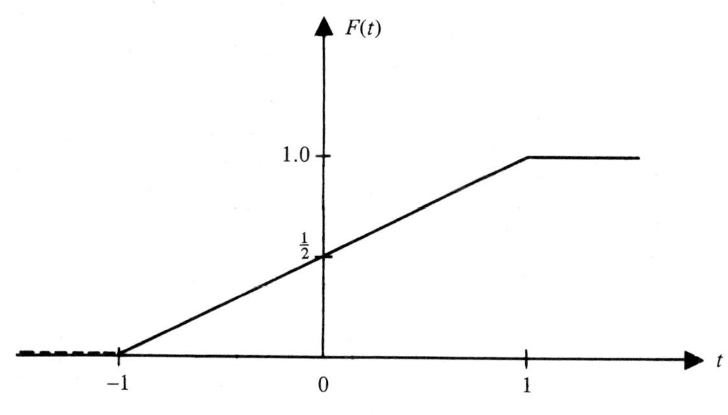

Figure 3.11

Example 3.13. If the pdf of an r.v. is given by

$$f(x) = \begin{cases} \frac{1}{2}\cos x, & -\pi/2 < x < \pi/2 \\ 0, & \text{elsewhere} \end{cases}$$

find the D.F.

Solution. We have

$$F(u) = \int_{-\infty}^{u} f(x)\,dx$$

$$= \begin{cases} \int_{-\infty}^{u} 0\,dx, & u < -\pi/2 \\ \int_{-\infty}^{-\pi/2} 0\,dx + \int_{-\pi/2}^{u} \frac{1}{2}\cos x\,dx, & -\pi/2 \leqslant u < \pi/2 \\ \int_{-\infty}^{-\pi/2} 0\,dx + \int_{-\pi/2}^{\pi/2} \frac{1}{2}\cos x\,dx + \int_{\pi/2}^{u} 0\,dx, & u \geqslant \pi/2 \end{cases}$$

Hence, integrating,

$$F(u) = \begin{cases} 0, & u < -\pi/2 \\ \frac{1}{2}[1 + \sin u], & -\pi/2 \leqslant u < \pi/2 \\ 1, & u \geqslant \pi/2 \end{cases}$$

Example 3.14. An r.v. has the following pdf f:

$$f(x) = \begin{cases} \frac{1}{2}, & -1 < x \leqslant 0 \\ \frac{1}{4}(2-x), & 0 < x < 2 \\ 0, & \text{elsewhere} \end{cases}$$

(*a*) Find the D.F.

(*b*) Find:

(*i*) $P(-\frac{1}{2} < X \leqslant 1)$ (*ii*) $P(X > 1)$ (*iii*) $P(X > -\frac{1}{2} | X \leqslant 1)$

Solution

(*a*) Following the argument of the previous examples,

$$F(u) = \int_{-\infty}^{u} f(x)\,dx$$

$$= \begin{cases} \int_{-\infty}^{u} 0\,dx, & u < -1 \\ \int_{-\infty}^{-1} 0\,dx + \int_{-1}^{u} \frac{1}{2}\,dx, & -1 \leqslant u < 0 \\ \int_{-\infty}^{-1} 0\,dx + \int_{-1}^{0} \frac{1}{2}\,dx + \int_{0}^{u} \frac{1}{4}(2-x)\,dx, & 0 \leqslant u < 2 \\ \int_{-\infty}^{-1} 0\,dx + \int_{-1}^{0} \frac{1}{2}\,dx + \int_{0}^{2} \frac{1}{4}(2-x)\,dx + \int_{2}^{u} 0\,dx, & u \geqslant 2 \end{cases}$$

$$= \begin{cases} 0, & u < -1 \\ \dfrac{1}{2}(1+u), & -1 \leqslant u < 0 \\ \dfrac{1}{2} + \dfrac{u}{4}\left(2 - \dfrac{u}{2}\right), & 0 \leqslant u < 2 \\ 1, & u \geqslant 2 \end{cases}$$

(*b*) (*i*) $P(-\frac{1}{2} < X \leqslant 1) = F(1) - F(-\frac{1}{2}) = [\frac{1}{2} + \frac{1}{4}(2 - \frac{1}{2})] - \frac{1}{2}(1 - \frac{1}{2}) = \frac{5}{8}$.

(*ii*) $P(X > 1) = 1 - F(1) = 1 - [\frac{1}{2} + \frac{1}{4}(2 - \frac{1}{2})] = \frac{1}{8}$.

(*iii*) $P(X > -\frac{1}{2} | X \leqslant 1) = \dfrac{P(-\frac{1}{2} < X \leqslant 1)}{P(X \leqslant 1)} = \dfrac{\frac{5}{8}}{\frac{7}{8}} = \frac{5}{7}$.

3.3 Mixed Distributions

The random variables that we encounter in practice are usually either discrete or absolutely continuous. However, there are distributions which combine the features of the two. Such distributions are called **mixed distributions**. See, for instance, the D.F. whose graph is given in Figure 3.12.

Here there is a positive probability that $X = -1$ and that $X = 2$. The rest of the probability mass namely $1 - P(X = -1) - P(X = 2)$ is spread smoothly along the remaining portion of the x-axis.

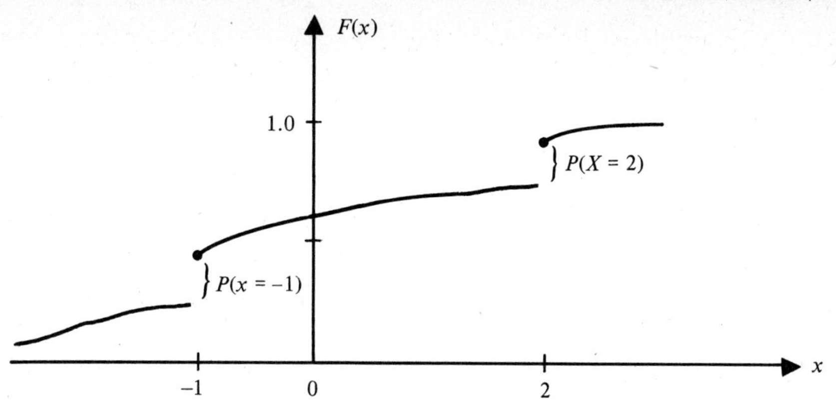

Figure 3.12

Example 3.15. Consider the following D.F. F:

$$F(x) = \begin{cases} 0, & x < 0 \\ x, & 0 \leqslant x < \frac{1}{2} \\ \frac{3}{4}, & \frac{1}{2} \leqslant x < 1 \\ \frac{1}{8}x + \frac{6}{8}, & 1 \leqslant x < 2 \\ 1, & x \geqslant 2 \end{cases}$$

(*a*) Plot the D.F.

(*b*) Find:

(*i*) $P(X = \frac{1}{2})$ (*ii*) $P(0.5 < X \leqslant 1.5)$ (*iii*) $P(0.5 \leqslant X \leqslant 1.5)$

(*iv*) $P(\frac{1}{4} < X \leqslant \frac{3}{4})$ (*v*) $P(X > \frac{3}{4} \mid X \leqslant 1.5)$

Solution

(*a*) The graph of the D.F. is given in Figure 3.13.

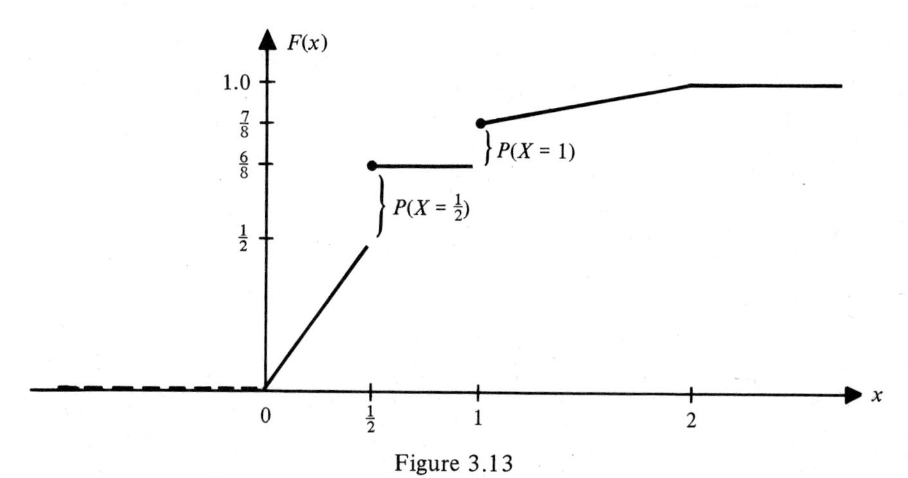

Figure 3.13

(*b*) (*i*) From the graph, we see that $P(X = \frac{1}{2}) = \frac{1}{4}$. We can get this answer without the aid of the graph by simply noting that $P(X = \frac{1}{2}) = F(\frac{1}{2}) - F(\frac{1}{2}^-) = \frac{3}{4} - \lim_{x \to 1/2} x = \frac{1}{4}$.

(*ii*) $P(0.5 < X \leqslant 1.5) = F(1.5) - F(0.5) = [\frac{1}{8}(1.5) + \frac{6}{8}] - \frac{3}{4} = \frac{3}{16}$.

(*iii*) $P(0.5 \leqslant X \leqslant 1.5) = [F(1.5) - F(0.5)] + P(X = 0.5) = \frac{3}{16} + \frac{1}{4} = \frac{7}{16}$.

(*iv*) $P(\frac{1}{4} < X \leqslant \frac{3}{4}) = F(\frac{3}{4}) - F(\frac{1}{4}) = \frac{3}{4} - \frac{1}{4} = \frac{1}{2}$.

(*v*)
$$P(X > \tfrac{3}{4} | X \leqslant 1.5) = \frac{P(\{X > \frac{3}{4}\} \cap \{X \leqslant 1.5\})}{P(X \leqslant 1.5)}$$
$$= \frac{P(\frac{3}{4} < X \leqslant 1.5)}{F(1.5)} = \frac{F(1.5) - F(\frac{3}{4})}{F(1.5)}$$
$$= \frac{[\frac{1}{8}(1.5) + \frac{6}{8}] - \frac{3}{4}}{\frac{1}{8}(1.5) + \frac{6}{8}} = \frac{1}{5}$$

Example 3.16. Suppose the D.F. of a random variable is given as follows:

$$F(x) = \begin{cases} 0, & x < 0 \\ 1 - (\frac{1}{2})^{x+1} - (\frac{1}{2})^{[x]+1}, & x \geqslant 0 \end{cases}$$

(*a*) Make a general sketch of the D.F.

(*b*) Find:

(*i*) $P(1 < X \leqslant 3.5)$ (*ii*) $P(1 < X < 4)$ (*iii*) $P(1 \leqslant X < 4)$

(*iv*) $P(1 \leqslant X \leqslant 4)$ (*v*) $P(1.5 \leqslant X < 4.5)$

Solution

(*a*) We see that

if $0 \leqslant x < 1$, $[x] = 0$ and therefore $F(x) = 1 - (\frac{1}{2})^{x+1} - \frac{1}{2}$

if $1 \leqslant x < 2$, $[x] = 1$ and therefore $F(x) = 1 - (\frac{1}{2})^{x+1} - (\frac{1}{2})^2$

Continuing the argument, for any positive integer i,

if $i \leqslant x < i + 1$, $[x] = i$ and therefore $F(x) = 1 - (\frac{1}{2})^{x+1} - (\frac{1}{2})^{i+1}$

We make the observation that F is not continuous at $i = 1, 2, \ldots$. It is continuous everywhere else. We have, moreover,

$$\begin{aligned} P(X = i) &= F(i) - F(i^-) \\ &= [1 - (\tfrac{1}{2})^{i+1} - (\tfrac{1}{2})^{i+1}] - \lim_{x \to i^-} [1 - (\tfrac{1}{2})^{x+1} - (\tfrac{1}{2})^{[x]+1}] \\ &= [1 - (\tfrac{1}{2})^{i}] - [1 - (\tfrac{1}{2})^{i+1} - (\tfrac{1}{2})^{(i-1)+1}] \\ &= (\tfrac{1}{2})^{i+1} \end{aligned}$$

Hence,

$$P(X = i) = (\tfrac{1}{2})^{i+1}, \qquad i = 1, 2, \ldots$$

We note that

$$\sum_{i=1}^{\infty} P(X = i) = \sum_{i=1}^{\infty} (\tfrac{1}{2})^{i+1} = \tfrac{1}{2}$$

Therefore, the remaining probability mass, namely $\frac{1}{2}$, is distributed smoothly along the x-axis. The D.F. is sketched in Figure 3.14.

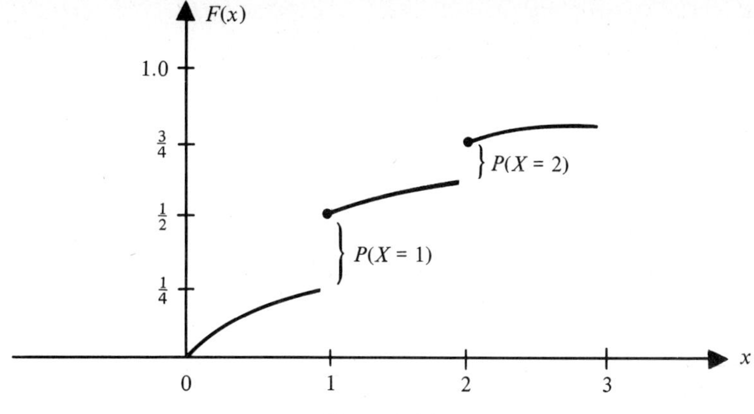

Figure 3.14

(b) (i) $P(1 < X \leqslant 3.5) = F(3.5) - F(1)$
$= [1 - (\frac{1}{2})^{3.5+1} - (\frac{1}{2})^{3+1}] - [1 - (\frac{1}{2})^{1+1} - (\frac{1}{2})^{1+1}]$
$= \frac{7}{16} - (\frac{1}{2})^{4.5}$

(ii) $P(1 < X < 4) = [F(4) - F(1)] - P(X = 4)$
$= [(1 - (\frac{1}{2})^{4+1} - (\frac{1}{2})^{4+1}) - (1 - (\frac{1}{2})^{1+1} - (\frac{1}{2})^{1+1})] - (\frac{1}{2})^{4+1}$
$= \frac{13}{32}$

(iii) $P(1 \leqslant X < 4) = P(1 < X < 4) + P(X = 1)$
$= \frac{13}{32} + (\frac{1}{2})^{1+1} = \frac{21}{32}$

(iv) $P(1 \leqslant X \leqslant 4) = P(1 \leqslant X < 4) + P(X = 4)$
$= \frac{21}{32} + (\frac{1}{2})^{4+1} = \frac{11}{16}$

(v) $P(1.5 \leqslant X < 4.5) = P(1.5 < X \leqslant 4.5) + P(X = 1.5) - P(X = 4.5)$
$= F(4.5) - F(1.5)$, since $P(X = 1.5) = P(X = 4.5) = 0$
$= [1 - (\frac{1}{2})^{4.5+1} - (\frac{1}{2})^{4+1}] - [1 - (\frac{1}{2})^{1.5+1} - (\frac{1}{2})^{1+1}]$
$= \frac{7}{32} - (\frac{1}{2})^{5.5} + (\frac{1}{2})^{2.5}$

3.4 Singular Distributions

We shall now briefly discuss distribution functions which are only of theoretical interest and do not arise in practical applications. Specifically, we shall consider random variables whose distribution functions are continuous, but whose derivative is zero "just about everywhere." Such distributions are called **singular distributions.** (We do not propose to define mathematically what is meant by "just about everywhere" and shall avoid all of the mathematical abstruseness. The reader should encounter no difficulty in understanding the spirit of the discussion.) Since the derivative is zero just about everywhere, it is impossible to recover F by writing $F(x) = \int_{-\infty}^{x} f(u)\, du$. It is this feature that distinguishes singular distributions from those that are absolutely continuous.

Consider a D.F. whose graph is partially indicated in Figure 3.15.

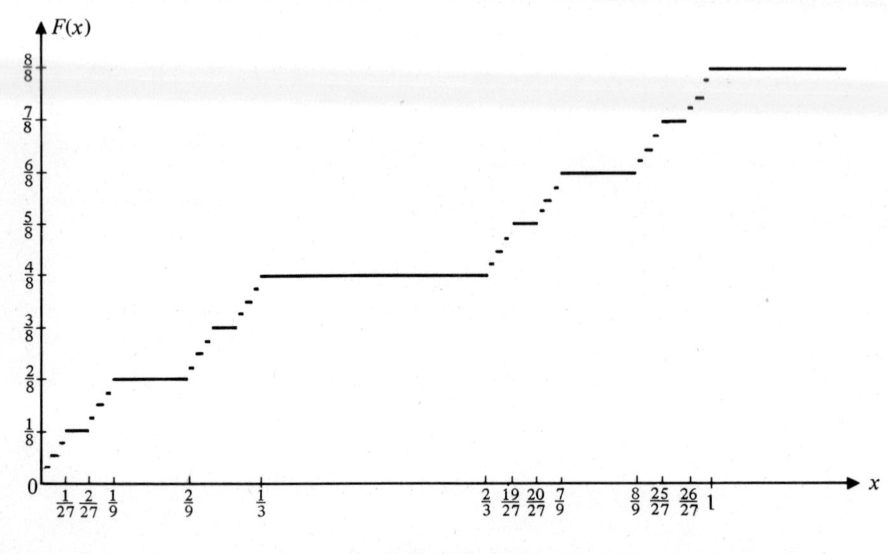

Figure 3.15

The D.F. is $\frac{1}{2}$ on the middle one-third of $[0, 1]$, namely $(\frac{1}{3}, \frac{2}{3})$, $\frac{3}{4}$ on the middle one-third of $(\frac{2}{3}, 1)$, $\frac{1}{4}$ on the middle one-third of $(0, \frac{1}{3})$, and so on. If we continue the pattern of drawing the graph as indicated in the figure, it is possible to obtain a non-decreasing function which is continuous everywhere and satisfies all the criteria for a D.F. (The reader might have to accept this on faith.)

We make the following observations:

(*i*) We see that $F(1) = 1$, and $F(0) = 0$ so that there is a definite increase in F as x goes from 0 to 1. Let us find where this increase might have taken place.

F does not increase in $(\frac{1}{3}, \frac{2}{3})$, or in $(\frac{1}{9}, \frac{2}{9}) \cup (\frac{7}{9}, \frac{8}{9})$, or in $(\frac{1}{27}, \frac{2}{27}) \cup (\frac{7}{27}, \frac{8}{27}) \cup (\frac{19}{27}, \frac{20}{27}) \cup (\frac{25}{27}, \frac{26}{27})$, and so on. (The lengths of these regions are respectively $\frac{1}{3}, \frac{2}{3^2}, \frac{4}{3^3}$, and so on.) Thus F does not increase over a subset of $[0, 1]$ whose length is $\frac{1}{3} + \frac{2}{3^2} + \frac{4}{3^3} + \ldots$. But this series converges to 1. Thus, *the increase in F from a low of 0 to a high of 1 takes place on a set whose length is zero.*

(*ii*) The derivative of F is zero in $(\frac{1}{3}, \frac{2}{3})$, $(\frac{1}{9}, \frac{2}{9}) \cup (\frac{7}{9}, \frac{8}{9})$, and so on. Thus, arguing as in (*i*), the derivative of F is zero on a subset of $[0, 1]$ whose length is 1. That is, *the derivative of F is zero just about everywhere on [0, 1].*

On the basis of (*ii*), the continuous distribution function F is *singular,* since the derivative of F exists and is equal to zero "just about everywhere."

We do not intend to pursue the topic on singular distributions any further at this stage. We shall mention them again, briefly, when we discuss joint distributions.

We conclude this chapter having introduced the concepts of random variable, distribution function, probability (mass) function, and probability density function. The probability distribution, or, simply, the distribution, of a random variable can be described completely by giving its distribution function (whether the random variable is discrete or continuous or of any other type). The distribution can also be completely specified for a discrete random variable by giving the probability function, and for a continuous random variable by giving the probability density function.

EXERCISES–SECTION 3

1. A number is picked at random from the set of integers $\{1, 2, 3, \ldots, 100\}$. Suppose X represents the remainder after dividing the number by 7. Find the distribution of X.
2. Suppose the sample space S consists of positive integers and $P(\{i\}) = 1/2^i$, $i = 1, 2, \ldots$. For any positive integer i, let $X(i)$ represent the remainder when i is divided by 10. Find the distribution of X.
3. Let X be a random variable with the following probability function:

x	-5	-3	-1	0	1	2	3	8
$P(X = x)$	0.2	0.1	0.15	0.05	0.1	0.2	0.15	0.05

Find:
(a) $P(X \text{ is even})$ (b) $P(X \text{ is a multiple of } 3)$
(c) $P(|X| < 3)$ (d) $P(X \leqslant -3 \mid X \leqslant 0)$

4. A box contains three defective and four nondefective items. If the items are picked one by one at random, find the distribution of X, where X represents the number of items that need to be picked to remove all the defective items.
5. A number is picked at random in the interval $[0, 0.5)$. If X denotes the first digit of the number picked when expressed in its decimal expansion, find the distribution of X. *Hint:* Any number in the interval $[0, 0.5)$ can be expressed as $0.x_1x_2\ldots$. What are the possible values of x_1?
6. Answer exercise 5 if we consider the interval $[0, 0.46)$.
7. Determine the constant k in each of the following cases if the probability function of a random variable X is given by—

(a) $p(x) = k(x-2), \quad x = 3, 4, 5, 6$
(b) $p(x) = k\,|x-2|, \quad x = -1, 0, 1, 3$
(c) $p(x) = k\dfrac{2^x}{x!}, \quad x = 0, 1, 2, \ldots$

8. Suppose a random variable has the probability function p given in each part below. In each case, find the D.F. of X and sketch its graph.

(a) $p(x) = 1, \quad x = 5$

(b)
$$p(x) = \begin{cases} \frac{1}{4}, & x = -2 \\ \frac{1}{2}, & x = 1 \\ \frac{1}{4}, & x = 4 \end{cases}$$

(c) $p(x) = \dfrac{x-2}{10}, \quad x = 3, 4, 5, 6$

(d) $p(x) = \dfrac{|x-2|}{7}, \quad x = -1, 0, 1, 3$

9. The number X of pies sold by a grocery store on any day is a random variable with the following probability function:

$$P(X = x) = \begin{cases} cx, & x \in \{0, 1, \ldots, 20\} \\ c, & x \in \{21, 22, \ldots, 40\} \\ c(60 - x), & x \in \{41, 42, \ldots, 60\} \end{cases}$$

Find:

(a) the constant c

(b) the probability of selling at most 20 pies

(c) the probability of selling at least 35 pies

10. Suppose X is a random variable whose probability function is given by

$$P(X = 0) = \frac{x+2}{3}, \quad P(X = 3) = \frac{2x+1}{4}, \quad \text{and} \quad P(X = 5) = \frac{1 - 10x}{12}$$

Determine for what values of x the above is a probability function.

11. Find the probability function of a random variable X if it has the following distribution function:

$$F(x) = \begin{cases} 0, & x < -5 \\ \frac{1}{12}, & -5 \leqslant x < -1 \\ \frac{1}{6}, & -1 \leqslant x < 1.5 \\ \frac{1}{3}, & 1.5 \leqslant x < 3.5 \\ 1, & x \geqslant 3.5 \end{cases}$$

12. Suppose X is a discrete random variable with the following probability function:

$$P(X = i) = \frac{C}{(i+1)(i+2)}, \quad i = 1, 2, \ldots$$

where C is a constant.

(a) Determine C.

(b) Find the D.F. of X.

(c) Find $P(r \leqslant X \leqslant s)$, where r, s are positive integers.

13. The equilateral triangle of side length a and the semicircle of radius r given in Figure 3.16 are the graphs of pdf's. Determine a and r.

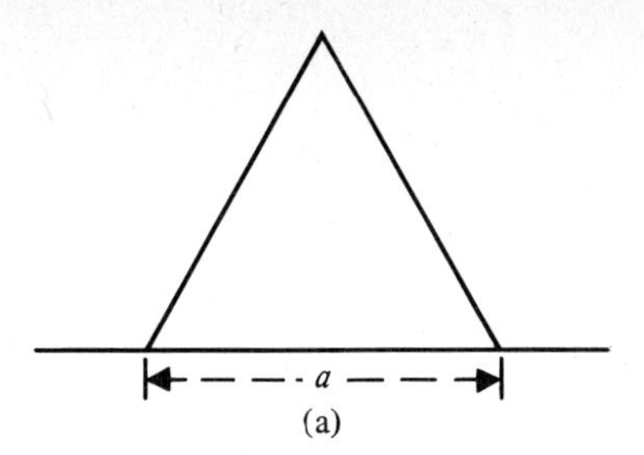

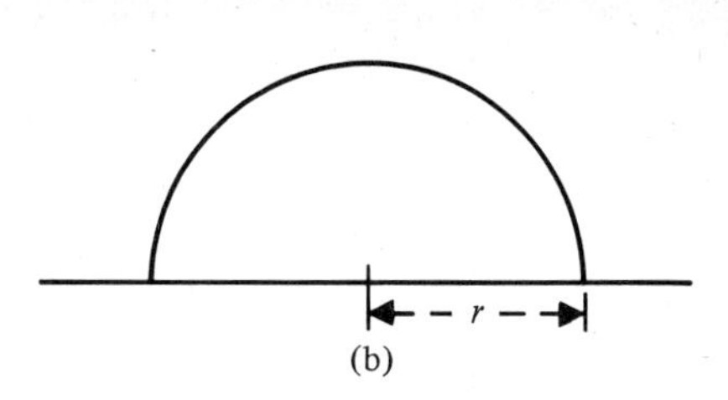

Figure 3.16

14. Indicate which of the functions given below represent genuine pdf's.

(a)
$$f(x) = \begin{cases} \dfrac{\ln x}{\int_{1/2}^{3} \ln x \, dx}, & \frac{1}{2} < x < 3 \\ 0, & \text{elsewhere} \end{cases}$$

(b)
$$f(x) = \begin{cases} \frac{1}{2} \cos x, & -\pi/2 < x < \pi/2 \\ 0, & \text{elsewhere} \end{cases}$$

(c)
$$f(x) = \begin{cases} \sin x, & -\pi/2 < x < \pi \\ 0, & \text{elsewhere} \end{cases}$$

(d)
$$f(x) = \begin{cases} \frac{1}{2} |\sin x|, & -\pi/2 < x < \pi/2 \\ 0, & \text{elsewhere} \end{cases}$$

(e)
$$f(x) = \begin{cases} \frac{1}{5} |x|, & -1 < x < 4 \\ 0, & \text{elsewhere} \end{cases}$$

15. Determine the constant k in each of the following functions f, where each f is the pdf of an r.v. with a continuous distribution.

(a)
$$f(x) = \begin{cases} kx^2, & 0 < x < 2 \\ 0, & \text{elsewhere} \end{cases}$$

(b)
$$f(x) = \begin{cases} k|x|, & -1 < x < 2 \\ 0, & \text{elsewhere} \end{cases}$$

(c)
$$f(x) = \begin{cases} k(1-x)^2, & 0 < x < 1 \\ kx^2, & 2 < x < 3 \\ 0, & \text{elsewhere} \end{cases}$$

16. If f and g are two pdf's, show that $af + bg$ is also a pdf, if a and b are nonnegative real numbers with $a + b = 1$.

17. Let f be the pdf of a continuous r.v. X. In each of the following cases, obtain the D.F. of X and plot its graph.

(a)
$$f(x) = \begin{cases} \frac{3}{8}x^2, & 0 \leqslant x \leqslant 2 \\ 0, & \text{elsewhere} \end{cases}$$

(b)
$$f(x) = \begin{cases} \frac{2}{5}|x|, & -1 < x < 2 \\ 0, & \text{elsewhere} \end{cases}$$

(c)
$$f(x) = \begin{cases} \frac{1}{x^2}, & x \geqslant 1 \\ 0, & \text{elsewhere} \end{cases}$$

(d)
$$f(x) = \begin{cases} \frac{1}{4}, & -1 < x < 1 \\ \frac{1}{6}, & 2 < x < 5 \\ 0, & \text{elsewhere} \end{cases}$$

18. In each of the following cases, find the pdf of X from the given distribution function.

(a)
$$F(x) = \begin{cases} 0, & x < 0 \\ \frac{x}{2}, & 0 \leqslant x < 1 \\ 1 - \frac{1}{2x}, & x \geqslant 1 \end{cases}$$

(b)
$$F(x) = \begin{cases} 0, & x < 0 \\ (1 - e^{-x})^2, & x \geqslant 0 \end{cases}$$

(c)
$$F(x) = \begin{cases} 0, & x < 0 \\ 140\left(\frac{x^4}{4} - \frac{3x^5}{5} + \frac{x^6}{2} - \frac{x^7}{7}\right), & 0 \leqslant x < 1 \\ 1, & x \geqslant 1 \end{cases}$$

(d)
$$F(x) = \begin{cases} 0, & x < 0 \\ 1 - \frac{1}{1+x}, & x \geqslant 0 \end{cases}$$

(e)
$$F(x) = \begin{cases} 0, & x < 0 \\ x - x \ln x, & 0 \leqslant x < 1 \\ 1, & x \geqslant 1 \end{cases}$$

19. Determine the median of the distribution of X if the pdf of X is given by

$$f(x) = \begin{cases} c(1-x), & 0 < x < 1 \\ 0, & \text{elsewhere} \end{cases}$$

20. If X has a continuous distribution with the pdf

$$f(x) = \begin{cases} 3x^2 e^{-x^3}, & x > 0 \\ 0 & \text{elsewhere} \end{cases}$$

find:

(a) $P(X > 2)$ (b) $P(X = 0)$ (c) $P(X = 2)$
(d) $P(1 \leqslant X \leqslant 2)$ (e) $P(X > 2 \mid 1 < X < 3)$

21. A random variable X with a mixed distribution has the following D.F.:

$$F(x) = \begin{cases} 0, & x < -1 \\ k(x+2), & -1 \leqslant x < 1 \\ 1, & x \geqslant 1 \end{cases}$$

If it is given that $P(X = 1) = \frac{1}{3}$, determine:

(a) the constant k (b) $P(X = -1)$
(c) $P(-1 \leqslant X < 1)$ (d) $P(-1 < X < 1)$
(e) $P(X \geqslant -1)$ (f) $P(X > -1)$
(g) $P(X \geqslant \frac{1}{2} | X > \frac{1}{3})$ (h) $P(X \geqslant \frac{1}{2} | X > -1)$

22. The length of time (measured in hours) that it takes a person to reach his office has a *triangular distribution* with the following pdf:

$$f(x) = \begin{cases} \frac{32}{3}(2x - 1), & \frac{1}{2} < x \leqslant \frac{3}{4} \\ \frac{32}{3}(-4x + \frac{7}{2}), & \frac{3}{4} < x < \frac{7}{8} \end{cases}$$

If the person starts from home at 8:20, what is the probability that he will be late at the office, which opens at 9:00?

23. A point is picked in the interval [0, 3] according to the law specified by the following pdf:

$$f(x) = \begin{cases} x, & 0 \leqslant x < 1 \\ \frac{1}{4}(3 - x), & 1 \leqslant x < 3 \\ 0, & \text{elsewhere} \end{cases}$$

Find the probability that the ratio of the short segment to the large segment is less than $\frac{1}{3}$.

24. A random variable X has the following pdf:

$$f(x) = \begin{cases} (0.003)x^2, & 0 < x < 10 \\ 0, & \text{elsewhere} \end{cases}$$

Find the probability that the roots of the quadratic equation in t, $t^2 + 2Xt + 5X - 4 = 0$, are real. *Hint:* The roots are real if the discriminant is positive.

25. The *logistic distribution* is defined as one whose distribution function is given by

$$F(x) = [1 + e^{-(ax + b)}]^{-1}, \quad -\infty < x < \infty$$

$(a > 0, -\infty < b < \infty)$. Show that its pdf is given by

$$f(x) = aF(x)[1 - F(x)]$$

26. Show that the Weibull distribution given by the D.F.

$$F(x) = \begin{cases} 0, & x < 0 \\ 1 - e^{-\lambda x^a}, & x \geqslant 0 \end{cases}$$

where $\lambda > 0$ and $a > 0$, may be written as

$$F(x) = \begin{cases} 0, & x < 0 \\ 1 - 2^{-(x/m)^a}, & x \geqslant 0 \end{cases}$$

where m represents the unique median. *Hint:* $e^{-\lambda m^a} = \frac{1}{2}$.

5 Some Special Distributions

INTRODUCTION

In Chapter 4 we discussed at length the basic properties of random variables and their distributions. We now propose to study some special random variables which find frequent applications in physical problems. We shall first study certain discrete distributions. Of these, the binomial, the hypergeometric, and the Poisson should be considered the principal ones. Of the continuous distributions, the normal distribution unquestionably enjoys a place of total prominence.

1. DISCRETE DISTRIBUTIONS

1.1 Bernoulli Distribution

A trial is said to be a *Bernoulli trial* if its performance entails two possible outcomes. One of the outcomes is called a "success"; the other a "failure."

For example:

(*i*) When a coin is tossed, either heads shows up, or tails.

(*ii*) If a basketball player attempts a free throw, either he makes the basket or he does not.

(*iii*) On any birth, as far as the sex of the child is concerned, it is either a boy or a girl.

If we let $X = 1$ when the outcome is a success, and $X = 0$ when the outcome is a failure, then the probability function of X is given by

$$P(X = 0) = 1 - p \quad \text{and} \quad P(X = 1) = p$$

where p, $0 \leqslant p \leqslant 1$, represents the probability that a trial results in a success. The probability function can be expressed compactly as

$$P(X = r) = p^r(1 - p)^{1-r}, \qquad r = 0, 1$$

A random variable with this type of probability function is called a **Bernoulli random variable**, with parameter p. (In general, *parameters* are the constants that characterize a distribution.) The probability function and the distribution function of a Bernoulli random variable are pictured in Figure 1.1(a) and Figure 1.1(b) below.

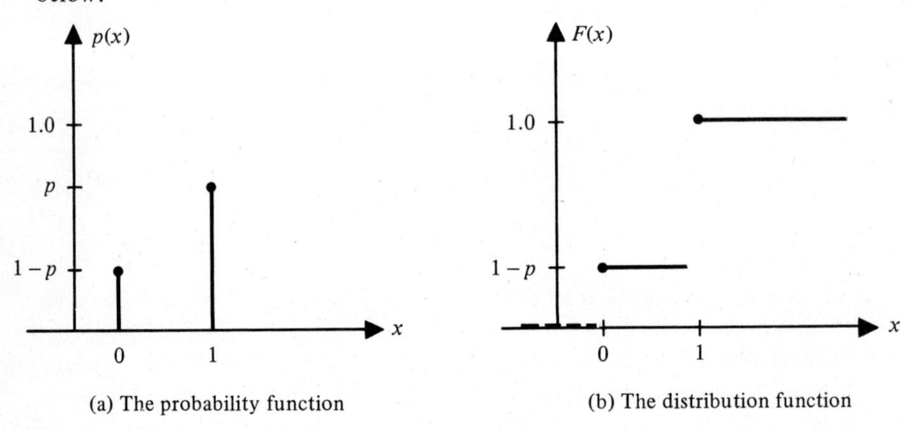

(a) The probability function

(b) The distribution function

Figure 1.1

1.2 The Binomial Distribution

The binomial distribution is by far the most important discrete distribution. It is characterized by the following features:

(*i*) An experiment consists of n independent trials.

(*ii*) On each trial, there are only two possible outcomes—either an event E occurs or it does not. If E occurs on a trial, we call it a "success"; if E does not occur (that is, E' occurs), we call it a "failure."

(*iii*) The probability of the occurrence of E is known (say, equal to p) and *remains the same from trial to trial.*

In short, the experiment consists of n independent and identical Bernoulli trials. *If X represents the number of successes among n trials, then X is said to be a* **binomial random variable** *with parameters n and p.* For brevity, we shall say that "X is binomial n, p" and write "X is $B(n, p)$." Of course, the parameter n which represents the number of trials is predetermined and is under the experimenter's control so that, in essence, p is the more important parameter. As some examples of binomial random variables, we cite: (*a*) the number of heads when a coin is tossed n times, (*b*) the number of baskets a player makes in n free throws, and (*c*) the number of sons among n children in a family.

On any trial, either E occurs, or E' occurs. As a result, the sample space S representing the outcomes of the n trials can be written as a collection of 2^n ordered n-tuples, the ith component being E or E' depending upon whether the ith trial resulted in a success or a failure.

The nature of the mapping by X is shown in Figure 1.2. The possible values of

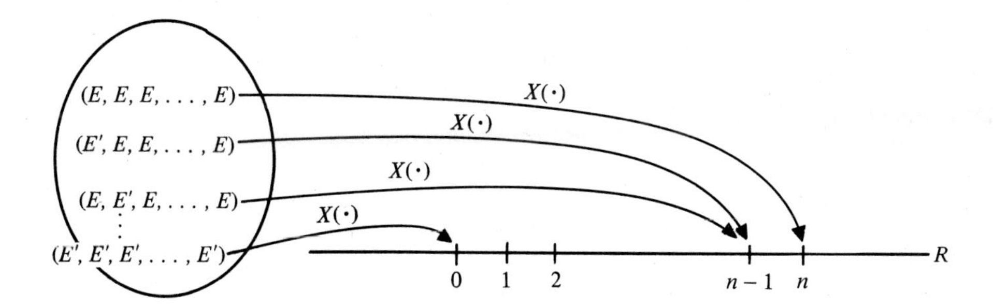

Figure 1.2

X are $k = 0, 1, 2, \ldots, n$, and we want to find $P(X = k)$—that is, the probability of k successes (and, consequently, $n - k$ failures) in n Bernoulli trials.

One of the possible ways of obtaining k successes is to have the outcome

$$s = (\underbrace{E, E, \ldots, E}_{k \text{ trials}}, \underbrace{E', E', \ldots, E'}_{n-k \text{ trials}})$$

where the first k trials are successes and the last $n - k$ are failures. Recall from our discussion on independent trials in Chapter 3 that, if the trials are independent, then the probabilities are assigned to the outcomes of the composite experiment by $P(s_1, s_2, \ldots, s_n) = P_1(s_1)P_1(s_2) \ldots P_1(s_n)$, where P_1 is the probability measure on the outcomes of the basic experiment. Therefore, it follows that

$$\begin{aligned} P(\{s\}) &= P_1(E) \cdot P_1(E) \ldots P_1(E)P_1(E') \ldots P_1(E') \\ &= p^k(1-p)^{n-k} \end{aligned}$$

As a matter of fact, it is easy to see that any sample point with k successes and $n - k$ failures is assigned a probability of $p^k(1-p)^{n-k}$. How many sample points, that is, n-tuples, are there in the sample space which have k successes and $n - k$ failures? This is simply the number of ways of picking k trials out of n so that E occurs on these trials. As we know, this number is equal to $\binom{n}{k}$. Therefore, in conclusion,

$$P(X=k) = \underbrace{\binom{n}{k}}_{\text{the number of outcomes with } k \text{ successes and } n-k \text{ failures}} \underbrace{p^k(1-p)^{n-k}}_{\text{the probability of each outcome with } k \text{ successes and } n-k \text{ failures}}, \qquad k = 0, 1, \ldots, n$$

Note that

(*i*) $P(X=k) \geqslant 0$

(*ii*) $\sum_{k=0}^{n} P(X=k) = \sum_{k=0}^{n} \binom{n}{k} p^k(1-p)^{n-k} = [p + (1-p)]^n = 1$

Hence $\binom{n}{k} p^k(1-p)^{n-k}$, $k = 0, 1, \ldots, n$, indeed represents a genuine probability function.

In summary, *the number of successes, X, in n independent trials, with constant probability p of success at each trial, has the binomial distribution with*

$$P(X=k) = \binom{n}{k} p^k(1-p)^{n-k}, \qquad k = 0, 1, 2, \ldots, n$$

Example 1.1. Two fair dice are rolled ten times. What is the probability that a total of 9 or 7 is rolled k times, where $k = 0, 1, 2, \ldots, 10$.

Solution. We have 10 independent trials with

E = Total of 9 or 7 on two dice

Now the probability of getting a total of 9 or 7 on two dice is equal to $\frac{10}{36} = \frac{5}{18}$. Hence $p = \frac{5}{18}$. Letting

X = Number of times a total of 9 or 7 is rolled in ten attempts

we have

$$P(X=k) = \binom{10}{k}\left(\frac{5}{18}\right)^k\left(1 - \frac{5}{18}\right)^{10-k}, \qquad k = 0, 1, \ldots, 10$$

Example 1.2. The probability of hitting a target on a shot is $\frac{2}{3}$. If a person fires 8 shots at a target, let X denote the number of times he hits the target, and find:

(*a*) $P(X=3)$ (*b*) $P(1 < X \leqslant 6)$ (*c*) $P(X > 3)$

Solution. The probability function of X is given by

$$P(X=k) = \binom{8}{k}\left(\frac{2}{3}\right)^k\left(\frac{1}{3}\right)^{8-k}, \qquad k = 0, 1, \ldots, 8$$

Hence

(*a*) $P(X=3) = \binom{8}{3}\left(\frac{2}{3}\right)^3\left(\frac{1}{3}\right)^5 = 0.0683$

(*b*) $P(1 < X \leqslant 6) = \sum_{k=2}^{6} \binom{8}{k}\left(\frac{2}{3}\right)^k \left(\frac{1}{3}\right)^{8-k} = 0.8023$

(*c*) $P(X > 3) = \sum_{k=4}^{8} \binom{8}{k}\left(\frac{2}{3}\right)^k \left(\frac{1}{3}\right)^{8-k} = 0.912$ ■

The reader will recall from Chapter 3, Section 3, that $P(X = k)$, $k = 0, 1, \ldots, n$, are simply the binomial probabilities $b(k; n, p)$. Extensive tables giving $b(k; n, p)$, $k = 0, 1, 2, \ldots, n$, are available for various values of p and n. We reproduce a short table giving $b(k; n, p)$ for $n = 2, 3, \ldots, 10$ and various values of p as Table B in the Appendix. Notice that Table B lists $b(k; n, p)$ only for values of $p \leqslant 0.5$. For values of $p > 0.5$, the quantities $b(k; n, p)$ can be obtained from the table by observing that $b(k; n, p) = b(n-k; n, 1-p)$. This follows since

$$b(k; n, p) = \binom{n}{k} p^k (1-p)^{n-k} = \binom{n}{n-k}(1-p)^{n-k}[1-(1-p)]^{n-(n-k)}$$
$$= b(n-k; n, 1-p)$$

Thus, for example, $b(6; 10, 0.7) = b(4; 10, 0.3) = 0.2001$.

Example 1.3. Use Table B to answer the following:

(*a*) Find $b(3; 7, 0.9)$.
(*b*) If X is $B(10, 0.3)$, find $P(X \geqslant 6)$.
(*c*) If X is $B(10, 0.8)$, find $P(X \geqslant 6)$.
(*d*) If X is $B(10, 0.7)$, find an integer r such that $P(X \geqslant r) = 0.8497$.

Solution

(*a*) We have $b(3; 7, 0.9) = b(4, 7, 0.1) = 0.0026$.

(*b*) $P(X \geqslant 6) = \sum_{k=6}^{10} b(k; 10, 0.3) = 0.0368 + 0.0090 + 0.0014 + 0.001 = 0.0473$.

(*c*) In this case, since $p > 0.5$, we use the relation $b(k; n, p) = b(n-k; n, 1-p)$. We get

$$P(X \geqslant 6) = \sum_{k=6}^{10} b(k; 10, 0.8) = \sum_{r=0}^{10-6} b(r; 10, 0.2)$$

(Verify this; also see exercise 9 at the end of this section.) Hence

$$P(X \geqslant 6) = 0.1074 + 0.2684 + 0.3020 + 0.2013 + 0.0881 = 0.9672$$

(*d*) We want to find an integer r for which $P(X \geqslant r) = 0.8497$; that is, $\sum_{k=r}^{10} b(k; 10, 0.7) = 0.8497$; that is, $\sum_{k=0}^{10-r} b(k; 10, 0.3) = 0.8497$. Now, from the table we see that $b(0; 10, 0.3) + b(1; 10, 0.3) + b(2; 10, 0.3) + b(3; 10, 0.3) + b(4; 10, 0.3) = 0.8497$. Hence $10 - r = 4$ and, consequently, $r = 6$.

Example 1.4. A drug manufacturing company is debating whether a vaccine is effective enough to be marketed. The company claims that the vaccine is 90 percent effective—that is, when the vaccine is administered to a person, the chance that the person will develop immunity is 0.9. The federal drug agency, however, believes that the claim is exaggerated and that the drug is 40 percent effective. To test the company claim, the following procedure is devised: The vaccine will be tried on ten people. If eight or more people develop immunity, the company claim will be granted. Find the probability that: (*a*) the company claim will be granted incorrectly (that is, when the federal drug agency is correct in its assertion); (*b*) the company claim will be denied incorrectly (that is, when the vaccine is indeed 90 percent effective).

Solution. Let X denote the number of people among the ten who develop immunity. If the company claim is valid, then the distribution of X is binomial with $n = 10, p = 0.9$. On the other hand, if the federal agency claim is valid, then X is binomial with $n = 10, p = 0.4$.

(*a*) The company claim will be granted if and only if $X \geqslant 8$. The probability that the claim will be granted incorrectly is equal to the probability that $X \geqslant 8$ when $p = 0.4$, which is equal to $\sum_{k=8}^{10} b(k; 10, 0.4) = 0.0106 + 0.0016 + 0.0001 = 0.0123$.

(*b*) The company claim will be denied if and only if $X < 8$. The probability that the claim will be denied incorrectly is equal to the probability that $X < 8$ when $p = 0.9$, which is equal to $\sum_{k=0}^{7} b(k; 10, 0.9) = \sum_{k=3}^{10} b(k; 10, 0.1) = 0.0702$.

Example 1.5. The probability that a basketball player scores at least once in six free throws is equal to 0.999936. Find: (*a*) the probability function of X, the number of times he scores; (*b*) the probability that he makes at least three baskets.

Solution. If p represents the probability of making the basket on any throw, then the probability of scoring at least once in 6 free throws is equal to

$$1 - (1 - p)^6$$

(Why?) Hence we are given that

$$1 - (1 - p)^6 = 0.999936$$

That is,

$$(1 - p)^6 = 0.000064$$

Hence $1 - p = 0.2$ and, consequently, $p = 0.8$.

(*a*) The probability function of X is given as

$$P(X = k) = \binom{6}{k}(0.8)^k(0.2)^{6-k}$$

(*b*) Here we want $P(X \geqslant 3)$. We have

$$\begin{aligned} P(X \geqslant 3) &= \sum_{k=3}^{6} \binom{6}{k}(0.8)^k(0.2)^{6-k} \\ &= 0.9830 \end{aligned}$$

using the table. ■

The most probable number

For fixed n and p, the binomial probabilities $b(k; n, p)$ depend on k. We now propose to investigate the behavior of these probabilities as k goes from 0 to n. Towards this we shall use the following identity which can be verified without much difficulty:

$$\frac{b(k; n, p)}{b(k-1; n, p)} = \frac{(n-k+1)p}{k(1-p)} = 1 + \frac{(n+1)p-k}{k(1-p)}$$

Now:

(i) If $k < (n+1)p$, then $\dfrac{(n+1)p-k}{k(1-p)} > 0$, so that $\dfrac{b(k; n, p)}{b(k-1; n, p)} > 1$.
Hence, if $k < (n+1)p$, the terms $b(k; n, p)$ increase with k.

(ii) If $k > (n+1)p$, then $\dfrac{(n+1)p-k}{k(1-p)} < 0$, so that $\dfrac{b(k; n, p)}{b(k-1; n, p)} < 1$.
Consequently, if $k > (n+1)p$, the terms $b(k; n, p)$ decrease with k.

(iii) If $(n+1)p$ is an integer, then $(n+1)p - k = 0$ for some k, say $k = m$.
For such m we then have $\dfrac{b(m; n, p)}{b(m-1; n, p)} = 1$.

In conclusion:

$b(k; n, p)$ increases with k if $k < (n+1)p$ and decreases with k if $k > (n+1)p$. If $(n+1)p$ is an integer, say equal to m, then $b(m-1; n, p) = b(m; n, p)$. *The integral part of the number $(n+1)p$ represents the most probable number of successes.* If $(n+1)p$ is an integer m, the largest value of the probability $b(k; n, p)$ is attained for two integers $m-1$ and m.

For instance:

(a) Suppose $n = 20$ and $p = 0.30$. Then $(n+1)p = 6.3$, so that $b(k; 20, 0.3)$ increases monotonically as k goes from 0 to 6 and then decreases as k goes from 7 to 20.

(b) Suppose $n = 24$ and $p = 0.4$. Since $(n+1)p = 10$, an integer, $b(k; 24, 0.4)$ increases as k goes from 0 to 9 and decreases as k goes from 10 to 24, with $b(9; 24, 0.4) = b(10; 24, 0.4)$.

(c) Consider the graphs of binomial probabilities in Figure 1.3. Figure 1.3(a) corresponds to $n = 10, p = 0.25$; in this case, $(n+1)p = 2.75$ and the maximum value is attained for $k = 2$, the integral part of 2.75. Figure 1.3(b) corresponds to $n = 5, p = 0.50$. Here $(n+1)p = 3.0$, an integer, and the maximum value is attained for $k = 2$ and $k = 3$.

Example 1.6. Thirteen machines are in operation. The probability that, at the end of one day, a machine is still in operation is 0.6. If the machines function independently, find the most probable number of machines in operation at the end of that day and the probability that these many machines are operating.

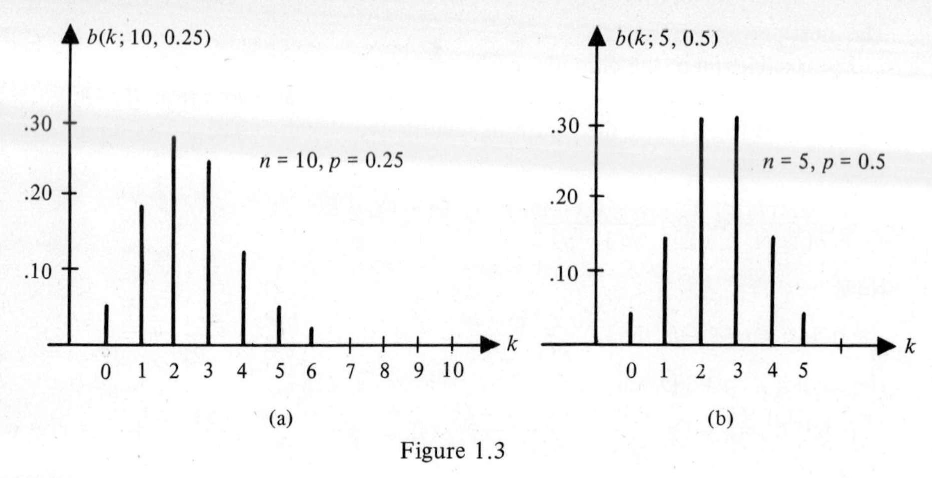

Figure 1.3

Solution. We know that $n = 13$ and $p = 0.6$. Hence $(n + 1)p = 8.4$. The most probable number of machines in operation is the integral part of 8.4, that is, 8. Of course, the probability that 8 machines of the thirteen are in operation is equal to $\binom{13}{8}(0.6)^8(0.4)^5$.

1.3 The Hypergeometric Distribution

The reader is already familiar with the hypergeometric probabilities described in Section 3 of Chapter 2. The setup here is exactly the same. A lot contains M objects, of which D are of one kind, say type I, and $M - D$ of another kind, say type II. Suppose n objects are picked at random without replacement and the random variable X is defined by

$$X = \text{The number of objects of type I}$$

We keep in mind the fact that a sample of n items cannot contain more than n items of type I. Also, since the sample is drawn without replacement, it cannot contain more items of type I than there are in the original lot, namely D. As mentioned in Chapter 2 in this context, these aspects are taken care of in the formula for hypergeometric probabilities by the fact that $\binom{a}{b} = 0$ if $b < 0$ or $b > a$.

We want to find the probability function of X, namely $P(X = k)$, $k = 0, 1, \ldots, n$. Now,

$$X = k \iff \begin{pmatrix}\text{we pick } k \text{ objects of type I} \\ \text{and } n - k \text{ objects of type II}\end{pmatrix}$$

Hence $P(X = k)$ is given by the hypergeometric probabilities as

$$P(X = k) = \frac{\binom{D}{k}\binom{M-D}{n-k}}{\binom{M}{n}}, \qquad k = 0, 1, 2, \ldots, n$$

Let us write

$$p = \frac{D}{M} \quad \text{and} \quad q = \frac{M-D}{M} = 1 - p$$

Thus p denotes the proportion of objects of type I in the original lot. Then

$$P(X = k) = \frac{\binom{Mp}{k}\binom{Mq}{n-k}}{\binom{M}{n}}, \qquad k = 0, 1, 2, \ldots, n$$

This form is often referred to as the probability function of a hypergeometric distribution with parameters M, n, and p.

Using the combinatorial result

$$\sum_{k=0}^{n} \frac{\binom{D}{k}\binom{M-D}{n-k}}{\binom{M}{n}} = 1$$

it follows that we indeed have a genuine probability function.

Comment. As in the binomial case, the experiment consists of n trials. However, there is a basic difference in that the trials are not independent.

Example 1.7. A box contains eight defective and twelve nondefective tubes. Six tubes are picked at random without replacement. Let X = The number of defective tubes picked. Find:

(*a*) the probability function of X
(*b*) the probability of picking no defective tubes
(*c*) the probability of picking at least one defective tube.

Solution

(*a*) The random variable takes integral values 0, 1, 2, 3, 4, 5, 6. (*Note:* If, for example, we had picked 10 tubes, then the random variable would have taken integral values 0, 1, 2, . . . , 6, 7, 8.)

We have

$$P(X = k) = \frac{\binom{8}{k}\binom{12}{6-k}}{\binom{20}{6}}, \qquad k = 0, 1, 2, 3, 4, 5, 6$$

(*b*) We want to find $P(X = 0)$; this is equal to

$$\frac{\binom{8}{0}\binom{12}{6}}{\binom{20}{6}} = \frac{\binom{12}{6}}{\binom{20}{6}}$$

(*c*) The probability of at least one defective tube is

$$P(X \geqslant 1) = \sum_{k=1}^{6} \frac{\binom{8}{k}\binom{12}{6-k}}{\binom{20}{6}}$$

However, an easier way to compute this is to note that

$$P(X \geqslant 1) = 1 - P(X = 0) = 1 - \frac{\binom{12}{6}}{\binom{20}{6}}$$

Example 1.8. A bowl contains M beads of which W are white and $M - W$ are black. Suppose n beads are picked. Let X denote the number of white beads in the sample. Find the distribution of X assuming that–

(*a*) the sample is drawn *without replacement*

(*b*) the sample is drawn *with replacement.*

Solution

(*a*) In this case the distribution of X is clearly hypergeometric and is given by

$$P(X = k) = \frac{\binom{W}{k}\binom{M-W}{n-k}}{\binom{M}{n}}, \qquad k = 0, 1, \ldots, n$$

(*b*) Here n beads are picked with replacement. As a result, we have n independent trials each with the probability of sucess equal to W/M. ("Success" stands for "getting a white bead on a pick.") Hence the distribution of X is binomial with

$$P(X = k) = \binom{n}{k}\left(\frac{W}{M}\right)^k \left(1 - \frac{W}{M}\right)^{n-k}, \qquad k = 0, 1, \ldots, n$$

This result was derived in Chapter 2 (see page 60) using a purely combinatorial argument.

1.4 The Geometric Distribution

The geometric distribution finds applications in situations of the following nature: A person tosses a coin until heads show up for the first time; or a basketball player attempts a basket until he scores one; or a billiards player keeps shooting until he misses a shot. Thus, as an idealization describing these situations, the experiment consists of a sequence of independent Bernoulli trials with probability of success p on any trial, where $0 < p < 1$, and the random variable X represents the number of trials required for the first success to occur. The random variable is commonly called a **geometric random variable**; it is also referred to as the *waiting time for the first success.* It should be realized that—unlike the binomial distribution, where the number of trials is *fixed*—in the present case, the number of trials is the random variable of interest.

The possible values of X are obviously $1, 2, 3, \ldots$, and

$$X = r \iff \begin{pmatrix} \text{the first } r-1 \text{ trials are failures} \\ \textit{and} \text{ the } r\text{th trial is a success} \end{pmatrix}$$

Therefore, since the trials are independent,

$$P(X = r) = (1 - p)^{r-1}p, \qquad r = 1, 2, 3, \ldots$$

The distribution is called the geometric distribution because the terms $p(1 - p)^{r-1}$, $r = 1, 2, 3, \ldots$, represent the successive terms of a geometric series.

Observe that we have a genuine assignment of probabilities because:

(*i*) For $r = 1, 2, 3, \ldots, P(X = r) = (1 - p)^{r-1}p \geqslant 0$.

(*ii*) Since the series $\sum_{r=1}^{\infty} (1 - p)^{r-1}$ is a geometric series with $0 < 1 - p < 1$,

$$\sum_{r=1}^{\infty} P(X = r) = p \sum_{r=1}^{\infty} (1 - p)^{r-1} = p \sum_{s=0}^{\infty} (1 - p)^s$$

$$= \frac{p}{1 - (1 - p)} = 1$$

Example 1.9. In order to attract customers, a grocery store has started a *SAVE* game. Any person who collects all four letters of the word *SAVE* gets a prize. A diligent Mrs. Y who has three letters *S, A,* and *E* keeps going to the store until she gets the fourth letter *V*. The probability that she gets the letter *V* on any visit is 0.002 and remains the same from visit to visit. Let X denote the number of times she visits the store until she gets the letter *V* for the first time. Find:

(*a*) the probability function of X

(*b*) the probability that she gets the letter *V* for the first time on the twentieth visit

(*c*) the probability that she will not have to visit more than three times

Solution

(*a*) The distribution of X is clearly geometric. Since $p = 0.002$, we have

$$P(X = r) = (1 - 0.002)^{r-1}(0.002)$$
$$= (0.998)^{r-1}(0.002), \qquad r = 1, 2, \ldots$$

(*b*) $P(X = 20) = (0.998)^{19}(0.002) = 0.0019$.

(*c*) We want $P(X \leqslant 3) = \sum_{r=1}^{3} (0.002)(0.998)^{r-1} = 0.006$.

1.5 The Negative Binomial Distribution

This distribution is also called the **Pascal distribution**. The experiment consists of independent Bernoulli trials with a constant probability of success p, where $0 < p < 1$, and the random variable X is defined as the number of trials needed for r successes to occur, where r is a fixed positive integer.

We are interested in finding how long it will take for the rth success to occur. Since success has to occur r times, the number of trials would have to be at least r. Hence X takes values $r, r+1, \ldots$. Now

$$X = r + k \iff (r\text{th success occurs on the } (r+k)\text{th trial})$$

that is

$$X = r + k \iff \begin{pmatrix}\text{success occurs } (r-1) \text{ times in the first } r+k-1 \\ \text{trials } \textit{and} \text{ success occurs on the } (r+k)\text{th trial}\end{pmatrix}$$

Therefore,

$$P(X = r + k) = P\begin{pmatrix}\text{success occurs } r-1 \text{ times} \\ \text{in the first } r+k-1 \text{ trials}\end{pmatrix} \cdot P\begin{pmatrix}\text{success occurs on} \\ \text{the } (r+k)\text{th trial}\end{pmatrix}$$

(Why?) The first term on the right represents the probability of $r-1$ successes in $r+k-1$ independent, identical Bernoulli trials. Hence, it follows that

$$\begin{aligned} P(X = r + k) &= \binom{r+k-1}{r-1} p^{r-1}(1-p)^k \cdot p \\ &= \binom{r+k-1}{r-1} p^r (1-p)^k, \quad k = 0, 1, 2, \ldots \end{aligned}$$

Hence, the probability function of X is given by

$$P(X = r + k) = \binom{r+k-1}{r-1} p^r (1-p)^k, \quad k = 0, 1, 2, \ldots$$

There is an alternate way of expressing the above probabilities; this version explains why the random variable is called negative binomial. We have

$$\begin{aligned} \binom{r+k-1}{k} &= \frac{(r+k-1)(r+k-2)\ldots(r+k-1-k+1)}{k!} \\ &= \frac{(r+k-1)(r+k-2)\ldots r}{k!} \\ &= \frac{(-r)(-r-1)\ldots(-r-k+1)}{k!}(-1)^k \\ &= \binom{-r}{k}(-1)^k \end{aligned}$$

Therefore,

$$P(X = r + k) = \binom{-r}{k}(-1)^k p^r (1-p)^k, \quad k = 0, 1, \ldots$$

The distribution is called the negative binomial distribution because the probabilities are given as typical terms of the expansion $p^r[1-(1-p)]^{-r}$ in a binomial series

$$p^r[1-(1-p)]^{-r} = p^r \sum_{k=0}^{\infty} (-1)^k \binom{-r}{k} (1-p)^k$$

Comment. If we take $r = 1$, then we get the geometric distribution as a special case of the negative binomial.

Example 1.10. At the end of the 1973 baseball season, Hammering Hank had scored 713 home runs. From past experience, it is reasonable to assume that the probability that he hits a home run when at bat is 0.1. Find the probability that:

(*a*) Hank would need to be at bat exactly five times to tie Babe Ruth's record

(*b*) Hank would need to be at bat exactly five times to break Ruth's record.

(Babe Ruth's record is 714 home runs.)

Solution

(*a*) Let X represent the number of times needed at bat to tie the record. Assuming independence, X has a geometric distribution with probability function

$$P(X = k) = (0.9)^{k-1}(0.1), \qquad k = 1, 2, \ldots$$

Hence $P(X = 5) = (0.9)^4(0.1) = 0.06561$.

(*b*) Let Y represent the number of times at bat needed to break the record, that is, the number of times needed to score two runs. Once again assuming independence, the distribution of Y is negative binomial, given by

$$P(Y = k) = \binom{k-1}{1}(0.9)^{k-2}(0.1)^2, \qquad k = 2, 3, \ldots$$

Therefore, $P(Y = 5) = \binom{4}{1}(0.9)^3(0.1)^2 = 0.02916$.

1.6 The Poisson Distribution

The Poisson distribution provides a probabilistic model for a wide class of phenomena. Typical examples are the number of telephone calls during a given period of time, the number of particles emitted from a radioactive source in a given period of time, and the number of bacteria on a plate of a given size.

In all the cases cited above, an event occurs repeatedly but in a haphazard way in an interval of time or space. If an event occurs, we shall say that "a change has taken place." Let

$$X_t = \text{The number of changes in an interval } (0, t)$$

Whether we consider the number of changes in time, or in space, or what have you, we shall refer to the interval $(0, t)$ as the *time interval.*

The following assumptions are basic for the distribution of X_t:

(*i*) The numbers of changes in two nonoverlapping intervals are independent.

(*ii*) The probability of the number of changes in an interval depends on the length of the interval, but does not depend upon the location of the interval. For instance, the probability of k changes in the interval $(0, t)$ is the same as the probability of k changes in the interval $(a, a + t)$ for any positive real number a.

(*iii*) The probability of one change in an interval of length h is approximately proportional to h, when h is small. Let λ (lambda) be the constant of proportionality. (Notice that λ has to be positive.)

(*iv*) The probability that two or more changes occur in an interval of length h is extremely small in comparison with the probability of one change, when h is small.

Under these basic assumptions, we can derive the probability law associated with the random variable X_t, the number of changes in the time interval of length t. It can be shown that

$$P(X_t = k) = \frac{e^{-\lambda t}(\lambda t)^k}{k!}, \qquad k = 0, 1, 2, \ldots$$

A random variable with the above probability function is called a **Poisson random variable** with parameter λt. In particular, the probability of no change in an interval of length t is $e^{-\lambda t}$. Thus, for a given interval $(0, t)$, the larger λ is, the smaller the probability of no change is. The reader will find this statement intuitively more meaningful after we attach a physical meaning to the parameter λ in Chapter 7.

If we consider a Poisson random variable over an interval of unit length, then writing X for X_1, we get

$$P(X = k) = \frac{e^{-\lambda}\lambda^k}{k!}, \qquad k = 0, 1, \ldots$$

Note

(*i*) $\dfrac{e^{-\lambda t}(\lambda t)^k}{k!} \geqslant 0, \qquad k = 0, 1, \ldots$

(*ii*)
$$\sum_{k=0}^{\infty} \frac{e^{-\lambda t}(\lambda t)^k}{k!} = e^{-\lambda t} \sum_{k=0}^{\infty} \frac{(\lambda t)^k}{k!}$$
$$= e^{-\lambda t} \cdot e^{\lambda t}, \qquad \text{since } \sum_{r=0}^{\infty} \frac{x^r}{r!} = e^x$$
$$= 1$$

Hence we have a genuine probability function.

Example 1.11. Suppose X has a Poisson distribution over an interval of unit length. If $P(X = 0) = 1/e^2$, find $P(X > 2)$.

Solution. To find the probabilities associated with a Poisson law, we must know the parameter λ. Now,

$$P(X = 0) = \frac{e^{-\lambda}\lambda^0}{0!} = e^{-\lambda}$$

and this is given to be e^{-2}. Therefore, $\lambda = 2$. Consequently,

$$P(X > 2) = \sum_{k=3}^{\infty} \frac{e^{-2}2^k}{k!}$$

or, equivalently,

$$\begin{aligned} P(X > 2) &= 1 - P(X = 0) - P(X = 1) - P(X = 2) \\ &= 1 - e^{-2}\left[1 + \frac{2}{1!} + \frac{2^2}{2!}\right] \\ &= 1 - 5e^{-2} = 0.323 \end{aligned}$$

Example 1.12. The number of telephone calls arriving at an exchange during any one hour period is a Poisson random variable with parameter $\lambda = 10$. What is the probability that during a fifteen-minute period there will be 4 calls?

Solution. Since we are interested in the number of calls in a fifteen-minute period, $t = \frac{1}{4}$. Therefore,

$$X_t = \text{The number of calls in a fifteen-minute period}$$

has the probability function given by

$$P(X_t = k) = \frac{e^{-10(1/4)}(10 \cdot \frac{1}{4})^k}{k!}, \qquad k = 0, 1, \ldots$$

Hence

$$P(X_t = 4) = \frac{e^{-2.5}(2.5)^4}{4!} = 0.1336$$

Example 1.13. Suppose X has a Poisson distribution with parameter λ.

(*a*) Show that $p(k + 1) = \dfrac{\lambda}{k + 1}p(k)$, where $p(k) = P(X = k)$.

(*b*) If $\lambda = 2$, compute $p(0)$, and then use the recursive relation in (*a*) to compute $p(1), p(2), p(3)$, and $p(4)$.

Solution

(*a*) We have

$$p(k + 1) = \frac{e^{-\lambda}\lambda^{k+1}}{(k+1)!} = \frac{\lambda}{k+1} \cdot \frac{e^{-\lambda}\lambda^k}{k!} = \frac{\lambda}{k+1} \cdot p(k)$$

(*b*) Since $\lambda = 2$, $p(0) = e^{-2} = 0.1353$. Now

$$p(1) = \lambda p(0) = 0.2706$$

$$p(2) = \frac{\lambda}{2}p(1) = p(1) = 0.2706$$

$$p(3) = \frac{\lambda}{3} p(2) = \frac{2}{3}(0.2706) = 0.1804$$

$$p(4) = \frac{\lambda}{4} p(3) = \frac{2}{4}(0.1804) = 0.0902$$

Example 1.14. Suppose X_t, the number of phone calls that arrive at an exchange during a period of length t, has a Poisson distribution with parameter λt. The probability that an operator answers any given phone call is equal to p, $0 \leqslant p \leqslant 1$. If Y_t denotes the number of phone calls answered, find the distribution of Y_t.

Solution. We want to find $P(Y_t = k)$, $k = 0, 1, 2, \ldots$. First of all we observe that, given $X_t = r$, the distribution of Y_t is binomial with r trials and the probability of success p. Hence

$$P(Y_t = k \mid X_t = r) = \binom{r}{k} p^k (1-p)^{r-k}, \qquad k = 0, 1, \ldots, r$$

Now, by the law of total probability,

$$\begin{aligned} P(Y_t = k) &= \sum_{r=k}^{\infty} P(Y_t = k \mid X_t = r) P(X_t = r) \\ &= \sum_{r=k}^{\infty} \binom{r}{k} p^k (1-p)^{r-k} \cdot \frac{e^{-\lambda t}(\lambda t)^r}{r!} \\ &= \frac{e^{-\lambda t}(\lambda t)^k p^k}{k!} \sum_{r=k}^{\infty} \frac{(1-p)^{r-k}(\lambda t)^{r-k}}{(r-k)!} \end{aligned}$$

Letting $r - k = i$, we see that

$$\sum_{r=k}^{\infty} \frac{(1-p)^{r-k}(\lambda t)^{r-k}}{(r-k)!} = \sum_{i=0}^{\infty} \frac{(1-p)^i(\lambda t)^i}{i!} = e^{(1-p)\lambda t}$$

Hence

$$P(Y_t = k) = \frac{e^{-\lambda t}(\lambda t)^k p^k}{k!} \cdot e^{(1-p)\lambda t} = \frac{e^{-\lambda p t}(\lambda p t)^k}{k!}$$

In conclusion, we see that Y_t has a Poisson distribution with parameter $p\lambda t$.

EXERCISES–SECTION 1

1. A fair die is rolled ten times. What is the probability of getting a multiple of 3 six times?

2. Assuming that a basketball player has a probability of 0.6 of making a basket, find the probability function of X, the number of baskets made in ten throws, and compute the following probabilities:
   (a) $P(X = 3)$ (b) $P(X > 0)$
   (c) $P(X \leqslant 3)$ (d) $P(X$ is even$)$

3. (a) What is the probability that the same number will appear on two dice at least once if the two dice are thrown four times?
   (b) How many times should the two dice be thrown if the probability in part (a) has to be approximately 0.9?

4. Suppose a penny, a dime, and a nickel with respective probabilities of heads equal to 0.2, 0.3, 0.4 are tossed. Find the distribution of X where X represents the number of heads.
5. A genetic pool contains N alleles of which i are of type A and $N - i$ of type a. If N alleles are picked with replacement, find the distribution of X, the number of alleles of type A.
6. A machine manufactures a certain type of bolts. From past experience, it is known that if ten bolts are tested, the probability of finding four defective bolts is 0.0112. Find the distribution of X, the number of defectives among ten bolts. *Hint:* Use the appropriate binomial table.
7. If the probability of success on a trial is 0.5, show that in n independent trials the probability of r successes is equal to the probability of r failures.
8. If X is $B(n, p)$, show that

$$\frac{b(k; n, p)}{b(k-1; n, p)} = 1 + \frac{(n+1)p - k}{k(1-p)}$$

9. Suppose X and Y each have a binomial distribution, where X is $B(n, p)$ and Y is $B(n, 1-p)$. Show that $P(X \geqslant r) = P(Y \leqslant n - r)$, $r = 0, 1, 2, \ldots, n$.
10. It is known that 80 percent of all families own a television set. If ten families are interviewed at random, use the table of binomial probabilities to find the probability that—
    (a) seven families own a television
    (b) at least seven families own a television
    (c) at most five families own a television.
11. The probability that a person contracts a certain type of influenza is 0.2. Past experience has revealed that the probability that a person who gets the influenza does not recover is 0.6. If there are n people in a group, find the distribution of X, the number of people who survive. (Assume that people respond to the disease independently.)
12. If a fair die is rolled fifteen times, find the most probable number of 6s. What if the die is rolled twenty-three times?
13. The probability that a child is a son is 0.4. If there are ten children in a family, find the most probable number of sons and the corresponding probability.
14. Suppose two fair coins are tossed until both coins show heads simultaneously. Find the distribution of the number of tosses required to accomplish this.
15. A billiards player keeps playing until he misses a shot. The probability that he misses any shot is 0.1. Let X denote the number of balls pocketed. Assuming that the player's performance on one shot does not influence his performance on other shots, find the distribution of X.
16. Suppose X has a geometric distribution with parameter p; that is, $P(X = r) = (1-p)^{r-1}p$, $r = 1, 2, \ldots$. Show that for any nonnegative integers s, t we have
    (a) $P(X > s) = (1-p)^s$
    (b) $P(X > s + t \mid X > s) = P(X > t)$

Comment on the result in (b).

17. A bowl contains w white beads and b black beads. If the beads are picked successively with replacement, find the distribution of the number of draws needed to pick the first white bead.

18. A person shoots arrows at a circular target which is 2 feet in radius. The distance from the bull's-eye to the point of impact has the pdf

$$f(x) = \begin{cases} x, & 0 \leqslant x \leqslant 1 \\ 2-x, & 1 \leqslant x \leqslant 2 \\ 0, & \text{elsewhere} \end{cases}$$

(a) Find the probability that the point of impact of an arrow is less than 0.9 foot from the bull's-eye. (Give your answer to one decimal place.)
(b) A person makes ten independent shots. Find the probability that he scores within 0.9 foot of the bull's-eye five times. (Use the binomial table.)
(c) If a person makes n independent shots, find the smallest n needed to have a probability of at least 0.98 of hitting within 0.9 foot of the bull's-eye at least once.

19. The pdf of a random variable X is given by

$$f(x) = \begin{cases} \dfrac{x^2}{9}, & 0 < x < 3 \\ 0, & \text{elsewhere} \end{cases}$$

(a) In ten independent observations, find the probability that X will take a value less than 2 six times.
(b) If independent observations are made, find the probability that ten observations will be needed to get a value less than 2 for the first time.

20. A random variable X has a pdf f.
(a) If n independent observations are made, find the probability that X will take a value less than a number c precisely k times, $k = 0, 1, 2, \ldots, n$.
(b) If independent observations are made, find the probability that k observations will be needed in order to get a value less than c for the first time.

21. The number of accidents in a certain locale on any given day is known to have a Poisson distribution. It is also known from past experience that the probability of no accidents is 0.1353. Find the probability that there will be at least three accidents on a given day.

22. The number of telephone calls arriving in an office obeys a Poisson distribution. The probabilities of two calls and four calls during a ten-minute period are respectively 0.1464 and 0.1952. Find the probability that there are three phone calls in a fifteen-minute period.

23. If X has a Poisson distribution with parameter λ, and λ is an integer, show that $P(X = i) = P(X = i + 1)$ for some integer $i \geqslant 0$.

24. Consider a Poisson random variable X with parameter λ. Show that $P(X = k)$ increases monotonically and then decreases monotonically as k increases, and that in the process it reaches its maximum when k is the largest integer not exceeding λ. *Hint:* Consider $P(X = k)/P(X = k - 1)$.

25. The number of misprints on a page is a Poisson random variable with parameter $\lambda = 0.011$. In a 400-page volume, what is the number of misprints that has the highest probability? Compute this probability.

2. ABSOLUTELY CONTINUOUS DISTRIBUTIONS

In this section we shall study some continuous probability distributions which often arise in physical applications.

2.1 The Uniform Distribution

The simplest example of a random variable of the continuous type is the random variable with the uniform distribution. A random variable has the uniform distribution over an interval $[a, b]$ if it has the following probability density function.

$$f(x) = \begin{cases} \dfrac{1}{b-a}, & a \leqslant x \leqslant b \\ 0, & \text{elsewhere} \end{cases}$$

The real numbers a and b are the parameters of the distribution. The pdf is sketched in Figure 2.1 below. Because of the shape of its graph, the uniform distribution is often referred to as the **rectangular distribution**.

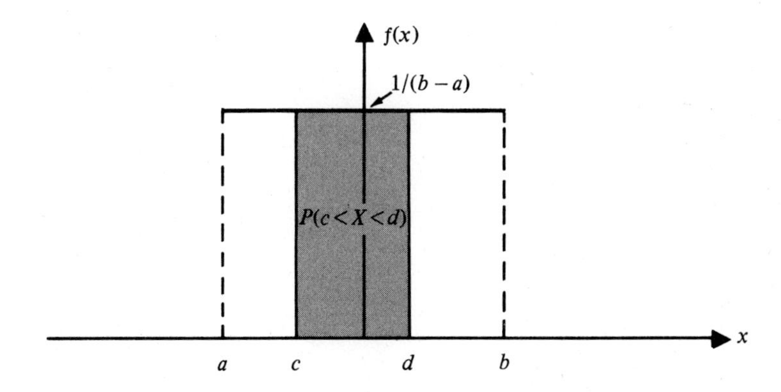

Figure 2.1 The Uniform Density Function

The shape of the distribution can be used to advantage to find the areas and, thereby, the probabilities associated with the events. For example, if (c, d) is a subinterval of (a, b), then $P(c < X < d)$ is the area of the rectangle whose sides are of length $\dfrac{1}{b-a}$ and $d - c$. Hence

$$P(c < X < d) = \frac{1}{b-a} \cdot (d - c)$$

if (c, d) is a subinterval of (a, b). That is, *the probability that X is in any subinterval of (a, b) is proportional to the length of the subinterval,* the constant of proportionality being equal to $1/(b-a)$. It is this property that characterizes a rectangular distribution.

This characterization implies that the probability that X falls in a subinterval of $[a, b]$ is the same for all subintervals as long as they have the same length. This is what we shall mean in our future reference when we say, for instance, that a point or a value is picked "at random" in the interval $[a, b]$, thereby indicating that the underlying random variable has a uniform distribution over $[a, b]$.

The distribution function of X is given by the formula $F(t) = \int_{-\infty}^{t} f(x)\,dx$ and can be easily verified to be

$$F(t) = \begin{cases} 0, & t < a \\ \dfrac{t-a}{b-a}, & a \leqslant t < b \\ 1, & t \geqslant b \end{cases}$$

The graph of the distribution function is given in Figure 2.2 below.

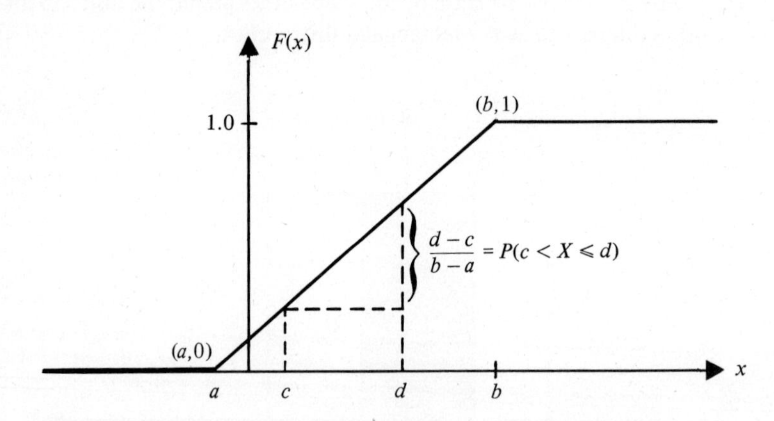

Figure 2.2 The D.F. of the Uniform Distribution

Let us now discuss a situation that provides an illustration of a phenomenon which can be regarded as having the uniform distribution.

Consider a circular horizontal board with a rotating pointer. When the pointer is spun, it can come to rest at any position on the board. We can therefore regard the sample space S as consisting of all the positions of the pointer. It seems safe to make the following assignment of probabilities:

$$\begin{pmatrix}\text{the probability that the pointer} \\ \text{comes to rest in any sector } QOR\end{pmatrix} = \frac{\text{area of the sector } QOR}{\text{area of the circle}}$$

(See Figure 2.3.)

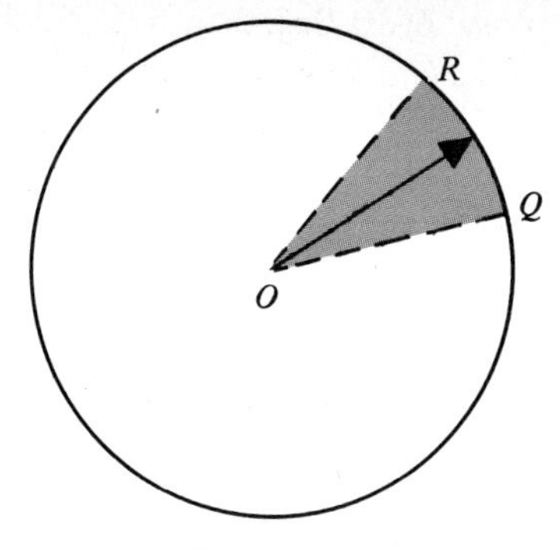

Figure 2.3

Next, let X represent the angle in radians that the pointer makes with OP, an arbitrarily chosen line of reference as shown in Figure 2.4(a). Then the range of X is the interval $[0, 2\pi]$.

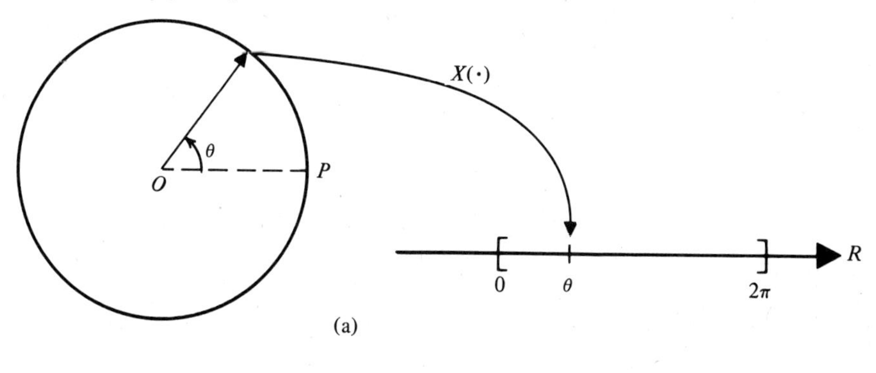

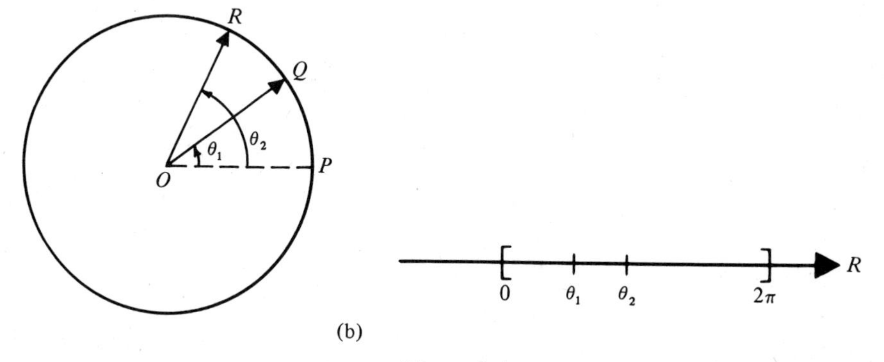

Figure 2.4

Also, if (θ_1, θ_2) is a subinterval of $[0, 2\pi]$, then

$$P^X[(\theta_1, \theta_2)] = P(\theta_1 < X < \theta_2)$$
$$= \frac{\text{area of the sector } QOR}{\text{area of the circle}}$$
$$= \frac{\frac{r^2}{2}(\theta_2 - \theta_1)}{\pi r^2}, \quad \text{assuming } r \text{ is the length of the pointer}$$
$$= \frac{\theta_2 - \theta_1}{2\pi}$$

(See Figure 2.4(b).)

In other words, $P(\theta_1 < X < \theta_2)$ is proportional to the length of the interval, $\theta_2 - \theta_1$. Consequently, the distribution of X is uniform over the interval $[0, 2\pi]$.

Example 2.1. A random variable X is uniformly distributed over the interval $[0, 2\pi]$. Find:

(*a*) the pdf of X
(*b*) the D.F. of X
(*c*) $P(\pi/6 < X \leqslant \pi/2)$
(*d*) $P(-\pi/6 < X \leqslant \pi/2)$

Solution

(*a*) The pdf of X is given as

$$f(x) = \begin{cases} \dfrac{1}{2\pi}, & 0 \leqslant x \leqslant 2\pi \\ 0, & \text{elsewhere} \end{cases}$$

(*b*) The D.F. is given by

$$F(u) = \begin{cases} 0, & u < 0 \\ \dfrac{u}{2\pi}, & 0 \leqslant u < 2\pi \\ 1, & u \geqslant 2\pi \end{cases}$$

(*c*) We have

$$P(\pi/6 < X \leqslant \pi/2) = F(\pi/2) - F(\pi/6)$$
$$= \frac{\pi/2}{2\pi} - \frac{\pi/6}{2\pi} = \frac{1}{6}$$

(*d*) In this case,

$$P(-\pi/6 < X \leqslant \pi/2) = F(\pi/2) - F(-\pi/6)$$
$$= \frac{\pi/2}{2\pi} - 0 = \frac{1}{4}$$

2.2 The Normal Distribution

This distribution is also called the **Gaussian distribution**. It is by far the most celebrated of the continuous distributions because a wide class of phenomena can be described in terms of this distribution.

A random variable X has the normal distribution if its probability density function is given by

$$f(x) = \frac{1}{b\sqrt{2\pi}} e^{-(x-a)^2/(2b^2)}, \quad -\infty < x < \infty$$

Here a and b are constants with $-\infty < a < \infty$ and $b > 0$. (The reason b cannot be negative or zero is obvious.) These constants are the parameters of the distribution and they completely describe the normal distribution. We shall investigate these constants more thoroughly when we study mathematical expectation, at which stage we shall replace a and b by the customary Greek letters μ and σ. It is a common practice to write the normal distribution with parameters a, b as $N(a, b^2)$.

(*i*) The normal distribution $N(a, b^2)$ is symmetric about a.

This follows because

$$f(a+t) = \frac{1}{b\sqrt{2\pi}} e^{-t^2/(2b^2)} = f(a-t)$$

for every t.

On account of this property we have, for example,

$$P(X \leqslant a - t) = P(X \geqslant a + t) \qquad \text{for any } t$$
$$P(a \leqslant X \leqslant a + t) = P(a - t \leqslant X \leqslant a) \qquad \text{for any } t > 0$$

and so on.

(*ii*) The parameter a represents the location where the maximum of f is attained.

If we differentiate f and set the derivative equal to zero, we get

$$f'(x) = \frac{1}{b\sqrt{2\pi}} e^{-(x-a)^2/(2b^2)} \cdot \frac{-2(x-a)}{2b^2} = 0$$

This equation is satisfied for $x = a$. Therefore, the maximum of f occurs at a and the maximum value is equal to $f(a) = 1/(b\sqrt{2\pi})$. It is therefore important to note that a normally distributed random variable has the highest probability of taking on a value in the neighborhood of a, the point of symmetry. More precisely

$$P(a - h < X < a + h) > P(c - h < X < c + h)$$

for any number c different from a.

(*iii*) The normal curve has two points of inflection. They correspond to $x = a - b$ and $x = a + b$.

The points of inflection are found by setting the second derivative of f equal to zero and then solving for x. We get

$$f''(x) = -\frac{1}{b^3\sqrt{2\pi}}\left[e^{-(x-a)^2/(2b^2)} - \frac{2(x-a)^2}{2b^2}e^{-(x-a)^2/(2b^2)}\right]$$

Setting this equal to zero gives

$$1 - \frac{(x-a)^2}{b^2} = 0$$

that is,

$$x = a + b \quad \text{and} \quad x = a - b$$

The significance of the constant b is to be seen in the fact that it determines the distance of the points of inflection from the line of symmetry and as such provides a measure of the flatness of the graph of the pdf. The larger the value of b, the flatter the curve.

In summary: If the pdf is sketched, we observe the following: *The graph is symmetric about a; the value at a is* $1/(b\sqrt{2\pi})$, *and this is the maximum value of f; and the graph has two points of inflection at a distance of b units from a, on either side of the line of symmetry.* (See Figure 2.5.)

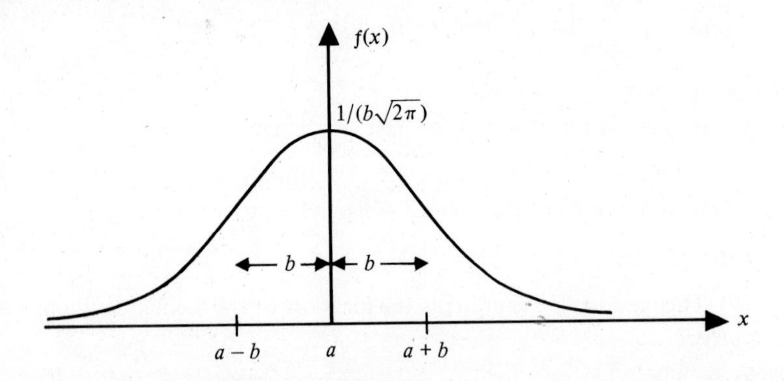

Figure 2.5 The Normal pdf

Figure 2.6 gives the graphs of some normal pdf's for different values of a and a fixed value of b ($b = 1$). Figure 2.7 gives the graphs for a fixed value of a ($a = 8$) and different values of b.

The particular normal distribution for which $b = 1$ and $a = 0$ is called the **standard normal distribution** and, following the convention, is denoted by $N(0, 1)$. Its pdf is

$$f(x) = \frac{1}{\sqrt{2\pi}}e^{-x^2/2}, \quad -\infty < x < \infty$$

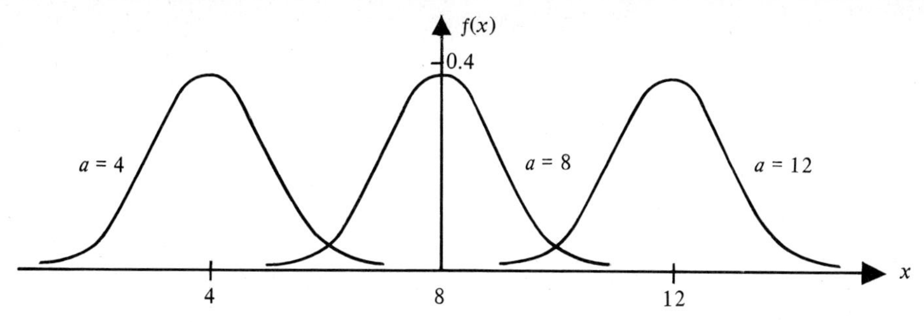

Figure 2.6 Normal Density Functions with $b = 1$ and $a = 4, 8, 12$

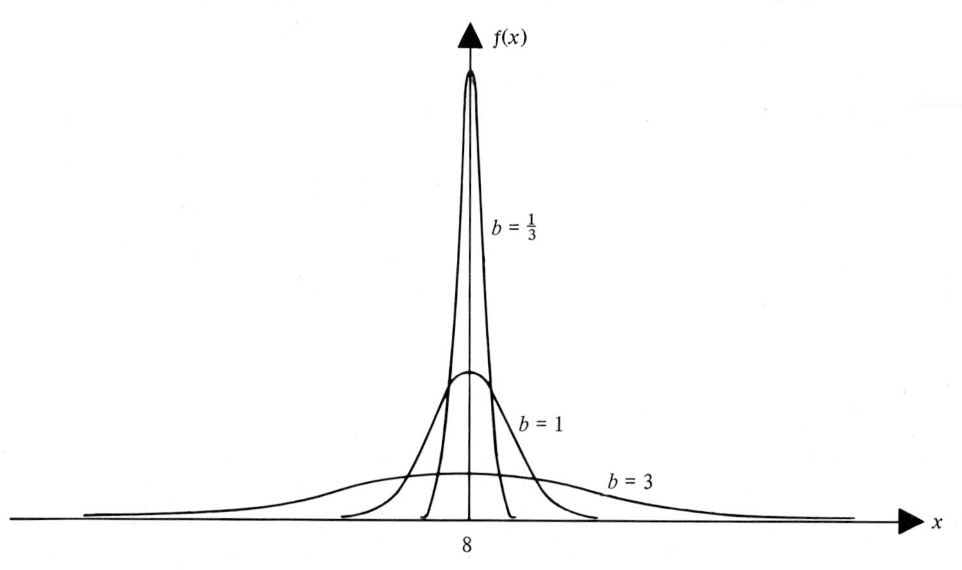

Figure 2.7 Normal Density Functions with $a = 8$ and $b = 3, 1$, and $\frac{1}{3}$

Since $-x^2/2$ is in the exponent, the pdf decreases to 0 very rapidly as $|x| \to \infty$.

Because of the importance of the standard normal distribution, we shall use the special symbol Φ throughout this text to represent its distribution function. We define

$$\Phi(u) = \frac{1}{\sqrt{2\pi}} \int_{-\infty}^{u} e^{-x^2/2}\, dx, \qquad \text{for any real number } u$$

It is not possible to express $\Phi(u)$ in a closed form. However, tables giving the values of Φ for different values of u are available. We reproduce such a table as Table C in the Appendix. The graph of Φ is given in Figure 2.8.

At the bottom of Table C are given the values of $\Phi(u)$ for values of $u = 1.282$, $1.645, \ldots, 4.417$. Notice how quickly Φ approaches 1. One can get an idea of this from the following bound on Φ.

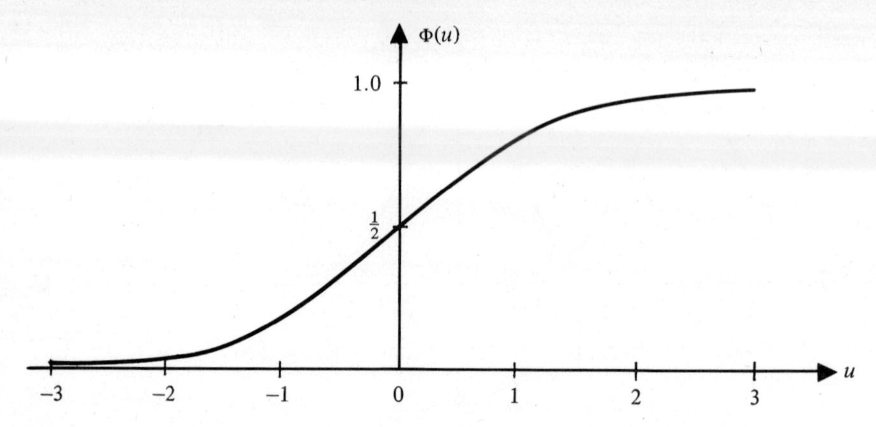

Figure 2.8 The D.F. of the Standard Normal Variable

$$\Phi(u) \geqslant 1 - \frac{1}{\sqrt{2\pi}} \cdot \frac{e^{-u^2/2}}{u}$$

To prove this let, $g(x) = (1/\sqrt{2\pi})e^{-x^2/2}$. Then, differentiating $g(x)$, it can be seen that $g'(x) = -xg(x)$, so that if $u > 0$ we have

$$\begin{aligned} 1 - \Phi(u) &= \int_u^\infty 1 \cdot g(x)\,dx \leqslant \int_u^\infty \frac{x}{u} g(x)\,dx, \qquad \text{since } \frac{x}{u} > 1 \text{ for } x > u \\ &= -\frac{1}{u}\int_u^\infty g'(x)\,dx \\ &= -\frac{1}{u} g(x)\Big|_u^\infty = \frac{g(u)}{u} = \frac{1}{\sqrt{2\pi}} \cdot \frac{e^{-u^2/2}}{u} \end{aligned}$$

Hence the bound.

How does one go about finding probabilities in the case of a normal distribution with arbitrary values of a and b ($b > 0$)? The standard normal table can be used to advantage for this purpose. This is accomplished as follows:

In the general case the D.F. is given by

$$F(u) = \frac{1}{b\sqrt{2\pi}} \int_{-\infty}^{u} e^{-(x-a)^2/(2b^2)}dx, \qquad -\infty < u < \infty$$

Let us make the substitution $(x - a)/b = y$. Then

$$dx = b\,dy \text{ and when } \begin{cases} x = u, & y = \dfrac{u-a}{b} \\ x = -\infty, & y = -\infty \end{cases}$$

Therefore,

$$F(u) = \frac{1}{\sqrt{2\pi}} \int_{-\infty}^{(u-a)/b} e^{-y^2/2}\,dy = \Phi\left(\frac{u-a}{b}\right)$$

Hence we get the following rule: *If X is $N(a, b^2)$, then to compute the value of F at u, first find $(u - a)/b$ and then refer to the standard normal table and find the value of Φ at $(u - a)/b$.*

Example 2.2. A random variable X has the following pdf:

$$f(x) = \frac{1}{\sqrt{18\pi}} e^{-(x^2 - 10x + 25)/18}, \qquad -\infty < x < \infty$$

Show that the distribution is normal and find the constants a and b. Also find the maximum of f.

Solution. The pdf can be rewritten as

$$f(x) = \frac{1}{3\sqrt{2\pi}} e^{-(x-5)^2/(2\cdot 3^2)}, \qquad -\infty < x < \infty$$

The distribution is therefore normal with $a = 5$ and $b = 3$. The maximum of f is equal to $1/(3\sqrt{2\pi})$.

Example 2.3. Suppose a random variable X has the following pdf:

$$f(x) = \frac{1}{2\sqrt{2\pi}} e^{-(x+4)^2/8}, \qquad -\infty < x < \infty$$

Compute:

(*a*) $P(X \leqslant -2)$ (*b*) $P(-5 < X \leqslant -2)$
(*c*) $P(|X + 3| \leqslant 1)$ (*d*) $P(X \geqslant -6)$

Solution. Note that the distribution is normal and $a = -4$, $b = 2$. Therefore, the point of symmetry is located at -4.

$$(a)\quad P(X \leqslant -2) = \Phi\left(\frac{-2-(-4)}{2}\right) = \Phi(1) = 0.8413$$

$$\begin{aligned}(b)\quad P(-5 < X \leqslant -2) &= P(X \leqslant -2) - P(X \leqslant -5)\\ &= \Phi\left(\frac{-2-(-4)}{2}\right) - \Phi\left(\frac{-5-(-4)}{2}\right)\\ &= \Phi(1) - \Phi(-\tfrac{1}{2})\\ &= 0.8413 - 0.3085 = 0.5328\end{aligned}$$

(*Note:* By symmetry, $\Phi(-\frac{1}{2}) = 1 - \Phi(\frac{1}{2}) = 1 - 0.6915 = 0.3085$.)

$$\begin{aligned}(c)\quad P(|X+3| \leqslant 1) &= P(-1 \leqslant X + 3 \leqslant 1)\\ &= P(-4 \leqslant X \leqslant -2)\\ &= P(X \leqslant -2) - P(X \leqslant -4)\\ &= \Phi(1) - \Phi(0) = 0.8413 - 0.5\\ &= 0.3413\end{aligned}$$

$$(d)\quad P(X \geqslant -6) = 1 - P(X \leqslant -6) = 1 - \Phi(-1) = 1 - 0.1587 = 0.8413.$$

Notice that the answer to part (*a*) is the same as the answer to (*d*). This is not surprising, because the distribution is symmetric about −4.

Example 2.4. Under the assumption that a random variable X is normally distributed with the following pdf,

$$f_1(x) = \frac{1}{6\sqrt{2\pi}}\, e^{-(x-60)^2/72}, \qquad -\infty < x < \infty$$

it was found that $P(X \leqslant u_o) = 0.8413$, where u_o is a real number. However, it was later discovered that the pdf should have been

$$f_2(x) = \frac{1}{6\sqrt{2\pi}}\, e^{-(x-69)^2/72}, \qquad -\infty < x < \infty$$

Find $P(X \leqslant u_o)$ when the correction is effected.

Solution. If the pdf is f_1, then $a = 60$ and $b = 6$. Since, under $f_1, P(X \leqslant u_o) =$ 0.8413, we have

$$\Phi\left(\frac{u_o - 60}{6}\right) = 0.8413$$

Referring to the standard normal table, we find that $\Phi(1) = 0.8413$. Therefore, we must have $(u_o - 60)/6 = 1$, that is, $u_o = 66$.

Since the correct distribution is as given by f_2, the corrected value of a should be 69 (with b still equal to 6). Therefore,

$$\begin{aligned} P(X \leqslant u_o) &= P(X \leqslant 66) = \Phi\left(\frac{66 - 69}{6}\right) = \Phi\left(-\frac{1}{2}\right) \\ &= 0.3085 \end{aligned}$$

Example 2.5. Suppose the weights of adult males are normally distributed and that 6.68 percent are under 130 lbs in weight, and 77.45 percent are between 130 and 180 lbs. Find the parameters of the distribution.

Solution. Let X represent the weight of a random male. Then we are given that

$$P(X \leqslant 130) = 0.0668 \quad \text{and} \quad P(130 < X \leqslant 180) = 0.7745$$

or, equivalently,

$$P(X \leqslant 130) = 0.0668 \quad \text{and} \quad P(X \leqslant 180) = 0.8413$$

(See Figure 2.9.) That is,

$$\Phi\left(\frac{130 - a}{b}\right) = 0.0668 \quad \text{and} \quad \Phi\left(\frac{180 - a}{b}\right) = 0.8413$$

However, from the standard normal table we know that

$$\Phi(-1.5) = 0.0668 \quad \text{and} \quad \Phi(1) = 0.8413$$

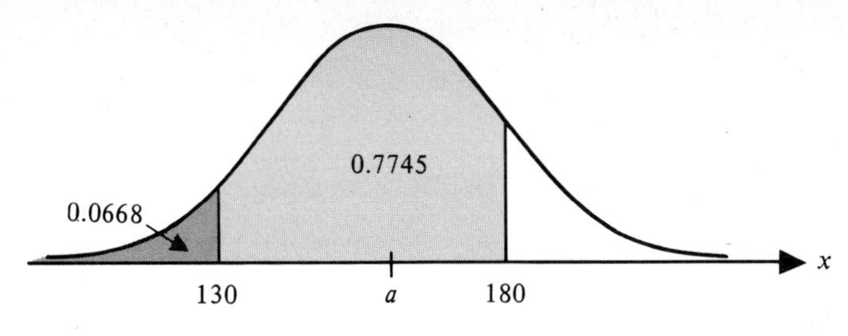

Figure 2.9

Therefore,

$$\frac{130-a}{b}=-1.5$$

$$\frac{180-a}{b}=1$$

Solving this system of equations, we get $a = 160$ and $b = 20$.

2.3 The Gamma Distribution

Before discussing the gamma distribution, let us introduce a mathematical function called the **gamma function** Γ. If p is a real number such that $p > 0$, then the gamma function is defined by

$$\Gamma(p) = \int_0^\infty x^{p-1}e^{-x}\,dx$$

We shall establish the following recursive relation:

$$\Gamma(p) = (p-1)\Gamma(p-1) \qquad \text{if} \quad p > 1$$

To see this, let us make the following substitution in the integral:

$$u = x^{p-1} \quad \text{and} \quad dv = e^{-x}\,dx$$

Then

$$du = (p-1)x^{p-2}\,dx \quad \text{and} \quad v = -e^{-x}$$

Integration by parts then yields

$$\Gamma(p) = -e^{-x}x^{p-1}\Big|_0^\infty + \int_0^\infty (p-1)x^{p-2}e^{-x}\,dx$$

The first term on the right is zero (use l'Hospital's rule). Therefore,

$$\Gamma(p) = (p-1)\int_0^\infty x^{(p-1)-1}e^{-x}\,dx = (p-1)\Gamma(p-1)$$

If $p > 2$, we can use the recursive relation one more time to write

$$\Gamma(p) = (p-1)(p-2)\Gamma(p-2)$$

and so on. In particular, if n is a positive integer, we can write

$$\Gamma(n) = (n-1)(n-2)\ldots 3 \cdot 2 \cdot \Gamma(1)$$

Now,

$$\Gamma(1) = \int_0^\infty x^{1-1} e^{-x}\, dx = 1$$

In conclusion, we get

$$\Gamma(n) = (n-1)! \qquad \text{if } n \text{ is a positive integer.}$$

The fact that $\frac{1}{\sqrt{2\pi}} \int_{-\infty}^{\infty} e^{-u^2/2}\, du = 1$ can be used to show that

$$\Gamma(\tfrac{1}{2}) = \sqrt{\pi}$$

Since $\int_{-\infty}^{\infty} e^{-u^2/2}\, du = \sqrt{2\pi}$, we get

$$2\int_0^\infty e^{-u^2/2}\, du = \sqrt{2\pi}$$

because the integrand is an even function. Let $u^2/2 = y$; then $u\, du = dy$ and $u = \sqrt{2y}$; that is,

$$du = \frac{dy}{u} = \frac{dy}{\sqrt{2y}}$$

Therefore,

$$2\int_0^\infty e^{-y} \frac{dy}{\sqrt{2y}} = \sqrt{2\pi}$$

Simplifying,

$$\int_0^\infty y^{(1/2)-1} e^{-y}\, dy = \sqrt{\pi}$$

Consequently,

$$\Gamma(\tfrac{1}{2}) = \sqrt{\pi}$$

Having discussed the gamma function, we shall now introduce the gamma distribution. The **gamma distribution** is characterized by the following pdf:

$$f(x) = \begin{cases} \dfrac{\lambda^p}{\Gamma(p)} x^{p-1} e^{-\lambda x}, & x > 0 \\ 0, & \text{elsewhere} \end{cases}$$

The parameters λ and p satisfy $\lambda > 0$ and $p > 0$.

Notice that

$$f(x) \geqslant 0$$

since $\lambda > 0$ and $x > 0$, and

$$\int_{-\infty}^{\infty} f(x)\, dx = 1$$

in view of the definition of the gamma function.

Gamma distributions for a fixed value of λ ($\lambda = 1$) and $p = 1, 2, 3$ are graphed in Figure 2.10, and gamma distributions for a fixed value of p ($p = 2$) and $\lambda = 2, 1, \frac{1}{2}$ are graphed in Figure 2.11.

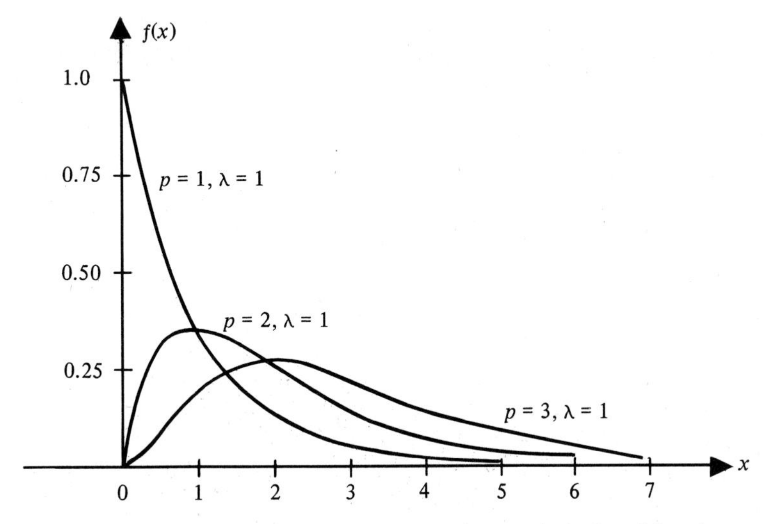

Figure 2.10 Gamma Density Functions for $p = 1, 2, 3$ and $\lambda = 1$

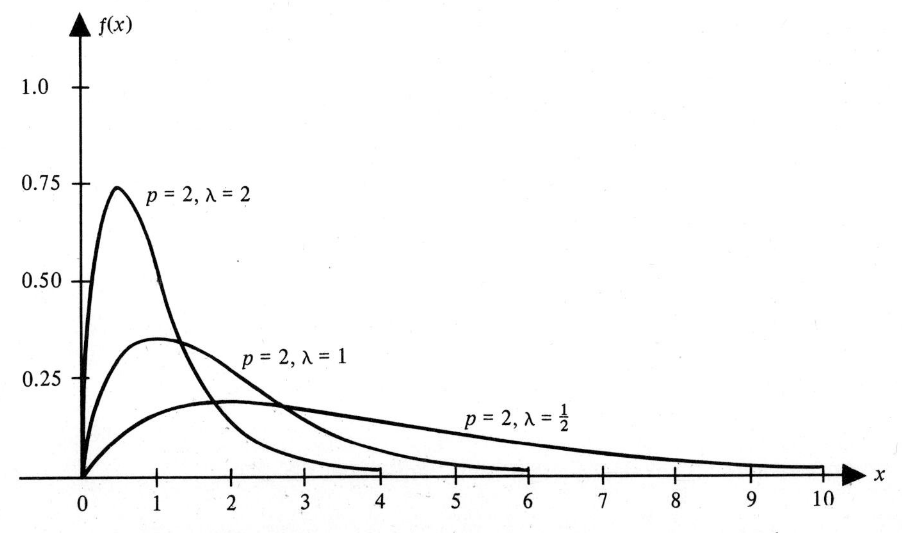

Figure 2.11 Gamma Density Functions for $p = 2$ and $\lambda = 2, 1, \frac{1}{2}$

Special cases of the gamma distribution are obtained by assigning specific values to the parameters λ and p.

The exponential distribution

If we set $p = 1$, we get a very important distribution with the following pdf

$$f(x) = \begin{cases} \lambda e^{-\lambda x}, & x > 0 \\ 0, & \text{elsewhere} \end{cases}$$

The random variable with this pdf is said to have the **exponential distribution** with the parameter λ ($\lambda > 0$).

The D.F. of the exponential distribution with parameter λ can be easily obtained as

$$F(x) = \begin{cases} 0, & x < 0 \\ 1 - e^{-\lambda x}, & x \geqslant 0 \end{cases}$$

The graph of the pdf and that of the D.F. will depend on the value of the parameter λ. Figures 2.12(a) and (b) indicate their typical shapes.

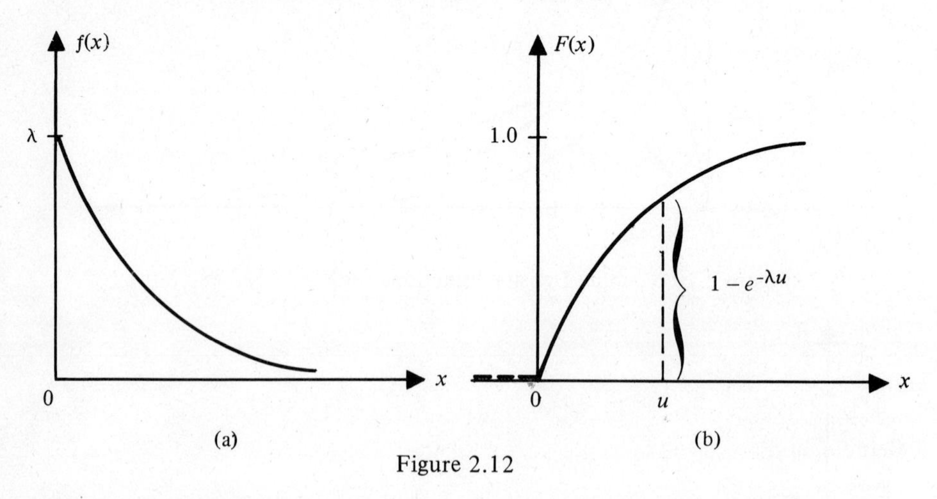

Figure 2.12

Example 2.6. Suppose the length of time (measured in hours) an electric bulb lasts is a random variable with the following D.F.:

$$F(x) = \begin{cases} 0, & x < 0 \\ 1 - e^{-(x/500)}, & x \geqslant 0 \end{cases}$$

Find the probability that the bulb lasts–

(*a*) between 100 to 200 hours

(*b*) beyond 300 hours

(*c*) beyond 400 hours, given that it was working at the end of 100 hours.

Solution. Denoting the life span of the bulb by X, we see that it has an exponential distribution with parameter $\lambda = \frac{1}{500}$. Then:

(*a*) $P(100 < X < 200) = F(200) - F(100) = e^{-1/5} - e^{-2/5} = 0.1484$.

(*b*) $P(X > 300) = 1 - F(300) = e^{-3/5} = 0.5488$.

(*c*) In this case we want to find $P(X > 400 \mid X \geqslant 100)$. We have

$$\begin{aligned} P(X > 400 \mid X \geqslant 100) &= \frac{P(\{X > 400\} \cap \{X \geqslant 100\})}{P(X \geqslant 100)} \\ &= \frac{P(X > 400)}{P(X \geqslant 100)} = \frac{e^{-4/5}}{e^{-1/5}} = e^{-3/5} = 0.5488 \end{aligned}$$

■

Parts (*b*) and (*c*) of the above example bring out a very interesting property of the exponential distribution which is often referred to as *the lack of memory property*. To see what this means, let us pursue the problem in a general setting.

Suppose the bulb has been functioning for at least T units of time. Given this information, the probability that it will function for another t units is

$$\begin{aligned} P(X \geqslant T + t \mid X \geqslant T) &= \frac{P(X \geqslant T + t)}{P(X \geqslant T)} \\ &= \frac{e^{-\lambda(T+t)}}{e^{-\lambda T}} = e^{-\lambda t} \end{aligned}$$

However,

$$e^{-\lambda t} = P(X \geqslant t)$$

Hence

$$P(X \geqslant T + t \mid X \geqslant T) = P(X \geqslant t)$$

The implication is that, given that the bulb has lasted for T hours, the probability that it will last for another t hours is the same as though this information is not available. The lightbulb does not "remember" how long it has burned. Because of this remarkable property, the exponential distribution finds wide applications in reliability theory and renewal theory.

Comment. As we have seen, the negative binomial distribution (and consequently its special case, the geometric distribution) represents the number of independent Bernoulli trials one has to perform in order to get r successes for the first time. In other words, the underlying random variable represents the waiting time, measured in terms of the number of trials, until the rth success occurs. Analogously, a random variable which has a gamma distribution (or its special case, the exponential distribution) represents the extent of wait for r changes to occur when the number of changes takes place according to the Poisson probability law. (See exercise 23.)

The chi-square distribution

Another special case of the gamma distribution is obtained if we set $\lambda = \frac{1}{2}$ and $p = n/2$, where n is a positive integer. In this case the gamma density becomes

$$f(x) = \begin{cases} \dfrac{1}{2^{n/2}\Gamma(n/2)} x^{(n/2)-1} e^{-x/2}, & x > 0 \\ 0, & x \leqslant 0 \end{cases}$$

A random variable having such a pdf is said to have a **chi-square distribution** *with n degrees of freedom* and is denoted by χ_n^2. In subsequent chapters, we shall investigate an interesting relationship between the chi-square distribution and the standard normal distribution.

For the sake of completeness, we now give some classical distributions which are of some interest.

2.4 The Cauchy Distribution

A random variable has the **Cauchy distribution** with parameters a and b if it has the following pdf:

$$f(x) = \frac{1}{\pi} \cdot \frac{a}{a^2 + (x-b)^2}, \qquad -\infty < x < \infty$$

The parameters a and b satisfy $-\infty < b < \infty$ and $a > 0$. Observe that

(*i*) $f(x) \geqslant 0$

(*ii*) $\int_{-\infty}^{\infty} f(x)\, dx = 1$

To see (*ii*), let $y = (x-b)/a$; then $dy = dx/a$ and

$$\int_{-\infty}^{\infty} f(x)\, dx = \frac{1}{\pi} \int_{-\infty}^{\infty} \frac{dy}{1+y^2} = \frac{1}{\pi} \arctan y \Big|_{-\infty}^{\infty} = 1$$

(*iii*) $f(b+x) = f(b-x) = \dfrac{1}{\pi} \cdot \dfrac{a}{a^2 + x^2}$

Hence f is symmetric about b.

(*iv*) The D.F. is given by

$$\begin{aligned} F(u) &= \frac{1}{\pi a} \int_{-\infty}^{u} \frac{1}{1 + [(x-b)^2/a^2]}\, dx \\ &= \frac{1}{\pi} \int_{-\infty}^{(u-b)/a} \frac{1}{1+y^2}\, dy \\ &= \frac{1}{2} + \frac{1}{\pi} \arctan\left(\frac{u-b}{a}\right) \end{aligned}$$

Hence

$$F(u) = \frac{1}{2} + \frac{1}{\pi}\arctan\left(\frac{u-b}{a}\right), \qquad -\infty < u < \infty$$

The pdf and the D.F. for the Cauchy distribution with $b = 0$ and $a = 1$ are drawn in Figure 2.13. The reader will see a close resemblance between the above graphs and those for the normal distribution. However, it should be realized that the two distributions are quite different.

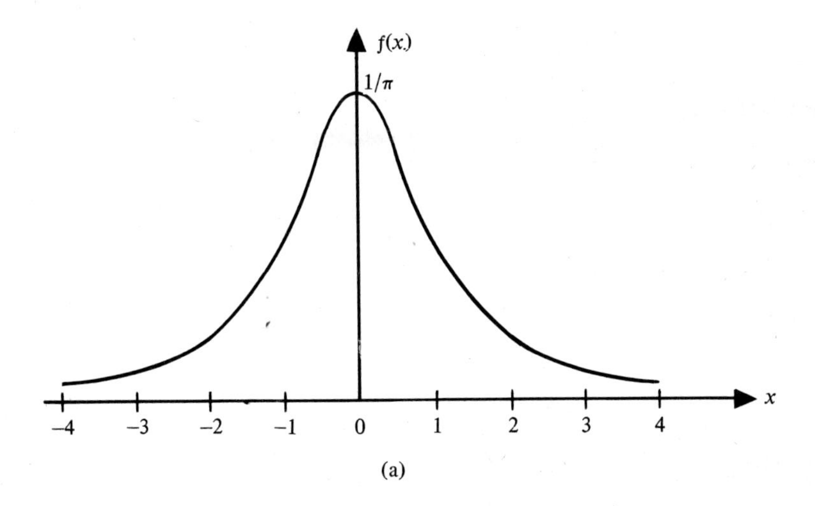

(a)

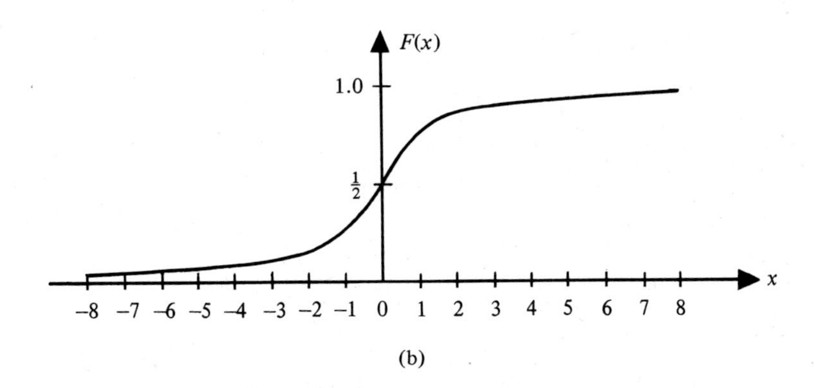

(b)

Figure 2.13

2.5 The Laplace Distribution

The pdf of the **Laplace distribution** is given by

$$f(x) = \frac{1}{2b} e^{-|x-a|/b}, \qquad -\infty < x < \infty$$

where a is any real number and $b > 0$.

The following can be established easily and are left as exercises:

(*i*) The pdf is symmetric about a.

(*ii*) $\int_{-\infty}^{\infty} f(x)\,dx = 1$.

(*Hint:* Write

$$\int_{-\infty}^{\infty} f(x)\,dx = \frac{1}{2b}\left[\int_{-\infty}^{a} e^{(x-a)/b}\,dx + \int_{a}^{\infty} e^{-(x-a)/b}\,dx\right]$$

and proceed.)

(*iii*) The D.F. is given as

$$F(x) = \begin{cases} \frac{1}{2} e^{(x-a)/b}, & \text{if } x < a \\ 1 - \frac{1}{2} e^{-(x-a)/b}, & \text{if } x \geqslant a \end{cases}$$

EXERCISES–SECTION 2

1. A number X is picked at random in the interval $[-3, 5]$. Find:
   (a) the pdf of X
   (b) the probability that the number is negative and greater than -2
2. A train arrives at a station at any random time between 2:00 p.m. and 4:30 p.m. If X represents the time (measured in hours) when the train arrives, find–
   (a) the pdf of X
   (b) $P(2.5 < X < 3.5)$
   (c) $P(|X - 3.5| > 0.5)$.
3. Suppose X is uniformly distributed over $[a, 9]$. If $P(3 < X < 5) = \frac{2}{7}$, find–
   (a) the D.F. of X
   (b) $P(|X - 5| < 2)$
   (c) a real number d for which $P(X > d) = \frac{1}{4}$.
4. If X is uniformly distributed over the interval $[-b, b]$, determine b if $P(|X| > 2) = \frac{3}{4}$, and give the pdf of X.
5. Show that $\int_{-\infty}^{\infty} e^{-x^2/2}\,dx = \sqrt{2\pi}$.

Hint: $\left[\int_{-\infty}^{\infty} e^{-x^2/2}\,dx\right]^2 = \int_{-\infty}^{\infty}\int_{-\infty}^{\infty} e^{-(x^2+y^2)/2}\,dx\,dy$

Use polar coordinates to show that the integral on the right-hand side is 2π.

6. Suppose the pdf of a random variable is given by

$$f(x) = ce^{-(x^2/2) + 3x}, \qquad -\infty < x < \infty$$

Determine the constant c without integrating.

7. If X has a standard normal distribution, show that for all $x > 0$ we have:
   (a) $\Phi(-x) + \Phi(x) = 1$
   (b) $P(|X| > x) = 2[1 - \Phi(x)] = 2\Phi(-x)$
   (c) $P(-x < X < x) = 2\Phi(x) - 1$.

8. If X is a standard normal variable, find c such that:
   (a) $P(X \leqslant c) = 0.1151$ (b) $P(|X| \geqslant c) = 0.5962$
   (c) $P(1 \leqslant X < c) = 0.1525$ (d) $P(X \geqslant -c) = 0.1539$.

9. Suppose X is $N(20, 16)$. Find:
   (a) $P(X < 22)$ (b) $P(18 < X < 24)$
   (c) $P(X > 18)$ (d) $P(|X - 20| > 6)$
   (e) $P(|X - 18| > 4)$

10. If X is normally distributed and is $N(20, 16)$, find a real number x for which:
    (a) $P(X > x) = 0.8944$ (b) $P(|X - 20| < x) = 0.9876$
    (c) $P(|X - 20| > x) = 0.0892$.

11. If X is $N(a, b^2)$, find $P(|X - a| < 1.5b)$.

12. Suppose X is $N(2, 9)$. Find
    (a) $P(|X| < 4)$ (b) $P(X^2 < 4)$

13. Suppose X is $N(0, 4)$ and $Y = X^2$. Find the positive real number c if:
    (a) $P(Y < c) = 0.6826$ (b) $P(1 < Y < c) = 0.4834$.

14. Suppose X is $N(2, 4)$. Find $P(X^2 - 4X < 5)$.

15. Suppose X is $N(a, b^2)$. If $P(X < 46) = 0.8413$ and $P(X > 31) = 0.9332$, find $P(49 < X < 55)$.

16. Show that

$$\int_1^3 e^{-(x^2 - 4x)/2}\, dx = (0.6826)\sqrt{2\pi}\, e^2$$

17. The length of life (measured in months) of an instrument produced by a machine has a normal distribution, $N(10, 4)$. Find the probability that an instrument produced by the machine will last–
    (a) less than six months
    (b) between seven and twelve months.

18. The diameter (measured in inches) of a tube produced by a machine has a normal distribution, $N(12, 0.04)$. A tube will be considered substandard if the diameter is outside the range 11.7 to 12.3. What is the probability that a tube will be considered substandard?

19. A random variable X is said to have a *log normal distribution* with parameters a, b^2 if its pdf is given by

$$f(x) = \frac{1}{xb\sqrt{2\pi}}\, e^{-((\ln x) - a)^2/2b^2}, \qquad 0 < x < \infty$$

Find:

(a) $P(X < e^{a-b})$ (b) $P(X > e^{a+2b})$

(c) $P(X < e^{a+b})$ (d) $P(X > e^{a-2b})$

20. The life of an electronic component (measured in hours) is an exponential random variable with $\lambda = \frac{1}{100}$. If ten components are functioning independently, find the probability that six of these will last beyond 100 hours.

21. A machine works on five electron tubes which operate independently. The time to failure (in hours) for any tube is an exponential random variable with parameter $\lambda = \frac{1}{60}$. Find the probability that the machine will function beyond 54 hours if–

(a) the machine works as long as *all* the tubes are operating

(b) the machine works as long as *at least one* tube is operating.

22. Consider a random variable X which has a gamma distribution given by

$$f(x) = \begin{cases} \dfrac{\lambda^p}{\Gamma(p)} x^{p-1} e^{-\lambda x}, & x > 0 \\ 0, & \text{elsewhere} \end{cases}$$

where p is a *positive integer*. Integrate by parts and show that

$$\frac{\lambda^p}{\Gamma(p)} \int_t^\infty x^{p-1} e^{-\lambda x}\, dx = \frac{e^{-\lambda t} \lambda t}{\Gamma(p)} + \frac{\lambda^{p-1}}{\Gamma(p-1)} \int_t^\infty x^{p-2} e^{-\lambda x}\, dx$$

Use this relationship recursively to show that

$$\frac{\lambda^p}{\Gamma(p)} \int_t^\infty x^{p-1} e^{-\lambda x}\, dx = \sum_{r=0}^{p-1} \frac{e^{-\lambda t} (\lambda t)^r}{r!}$$

What does the right-hand side represent? What can you conclude from this last relation?

23. Suppose the number of particles X_t emitted from a radioactive substance during an interval $(0, t)$ has a Poisson distribution with parameter λt. Let T represent the time needed to emit p particles. Show that the distribution of T is a gamma distribution. *Hint:* $T > t \Leftrightarrow X_t = 0$, or $X_t = 1$, or . . . , or $X_t = p - 1$. Use the result in exercise 22.

6 Functions of a Random Variable

INTRODUCTION

It often happens in applied problems that the distribution of a random variable is given, and it is necessary to find the distribution of a function of it. For instance, the velocity of a particle might be known to be a random variable with a given distribution, and we might actually be interested in the distance travelled by the particle or the kinetic energy of the particle. Looking at it purely from a non-mathematical point of view (without introducing the Borel fields into our discussion), if the velocity of a particle is a random variable, then so is the distance travelled or the kinetic energy. Thus, if we say that the probability that the velocity is between 300 miles/hour and 350 miles/hour is say, 0.95, then, of course, in talking about the distance travelled by the particle we will necessarily have to make a statement which is probabilistic in nature.

1. THE MATHEMATICAL FORMULATION*

We shall now briefly discuss the general mathematical setup for functions of a random variable.

Suppose X is a random variable on a probability space $(S, \mathcal{F}, P)$. By definition, X is a real-valued function on S. Therefore,

$$X : S \to \mathbf{R}$$

Now suppose h is a real-valued function of a real variable. That is,

$$h : \mathbf{R} \to \mathbf{R}$$

*The reader might find this section slightly abstruse. It is not essential in the sequel and can be omitted.

Since for any $s \in S$, $X(s)$ is a real number, $h(X(s))$ is well defined and is a real number. It is interesting that in the process we have a new function from S to $\mathbf{R}$, which assigns to any point $s \in S$ a real number $h(X(s))$. We shall denote this function by $h(X)$. For any point $s \in S$, it is defined by

$$h(X)(s) = h(X(s))$$

That is, to find $h(X)$ at $s \in S$, first find the real number $X(s)$ and for this real number find h, as indicated in Figure 1.1.

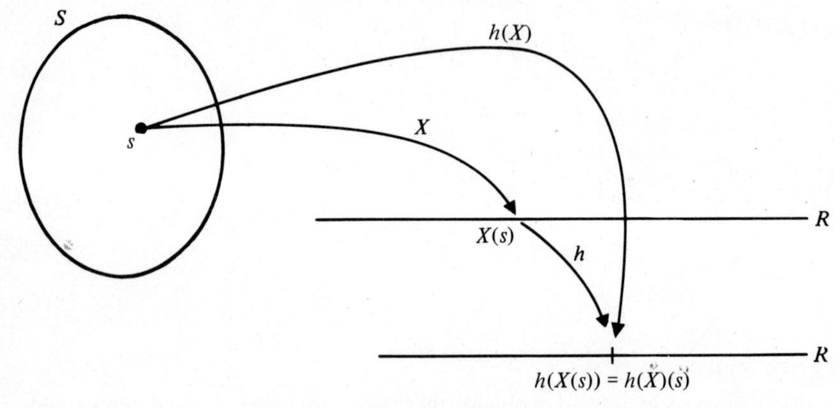

Figure 1.1

In short, $h(X)$ is what is called in mathematics a *composite function* defined on S, and

$$h(X) : S \to \mathbf{R}$$

With this definition of $h(X)$, it can be shown that if h is a piece-wise continuous function and X is a random variable, then for any Borel set C of the real line the set $\{s | h(X)(s) \in C\}$ is a member of $\mathcal{F}$, the sigma field of subsets of S. (See Figure 1.2.)

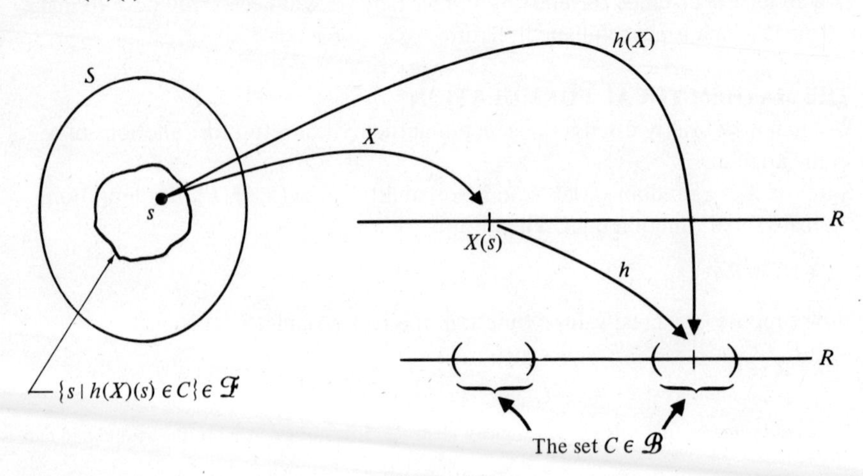

Figure 1.2

We shall not verify this assertion, but will accept it as such. Hence, by the definition of an r.v., it follows that $h(X)$ is a random variable.

Let us denote the random variable $h(X)$ by Y. Now, as we are aware, an r.v. induces a probability measure on the Borel sets of the real line. In our discussion, there are two random variables involved, namely X and Y, and these will induce two probability measures which, using our previous notation, we shall denote respectively by P^X and P^Y. Thus, for any Borel set B of the real line, we have

$$P^X(B) = P(\{s \mid X(s) \in B\})$$

and $$P^Y(B) = P(\{s \mid Y(s) \in B\}) = P(\{s \mid h(X)(s) \in B\})$$

The question is, how do the two probability measures P^X and P^Y relate to each other? To answer this, suppose C is a Borel set of the real line. Let

$$B = \{x \in \mathbf{R} \mid h(x) \in C\} \qquad A = \{s \in S \mid X(s) \in B\}$$

A is a subset of S that consists precisely of preimages (under X) of members of B which, in turn, is a subset of $\mathbf{R}$ and consists precisely of preimages (under h) of the members of C. (See Figure 1.3.)

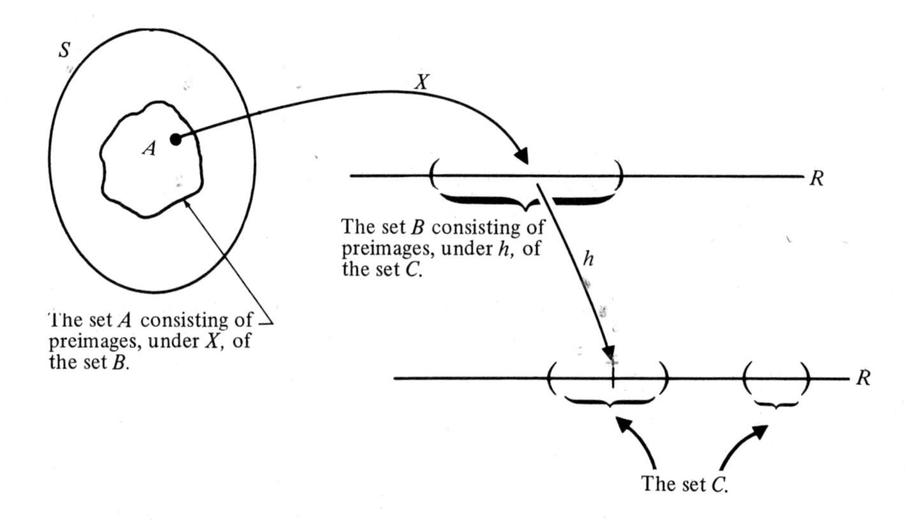

Figure 1.3

As a natural consequence, A consists precisely of preimages (under $h(X)$, that is, Y) of members of C. Thus

$$A = \{s \in S \mid h(X)(s) \in C\} = \{s \in S \mid Y(s) \in C\}$$

In summary, since $A = \{s \in S \mid X(s) \in B\} = \{s \in S \mid Y(s) \in C\}$, the probability measures P^X and P^Y are related to each other by the following assignment of probabilities:

$$P(A) = P^X(B) = P^Y(C)$$

2. THE DISTRIBUTION OF A FUNCTION OF A RANDOM VARIABLE

The discussion of Section 1 considers the problem in terms of broad generalities. It is not of much help in practical problems. Our main interest lies in finding the distribution of a function of X, say $Y = h(X)$, when the distribution of X is known. For example, if the distribution of X is known we might want to find the distribution of $\sin X$, or $X^2 + 3$, or $\log_e X$, and so on. The two cases, where X is a discrete random variable and where X is a continuous random variable, will be considered separately.

2.1 The Discrete Case

If X is a discrete random variable, then so is $Y = h(X)$. In this case the procedure for finding the distribution of Y is simple. *It is sufficient to give the set of values that Y assumes together with the corresponding probabilities.* To get an idea of the general procedure, let us start with the following example.

Example 2.1. Suppose X has the probability function given below.

x	$P(X = x)$
-1	$\frac{1}{3}$
0	$\frac{1}{6}$
1	$\frac{1}{6}$
2	$\frac{1}{3}$

Find the distribution of the random variable $Y = X^2$.

Solution. The function h is therefore defined by $h(x) = x^2$ for any real number x, and we have

x	-1	0	1	2
$h(x)$	$(-1)^2$	0^2	1^2	2^2

Now

$$Y = 1 \Leftrightarrow X = 1 \quad \text{or} \quad X = -1$$

Therefore,

$$P(Y = 1) = P(X = 1) + P(X = -1) = \tfrac{1}{6} + \tfrac{1}{3}$$

Similarly,

$$P(Y = 0) = \tfrac{1}{6} \quad \text{and} \quad P(Y = 4) = \tfrac{1}{3}$$

In summary, the probability function of Y is

y	$P(Y = y)$
0	$\frac{1}{6}$
1	$\frac{1}{6} + \frac{1}{3}$
4	$\frac{1}{3}$

■

In the general context, if $x_1, x_2, \ldots$ are the possible values of X, the possible values of $h(X)$ are $h(x_1), h(x_2), \ldots$. Let $x_{i_1}, x_{i_2}, \ldots$ be the values of X for which $h(x_{i_k}) = y_i$, for $k = 1, 2, \ldots$. Then

$$P(Y = y_i) = P(X = x_{i_1}) + P(X = x_{i_2}) + \ldots$$

This might be more clearly seen from the following tables.

The probability function of X

x	$P(X = x)$
⋮	⋮
x_{i_1}	$P(X = x_{i_1})$
⋮	⋮
x_{i_2}	$P(X = x_{i_2})$
⋮	⋮
x_{i_k}	$P(X = x_{i_k})$
⋮	⋮

The probability function of Y

y	$P(Y = y)$
⋮	⋮
y_i	$P(X = x_{i_1}) + P(X = x_{i_2}) + \cdots$
⋮	⋮

Example 2.2. A particle starts from the origin. Its position, X, along the line is given by the following probability function.

x	$P(X = x)$
−5	0.1
−3	0.2
−1	0.1
2	0.3
3	0.1
5	0.2

Find the distribution of the distance covered.

Solution. Letting Y denote the distance, $Y = |X|$. The probability function of Y is

y	$P(Y = y)$
1	0.1
2	0.3
3	$0.2 + 0.1 = 0.3$
5	$0.1 + 0.2 = 0.3$

Example 2.3. Suppose X has the Poisson distribution with parameter λ. Let

$$Y = \begin{cases} 1 & \text{if } X \text{ is even} \\ -1 & \text{if } X \text{ is odd} \end{cases}$$

Find the distribution of Y.

Solution. Y takes two values 1 and -1.

$$P(Y = 1) = \sum_{r=0}^{\infty} P(X = 2r) = \sum_{r=0}^{\infty} e^{-\lambda} \frac{\lambda^{2r}}{(2r)!} = e^{-\lambda} \sum_{r=0}^{\infty} \frac{\lambda^{2r}}{(2r)!}$$

and

$$P(Y = -1) = \sum_{r=0}^{\infty} P(X = 2r + 1) = \sum_{r=0}^{\infty} e^{-\lambda} \frac{\lambda^{2r+1}}{(2r + 1)!} = e^{-\lambda} \sum_{r=0}^{\infty} \frac{\lambda^{2r+1}}{(2r + 1)!}$$

Now

$$e^{\lambda} = \sum_{r=0}^{\infty} \frac{\lambda^r}{r!} \quad \text{and} \quad e^{-\lambda} = \sum_{r=0}^{\infty} (-1)^r \frac{\lambda^r}{r!}$$

Therefore,

$$e^{\lambda} + e^{-\lambda} = 2 \sum_{r=0}^{\infty} \frac{\lambda^{2r}}{(2r)!} \quad \text{and} \quad e^{\lambda} - e^{-\lambda} = 2 \sum_{r=0}^{\infty} \frac{\lambda^{2r+1}}{(2r + 1)!}$$

Hence

$$P(Y = 1) = e^{-\lambda}\left(\frac{e^{\lambda} + e^{-\lambda}}{2}\right) = \frac{1}{2}(1 + e^{-2\lambda})$$

$$P(Y = -1) = e^{-\lambda}\left(\frac{e^{\lambda} - e^{-\lambda}}{2}\right) = \frac{1}{2}(1 - e^{-2\lambda})$$

Example 2.4. X is a random variable with the probability function given by

$$P(X = i) = (\tfrac{1}{2})^i, \qquad i = 1, 2, \ldots$$

Find the distribution of $Y = \sin(\pi X/2)$.

Solution. We see right away that the possible values of Y are $0, 1, -1$.

$$Y = 0 \Leftrightarrow X \in \{2, 4, 6, \ldots\} = \{2k \mid k \text{ a positive integer}\}$$

$$Y = 1 \Leftrightarrow X \in \{1, 5, 9, \ldots\} = \{4k + 1 \mid k \text{ a nonnegative integer}\}$$

$$Y = -1 \Leftrightarrow X \in \{3, 7, 11, \ldots\} = \{4k + 3 \mid k \text{ a nonnegative integer}\}$$

Then

$$P(Y=0) = \sum_{k=1}^{\infty} P(X=2k) = \sum_{k=1}^{\infty} \left(\frac{1}{2}\right)^{2k} = \sum_{k=1}^{\infty} \left(\frac{1}{4}\right)^{k}$$

$$= \frac{1}{4} \cdot \sum_{k=0}^{\infty} \left(\frac{1}{4}\right)^{k} = \frac{1}{4} \cdot \frac{1}{1-\frac{1}{4}} = \frac{1}{3}$$

$$P(Y=1) = \sum_{k=0}^{\infty} P(X=4k+1) = \sum_{k=0}^{\infty} \left(\frac{1}{2}\right)^{4k+1} = \frac{1}{2} \sum_{k=0}^{\infty} \left(\frac{1}{16}\right)^{k}$$

$$= \frac{1}{2} \cdot \frac{1}{1-\frac{1}{16}} = \frac{8}{15}$$

$$P(Y=-1) = \sum_{k=0}^{\infty} P(X=4k+3) = \sum_{k=0}^{\infty} \left(\frac{1}{2}\right)^{4k+3} = \frac{1}{8} \sum_{k=0}^{\infty} \left(\frac{1}{16}\right)^{k} = \frac{2}{15}$$

2.2 The Continuous Case

In the continuous case, when we want to find the distribution of $Y = h(X)$, it is best to obtain the distribution function of Y. The procedure will therefore consist of finding the D.F. of X first and then using this in some way to find the D.F. of Y. Identifying the distribution functions by using subscripts, let F_X and F_Y denote the distribution functions of X and Y, respectively. Then, by definition, for any real number t,

$$F_Y(t) = P(Y \leqslant t) = P(h(X) \leqslant t)$$

Now suppose the graph of the function h is as given in Figure 2.1.

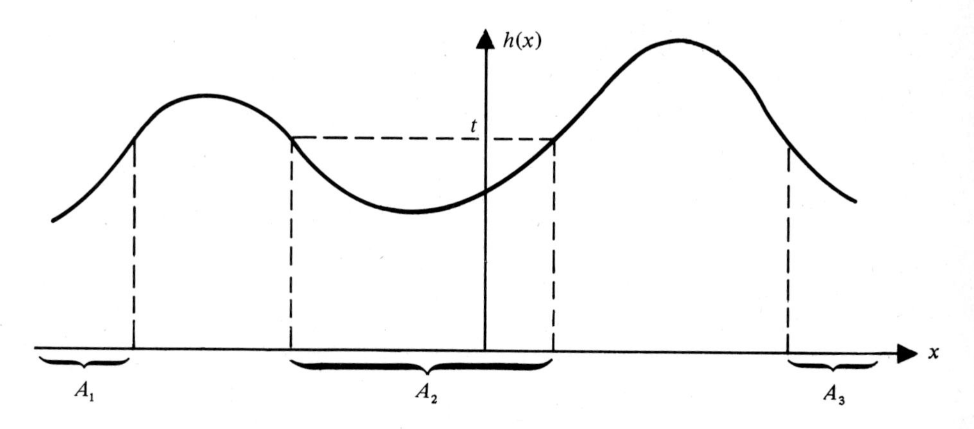

Figure 2.1

From the graph it follows that

$$h(x) \leqslant t \quad \text{if and only if} \quad x \in A_1 \cup A_2 \cup A_3$$

Therefore,

$$F_Y(t) = P(X \in A_1 \cup A_2 \cup A_3)$$

Since the distribution of X is known, it should be possible to obtain the probability $P(X \in A_1 \cup A_2 \cup A_3)$ and thereby $F_Y(t)$.

The following examples are designed to illustrate the essential ingredients of the procedure.

(*Note:* In working with the inequalities below, the following results would be worth keeping in mind:

(*i*) If $a \leqslant b$, then $a + c \leqslant b + c$ for any real number c.

(*ii*) If $a \leqslant b$, then $ac \leqslant bc$ if c is a positive number and $ac \geqslant bc$ if c is a negative number.)

Example 2.5. Suppose a continuous random variable X has the pdf given by

$$f(x) = \begin{cases} e^{-x}, & x > 0 \\ 0, & x \leqslant 0 \end{cases}$$

Find the distribution of–

(*a*) $Y = X^3$

(*b*) $Z = e^X$

(*c*) $W = \ln X$.

Solution. We shall need the D.F. of X, which is given by

$$F_X(u) = \begin{cases} 0, & u < 0 \\ 1 - e^{-u}, & u \geqslant 0 \end{cases}$$

(*a*) $Y = X^3$. Let us find the D.F. of Y.

$$F_Y(y) = P(Y \leqslant y) = P(X^3 \leqslant y)$$

From the graph of h in Figure 2.2, we see that

$$x^3 \leqslant y \iff x \leqslant \sqrt[3]{y}$$

Therefore,

$$F_Y(y) = P(X^3 \leqslant y) = P(X \leqslant \sqrt[3]{y}) = F_X(\sqrt[3]{y})$$

$$= \begin{cases} 0, & \sqrt[3]{y} < 0 \\ 1 - e^{-\sqrt[3]{y}}, & \sqrt[3]{y} \geqslant 0 \end{cases}$$

$$= \begin{cases} 0, & y < 0 \\ 1 - e^{-\sqrt[3]{y}}, & y \geqslant 0 \end{cases}$$

The pdf of Y is given by differentiating as

$$f_Y(y) = \frac{d}{dy} F_Y(y) = \begin{cases} 0, & y \leqslant 0 \\ \frac{1}{3} y^{-2/3} e^{-\sqrt[3]{y}}, & y > 0 \end{cases}$$

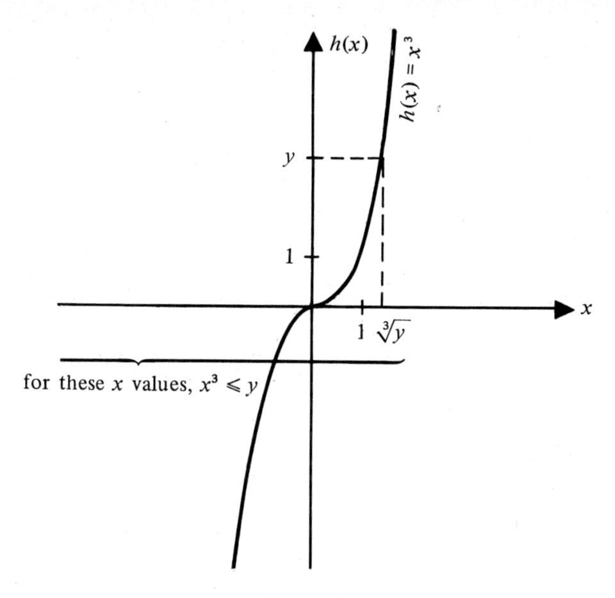

Figure 2.2

(*b*) $Z = e^X$. Here we want to obtain

$$F_Z(u) = P(Z \leqslant u) = P(e^X \leqslant u)$$

The graph of $h(x) = e^x$ is given in Figure 2.3.

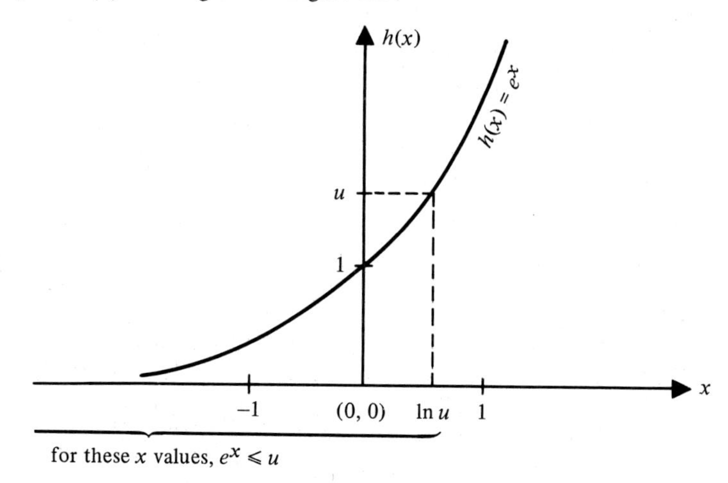

Figure 2.3

From this graph, it follows that the event $\{e^X \leqslant u\}$ is given for different values of u by

$$\{e^X \leqslant u\} = \begin{cases} \emptyset, & u \leqslant 0 \\ \{X \leqslant \ln u\}, & u > 0 \end{cases}$$

Therefore,

$$F_Z(u) = \begin{cases} 0, & u \leqslant 0 \\ P(X \leqslant \ln u), & u > 0 \end{cases}$$

$$= \begin{cases} 0, & u \leqslant 0 \\ F_X(\ln u), & u > 0 \end{cases}$$

$$= \begin{cases} 0, & u \leqslant 0 \\ 0, & \ln u < 0 \quad \text{(that is, } 0 < u < 1) \\ 1 - e^{-\ln u}, & \ln u \geqslant 0 \quad \text{(that is, } u \geqslant 1) \end{cases}$$

$$= \begin{cases} 0, & u < 1 \\ 1 - \dfrac{1}{u}, & u \geqslant 1 \end{cases}$$

The resulting pdf is

$$f_Z(u) = \begin{cases} 0, & u < 1 \\ \dfrac{1}{u^2}, & u \geqslant 1 \end{cases}$$

(*c*) $W = \ln X$. We shall find $F_W(u) = P(\ln X \leqslant u)$ as a function of u. From the graph in Figure 2.4, we see that for any u, $-\infty < u < \infty$,

$$\ln x \leqslant u \Leftrightarrow 0 < x \leqslant e^u$$

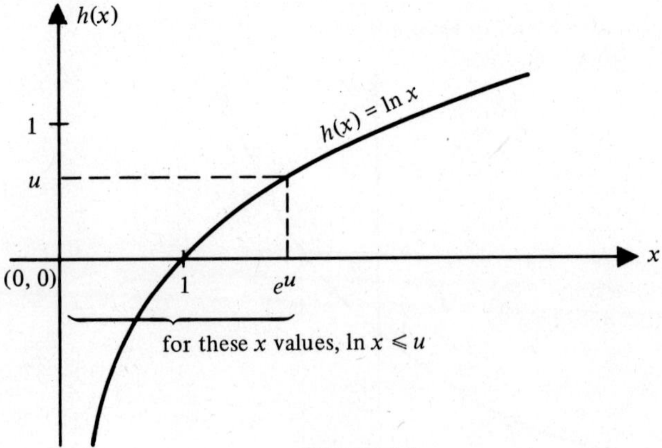

Figure 2.4

Therefore,

$$F_W(u) = P(0 < X \leqslant e^u) = F_X(e^u) = 1 - e^{-e^u}, \quad -\infty < u < \infty$$

The corresponding pdf is

$$f_W(u) = e^{(u - e^u)}, \quad -\infty < u < \infty$$

Example 2.6. Let X be a continuous r.v. whose pdf is given by

$$f_X(x) = \begin{cases} \frac{1}{6}(x^3 + 1), & 0 < x < 2 \\ 0, & \text{elsewhere} \end{cases}$$

Find the distribution of $Y = 2X + 1$.

Solution. Our first step is to obtain the D.F. of X. This is given by

$$F_X(u) = \begin{cases} 0, & u < 0 \\ \int_0^u \frac{1}{6}(x^3 + 1)\, dx, & 0 \leqslant u < 2 \\ 1, & u \leqslant 2 \end{cases}$$

$$= \begin{cases} 0, & u < 0 \\ \dfrac{u^4 + 4u}{24}, & 0 \leqslant u < 2 \\ 1, & u \geqslant 2 \end{cases}$$

We shall now find the distribution of Y by finding its distribution function.

$$F_Y(y) = P(Y \leqslant y) = P(2X + 1 \leqslant y)$$

From the graph in Figure 2.5, it follows that

$$2x + 1 \leqslant y \iff x \leqslant \frac{y - 1}{2}$$

(*Note:* We could also arrive at this conclusion from purely algebraic considerations without actually plotting the graph.)

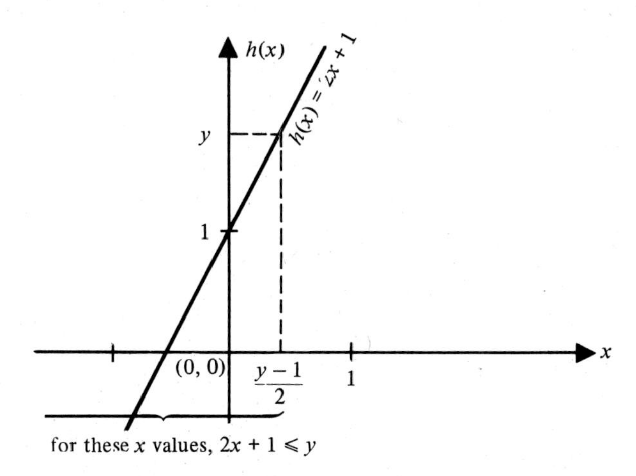

Figure 2.5

Hence

$$F_Y(y) = P\left(X \leqslant \frac{y-1}{2}\right) = F_X\left(\frac{y-1}{2}\right)$$

$$= \begin{cases} 0, & \dfrac{y-1}{2} < 0 \\ \dfrac{\left(\dfrac{y-1}{2}\right)^4 + 4\left(\dfrac{y-1}{2}\right)}{24}, & 0 \leqslant \dfrac{y-1}{2} < 2 \\ 1, & \dfrac{y-1}{2} \geqslant 2 \end{cases}$$

$$= \begin{cases} 0, & y < 1 \\ \dfrac{\left(\dfrac{y-1}{2}\right)^4 + 2(y-1)}{24}, & 1 \leqslant y < 5 \\ 1, & y \geqslant 5 \end{cases}$$

The pdf can now be obtained by differentiating.

Example 2.7. Suppose X has pdf given by

$$f(x) = \begin{cases} 2x, & 0 < x < 1 \\ 0, & \text{elsewhere} \end{cases}$$

Find the distribution of: (*a*) $Y = -2X + 3$; (*b*) $Z = \dfrac{1}{X+1}$.

Solution. The D.F. of X can be shown to be

$$F_X(x) = \begin{cases} 0, & x < 0 \\ x^2, & 0 \leqslant x < 1 \\ 1, & x \geqslant 1 \end{cases}$$

(*a*) To find the distribution of Y, we note that

$$F_Y(y) = P(Y \leqslant y) = P(-2X + 3 \leqslant y)$$

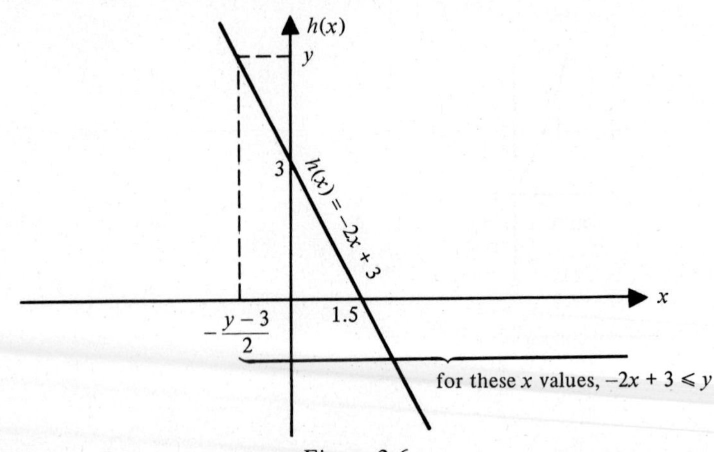

Figure 2.6

From the graph in Figure 2.6, or from purely algebraic considerations, we see that

$$-2x + 3 \leqslant y \Leftrightarrow x \geqslant -\frac{y-3}{2}$$

Therefore,

$$F_Y(y) = P\left(X \geqslant -\frac{y-3}{2}\right) = 1 - P\left(X \leqslant -\frac{y-3}{2}\right) = 1 - F_X\left(-\frac{y-3}{2}\right)$$

$$= \begin{cases} 1-0, & -\frac{y-3}{2} < 0 \\ 1-\left(-\frac{y-3}{2}\right)^2, & 0 \leqslant -\frac{y-3}{2} < 1 \\ 1-1, & -\frac{y-3}{2} \geqslant 1 \end{cases}$$

$$= \begin{cases} 1, & y > 3 \\ 1-\left(\frac{y-3}{2}\right)^2, & 1 < y \leqslant 3 \\ 0, & y \leqslant 1 \end{cases}$$

The pdf of Y is

$$f_Y(y) = \begin{cases} \frac{3-y}{2}, & 1 < y < 3 \\ 0, & \text{elsewhere} \end{cases}$$

(*b*) Since $Z = \frac{1}{X+1}$, $F_Z(u) = P\left(\frac{1}{X+1} \leqslant u\right)$.

From the graph in Figure 2.7, it follows that

$$\left\{\frac{1}{X+1} \leqslant u\right\} = \begin{cases} \emptyset, & u \leqslant 0 \\ \left\{X \geqslant \frac{1-u}{u}\right\}, & u > 0 \end{cases}$$

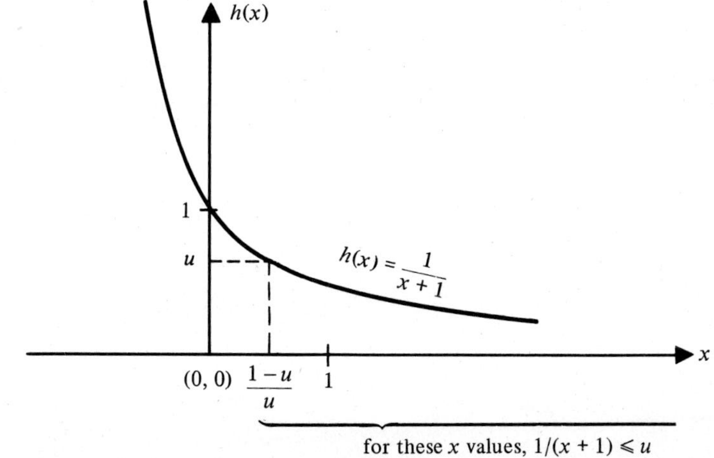

Figure 2.7

Therefore,

$$F_Z(u) = \begin{cases} 0, & u \leqslant 0 \\ P\left(X \geqslant \dfrac{1-u}{u}\right), & u > 0 \end{cases}$$

$$= \begin{cases} 0, & u \leqslant 0 \\ 1 - F_X\left(\dfrac{1-u}{u}\right), & u > 0 \end{cases}$$

Now if $u > 0$,

$$1 - F_X\left(\frac{1-u}{u}\right) = \begin{cases} 1 - 0, & \dfrac{1-u}{u} < 0 \quad (\text{that is, } u > 1) \\ 1 - \left(\dfrac{1-u}{u}\right)^2, & 0 \leqslant \dfrac{1-u}{u} < 1 \quad (\text{that is, } \tfrac{1}{2} < u \leqslant 1) \\ 1 - 1, & \dfrac{1-u}{u} \geqslant 1 \quad (\text{that is, } 0 < u \leqslant \tfrac{1}{2}) \end{cases}$$

Substituting, we get

$$F_Z(u) = \begin{cases} 0, & u \leqslant 0 \\ 1, & u > 1 \\ 1 - \left(\dfrac{1-u}{u}\right)^2, & \tfrac{1}{2} < u \leqslant 1 \\ 0, & 0 < u \leqslant \tfrac{1}{2} \end{cases}$$

In other words,

$$F_Z(u) = \begin{cases} 0, & u \leqslant \tfrac{1}{2} \\ 1 - \left(\dfrac{1-u}{u}\right)^2, & \tfrac{1}{2} < u \leqslant 1 \\ 1, & u > 1 \end{cases}$$

The pdf of Z can then be found by differentiation as

$$f_Z(u) = \begin{cases} 2\left(\dfrac{1-u}{u^3}\right), & \tfrac{1}{2} < u < 1 \\ 0, & \text{elsewhere} \end{cases}$$

Example 2.8. Suppose X is uniformly distributed over the interval $[-\pi, \pi]$. Find the distribution of–

(*a*) $Y = \cos X$ (*b*) $Z = \sin X$ (*c*) $W = |X|$.

Solution. The pdf of X is

$$f_X(x) = \begin{cases} \dfrac{1}{2\pi}, & -\pi \leqslant x \leqslant \pi \\ 0, & \text{elsewhere} \end{cases}$$

and its D.F. is

$$F_X(x) = \begin{cases} 0, & x < -\pi \\ \dfrac{x+\pi}{2\pi}, & -\pi \leqslant x < \pi \\ 1, & x \geqslant \pi \end{cases}$$

(*a*) $F_Y(y) = P(\cos X \leqslant y)$. From the graph of the cosine function shown in Figure 2.8, it follows that

$$\{\cos X \leqslant y\} = \begin{cases} \emptyset, & y < -1 \\ \{-\pi < X \leqslant -\cos^{-1} y\} \cup \{\cos^{-1} y < X \leqslant \pi\}, & -1 \leqslant y \leqslant 1 \\ S, & y > 1 \end{cases}$$

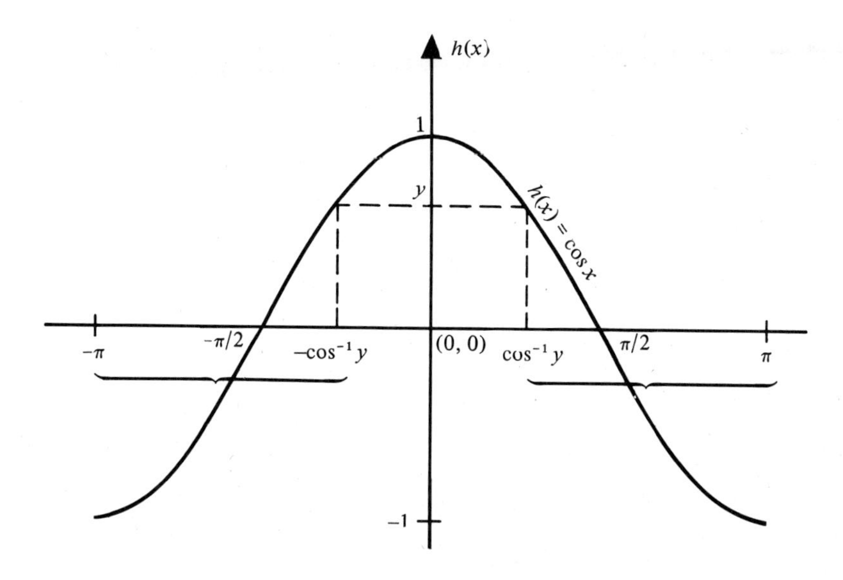

Figure 2.8 $\cos x \leqslant y$ precisely for $x \in (-\pi, -\cos^{-1} y) \cup (\cos^{-1} y, \pi)$

Therefore,

$$F_Y(y) = \begin{cases} 0, & y < -1 \\ P(-\pi < X \leqslant -\cos^{-1} y) + P(\cos^{-1} y < X \leqslant \pi), & -1 \leqslant y \leqslant 1 \\ 1, & y > 1 \end{cases}$$

$$= \begin{cases} 0, & y < -1 \\ \dfrac{-\cos^{-1} y + \pi}{2\pi} + \dfrac{\pi - \cos^{-1} y}{2\pi}, & -1 \leqslant y \leqslant 1 \\ 1, & y > 1 \end{cases}$$

In summary, simplifying,

$$F_Y(y) = \begin{cases} 0, & y < -1 \\ 1 - \dfrac{1}{\pi} \cos^{-1} y, & -1 \leqslant y \leqslant 1 \\ 1, & y > 1 \end{cases}$$

(*b*) $F_Z(u) = P(\sin X \leqslant u)$. From Figures 2.9(a) and (b), it can be seen that

$$\{\sin X \leqslant u\} = \begin{cases} \emptyset, & u < -1 \\ \{-\pi - \sin^{-1} u \leqslant X \leqslant \sin^{-1} u\}, & -1 \leqslant u < 0 \\ \{-\pi \leqslant X \leqslant \sin^{-1} u\} \cup \{\pi - \sin^{-1} u \leqslant X \leqslant \pi\}, & 0 \leqslant u \leqslant 1 \\ S, & u > 1 \end{cases}$$

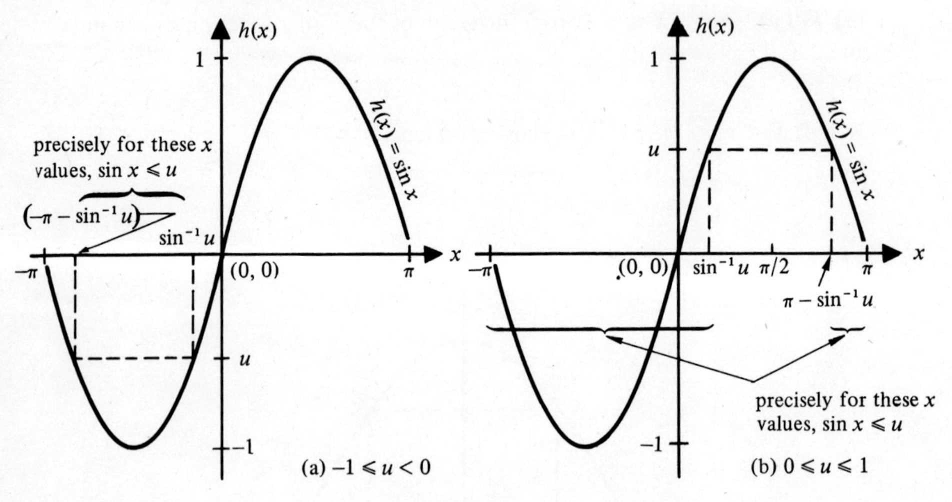

Figure 2.9

(The case $-1 \leqslant u < 0$ is shown in Figure 2.9(a) and the case of $0 \leqslant u \leqslant 1$ in Figure 2.9(b).)

Therefore,

$$F_Z(u) = P(\sin X \leqslant u)$$
$$= \begin{cases} 0, & u < -1 \\ P(-\pi - \sin^{-1} u \leqslant X \leqslant \sin^{-1} u), & -1 \leqslant u < 0 \\ P(-\pi \leqslant X \leqslant \sin^{-1} u) + P(\pi - \sin^{-1} u \leqslant X \leqslant \pi), & 0 \leqslant u \leqslant 1 \\ 1, & u > 1 \end{cases}$$

After simplifying this yields

$$F_Z(u) = \begin{cases} 0, & u < -1 \\ \dfrac{2 \sin^{-1} u + \pi}{2\pi}, & -1 \leqslant u \leqslant 1 \\ 1, & u > 1 \end{cases}$$

(*c*) $F_W(u) = P(|X| \leqslant u)$. From the definition of the absolute value function it follows that

$$\{|X| \leqslant u\} = \begin{cases} \emptyset, & u < 0 \\ \{-u \leqslant X \leqslant u\}, & u \geqslant 0 \end{cases}$$

(The case $u \geqslant 0$ is shown in Figure 2.10.)

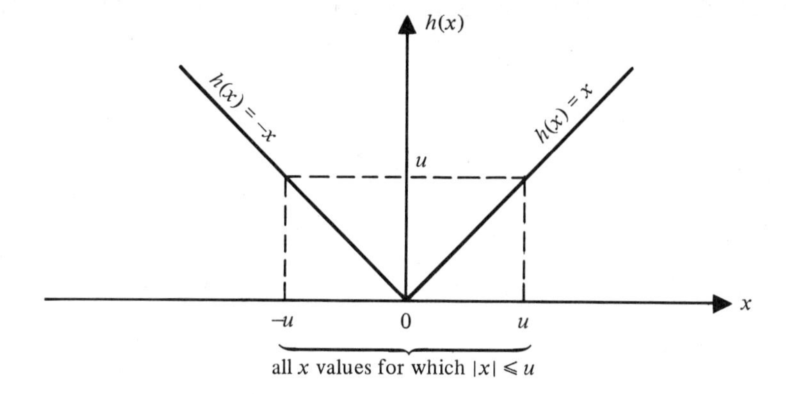

Figure 2.10

Therefore,

$$F_W(u) = \begin{cases} 0, & u < 0 \\ F_X(u) - F_X(-u), & u \geqslant 0 \end{cases}$$

$$= \begin{cases} 0, & u < 0 \\ \dfrac{u}{\pi}, & 0 \leqslant u < \pi \\ 1, & u \geqslant \pi \end{cases}$$

In conclusion, $|X|$ has a uniform distribution over the interval $[0, \pi]$ if X is uniformly distributed over the interval $[-\pi, \pi]$. This is in keeping with what we might anticipate intuitively.

- *Example 2.9.* Suppose a pointer is supported at the point (0, 1) on the y-axis as shown in Figure 2.11. When the pointer is spun it comes to rest making an angle X with the y-axis, where $-\pi/2 \leqslant X \leqslant \pi/2$. Assuming that X has a uniform distribution over $[-\pi/2, \pi/2]$, find the distribution of Y, the point on the x-axis at which the pointer is pointing.

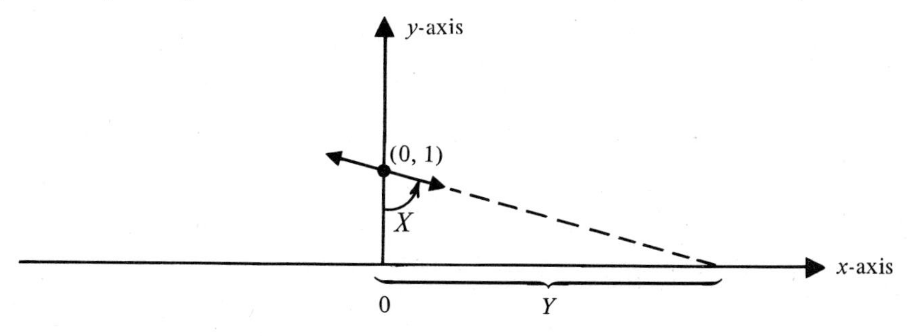

Figure 2.11

Solution. The D.F. of X is

$$F_X(x) = \begin{cases} 0, & x < -\pi/2 \\ \dfrac{x + (\pi/2)}{\pi}, & -\pi/2 \leqslant x < \pi/2 \\ 1, & x \geqslant \pi/2 \end{cases}$$

It follows that the relation connecting X and Y is

$$Y = \tan X$$

Now from Figure 2.12 it can be seen that for any u, $-\infty < u < \infty$,

$$\tan x \leqslant u \Leftrightarrow -\pi/2 < x \leqslant \tan^{-1} u$$

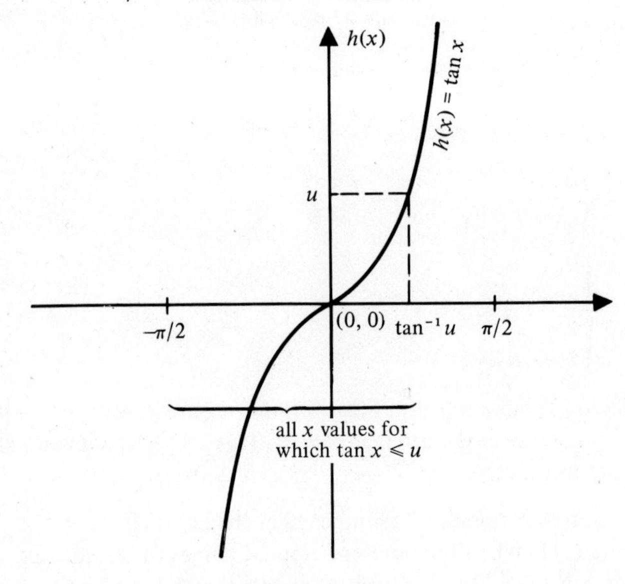

Figure 2.12

Therefore,

$$F_Y(u) = P(\tan X \leqslant u) = P(-\pi/2 \leqslant X \leqslant \tan^{-1} u) = \frac{\tan^{-1} u + (\pi/2)}{\pi}$$

for $-\infty < u < \infty$. Hence the pdf of Y is

$$f_Y(u) = \frac{1}{\pi(1 + u^2)}, \qquad -\infty < u < \infty$$

That is, Y has a Cauchy distribution. This example shows how the Cauchy distribution is related to the uniform distribution.

- *Example 2.10.* Suppose X has the standard normal distribution. Find the distribution of $Y = X^2$.

Solution. It can be seen immediately from Figure 2.13 that

$$F_Y(u) = \begin{cases} 0, & u \leqslant 0 \\ P(-\sqrt{u} \leqslant X \leqslant \sqrt{u}), & u > 0 \end{cases}$$

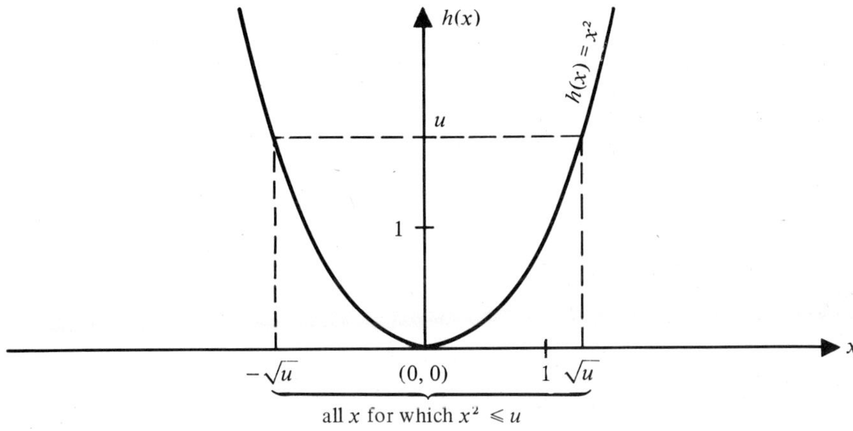

Figure 2.13

Therefore,

$$F_Y(u) = \begin{cases} 0, & u \leqslant 0 \\ \Phi(\sqrt{u}) - \Phi(-\sqrt{u}), & u > 0 \end{cases}$$

(Recall that Φ represents the D.F. of the standard normal variable.)

Since an explicit expression for $\Phi(x)$ is not available in a closed form, we cannot find an expression for $F_Y(u)$. However, since $\Phi'(x) = \frac{d}{dx}\Phi(x) = \frac{1}{\sqrt{2\pi}} e^{-(x^2/2)}$, we can find the pdf of Y. We have

$$f_Y(u) = \frac{d}{du}F_Y(u) = \begin{cases} 0, & u \leqslant 0 \\ \frac{d}{du}\Phi(\sqrt{u}) - \frac{d}{du}\Phi(-\sqrt{u}), & u > 0 \end{cases}$$

$$= \begin{cases} 0, & u \leqslant 0 \\ \frac{1}{2\sqrt{u}}\Phi'(\sqrt{u}) + \frac{1}{2\sqrt{u}}\Phi'(-\sqrt{u}), & u > 0 \end{cases}$$

$$= \begin{cases} 0, & u \leqslant 0 \\ \frac{1}{\sqrt{u}} \cdot \frac{1}{\sqrt{2\pi}} e^{-u/2}, & u > 0 \end{cases}$$

$$= \begin{cases} 0, & u \leqslant 0 \\ \frac{1}{2^{1/2}\Gamma(\frac{1}{2})} u^{(1/2)-1} e^{-u/2}, & u > 0 \end{cases}$$

But this is the chi-square distribution with one degree of freedom. Thus we see an important connection between the standard normal distribution and the chi-square distribution: *If X is $N(0, 1)$, then X^2 has a chi-square distribution with one degree of freedom.*

Example 2.11. Suppose X has the exponential distribution

$$F_X(u) = \begin{cases} 0, & u<0 \\ 1-e^{-u}, & u \geqslant 0 \end{cases}$$

Let

$$Y = \begin{cases} X & \text{if } X \leqslant 2 \\ \dfrac{1}{X} & \text{if } X > 2 \end{cases}$$

Find the distribution of Y.

Solution. (Note that the function h is not continuous, but is piecewise continuous.)

To find the D.F. of Y, we observe from the graphs in Figures 2.14(a) and (b) that

$$\{Y \leqslant u\} = \begin{cases} \emptyset, & u<0 \\ \{0 \leqslant X \leqslant u\} \cup \left\{X \geqslant \dfrac{1}{u}\right\}, & 0<u<\frac{1}{2} \qquad \text{(see Figure 2.14(a))} \\ \{0 \leqslant X \leqslant u\} \cup \{X \geqslant 2\}, & \frac{1}{2} \leqslant u \leqslant 2 \qquad \text{(see Figure 2.14(b))} \\ S, & u>2 \end{cases}$$

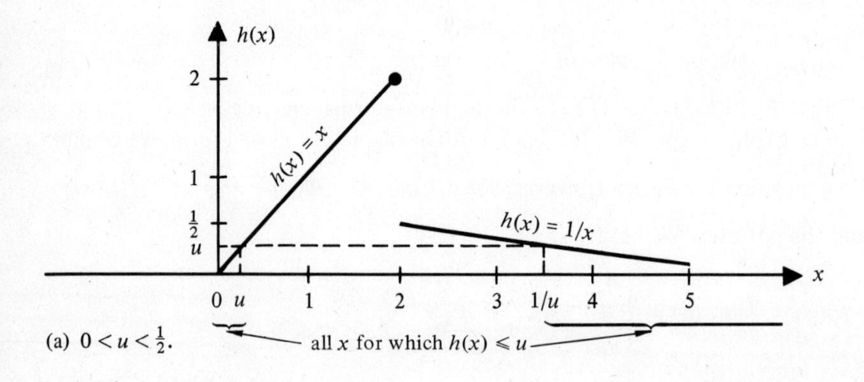

(a) $0<u<\frac{1}{2}$.

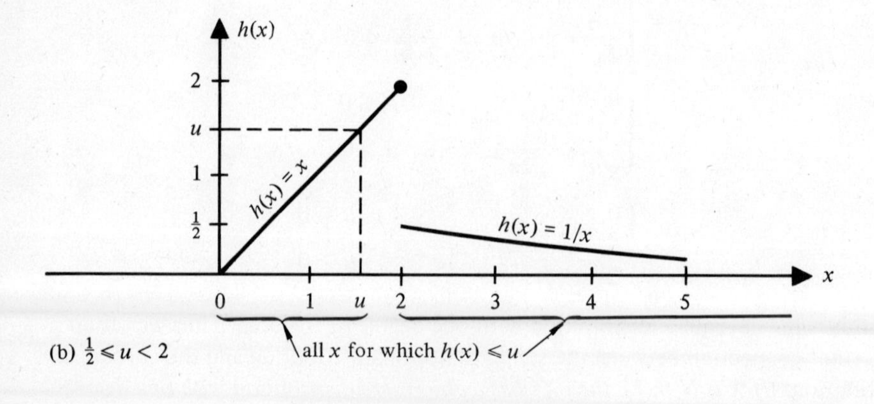

(b) $\frac{1}{2} \leqslant u < 2$

Figure 2.14

Therefore,

$$F_Y(u) = \begin{cases} 0, & u \leqslant 0 \\ P(0 \leqslant X \leqslant u) + P\left(X \geqslant \dfrac{1}{u}\right), & 0 < u < \frac{1}{2} \\ P(0 \leqslant X \leqslant u) + P(X \geqslant 2), & \frac{1}{2} \leqslant u \leqslant 2 \\ 1, & u > 2 \end{cases}$$

$$= \begin{cases} 0, & u \leqslant 0 \\ (1 - e^{-u}) + [1 - (1 - e^{-1/u})], & 0 < u < \frac{1}{2} \\ (1 - e^{-u}) + [1 - (1 - e^{-2})], & \frac{1}{2} \leqslant u \leqslant 2 \\ 1, & u > 2 \end{cases}$$

Simplifying, this gives

$$F_Y(u) = \begin{cases} 0, & u \leqslant 0 \\ 1 - e^{-u} + e^{-1/u}, & 0 < u < \frac{1}{2} \\ 1 - e^{-u} + e^{-2}, & \frac{1}{2} \leqslant u \leqslant 2 \\ 1, & u > 2 \end{cases}$$

- *Example 2.12.* Suppose X is an absolutely continuous random variable with a distribution function F which is *strictly increasing.* Define a random variable Y by $Y = F(X)$. Find the distribution of Y.

Solution. $F_Y(y) = P(F(X) \leqslant y)$. Since F is a D.F.,

$$\{F(X) \leqslant y\} = \begin{cases} \emptyset, & y < 0 \\ \{X \leqslant F^{-1}(y)\}, & 0 \leqslant y \leqslant 1 \\ S, & y > 1 \end{cases}$$

(See Figure 2.15. Here $F^{-1}(y)$ is a number which is the preimage of y. Since F is strictly increasing, there is precisely one preimage for any y.)

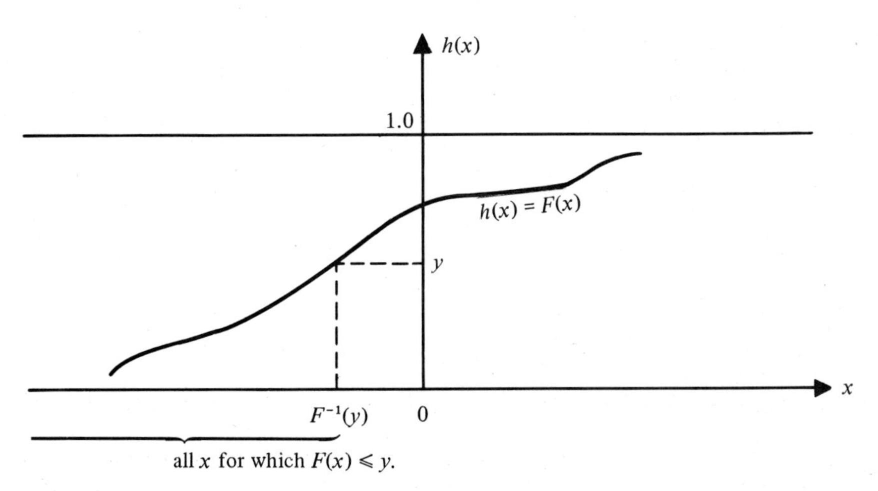

Figure 2.15

Therefore,

$$F_Y(y) = \begin{cases} 0, & y < 0 \\ F(F^{-1}(y)), & 0 \leqslant y \leqslant 1 \\ 1, & y > 1 \end{cases}$$

$$= \begin{cases} 0, & y < 0 \\ y, & 0 \leqslant y \leqslant 1 \\ 1, & y > 1 \end{cases}$$

■

Comment. Example 2.12 shows that *for any absolutely continuous random variable X with a strictly increasing distribution function F, the distribution of Y = F(X) is uniform over the interval [0, 1].*

Example 2.13. The length of time X (measured in days) that a gadget lasts has the following pdf:

$$f(x) = \begin{cases} 2e^{-2x}, & x > 0 \\ 0, & \text{elsewhere} \end{cases}$$

If the gadget lasts more than three days the profit is \$1; otherwise it incurs a loss of \$2. Find the distribution of Y, the profit.

Solution. The random variable Y is defined by

$$Y = \begin{cases} 1 & \text{if } X > 3 \\ -2 & \text{if } X \leqslant 3 \end{cases}$$

We shall solve this example by two methods.

Method 1: Here, following the routine adopted so far, we shall find the D.F. of Y. From the graph in Figure 2.16, notice, for example, that if $-2 \leqslant u < 1$, then

$$y \leqslant u \Leftrightarrow x \leqslant 3$$

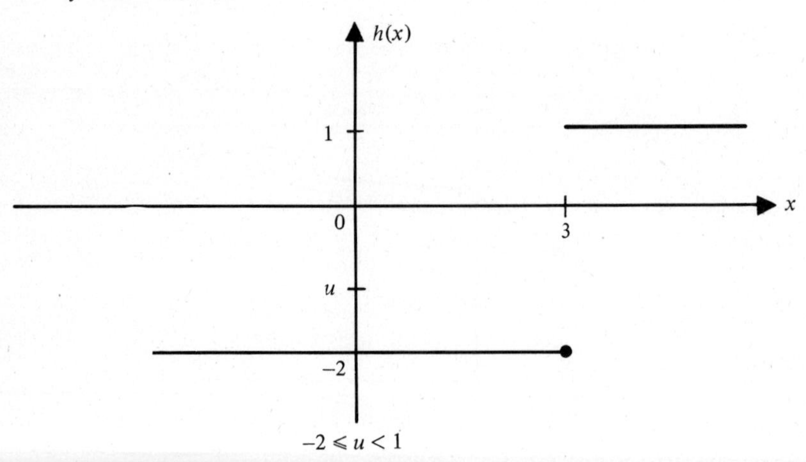

Figure 2.16

Continuing this argument, we get

$$F_Y(u) = P(Y \leqslant u) = \begin{cases} 0, & u < -2 \\ P(X \leqslant 3), & -2 \leqslant u < 1 \\ 1, & u \geqslant 1 \end{cases}$$

$$= \begin{cases} 0, & u < -2 \\ 1 - e^{-6}, & -2 \leqslant u < 1 \\ 1, & u \geqslant 1 \end{cases}$$

As can be seen immediately, this is the D.F. of a discrete random variable whose probability function is given by

$$P(Y = -2) = 1 - e^{-6} \quad \text{and} \quad P(Y = 1) = e^{-6}$$

Method 2: This method is simpler than Method 1. The first observation we make is that Y is a discrete r.v. which takes the values -2 and 1. We can therefore zero in and give the probability function of Y. Now

$$Y = -2 \Leftrightarrow X \leqslant 3$$

and $$Y = 1 \Leftrightarrow X > 3$$

Therefore,

$$P(Y = -2) = P(X \leqslant 3) = F_X(3) = 1 - e^{-6}$$

and $$P(Y = 1) = P(X > 3) = 1 - P(X \leqslant 3) = e^{-6}$$

EXERCISES–SECTION 2

1. Let $Y = h(X)$. The four cases on the basis of the nature of the distribution functions F_X and F_Y are listed below.

Case	F_X	F_Y
1.	absolutely continuous	absolutely continuous
2.	absolutely continuous	discrete
3.	discrete	absolutely continuous
4.	discrete	discrete

Show which of the cases are not possible.

2. If the distribution of X is symmetric about c, show that $X - c$ and $-X + c$ have the same distribution. (The distribution of a random variable X is symmetric about c if $P(X \leqslant c - x) = P(X \geqslant c + x)$ for any real number x.)

3. Suppose X is any random variable and $Y = |X|$. Show that

$$F_X(y) = \begin{cases} 0, & y \leqslant 0 \\ F_X(y) - F_X(-y) + P(X = -y), & y > 0 \end{cases}$$

4. Suppose X and Y are continuous random variables with respective pdf's f_X and f_Y.

   (a) If $Y = h(X)$, where h is a *monotone increasing* function, show that $f_Y(y) = f_X(g(y))g'(y)$, where g is a function inverse of h defined by $X = g(Y)$.

(b) If $Y = h(X)$, where h is a *monotonically decreasing* function, show that $f_Y(y) = -f_X(g(y))g'(y)$, where g is defined by $X = g(Y)$.

5. A random variable X has the following probability function:

$$P(X = x) = \tfrac{1}{35}(4 + x), \qquad x \in \{-3, -1, 0, 1, 2, 3, 5\}$$

Find the distribution of–

(a) $Y = 3X - 4$ (b) $Z = X^2 + 1$.

6. If $P(X = 2) = p$ and $P(X = 0) = 1 - p$, where $0 \leqslant p \leqslant 1$, find the probability function of $Y = X(2 - X)$.

7. (a) If X is $B(5, 0.4)$, find the probability function of $Y = (X - 2)^2$.
(b) If X is $B(5, 0.3)$ find the probability function of $Y = (X - 1.5)^2$.

8. If X is $B(n, p)$, find the probability function of $Y = (X - np)^2$.

9. Suppose X is $B(n, p)$. Let

$$Y = \begin{cases} 1 & \text{if } X \text{ is even} \\ -1 & \text{if } X \text{ is odd} \end{cases}$$

Show that the probability function of Y is given by

$$P(Y = 1) = \tfrac{1}{2}[1 + (1 - 2p)^n], \qquad P(Y = -1) = \tfrac{1}{2}[1 - (1 - 2p)^n]$$

Hint: $$(x + y)^n + (x - y)^n = 2 \sum_{r=0}^{n} \binom{n}{2r} x^{2r} y^{n-2r}$$

10. Suppose X has an exponential distribution with parameter λ. Define a random variable Y as follows:

$$Y = i \text{ if and only if } i - 1 \leqslant X < i, \qquad i = 1, 2, 3, \ldots$$

Find the distribution of Y.

11. The radius of a particle is a random variable with a pdf given by

$$f(x) = \begin{cases} 3x^2, & 0 < x < 1 \\ 0, & \text{elsewhere} \end{cases}$$

Find the distribution of the volume of the particle.

12. The length of the side of a cube is a random variable which is uniformly distributed over the interval $[0, b]$. Find the distribution of the volume of the cube.

13. A random variable X has the following distribution function:

$$F(x) = \begin{cases} 0, & x < 0 \\ 1 - e^{-x}, & x \geqslant 0 \end{cases}$$

Find the distribution of $Y = \sqrt{X}$.

14. Suppose X is a standard normal variable. Find the distribution of Y when Y is defined by

$$Y = \begin{cases} \sqrt{X} & \text{if } X \geqslant 0 \\ \sqrt{-X} & \text{if } X < 0 \end{cases}$$

15. The radius of a circle (measured in inches) is a random variable with a uniform distribution over the interval $[10, 14]$. Find the distribution of the area of the circle.

16. If X is uniformly distributed over the interval $[-\pi, 2\pi]$, find the distributions of–

(a) $Y = \cos X$ (b) $Z = \sin X$ (c) $W = |X|$

17. Suppose X is $N(\mu, \sigma^2)$. Show that the distribution of $Y = cX + d$, where $c \neq 0$, is $N(c\mu + d, c^2\sigma^2)$.
18. If Y is $N(\mu, \sigma^2)$, show that $X = (Y - \mu)/\sigma$ is $N(0, 1)$.
19. Suppose X is $N(a, b^2)$. Find the distribution of $Y = e^X$.
20. Suppose X is a standard normal variable. Show that

$$P(|X| < u) = \begin{cases} 0, & u < 0 \\ 2\Phi(u) - 1, & u \geq 0 \end{cases}$$

where Φ is the D.F. of a standard normal variable. Hence, show that the pdf of $Z = |X|$ is given by

$$f_Z(u) = \begin{cases} (\sqrt{2/\pi})e^{-u^2/2}, & u > 0 \\ 0, & \text{elsewhere} \end{cases}$$

21. If X is a random variable having the normal distribution $N(\mu, \sigma^2)$, find the distribution of $Y = |X - \mu|$.
22. If X has the pdf

$$f(x) = \tfrac{1}{2}e^{-|x|}, \quad -\infty < x < \infty$$

find the pdf of (a) $Y = \sqrt[3]{X}$, (b) $Z = |X|$.
23. The amount of time T (measured in hours) that it takes to repair a machine has the following pdf:

$$f_T(x) = \begin{cases} e^{-x}, & x > 0 \\ 0, & x \leq 0 \end{cases}$$

During the first hour there is a flat charge of \$12. There is an additional charge of \$6 per hour for any time beyond the first hour. If Y represents the amount charged, find the distribution of Y.
24. Suppose the pdf of a random variable X is

$$f(x) = \begin{cases} 2(1 - x), & 0 < x < 1 \\ 0, & \text{elsewhere} \end{cases}$$

Define Y as follows:

$$Y = \begin{cases} 2X & \text{if } 0 < X < \tfrac{1}{2} \\ 5 & \text{if } X \geq \tfrac{1}{2} \end{cases}$$

Find the distribution of Y.
25. Suppose X is a random variable with a known distribution. For a given $b > 0$, let $Y = h(X)$, where the function h is defined by

$$h(x) = \begin{cases} b, & x \geq b \\ x, & -b < x < b \\ -b, & x \leq -b \end{cases}$$

Show that

$$F_Y(y) = \begin{cases} 0, & y < -b \\ F_X(y), & -b \leqslant y < b \\ 1, & y \geqslant b \end{cases}$$

26. If X is a random variable, let $Y = h(X)$, where h is defined by

$$h(x) = \begin{cases} x + b, & x \geqslant 0 \\ x - b, & x < 0 \end{cases}$$

for a fixed given value $b > 0$. Show that

$$F_Y(y) = \begin{cases} F_X(y + b), & y < -b \\ F_X(0), & -b \leqslant y < b \\ F_X(y - b), & y \geqslant b \end{cases}$$

7 Expectation–A Single Variable

INTRODUCTION

The expected value of a random variable is a single number which gives us some information about the distribution of the random variable. There are basically two approaches to define the expected value of a random variable. Since a random variable is a real-valued function defined on the sample space S, we can formulate expectation in terms of the probability space $(S, \mathcal{F}, P)$. Such a treatment leads into the theory of abstract Lebesgue integration. Since we do not presuppose familiarity with this general concept of the integral, we shall formulate our definition equivalently in an alternate way.

Recall that a random variable induces a probability distribution on the real line and that this distribution can be expressed in terms of the distribution function of X. As a consequence, the concept of the expected value of a random variable can be developed by referring to the distribution of the probability mass on the real line. Thus the real line is our setting.

1. DEFINITIONS AND BASIC RESULTS

1.1 The Definition of Expectation

If X is a random variable, the **expected value of** X (or the **expectation of** X, or the **mean of** X), to be denoted by $E(X)$ or μ_X, is a real number defined by

$$E(X) = \begin{cases} \sum_i x_i P(X = x_i) & \text{in the discrete case} \\ \int_{-\infty}^{\infty} x f_X(x)\, dx & \text{in the continuous case} \end{cases}$$

(*Note:* In the discrete case, the summation is carried over all the possible values of X.)

If the context is well defined and there is no possibility of confusion, we shall simply write μ, omitting the subscript X. This policy will also apply to the subscripts X of the probability function p_X and the pdf f_X.

$E(X)$ is defined if the sum or the integral, as the case may be, converges absolutely; that is,

$$\sum_i |x_i| P(X = x_i) < \infty \quad \text{or} \quad \int_{-\infty}^{\infty} |x| f(x)\, dx < \infty$$

In the above definition of the expected value, we treated the discrete and the continuous cases separately. A unifying approach which gives one single expression for both the discrete and the continuous case is provided by the concept of the improper Riemann-Stieltjes integral $\int_{-\infty}^{\infty} x\, dF(x)$. For our treatment of the topic the knowledge of Stieltjes integral is not essential, and we do not assume that the reader is familiar with the concept. The propositions that we state will hold true for the discrete as well as the continuous case. However, in many instances, we shall provide proofs assuming that the distribution is absolutely continuous. In other instances we will assume a discrete distribution because of the delicate nature of proof for the continuous case.

The physical interpretation. We can give the following physical interpretation of the mathematical expectation of a random variable: In physics, if the mass $m(x_i)$ is located at a distance x_i, $i = 1, 2, \ldots,$ from the origin, then the center of gravity of the mass is located at $\sum_i x_i m(x_i) / \sum_i m(x_i)$. If we identify $m(x_i)$ with $p(x_i)$ then, since

$$E(X) = \sum_i x_i p(x_i) = \frac{\sum_i x_i p(x_i)}{\sum_i p(x_i)}$$

the expected value of a random variable can be regarded analogously as representing the center of gravity of the probability distribution. A similar explanation holds in the continuous case.

The definition of $E(X)$ states that $E(X)$ exists if $\int_{-\infty}^{\infty} |x| f(x)\, dx < \infty$ in the continuous case and $\sum_i |x_i| p(x_i) < \infty$ in the discrete case. Let us investigate this. We shall discuss the continuous case and our discussion will be purely intuitive.

We know that a pdf is always nonnegative. Therefore,

$$\begin{aligned} & xf(x) \geqslant 0 \quad \text{if} \quad x \geqslant 0 \\ \text{and} \quad & xf(x) \leqslant 0 \quad \text{if} \quad x \leqslant 0 \end{aligned}$$

The graph of $xf(x)$ might be as given in Figure 1.1.

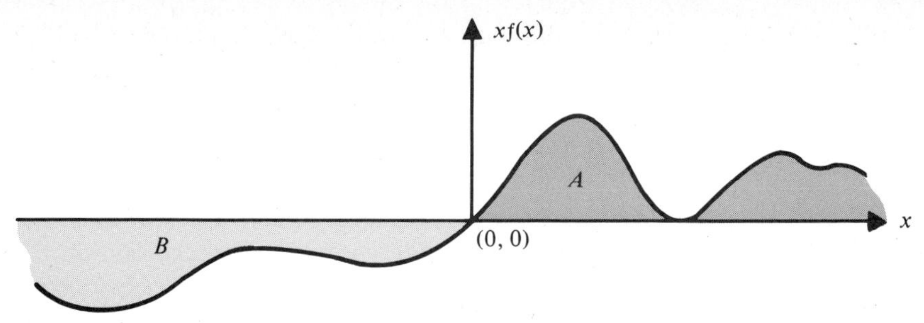

Figure 1.1

Let us denote by A the area which lies under the curve $xf(x)$ and is bounded below by the x-axis, and by B the area which lies above the curve $xf(x)$ and is bounded above by the x-axis. Then, geometrically,

$$\int_{-\infty}^{\infty} xf(x)\,dx = A - B$$

and $$\int_{-\infty}^{\infty} |x|f(x)\,dx = A + B$$

There are four possibilities:

(1) $A = \infty$ and $B = \infty$
(2) $A = \infty$ and $0 \leqslant B < \infty$
(3) $0 \leqslant A < \infty$ and $B = \infty$
(4) $0 \leqslant A < \infty$ and $0 \leqslant B < \infty$

In the first three cases, we see that $A + B = \infty$ and $A - B$ is not a finite number. (As far as the first possibility is concerned, we would get $\infty - \infty$ for $A - B$, which does not even make sense.) That is, $\int_{-\infty}^{\infty} |x|f(x)\,dx = \infty$, and $\int_{-\infty}^{\infty} xf(x)\,dx$ does not exist. In the fourth case, $A + B$ is finite and so is $A - B$. Therefore, $\int_{-\infty}^{\infty} xf(x)\,dx$ is defined whenever $\int_{-\infty}^{\infty} |x|f(x)\,dx < \infty$.

Example 1.1. Suppose the earnings X of a laborer are given by the following probability function:

x	0	8	12	16
$P(X = x)$	0.3	0.2	0.3	0.2

Find the laborer's expected earnings.

Solution. We have, applying the formula,

$$E(X) = 0(0.3) + 8(0.2) + 12(0.3) + 16(0.2) = 8.4$$ ■

Discussion. In Example 1.1, what does the number 8.4 mean in practical terms? Suppose we keep track of the laborer's earnings over N days, and say he earns nothing on f_1 days, \$8 on f_2 days, \$12 on f_3 days, and \$16 on f_4 days (so that $f_1 + f_2 + f_3 + f_4 = N$). Then the "ordinary average" of his earnings will be

$$\frac{0 \cdot f_1 + 8 \cdot f_2 + 12 \cdot f_3 + 16 \cdot f_4}{N} = 0\left(\frac{f_1}{N}\right) + 8\left(\frac{f_2}{N}\right) + 12\left(\frac{f_3}{N}\right) + 16\left(\frac{f_4}{N}\right)$$

Our empirical notion of probability tells us that if N is very large, then f_1/N, f_2/N, f_3/N, and f_4/N should be close to the corresponding theoretical probabilities, 0.3, 0.2, 0.3, and 0.2. Hence if we keep track of the laborer's earnings over a very long period of time, then the ordinary average of his earnings would be close to 8.4, the theoretical expectation. In general, the empirical concept of expectation entails taking the average over a large number of repeated performances of the experiment.

Example 1.2. A box contains three white beads and two black. If three beads are picked at random without replacement, find the expected number of white beads.

Solution. Let X represent the number of white beads in the sample. Then the probability function of X is

x	0	1	2	3
$P(X = x)$	0	0.3	0.6	0.1

Hence, by definition

$$E(X) = 1(0.3) + 2(0.6) + 3(0.1) = 1.8$$

Example 1.3. A town has N cars with license plates numbered 1 to N. If a person observes n cars at random, find $E(X)$, where X represents the highest number on the license plate. (Of course, it is possible that the person might observe a car more than once. Thus the sampling is carried out with replacement.)

Solution. Let us first find the probability function of X.

The event $\{X \leqslant k\}$ will occur precisely if the number on *each* of the n license plates is less than or equal to k. Now, the probability that a number drawn is less than or equal to k is k/N. Hence $P(X \leqslant k) = (k/N)^n$, since the numbers are sampled with replacement. Therefore,

$$\begin{aligned} P(X = k) &= P(X \leqslant k) - P(X \leqslant k-1) \\ &= \{k^n - (k-1)^n\}/N^n, \qquad k = 1, 2, \ldots, N \end{aligned}$$

This gives

$$E(X) = \sum_{k=1}^{N} kP(X = k) = \sum_{k=1}^{N} \{k^{n+1} - k(k-1)^n\}/N^n$$

Writing $k(k-1)^n = (k-1)^{n+1} + (k-1)^n$ and simplifying, we get

$$E(X) = \{N^{n+1} - \sum_{k=1}^{N} (k-1)^n\}/N^n$$

For large N, $\sum_{k=1}^{N} (k-1)^n$ is approximately equal to $\int_0^N x^n \, dx = \frac{N^{n+1}}{n+1}$. Therefore, if N is large, $E(X)$ is approximately equal to

$$\left(N^{n+1} - \frac{N^{n+1}}{n+1}\right)\Big/N^n = \frac{n}{n+1}N$$

■

Comment. The significance of the answer in Example 1.3 is not so much that it gives the expected highest license plate number, when n and N are given, but that it allows us to make a "rough guess" about N, the total number of cars in the town. For example, suppose a person picks a random day and finds that, among the 20 cars that were randomly observed, the highest license plate number was 1200. Then by setting $1200 = \frac{20}{21}N$, it might be inferred roughly that the total number of cars in the town is 1260. The discussion of problems of this nature comes under the broad heading of *statistical estimation.*

Example 1.4. A random variable X has the distribution whose pdf is given by

$$f(x) = \begin{cases} \dfrac{x}{2}, & 0 \leqslant x < 1 \\ \dfrac{1}{2}, & 1 \leqslant x < 2 \\ -\dfrac{x}{2} + \dfrac{3}{2}, & 2 \leqslant x < 3 \\ 0, & \text{elsewhere} \end{cases}$$

Find $E(X)$.

Solution. The distribution of X is continuous. Hence,

$$\begin{aligned} E(X) &= \int_{-\infty}^{\infty} x f(x)\,dx = \int_{-\infty}^{0} x \cdot 0\,dx + \int_0^1 x\left(\frac{x}{2}\right)dx + \int_1^2 x\left(\frac{1}{2}\right)dx \\ &\quad + \int_2^3 x\left(-\frac{x}{2} + \frac{3}{2}\right)dx + \int_3^{\infty} x \cdot 0\,dx \\ &= \frac{1}{6} + \frac{3}{4} + \frac{7}{12} = \frac{3}{2} \end{aligned}$$

Example 1.5. Suppose a random variable X has the following distribution function F:

$$F(x) = \begin{cases} 0, & x < -1 \\ \dfrac{x^3}{2} + \dfrac{1}{2}, & -1 \leqslant x < 1 \\ 1, & x \geqslant 1 \end{cases}$$

Find $E(X)$.

Solution. To apply the formula, first we must find the pdf of X, which is given by

$$f(x) = \begin{cases} \dfrac{3x^2}{2}, & -1 < x < 1 \\ 0, & \text{elsewhere} \end{cases}$$

Therefore, by definition,

$$E(X) = \int_{-\infty}^{\infty} xf(x)\,dx = \int_{-\infty}^{-1} x\cdot 0\,dx + \int_{-1}^{1} x\cdot \frac{3x^2}{2}\,dx + \int_{1}^{\infty} x\cdot 0\,dx = 0$$

- *Example 1.6.* If X has the Cauchy distribution with the pdf given by

$$f(x) = \frac{1}{\pi(1+x^2)}, \qquad -\infty < x < \infty$$

show that $E(X)$ does not exist.

Solution. The expected value exists provided $\int_{-\infty}^{\infty} |x| f(x)\,dx < \infty$. Now

$$\begin{aligned}\int_{-\infty}^{\infty} |x| f(x)\,dx &= \frac{1}{\pi}\int_{-\infty}^{\infty} |x| \frac{1}{1+x^2}\,dx \\ &= \frac{2}{\pi}\int_{0}^{\infty} \frac{x}{1+x^2}\,dx \qquad \text{(why?)} \\ &= \frac{2}{\pi}\lim_{y\to\infty}\int_{0}^{y} \frac{x}{1+x^2}\,dx \\ &= \frac{1}{\pi}\lim_{y\to\infty} \ln(1+y^2) = \infty\end{aligned}$$

Since the integral does not converge absolutely, *E(X) does not exist for the Cauchy distribution.*

Geometric interpretation of expectation

Under some general conditions, the expected value of a random variable can be given the following simple geometric interpretation. We shall first state and prove the result which leads to this interpretation for the continuous case. (The geometric interpretation also holds in the discrete case.)

If the distribution function F of X satisfies the conditions $xF(x) \to 0$ as $x \to -\infty$ and $x[1 - F(x)] \to 0$ as $x \to \infty$, then

$$E(X) = \int_{0}^{\infty} [1 - F(x)]\,dx - \int_{-\infty}^{0} F(x)\,dx$$

This can be proved as follows: We have

$$E(X) = \int_{-\infty}^{\infty} xf(x)\,dx = \int_{-\infty}^{0} xf(x)\,dx + \int_{0}^{\infty} xf(x)\,dx$$

Now consider the integral $\int_{-\infty}^{0} xf(x)\,dx$. Noting that $f(x)\,dx = dF(x)$, integration by parts yields

$$\int_{-\infty}^{0} xf(x)\,dx = xF(x)\Big|_{-\infty}^{0} - \int_{-\infty}^{0} F(x)\,dx = -\int_{-\infty}^{0} F(x)\,dx$$

since $xF(x) \to 0$ as $x \to -\infty$, and $xF(x) = 0$ at $x = 0$. Next, consider the integral $\int_{0}^{\infty} xf(x)\,dx$. We can write

$$\int_0^\infty xf(x)\,dx = -\int_0^\infty x\,d(1-F(x))$$

since $d(1-F(x)) = -f(x)\,dx$. Integration by parts then gives

$$\int_0^\infty xf(x)\,dx = -[x(1-F(x))]\Big|_0^\infty + \int_0^\infty (1-F(x))\,dx = \int_0^\infty (1-F(x))\,dx$$

since $x[1-F(x)] \to 0$ as $x \to \infty$ and $x[1-F(x)] = 0$ at $x = 0$. Combining the two integrals, the result follows.

To see the geometric interpretation, notice that $\int_{-\infty}^0 F(x)\,dx$ represents the area under the distribution function F from $-\infty$ to 0 (this is the area B shaded in Figure 1.2(a)) and $\int_0^\infty (1-F(x))\,dx$ the area under the graph of $1-F(x)$ from 0 to ∞ (this is the area A shaded in Figure 1.2(a)). The above result simply states that we can regard the expected value of a random variable as the difference $A-B$ of the two areas. The situation in the discrete case is shown in Figure 1.2(b).

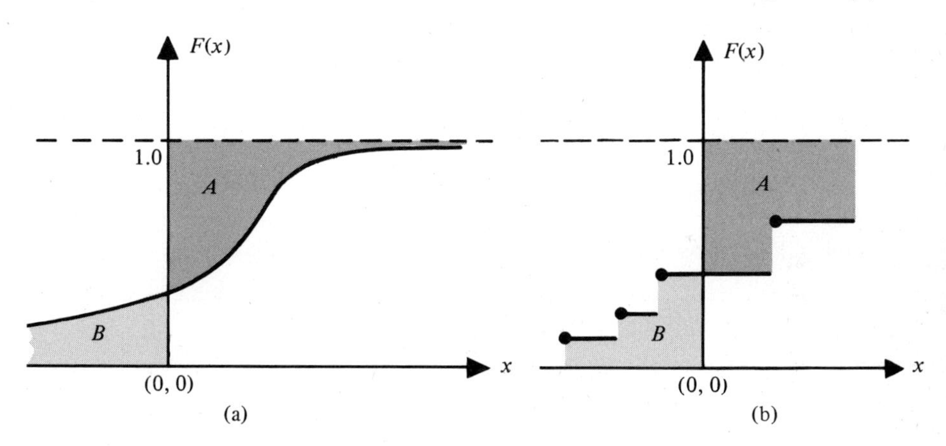

Figure 1.2

Example 1.7. Suppose the life span of an electronic component is a continuous random variable X with the following D.F.:

$$F(x) = \begin{cases} 0, & x < 200 \\ 1 - \left(\dfrac{200}{x}\right)^2, & x \geqslant 200 \end{cases}$$

Find the expected life span of the electronic component.

Solution. We could, of course, find $E(x)$ by first obtaining the pdf of X and then going the usual route. Instead, using the relation

$$E(X) = \int_{-\infty}^{0} F(x)\,dx + \int_{0}^{\infty} (1 - F(x))\,dx$$

it follows that, substituting for F appropriately,

$$\begin{aligned} E(X) &= \int_{-\infty}^{0} 0\,dx + \int_{0}^{200} (1-0)\,dx + \int_{200}^{\infty} \left[1 - \left(1 - \left(\frac{200}{x}\right)^2\right)\right] dx \\ &= \int_{0}^{200} dx + \int_{200}^{\infty} \left(\frac{200}{x}\right)^2 dx = 400 \end{aligned}$$

1.2 The Expectation of a Function of a Random Variable

Suppose Y is a function of a random variable X given by $Y = h(X)$ and we want to find the expected value of Y. We have seen in Chapter 6 how to find the distribution of Y if that of X is known. Now, if the distribution of Y is known, we can find $E(Y)$ by using the basic definition of the expected value of a random variable as follows:

$$E(Y) = \begin{cases} \sum_i y_i P(Y = y_i), & \text{in the discrete case} \\ \int_{-\infty}^{\infty} y f_Y(y)\,dy, & \text{in the continuous case} \end{cases}$$

This approach, of course, assumes the knowledge of the distribution of Y. The following equivalent formula shows that there is no need to derive the distribution of Y. We have the following result:

If $Y = h(X)$, then

$$E(Y) = \begin{cases} \sum_i h(x_i) P(X = x_i) & \text{in the discrete case} \\ \int_{-\infty}^{\infty} h(x) f_X(x)\,dx & \text{in the continuous case} \end{cases}$$

The proof in the continuous case being slightly involved, we shall prove the result only for the discrete case. To get an idea of the proof in the general discrete case, let us start with the following simple example.

Example 1.8. Suppose X has the following probability function, and we want to find the expected value of $Y = X^2$. (Here $h(x) = x^2$.)

x	-1	0	1	2
$P(X = x)$	$\frac{1}{3}$	$\frac{1}{6}$	$\frac{1}{6}$	$\frac{1}{3}$

Solution. To obtain $E(Y)$, we first find the probability function of Y, which is given as

y	0	1	4
$P(Y = y)$	$\frac{1}{6}$	$\frac{1}{3} + \frac{1}{6}$	$\frac{1}{3}$

Using the probability function of Y,

$$\begin{aligned} E(Y) &= 0(\tfrac{1}{6}) + 1(\tfrac{1}{3} + \tfrac{1}{6}) + 4(\tfrac{1}{3}) \\ &= 0^2(\tfrac{1}{6}) + (-1)^2(\tfrac{1}{3}) + 1^2(\tfrac{1}{6}) + 2^2(\tfrac{1}{3}) \\ &= 0^2 \cdot P(X = 0) + (-1)^2 \cdot P(X = -1) + 1^2 \cdot P(X = 1) + 2^2 \cdot P(X = 2) \\ &= \sum_i h(x_i)P(X = x_i) \end{aligned}$$

■

Now let us turn our attention to the general discrete case. We have

$$\begin{aligned} E(Y) &= \sum_j y_j P(Y = y_j), \qquad \text{by the basic definition} \\ &= \sum_j h(x_i) \cdot \sum_{\substack{\text{all } x_i \\ \text{such that } h(x_i) = y_j}} P(X = x_i) \\ &= \sum_i h(x_i)P(X = x_i) \end{aligned}$$

The Moral: To find the expectation of $Y = h(X)$, go ahead and use the distribution of X appropriately. There is no need to obtain the distribution of Y.

Example 1.9. Suppose X has the pdf f where

$$f(x) = \begin{cases} 2x, & 0 < x < 1 \\ 0, & \text{elsewhere} \end{cases}$$

Let $Y = -2X + 3$. Find $E(Y)$.

Solution. We shall find $E(Y)$ using two approaches.

Method 1: In Example 2.7 of Chapter 6, we have already found the pdf of $Y = -2X + 3$ as

$$f_Y(y) = \begin{cases} \dfrac{3-y}{2}, & 1 < y < 3 \\ 0, & \text{elsewhere} \end{cases}$$

Therefore,

$$E(Y) = \int_{-\infty}^{\infty} y f_Y(y)\, dy = \int_1^3 y\left(\frac{3-y}{2}\right) dy = \frac{5}{3}$$

Method 2: Using the alternate method,

$$E(Y) = \int_{-\infty}^{\infty} (-2x + 3) f_X(x)\, dx = \int_0^1 (-2x + 3)(2x)\, dx = \frac{5}{3}$$

Thus we get the same answer using the two methods.

Example 1.10. The length of time X (measured in days) that a gadget lasts has the following pdf:

$$f(x) = \begin{cases} 2e^{-2x}, & x > 0 \\ 0, & \text{elsewhere} \end{cases}$$

If the gadget lasts more than three days the profit is \$1; otherwise, it incurs a loss of \$2. Find the expected profit.

Solution. Letting Y represent the profit, we want to find $E(Y)$, where

$$Y = h(X) = \begin{cases} 1 & \text{if } X > 3 \\ -2 & \text{if } X \leqslant 3 \end{cases}$$

Method 1: In Example 2.13 of Chapter 6, we showed that Y has a discrete distribution with the probability function:

$$P(Y = -2) = 1 - e^{-6} \quad \text{and} \quad P(Y = 1) = e^{-6}$$

Therefore,

$$E(Y) = (-2)(1 - e^{-6}) + (1)e^{-6} = 3e^{-6} - 2$$

Method 2: Here we proceed by using the distribution of X. We have

$$\begin{aligned} E(Y) &= \int_{-\infty}^{\infty} h(x)f(x)\,dx \\ &= 2\int_0^{\infty} h(x) \cdot e^{-2x}\,dx, \qquad \text{since } f(x) = \begin{cases} 0, & x \leqslant 0 \\ 2e^{-2x}, & x > 0 \end{cases} \\ &= 2\left[\int_0^3 h(x)e^{-2x}\,dx + \int_3^{\infty} h(x)e^{-2x}\,dx\right] \\ &= 2\left[\int_0^3 (-2)e^{-2x}\,dx + \int_3^{\infty} (1)e^{-2x}\,dx\right] = 3e^{-6} - 2 \end{aligned}$$

Example 1.11. A random variable X has the following pdf:

$$f(x) = \begin{cases} e^{-x}, & x > 0 \\ 0, & \text{elsewhere} \end{cases}$$

Let

$$Z = h(X) = \begin{cases} X & \text{if } X > 3 \\ 2X + 3 & \text{if } X \leqslant 3 \end{cases}$$

Find $E(Z)$.

Solution. Using the distribution of X,

$$\begin{aligned} E(Z) &= \int_{-\infty}^{\infty} h(x)f(x)\,dx \\ &= \int_0^{\infty} h(x)(e^{-x})\,dx, \qquad \text{since } f(x) = \begin{cases} 0, & x \leqslant 0 \\ e^{-x}, & x > 0 \end{cases} \\ &= \int_0^3 (2x + 3)e^{-x}\,dx + \int_3^{\infty} xe^{-x}\,dx \end{aligned}$$

Upon simplification, this gives

$$E(Z) = 5 - 7e^{-3}$$

1.3 Some Properties of Expectation

We now give some very important properties of the operation of finding expected values. The reader will find them quite helpful in formal manipulations. These results will be proved in greater generality in Chapter 11, where we will consider the expectation of a function of several random variables. For the present, we shall state and prove the results for the functions of a *single* random variable.

(*i*) For any constant C, $E(C) = C$

The random variable in this case is the degenerate random variable $X \equiv C$, where

$$X(s) = C \qquad \text{for every } s \in S$$

Therefore,

$$P(X = C) = 1 \quad \text{and} \quad P(X \neq C) = 0$$

and, consequently, by definition,

$$E(X) = C \cdot P(X = C) = C$$

(*ii*) If h_1 and h_2 are real-valued functions, and if a and b are any real numbers, then

$$E[ah_1(X) + bh_2(X)] = aE(h_1(X)) + bE(h_2(X))$$

In the continuous case, this can be shown as follows:

$$\begin{aligned} E(ah_1(X) + bh_2(X)) &= \int_{-\infty}^{\infty} (ah_1(x) + bh_2(x)) f_X(x)\, dx \\ &= a \int_{-\infty}^{\infty} h_1(x) f_X(x)\, dx + b \int_{-\infty}^{\infty} h_2(x) f_X(x)\, dx \\ &= aE(h_1(X)) + bE(h_2(X)) \end{aligned}$$

In particular, if we set $b = 0$ in the above result, we get

$$E(ah_1(X)) = aE(h_1(X)) \qquad \text{for any constant } a$$

In other words, a constant multiplier can be pulled out.

An obvious extension, if there are r real-valued functions $h_1, h_2, \ldots, h_r$, is the following:

$$E(a_1h_1(X) + a_2h_2(X) + \ldots + a_rh_r(X)) = a_1E(h_1(X)) + \ldots + a_rE(h_r(X))$$

where $a_1, a_2, \ldots, a_r$ are constants.

Moments

Among the quantities which play an important role in probability theory are those called the *moments* of a distribution. (The term *moment* is borrowed from mechanics.) These are obtained by taking special functions $h(X)$ and computing their expected values.

For every nonnegative integer n, the ***n*th moment** of the distribution of an r.v. X, or simply of X, is defined to be the expected value $E(X^n)$ and, using the customary notation, is denoted by μ_n'. Hence

$$\mu_n' = E(X^n), \qquad n = 0, 1, \ldots$$

Observe that the zeroth moment of any distribution is 1 and the first moment is the mean of the distribution. The moments $E(X^n)$ are also referred to as the ***n*th raw moments.**

The *nth moment of X about any point* $x = a$, if it exists, is defined as $E[(X-a)^n]$.

If, in particular, a is taken as the mean μ of X, the moments are called nth moments about the mean μ or the ***n*th central moments** and are denoted by μ_n. Hence

$$\mu_n = E[(X-\mu)^n], \qquad n = 0, 1, \ldots$$

There are moments of another kind, called factorial moments, which are especially useful for discrete distributions. If X is a random variable, then for any positive integer n, the ***n*th factorial moment** of X is defined to be $E(X(X-1)\ldots(X-n+1))$. They are so called because of the factorial nature of the expression $(x)_n = x(x-1)\ldots(x-n+1)$.

If $E(X^n)$ exists for some positive integer n, then all the raw moments $E(X^k)$, $k = 1, 2, \ldots, n-1$, exist.

We shall prove this result only in the discrete case. Let us assume that $E(X^n)$ exists for some fixed positive integer n. Therefore, $E(|X|^n) < \infty$ and consequently $\sum_{x_i} |x_i|^n P(X = x_i) < \infty$. Now for any positive integer $k < n$,

$$\begin{aligned} E(|X|^k) &= \sum_{x_i} |x_i|^k P(X = x_i) = \sum_{|x_i|<1} |x_i|^k P(X = x_i) + \sum_{|x_i| \geq 1} |x_i|^k P(X = x_i) \\ &\leq \sum_{|x_i|<1} P(X = x_i) + \sum_{|x_i| \geq 1} |x_i|^k P(X = x_i) \\ &\leq 1 + \sum_{|x_i| \geq 1} |x_i|^n P(X = x_i) \end{aligned}$$

since $\sum_{|x_i|<1} P(X = x_i) \leq \sum_{x_i} P(X = x_i) = 1$, and $|x_i|^k \leq |x_i|^n$ if $|x_i| \geq 1$ and $k < n$.

Therefore,

$$
\begin{aligned}
E(|X|^k) &\leqslant 1 + \sum_{|x_i|<1} |x_i|^n P(X = x_i) + \sum_{|x_i|\geqslant 1} |x_i|^n P(X = x_i) \\
&= 1 + \sum_{x_i} |x_i|^n P(X = x_i) \\
&= 1 + E(|X|^n)
\end{aligned}
$$

In conclusion, $E(|X|^k) \leqslant 1 + E(|X|^n)$ if $k < n$. Therefore, $E(|X|^k) < \infty$ if $E(|X|^n) < \infty$, which in turn implies that $E(X^k)$ exists if $E(X^n)$ exists and $k < n$.

1.4 The Variance of a Random Variable

A very important constant associated with the distribution of a random variable is its *variance,* denoted by Var(X) or σ_X^2 (sigma squared).

The **variance** *of the distribution of a random variable X,* or simply the *variance of X,* is defined as

$$
\text{Var}(X) = E[(X - E(X))^2]
$$

In other words, the variance of X is the second central moment. Therefore, writing μ for $E(X)$,

$$
\text{Var}(X) = \begin{cases} \int_{-\infty}^{\infty} (x - \mu)^2 f_X(x)\, dx, & \text{in the continuous case} \\ \sum_i (x_i - \mu)^2 p(x_i), & \text{in the discrete case} \end{cases}
$$

Since $(x - \mu)^2 f_X(x) \geqslant 0$, we always have $\int_{-\infty}^{\infty} (x - \mu)^2 f_X(x)\, dx \geqslant 0$, and since $(x_i - \mu)^2 p(x_i) \geqslant 0$, we have $\sum_i (x_i - \mu)^2 p(x_i) \geqslant 0$. Hence, *Var(X)* $\geqslant$ *0 for any random variable X.*

The positive square root of the variance of X is called the **standard deviation.**

Example 1.12. Compute the variance and the standard deviation of the earnings X of the laborer as given in Example 1.1.

Solution. We have already seen that $E(X) = 8.4$. Therefore, by definition

$$
\begin{aligned}
\text{Var}(X) &= (0 - 8.4)^2(0.3) + (8 - 8.4)^2(0.2) + (12 - 8.4)^2(0.3) + (16 - 8.4)^2(0.2) \\
&= 36.64
\end{aligned}
$$

The standard deviation is $\sqrt{36.64} = 6.05$ (approximately).

Example 1.13. Suppose an r.v. X has the following pdf:

$$
f(x) = \begin{cases} \dfrac{x}{2}, & 0 < x < 2 \\ 0, & \text{elsewhere} \end{cases}
$$

Find the variance and standard deviation of X.

Solution. First we need $E(X)$, which is equal to

$$\int_0^2 x \cdot \frac{x}{2}\, dx = \frac{4}{3}$$

Next, by definition,

$$\mathrm{Var}(X) = \int_{-\infty}^{\infty} \left(x - \frac{4}{3}\right)^2 f(x)\, dx = \int_0^2 \left(x - \frac{4}{3}\right)^2 \cdot \frac{x}{2}\, dx = \frac{2}{9}$$

Hence the standard deviation is equal to $\sqrt{\frac{2}{9}} = 0.471$. ■

The following observation about the variance of a random variable is noteworthy:

$$\mathrm{Var}(X) = 0 \quad \text{if and only if} \quad P(X = \mu) = 1$$

We shall prove the result assuming that X has a discrete distribution. In fact, $\mathrm{Var}(X) = 0$ implies that $\sum_i (x_i - \mu)^2 P(X = x_i) = 0$, which can happen if and only if $x_i = \mu$ for every i. Consequently, $\mathrm{Var}(X) = 0$ if and only if the only possible value of X (that is, the only value assumed with positive probability) is μ; in other words, if and only if $P(X = \mu) = 1$.

The proof where X is assumed to have a continuous distribution will not be attempted since it calls for a more involved argument.

The second moment of a random variable X about any point a is, by definition, $E[(X - a)^2]$. As a changes, we get different values for $E[(X - a)^2]$, thus giving rise to a function which depends on a, and whose value at $a = \mu$ is the variance of the distribution.

The second moment $E[(X - a)^2]$ can be expressed in terms of the variance as follows:

$$\begin{aligned} E[(X-a)^2] &= E[((X-\mu) + (\mu - a))^2] \\ &= E[(X-\mu)^2] + 2(\mu - a)E(X - \mu) + (\mu - a)^2 \qquad \text{(why?)} \\ &= E[(X-\mu)^2] + (\mu - a)^2 \qquad \text{(why?)} \end{aligned}$$

Now $(\mu - a)^2 \geqslant 0$. Hence

$$E[(X-a)^2] \geqslant E[(X-\mu)^2] = \sigma^2$$

We thus arrive at a very important conclusion:

The variance of a random variable is the smallest second moment.

Using the properties of the expectation operation, it is possible to derive alternate expressions for the variance of X. These expressions turn out to be convenient for finding $\mathrm{Var}(X)$ in some instances. Specifically, we shall show that

(*i*) $\mathrm{Var}(X) = E(X^2) - [E(X)]^2$

(*ii*) $\mathrm{Var}(X) = E(X(X-1)) + E(X) - [E(X)]^2$

To prove (*i*), we see that

$$\begin{aligned}\mathrm{Var}(X) &= E[(X-\mu)^2] = E(X^2 - 2\mu X + \mu^2)\\ &= E(X^2) - 2\mu E(X) + E(\mu^2) \qquad \text{(why?)}\\ &= E(X^2) - \mu^2\end{aligned}$$

To prove (*ii*), we note that

$$E(X(X-1)) = E(X^2 - X) = E(X^2) - E(X)$$

Therefore,

$$E(X^2) = E(X(X-1)) + E(X)$$

Hence, from (*i*) it follows that

$$\mathrm{Var}(X) = E(X^2) - [E(X)]^2 = E(X(X-1)) + E(X) - [E(X)]^2$$

Example 1.14. If the probability function of X is given by

$$p(x) = \begin{cases} \frac{1}{5}, & x \in \{0, 2, 4, 8, 16\} \\ 0, & \text{elsewhere} \end{cases}$$

find the variance of X.

Solution. First we observe that X is a discrete random variable. Now,

$$E(X) = 0(\tfrac{1}{5}) + 2(\tfrac{1}{5}) + 4(\tfrac{1}{5}) + 8(\tfrac{1}{5}) + 16(\tfrac{1}{5}) = 6$$

and

$$E(X^2) = 0^2(\tfrac{1}{5}) + 2^2(\tfrac{1}{5}) + 4^2(\tfrac{1}{5}) + 8^2(\tfrac{1}{5}) + 16^2(\tfrac{1}{5}) = 68$$

Therefore,

$$\mathrm{Var}(X) = 68 - (6)^2 = 32$$

Example 1.15. If the pdf of X is given by

$$f(x) = \begin{cases} \dfrac{1}{\ln 3} \cdot \dfrac{1}{x}, & 1 < x < 3 \\ 0, & \text{elsewhere} \end{cases}$$

find $E(X)$, $E(X^2)$, and the variance of X.

Solution. Instead of finding $E(X)$, $E(X^2)$ separately, we shall find $E(X^n)$ for any positive integer n. We get

$$\begin{aligned}E(X^n) &= \int_{-\infty}^{\infty} x^n \cdot f(x)\,dx = \int_1^3 x^n \frac{1}{\ln 3} \cdot \frac{1}{x}\,dx = \frac{1}{\ln 3}\int_1^3 x^{n-1}\,dx\\ &= \frac{3^n - 1}{n \ln 3}\end{aligned}$$

Therefore,

$$E(X) = \frac{2}{\ln 3} \quad \text{and} \quad E(X^2) = \frac{4}{\ln 3}$$

and, consequently,

$$\text{Var}(X) = \frac{4}{\ln 3} - \left(\frac{2}{\ln 3}\right)^2 = \frac{4(\ln 3 - 1)}{(\ln 3)^2}$$ ■

The relation $\text{Var}(X) = E(X^2) - [E(X)]^2$ leads to some important consequences. First of all, since $\text{Var}(X) \geqslant 0$, we always have

$$E(X^2) \geqslant [E(X)]^2 \qquad \text{for any r.v. } X$$

For example, it is not possible to have a random variable X with a distribution having $E(X^2) = 15$ and $E(X) = 5$.

Another consequence is: *If $E(X^2)$ is finite, then the variance of X always exists.*

Example 1.16. Suppose X has a continuous distribution with the following pdf:

$$f(x) = \begin{cases} \dfrac{8}{x^3}, & x > 2 \\ 0, & \text{elsewhere} \end{cases}$$

Find the variance of X.

Solution. It can be easily seen that $E(X) = \int_2^\infty x \cdot \frac{8}{x^3}\, dx = 4$. We shall next find $E(X^2)$.

$$E(X^2) = \int_2^\infty x^2 \cdot \frac{8}{x^3}\, dx = \lim_{t\to\infty} \int_2^t \frac{8}{x}\, dx = 8 \lim_{t\to\infty} [\ln t - \ln 2] = \infty$$

In this case the variance of X does not exist because it is not finite.

1.5 Conditional Expectation

Let X be a random variable and A a (Borel) set of the real line such that $P(X \in A) > 0$. Our aim is to find the *conditional expectation of X given that X has taken a value in A*. The obvious first step in this direction would be to find the conditional distribution of X given $X \in A$. This distribution is often referred to as the **truncated distribution** of X.

Since we do not have the mathematical tools to treat the most general case where A is any Borel set, we shall derive the truncated distribution of X in the special case where A is an interval. This is the case which is most frequently encountered in applications. Specifically, we shall consider the case $A = (a, b]$.

Towards this end, let us first find the conditional distribution function of X given $a < X \leqslant b$, namely, $P(X \leqslant u \mid a < X \leqslant b)$ for any real number u. This is given by

$$P(X \leqslant u \mid a < X \leqslant b) = \frac{P(X \leqslant u,\ a < X \leqslant b)}{P(a < X \leqslant b)}$$

$$= \begin{cases} 0, & u < a \\ \dfrac{P(a < X \leqslant u)}{P(a < X \leqslant b)}, & a < u \leqslant b \\ 1, & u > b \end{cases}$$

$$= \begin{cases} 0, & u < a \\ \dfrac{F(u) - F(a)}{F(b) - F(a)}, & a < u \leqslant b \\ 1, & u > b \end{cases}$$

Hence:

If X is a continuous random variable, and if we denote the *conditional pdf of X given $a < X \leqslant b$* by $f_X(\ \mid a < X \leqslant b)$, then

$$f_X(u \mid a < X \leqslant b) = \begin{cases} \dfrac{f(u)}{\int_a^b f(x)\,dx}, & a < u \leqslant b \\ 0, & \text{elsewhere} \end{cases}$$

If X is a discrete random variable with possible values $x_1, x_2, \ldots,$ and if we denote *the conditional probability function of X given $a < X \leqslant b$* by $p_X(\ \mid a < X \leqslant b)$, then

$$p_X(x_i \mid a < X \leqslant b) = \begin{cases} \dfrac{p(x_i)}{\sum_{a < x_j \leqslant b} p(x_j)}, & a < x_i \leqslant b \\ 0, & \text{elsewhere} \end{cases}$$

If h is a real-valued function, then the conditional expectation of $h(X)$ given $a < X \leqslant b$ is now defined in an obvious way. If we denote this conditional expectation by $E(h(X) \mid a < X \leqslant b)$, and if X is a continuous random variable, then

$$E(h(X) \mid a < X \leqslant b) = \int_{-\infty}^{\infty} h(u) \cdot f_X(u \mid a < X \leqslant b)\,du$$

$$= \int_a^b h(u) \cdot \frac{f(u)}{\int_a^b f(x)\,dx}\,du$$

$$= \frac{\int_a^b h(u) \cdot f(u)\,du}{\int_a^b f(x)\,dx}$$

Hence

$$E(h(X) \mid a < X \leqslant b) = \frac{\int_a^b h(u)f(u)\,du}{\int_a^b f(x)\,dx}$$

An analogous definition holds in the discrete case. The reader should encounter no difficulty in formulating and deriving the appropriate expressions when A is any other type of interval, such as $[a, \infty)$, $(-\infty, a]$, and so on.

Example 1.17. If the probability that a child is a son is p, where $0 < p < 1$, find the expected number of sons in a family with n children, given that there is at least one son.

Solution. Letting X represent the number of sons, the conditional probability function of X given that there is at least one son is given by

$$P(X = r \mid X \geqslant 1) = \frac{P(X = r)}{P(X \geqslant 1)}, \qquad r = 1, 2, \ldots, n$$

$$= \frac{\binom{n}{r} p^r (1-p)^{n-r}}{1-(1-p)^n}, \qquad r = 1, 2, \ldots, n$$

Therefore,

$$E(X \mid X \geqslant 1) = \frac{\sum_{r=1}^{n} r \cdot \binom{n}{r} p^r (1-p)^{n-r}}{1-(1-p)^n}$$

$$= \frac{np}{1-(1-p)^n}$$

Example 1.18. Let X have a continuous distribution with the following pdf:

$$f(x) = \begin{cases} \frac{1}{6}(x^3+1), & 0 < x < 2 \\ 0, & \text{elsewhere} \end{cases}$$

Find $E(X^n \mid X < 1)$, where n is a nonnegative integer.

Solution. The conditional pdf of X given $X < 1$ is

$$f_X(u \mid X < 1) = \begin{cases} \dfrac{\frac{1}{6}(u^3+1)}{P(X<1)}, & 0 < u < 1 \\ 0, & \text{elsewhere} \end{cases}$$

$$= \begin{cases} \frac{4}{5}(u^3+1), & 0 < u < 1 \\ 0, & \text{elsewhere} \end{cases}$$

since

$$P(X < 1) = \int_0^1 \tfrac{1}{6}(x^3+1)\,dx = \tfrac{5}{24}$$

Therefore,

$$E(X^n \mid X<1) = \int_0^1 u^n \cdot \tfrac{4}{5}(u^3+1)\,du$$

$$= \tfrac{4}{5}\left[\frac{u^{n+4}}{n+4} + \frac{u^{n+1}}{n+1}\right]\Bigg|_{u=0}^{u=1}$$

$$= \tfrac{4}{5} \cdot \frac{2n+5}{(n+4)(n+1)}$$

EXERCISES–SECTION 1

1. If $S = \{-1, 1, 2, 3\}$ and $P(\{-1\}) = \frac{1}{6}$, $P(\{1\}) = \frac{1}{12}$, $P(\{2\}) = \frac{1}{3}$, find $E(X)$ where X is defined as follows:

   (a) $X(s) = 2s + 1$ for every $s \in S$.

   (b) $X(s) = 2s^2 + 1$ for every $s \in S$.

2. Consider two independent events A and B with $P(A) = \frac{1}{3}$ and $P(B) = \frac{2}{3}$. Define a random variable X as follows:

$$X(s) = \begin{cases} 0, & s \in (A \cup B)' \\ 2, & s \in AB \\ 3, & s \in A - B \\ 5, & s \in B - A \end{cases}$$

Find $E(X)$ and $\text{Var}(X)$.

3. In each of the following cases, find $E(X)$ and $\text{Var}(X)$ if X has a discrete distribution with the given probability function.

(a)
$$p(x) = \begin{cases} \frac{1}{4}, & x = -2 \\ \frac{1}{2}, & x = 1 \\ \frac{1}{4}, & x = 4 \end{cases}$$

(b) $p(x) = \dfrac{|x-2|}{7}, \qquad x = -1, 0, 1, 3$

(c) $p(x) = \frac{1}{6}x^2, \qquad x = -2, -1, 1$

(d) $p(x) = (\frac{1}{2})^x, \qquad x = 1, 2, 3, \ldots$

4. In each of the following cases, X has a continuous distribution with the given pdf. Find $E(X)$ and $\text{Var}(X)$ whenever they exist.

(a)
$$f(x) = \begin{cases} \frac{3}{8}x^2, & 0 \leqslant x \leqslant 2 \\ 0, & \text{elsewhere} \end{cases}$$

(b)
$$f(x) = \begin{cases} \frac{2}{5}|x|, & -1 < x < 2 \\ 0, & \text{elsewhere} \end{cases}$$

(c)
$$f(x) = \begin{cases} 1 - |1 - x|, & 0 < x < 2 \\ 0, & \text{elsewhere} \end{cases}$$

(d)
$$f(x) = \begin{cases} \dfrac{1}{x^2}, & x \geqslant 1 \\ 0, & \text{elsewhere} \end{cases}$$

(e)
$$f(x) = \begin{cases} \frac{1}{4}, & -1 < x < 1 \\ \frac{1}{6}, & 2 < x < 5 \\ 0, & \text{elsewhere} \end{cases}$$

5. Let X represent the number that shows up when a fair die is rolled. Find $E(X)$ and $\text{Var}(X)$.
6. In a raffle there is one first prize of \$3000, five second prizes of \$1000 each, and twenty third prizes of \$100 each. In all, ten thousand tickets are sold at \$1.50 each. What are the expected net winnings of a person who buys one ticket? Interpret the result.
7. A chip is picked at random from a box containing N chips numbered $1, 2, \ldots, N$. Let the random variable X denote the number on the chip. Find $E(X)$ and $\text{Var}(X)$.
8. A box contains five tubes of which three are defective. If three tubes are picked at random without replacement, find the expected number of defectives and the variance of this number.
9. Suppose cards numbered $1, 2, \ldots, n$ are arranged in a row. If card i is in the ith location, a match is said to have occurred. Let X denote the number of matches. Find the distribution of X and the expected number of matches if (a) $n = 3$, (b) $n = 4$, (c) $n = 5$. What is the conjecture for arbitrary n? *Hint:* For $n = 4$, see exercise 24 of Section 3 of Chapter 2. (This general result will be proved in Chapter 11 using the indicator random variables technique.)
10. Tom and John agree to play the following game: Tom picks two cards from a standard deck of cards. If the two cards are of the same suit, John pays Tom \$5; otherwise, Tom pays John \$2. In whose favor is the game tilted?
11. Tom has five similar keys of which only one key can open the lock. If he tries the keys one by one at random, discarding the unsuccessful ones, find the expected number of keys he must try in order to open the lock.
12. Is it true that $E(1/X) = 1/E(X)$? Give a reason for your answer.
13. A random variable X assumes only the values -2 and 5. If $E(X) = 2.9$, find $\text{Var}(X)$.
14. Suppose X is a discrete random variable which assumes three values, -3, 0, and 4, with $P(X = 0) = \frac{1}{2}$. If $E(X) = \frac{9}{8}$, find $P(X = -3)$ and $P(X = 4)$.
15. Consider the following probability function of a random variable X:

$$P(X = i) = \frac{6}{\pi^2} \frac{1}{i^2}, \qquad i = 1, -2, 3, -4, \ldots$$

Does $E(X)$ exist? Explain.
16. The pdf of a random variable X is given by

$$f(x) = \begin{cases} \frac{3}{2}x^2, & -1 \leqslant x \leqslant 1 \\ 0, & \text{elsewhere} \end{cases}$$

Find the variance of $Y = X^2$ by using the following two methods: first, by obtaining the distribution of Y; second, by using the distribution of X.

17. Suppose X is an integer-valued random variable with $P(X = i) = p_i$, $i = 1, 2, \ldots,$ and $\sum_{i=1}^{\infty} p_i = 1$. Let $P(X \geqslant i) = q_i$. If $E(X)$ exists, show that $E(X) = \sum_{i=1}^{\infty} q_i$.

18. Suppose the probability function of X is given by

$$P(X = i) = \frac{2i + 1}{i^2(i + 1)^2}, \qquad i = 1, 2, \ldots$$

Find $E(X)$. *Hint:* Use the result of exercise 17 and the fact that

$$\sum_{i=1}^{\infty} \frac{1}{i^2} = \frac{\pi^2}{6}$$

19. A random variable X has the factorial moments given by

$$E(X(X - 1) \ldots (X - n + 1)) = 3^n$$

for any positive integer n. Find the following raw moments of X:

(a) $E(X)$ (b) $E(X^2)$
(c) $E(X^3)$ (d) $E(X^4)$

20. The raw moments of a random variable are given by

$$E(X^n) = 3\left(\frac{2^n}{n + 3}\right)$$

for any nonnegative integer n. Find the following factorial moments:

(a) $E(X(X - 1))$ (b) $E(X(X - 1)(X - 2))$
(c) $E(X(X - 1)(X - 2)(X - 3))$

21. If X is a random variable with $E(X) = 5$ and $E(X^2) = 28$, find the mean and variance of $Y = 4X + 5$.

22. Suppose a random variable X has $E(X) = 8$ and $\text{Var}(X) = 10$. Find $E(2X^2 + 4X)$.

23. Show why the following facts about a random variable X are not compatible: $E(X(X - 1)) = 26$; $E(X) = 7$.

24. X is a random variable such that $E(X) = 5$ and $\text{Var}(X) = 16$. Find the constants a and b for which $Y = aX + b$ has expectation 18 and variance 64.

25. Discuss the raw moments of the distribution of a random variable whose pdf is given by

$$f(x) = \begin{cases} (n + 1)x^{-(n+2)}, & x \geqslant 1 \\ 0, & \text{elsewhere} \end{cases}$$

where n is a fixed positive integer.

26. The possible values of a discrete random variable are $-3, -1, 0, 2, 3$. If $P(X = 0) = \frac{1}{4}$, $P(X = -1) = \frac{1}{8}$, and $P(X = 2) = \frac{1}{4}$, find the following wherever possible:

(a) $E(X^2)$ (b) $E(X^3)$ (c) $E(X^4)$

27. The number of defective items X in a box has the following distribution:

x	0	1	2	3	4	5
$P(X = x)$	0.1	0.2	0.3	0.1	0.1	0.2

Find the expected number of defective items, given that there are at most three defective items.

28. If X has a continuous distribution with the pdf given by

$$f(x) = \begin{cases} \frac{1}{6}(x^3 + 1), & 0 < x < 2 \\ 0, & \text{elsewhere} \end{cases}$$

find $E(X^n \mid X > 1)$ for any nonnegative integer n.

29. The length of a telephone conversation (measured in minutes) has the following pdf:

$$f(x) = \begin{cases} \frac{1}{4}, & 0 < x \leqslant 2 \\ \frac{4}{x^3}, & x > 2 \\ 0, & \text{elsewhere} \end{cases}$$

(a) Find the expected length of a conversation.

(b) Find the expected length of a conversation, given that it lasts at least one minute.

30. Suppose X is a random variable. For any real number $a > 0$, define a new random variable Y as follows:

$$Y = \begin{cases} 0 & \text{if } |X| \leqslant a \\ a^2 & \text{if } |X| > a \end{cases}$$

Show that $E(Y) = a^2 P(|X| > a)$.

31. Let X be a continuous random variable with finite range $[a, b]$. Show that $E(X) = b - \int_a^b F(x)\,dx$.

32. If X is a continuous random variable, show that $E(2F(X) - 1) = 0$ where F is the D.F. of X.

33. Suppose X is an absolutely continuous random variable having a unique median m. If b is a real number, show that $E(|X - b|)$ is a minimum when $b = m$.

Hint: First show that $E(|X - b|) = E(|X - m|) + 2\int_m^b (b - x)f(x)\,dx$. Then consider the two cases $m < b$ and $m > b$, and show that $\int_m^b (b - x)f(x)\,dx \geqslant 0$ and that this integral is zero when $b = m$.

2. EXPECTATIONS OF SOME SPECIAL DISTRIBUTIONS

In Chapter 5 we discussed some important discrete distributions (the binomial, the Poisson, etc.) and continuous distributions (the uniform, the normal, etc.). We referred to the constants associated with these distributions as their *parameters.* We are now in a position to provide physical meanings to these constants. Before embarking on this, we shall prove a result which applies to symmetric distributions.

If the distribution of X is symmetric about a, and if $E(X)$ exists, then $E(X) = a$, the point of symmetry.

The proof depends on the easily proved fact that if the distribution of X is symmetric about a, then $X - a$ and $-X + a$ have the same distribution. (See Chapter 6, exercise 2 on page 217.) Therefore,

$$E(X - a) = E(-X + a)$$

That is,

$$E(X) - a = -E(X) + a$$

Consequently, since $E(X) < \infty$, we get $E(X) = a$.

Comment. The Cauchy distribution provides an example of a distribution which is symmetric, but the point of symmetry is not the mean. The mean does not exist for the Cauchy distribution.

The Bernoulli distribution

If X has the Bernoulli distribution, then

$$P(X = x) = p^x(1 - p)^{1-x}, \qquad x = 0, 1$$

where $0 \leqslant p \leqslant 1$. Therefore,

$$E(X) = 0 \cdot (1 - p) + 1 \cdot p = p$$
$$E(X^2) = 0^2 \cdot (1 - p) + 1^2 \cdot p = p$$

Hence,

$$\text{Var}(X) = E(X^2) - [E(X)]^2 = p - p^2 = p(1 - p)$$

Thus

$$E(X) = p$$
$$\text{Var}(X) = p(1 - p)$$

Since $\frac{d}{dp} p(1 - p) = 1 - 2p$, we see that the variance is the largest when $p = \frac{1}{2}$. This stands to reason in view of the fact that the outcome of the experiment is least predictable when $p = \frac{1}{2}$.

The binomial distribution

Suppose X has the binomial distribution consisting of n independent trials with probability of success equal to p, $0 \leqslant p \leqslant 1$. Then the probability function of X is

$$P(X = r) = \binom{n}{r} p^r(1 - p)^{n-r}, \qquad r = 0, 1, \ldots, n$$

Therefore,

$$E(X) = \sum_{r=0}^{n} r \cdot P(X = r)$$
$$= \sum_{r=0}^{n} r \cdot \binom{n}{r} p^r (1-p)^{n-r}$$
$$= np \sum_{r=1}^{n} \frac{(n-1)!}{(r-1)!(n-r)!} p^{r-1}(1-p)^{n-r}$$

(The reader should supply and justify the intervening steps.) Letting $s = r - 1$, we can write

$$E(X) = np \sum_{s=0}^{n-1} \frac{(n-1)!}{s!(n-1-s)!} p^s (1-p)^{n-1-s}$$
$$= np[p + (1-p)]^{n-1} = np$$

Hence $E(X) = np$.

This result is in conformity with our intuitive answer. For example, if we knew that the probability of making a basket was 0.3, we would expect to score about 30 times in 100 attempts.

To find $\mathrm{Var}(X)$, we shall obtain $E(X(X-1))$.

$$E(X(X-1)) = \sum_{r=0}^{n} r(r-1)P(X = r)$$
$$= \sum_{r=0}^{n} r(r-1) \cdot \binom{n}{r} p^r (1-p)^{n-r}$$
$$= p^2 n(n-1) \sum_{r=2}^{n} \frac{(n-2)!}{(r-2)!(n-r)!} p^{r-2}(1-p)^{n-r}$$

(Supply the intermediate steps.) Letting $s = r - 2$, the above gives

$$E(X(X-1)) = p^2 n(n-1) \sum_{s=0}^{n-2} \frac{(n-2)!}{s!(n-2-s)!} p^s (1-p)^{n-2-s}$$
$$= p^2 n(n-1)[p + (1-p)]^{n-2}$$
$$= p^2 n(n-1)$$

Using the relation $\mathrm{Var}(X) = E(X(X-1)) + E(X) - [E(X)]^2$, we now get

$$\mathrm{Var}(X) = p^2 n(n-1) + np - (np)^2 = np(1-p)$$

In summary, if X has the binomial distribution $B(n, p)$, $0 \leqslant p \leqslant 1$, then

$$E(X) = np$$
$$\mathrm{Var}(X) = np(1-p)$$

The geometric distribution

A random variable with the geometric distribution with parameter p has the probability function

$$P(X = r) = p(1-p)^{r-1}, \qquad r = 1, 2, \ldots$$

In this case,

$$E(X) = \sum_{r=1}^{\infty} r(1-p)^{r-1}p = p \sum_{r=1}^{\infty} r(1-p)^{r-1}$$

Using the fact that $\sum_{r=1}^{\infty} rx^{r-1} = 1/(1-x)^2$ if $|x| < 1$, this gives

$$E(X) = \frac{p}{[1-(1-p)]^2} = \frac{1}{p}$$

The expected number of trials is inversely related to p, the probability of success, as might have been anticipated. The expected number of trials will be large if p is small, and conversely.

We now find $E(X(X-1))$. We have

$$E(X(X-1)) = \sum_{r=1}^{\infty} r(r-1)(1-p)^{r-1}p$$
$$= p(1-p) \sum_{r=2}^{\infty} r(r-1)(1-p)^{r-2}$$

Now $\sum_{r=2}^{\infty} r(r-1)x^{r-2} = 2/(1-x)^3$ if $|x| < 1$. Therefore,

$$E(X(X-1)) = p(1-p) \cdot \frac{2}{[1-(1-p)]^3} = \frac{2(1-p)}{p^2}$$

Consequently,

$$\text{Var}(X) = E(X(X-1)) + E(X) - [E(X)]^2$$
$$= \frac{2(1-p)}{p^2} + \frac{1}{p} - \frac{1}{p^2} = \frac{1-p}{p^2}$$

In conclusion,

$$E(X) = \frac{1}{p}$$
$$\text{Var}(X) = \frac{1-p}{p^2}$$

In Example 1.9 of Chapter 5, the probability that Mrs. Y gets the letter V on any visit is 0.002. Therefore, the expected number of times she will have to go to the store until she gets the letter V is $1/(0.002) = 500$ visits.

Example 2.1. The cost of launching a missile is \$20 million. Any failure incurs an additional expenditure (for repairing the launching pad, and so on) of \$2 million per launch. The probability of a successful launch is 0.4. If the missiles are fired until a successful launch is accomplished, find the expected total cost.

Solution. Let X denote the total number of attempts needed for a successful launch. Since X has a geometric distribution (assuming the outcomes of the launches are independent), $E(X) = 1/(0.4) = 2.5$.

The total cost Y (in millions of dollars) for the first successful launch is given by

$$Y = 20X + 2(X - 1) = 22X - 2$$

Therefore, $E(Y) = 22E(X) - 2 = 53$. That is, the expected total cost is \$53 million.

The Poisson distribution

If X_t has the Poisson distribution with parameter λt, then its probability function is

$$P(X_t = r) = \frac{e^{-\lambda t}(\lambda t)^r}{r!}, \qquad r = 0, 1, 2, \ldots$$

Therefore,

$$E(X_t) = \sum_{r=0}^{\infty} r \cdot \frac{e^{-\lambda t}(\lambda t)^r}{r!}$$

$$= e^{-\lambda t}\lambda t \sum_{r=1}^{\infty} \frac{(\lambda t)^{r-1}}{(r-1)!}$$

Letting $r - 1 = s$ gives

$$E(X_t) = e^{-\lambda t}\lambda t \sum_{s=0}^{\infty} \frac{(\lambda t)^s}{s!} = e^{-\lambda t}\lambda t e^{\lambda t} = \lambda t$$

The parameter λ therefore represents the expected number of changes per unit interval of time.

To find the variance of X_t, we first obtain $E(X_t(X_t - 1))$.

$$E(X_t(X_t - 1)) = \sum_{r=0}^{\infty} r(r-1) \frac{e^{-\lambda t}(\lambda t)^r}{r!}$$

$$= e^{-\lambda t}(\lambda t)^2 \sum_{r=2}^{\infty} \frac{(\lambda t)^{r-2}}{(r-2)!}$$

Letting $r - 2 = s$ yields

$$E(X_t(X_t - 1)) = e^{-\lambda t}(\lambda t)^2 \sum_{s=0}^{\infty} \frac{(\lambda t)^s}{s!} = e^{-\lambda t}(\lambda t)^2 e^{\lambda t} = (\lambda t)^2$$

Hence

$$\text{Var}(X_t) = E(X_t(X_t - 1)) + E(X_t) - [E(X_t)]^2 = (\lambda t)^2 + \lambda t - (\lambda t)^2 = \lambda t$$

Summarizing, if X_t has the Poisson distribution with parameter λt, then

$$E(X_t) = \lambda t$$
$$\text{Var}(X_t) = \lambda t$$

Example 2.2. Past experience shows that an office receives telephone calls at the mean rate of thirty calls per hour. Assuming that the number of calls have the Poisson distribution, find the probability that–

(*a*) during a ten-minute coffee-break, no calls will be received
(*b*) during a twenty-minute coffee-break, at least two calls will be received.

Solution. Assuming that time is measured in hours, we are given that $\lambda = 30$.

(*a*) We want the number of calls in a period of 10 minutes, that is, during an interval of time $(0, \frac{1}{6})$. Therefore, $\lambda t = 30(\frac{1}{6}) = 5$, and the probability of r calls is

$$\frac{e^{-5}5^r}{r!}, \qquad r = 0, 1, 2, \ldots$$

Hence the probability of no calls is $e^{-5} = 0.0067$.

(*b*) In this case, $t = \frac{1}{3}$ and consequently $\lambda t = 30(\frac{1}{3}) = 10$. Therefore, the probability of r calls is

$$\frac{e^{-10}(10)^r}{r!}, \qquad r = 0, 1, \ldots$$

As a result, the probability of at least two calls is $1 - [e^{-10} + 10e^{-10}] = 0.9995$.

Example 2.3. Defects occur along the length of a cable at an average of six defects per 4000 feet. If the number of defects on any length of the cable has a Poisson distribution, find the probability that a 3000-foot cable will have at most two defects.

Solution. If we measure t in units of feet, we are given that $\lambda(4000) = 6$, so that λ, the expected number of defects per foot, is $\frac{6}{4000}$. Therefore, the expected number of defects on a 3000-foot cable is $\frac{6}{4000}(3000) = 4.5$. Hence, the probability of r defects on a 3000-foot cable is

$$\frac{e^{-4.5}(4.5)^r}{r!}, \qquad r = 0, 1, 2, \ldots$$

Consequently, the probability of at most two defects is

$$e^{-4.5} + e^{-4.5}(4.5) + \frac{e^{-4.5}(4.5)^2}{2!} = (15.625)e^{-4.5} = 0.1736$$

Example 2.4. Suppose "changes" take place according to the Poisson distribution at the mean rate of λ changes per unit interval. Given that during the interval $[0, 1]$ exactly one change has occurred, find the distribution of the time T when the change occurs.

Solution. We observe first of all that the distribution of T is continuous. Let $0 \leqslant t \leqslant 1$. Then

$$P(T \leqslant t \mid \text{one change in } [0, 1]) = \frac{P(T \leqslant t \textit{ and } \text{one change in } [0, 1])}{P(\text{one change in } [0, 1])}$$

(Recall that $P(A \mid B) = P(A \text{ and } B)/P(B)$). Now

$$\begin{aligned} &P(T \leqslant t \text{ and one change in } [0, 1]) \\ &\quad = P(\text{one change in } [0, t] \text{ and none in } (t, 1]) \\ &\quad = P(\text{one change in } [0, t])P(\text{none in } (t, 1]) \end{aligned}$$

since changes in nonoverlapping intervals are independent.

We further note that

$$P(\text{one change in } [0, 1]) = \frac{e^{-\lambda}\lambda^1}{1!} = \lambda e^{-\lambda}$$

$$P(\text{one change in } [0, t]) = \frac{e^{-\lambda t}(\lambda t)^1}{1!} = \lambda t e^{-\lambda t}$$

and $P(\text{none in } (t, 1]) = e^{-\lambda(1-t)} \dfrac{[\lambda(1-t)]^0}{0!} = e^{-\lambda(1-t)}$

since the length of the interval $(t, 1]$ is $1 - t$.

Substituting, it follows that

$$P(T \leqslant t \mid \text{one change in } [0, 1]) = \frac{\lambda t e^{-\lambda t} e^{-\lambda(1-t)}}{\lambda e^{-\lambda}} = t$$

Of course, it is obvious that

$$P(T \leqslant t \mid \text{one change in } [0, 1]) = \begin{cases} 0 & \text{if } t < 0 \\ 1 & \text{if } t > 1 \end{cases}$$

In conclusion, therefore, given that exactly one change has occurred in $[0, 1]$ the distribution of the time T when the change occurs is uniform.

The uniform distribution

If X is uniformly distributed over the interval $[a, b]$, its pdf is given by

$$f(x) = \begin{cases} \dfrac{1}{b-a}, & a < x < b \\ 0, & \text{elsewhere} \end{cases}$$

We shall first find $E(X^n)$ for any nonnegative integer n.

$$\begin{aligned} E(X^n) &= \int_{-\infty}^{\infty} x^n \cdot f(x)\, dx = \int_a^b \frac{x^n}{b-a} dx \\ &= \frac{1}{n+1}\left(\frac{b^{n+1} - a^{n+1}}{b-a}\right) \end{aligned}$$

In particular, setting $n = 1$, we get

$$E(X) = \frac{b+a}{2}$$

Notice that $(b+a)/2$ is the midpoint of the interval $[a, b]$, and this makes good sense in view of the fact that X is uniformly distributed over $[a, b]$. Also observe that $(a+b)/2$ is the point about which the distribution is symmetric.

Setting $n = 2$ gives

$$E(X^2) = \frac{1}{3}\left(\frac{b^3 - a^3}{b-a}\right) = \frac{1}{3}(b^2 + ab + a^2)$$

Therefore,

$$\begin{aligned}\text{Var}(X) &= E(X^2) - [E(X)]^2 \\ &= \frac{1}{3}(b^2 + ab + a^2) - \left(\frac{a+b}{2}\right)^2 = \frac{(b-a)^2}{12}\end{aligned}$$

In summary,

$$E(X) = \frac{a+b}{2}$$

$$\text{Var}(X) = \frac{(b-a)^2}{12}$$

Comment. If both the expected value and the variance of a uniformly distributed random variable are known, we can find its parameters a and b, thereby completely specifying the underlying distribution. For example, if X has a uniform distribution with $E(X) = 1$ and $\text{Var}(X) = 3$, then $(a+b)/2 = 1$ and $(b-a)^2/12 = 3$, so that $a = -2$ and $b = 4$. The underlying pdf is then obtained as

$$f(x) = \begin{cases} \frac{1}{6}, & -2 < x < 4 \\ 0, & \text{elsewhere} \end{cases}$$

The normal distribution

Suppose X is normally distributed with the pdf given by

$$f(x) = \frac{1}{b\sqrt{2\pi}} e^{-(x-a)^2/(2b^2)}, \quad -\infty < x < \infty$$

where $-\infty < a < \infty$, $b > 0$.

Instead of finding $E(X)$ directly, we shall first find $E(X-a)$. Before this can be done we have to make sure that $E(X-a)$ exists. Towards this end, let us consider $E(|X-a|)$:

$$\begin{aligned}E(|X-a|) &= \frac{1}{b\sqrt{2\pi}} \int_{-\infty}^{\infty} |x-a| e^{-(x-a)^2/(2b^2)} dx, \\ &= \frac{1}{b\sqrt{2\pi}} \left[\int_{-\infty}^{a} (a-x) e^{-(x-a)^2/(2b^2)} dx + \int_{a}^{\infty} (x-a) e^{-(x-a)^2/(2b^2)} dx\right]\end{aligned}$$

Now

$$\int_{-\infty}^{a} (a-x) e^{-(x-a)^2/(2b^2)}\, dx = \int_{a}^{\infty} (x-a) e^{-(x-a)^2/(2b^2)} dx \quad \text{(why?)}$$

Therefore,

$$E(|X-a|) = \frac{2}{b\sqrt{2\pi}} \int_a^\infty (x-a)e^{-(x-a)^2/(2b^2)}dx$$

Let $(x-a)^2/(2b^2) = u$; then $x - a = b\sqrt{2u}$ and $dx = (b\,du)/\sqrt{2u}$. Hence

$$E(|X-a|) = \frac{2}{b\sqrt{2\pi}} \int_0^\infty b\sqrt{2u}\, e^{-u} \cdot \frac{b\,du}{\sqrt{2u}} = \frac{2b}{\sqrt{2\pi}} \int_0^\infty e^{-u}\,du = \frac{2b}{\sqrt{2\pi}}$$

That is, $E(|X-a|) < \infty$. Consequently, $E(X-a)$ exists. Now that we have shown $E(X-a) < \infty$, that is, $E(X) - a < \infty$, it follows that $E(X) < \infty$. Since the distribution of X is symmetric around a and $E(X) < \infty$, we have

$$E(X) = a$$

by the result on symmetric distributions proved at the beginning of this section.

To find the variance of X, we shall employ the following integration formula:

$$\int_{-\infty}^\infty x^{2n}e^{-\alpha x^2}\,dx = \frac{1\cdot 3 \cdot \ldots \cdot (2n-1)}{(2\alpha)^n}\sqrt{\frac{\pi}{\alpha}}$$

where n is a positive integer and $\alpha > 0$. Now

$$\mathrm{Var}(X) = E[(X-a)^2] = \int_{-\infty}^\infty (x-a)^2 \cdot \frac{1}{b\sqrt{2\pi}}\, e^{-(x-a)^2/(2b^2)}dx$$

Let $(x-a)/b = y$; then $dx = b\,dy$, and we get

$$\mathrm{Var}(X) = \frac{b^2}{\sqrt{2\pi}} \int_{-\infty}^\infty y^2 e^{-y^2/2}\,dy = b^2$$

setting $n = 1$ and $\alpha = \frac{1}{2}$ in the formula.

We see therefore that *the parameter b which gives the distance of the points of inflection from the line of symmetry is actually the standard deviation of the random variable.*

Summarizing,

$$E(X) = a$$
$$\mathrm{Var}(X) = b^2$$

The normal distribution is completely determined by the knowledge of the expectation and variance of the random variable.

Having seen the significance of the constants a and b, from now on we shall feel free to use μ in place of a and σ^2 in place of b^2.

Example 2.5. (Do you grade on the curve?) Suppose the grades on a test are (approximately) normally distributed with a mean score of 70 points and a standard deviation of 10 points. It is decided to give A's to 10 percent of the students and B's to 25 percent of the students. Find what scores should be assigned A's and B's.

Solution. "The grades on a test" is in reality a discrete random variable, but the normal distribution provides a good approximation. Let us denote the random variable by X. The problem is reduced to the following: We want to find two numbers u and v for which

$$P(X \geqslant u) = 0.1$$
$$\text{and} \quad P(v \leqslant X < u) = 0.25$$

In other words, $P(X \leqslant u) = 0.9$ and $P(X \leqslant v) = 0.65$ (see Figure 2.1).

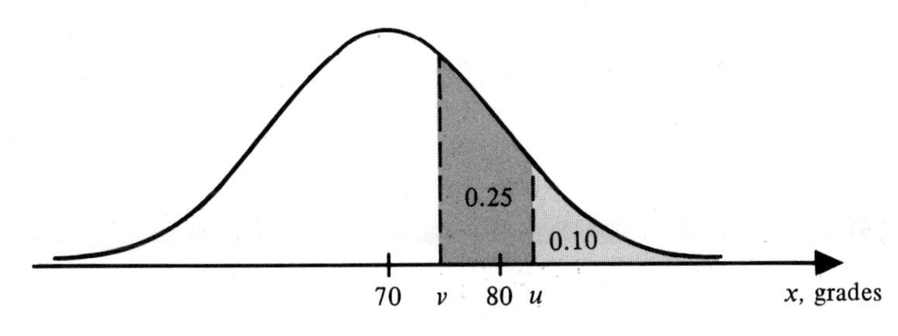

Figure 2.1

Therefore, since $\mu = 70$ and $\sigma = 10$,

$$\Phi\left(\frac{u-70}{10}\right) = 0.9 \quad \text{and} \quad \Phi\left(\frac{v-70}{10}\right) = 0.65$$

Hence, from the standard normal table (Table C, Appendix),

$$\frac{u-70}{10} = 1.28 \quad \text{and} \quad \frac{v-70}{10} = 0.44$$

Consequently, $u = 82.8$ and $v = 74.4$.

Thus students with the score of 83 points or over should be given an A; those with a score between 74 and 82 points should receive a B.

Example 2.6. The demand for meat at a grocery store during any week is approximately normally distributed with a mean demand of 5000 lbs and a standard deviation of 300 lbs.

(*a*) If the store has 5300 lbs of meat in stock, what is the probability that it is overstocked?

(*b*) How much meat should the store have in stock per week so as to not run short more than 10 percent of the time?

Solution

(*a*) Let X denote the demand for meat. The store will be overstocked if and only if $X < 5300$. Since $\mu = 5000$ and $\sigma = 300$, the required probability is

$$P(X < 5300) = \Phi\left(\frac{5300 - 5000}{300}\right) = \Phi(1) = 0.8413$$

Rephrased, if the store makes it a policy to keep 5300 lbs in stock per week, it will not run short approximately 84 percent of the time.

(*b*) Let q denote the amount of meat that should be stocked. The store will run short if and only if $X > q$, and we are given that the probability of this happening should not exceed 10 percent. Hence it is safe to say that q satisfies

$$P(X > q) = 0.10 \quad (\text{in reality}, \leqslant 0.10)$$

that is, $P(X \leqslant q) = 0.90$. Therefore, q satisfies

$$\Phi\left(\frac{q - 5000}{300}\right) = 0.90$$

But, from the standard normal table, $\Phi(1.28) = 0.8997$ (or approximately 0.90). Hence setting $(q - 5000)/300 = 1.28$ we get $q = 5384$. That is, 5384 lbs of meat should be stocked so that the store will not run out of meat more than 10 percent of the time.

The exponential distribution

The pdf of an exponentially distributed random variable X is

$$f(x) = \begin{cases} \lambda e^{-\lambda x}, & x > 0 \\ 0, & \text{elsewhere} \end{cases}$$

We shall first find $E(X^n)$ for any nonnegative integer n.

$$E(X^n) = \int_0^\infty x^n \cdot \lambda e^{-\lambda x}\, dx$$

Let $u = \lambda x$; then $dx = du/\lambda$. Hence

$$E(X^n) = \int_0^\infty \frac{u^n}{\lambda^n} \cdot \lambda e^{-u} \frac{du}{\lambda}$$

$$= \frac{1}{\lambda^n} \int_0^\infty u^{(n+1)-1} e^{-u}\, du = \frac{\Gamma(n+1)}{\lambda^n} = \frac{n!}{\lambda^n}$$

Consequently,

$$E(X^n) = \frac{n!}{\lambda^n}$$

In particular, $E(X) = 1/\lambda$ and $E(X^2) = 2/\lambda^2$. Therefore,

$$\text{Var}(X) = E(X^2) - [E(X)]^2 = \frac{2}{\lambda^2} - \frac{1}{\lambda^2} = \frac{1}{\lambda^2}$$

Summarizing,

$$E(X) = \frac{1}{\lambda}$$

$$\text{Var}(X) = \frac{1}{\lambda^2}$$

Example 2.7. (How long is the wait?) Telephone calls are received in an office at the mean rate of λ calls per unit period of time. Assuming a Poisson distribution of the number of calls, find the distribution of the length of time that a secretary has to wait for the first call.

Solution. Let X_t = Number of calls during the interval $(0, t)$, and let T = The length of time the secretary has to wait for the first call. (T is a random variable.)

The secretary will have to wait for at least t units of time if and only if during the period $(0, t)$ there is no call. That is,

$$T > t \Leftrightarrow X_t = 0$$

Therefore,

$$P(T > t) = P(X_t = 0)$$

Consequently,

$$1 - F_T(t) = e^{-\lambda t}$$

Hence

$$F_T(t) = 1 - e^{-\lambda t}, \qquad t > 0$$

Of course, $F_T(t) = 0$ if $t \leqslant 0$. Thus

$$F_T(t) = \begin{cases} 0, & t \leqslant 0 \\ 1 - e^{-\lambda t}, & t > 0 \end{cases}$$

The distribution of T is therefore exponential with parameter λ. ■

Comment. Since for an exponential distribution with parameter λ the expected value is $1/\lambda$, we see that the expected length of wait is equal to the reciprocal of the expected number of calls. The higher the mean rate of calls, the shorter the expected wait.

Example 2.8. Suppose the lifetime of an electronic tube has an exponential distribution with a mean lifetime of μ hours. Find the expected lifetime of a tube, given that the tube has lasted at least T hours.

Solution. Let X represent the lifetime of an electronic tube. Then, since X has an exponential distribution with mean μ, its pdf is given by

$$f(x) = \begin{cases} \frac{1}{\mu} e^{-x/\mu}, & x > 0 \\ 0, & \text{elsewhere} \end{cases}$$

Hence the conditional pdf of X given $X \geqslant T$ is

$$f_X(x \mid X \geqslant T) = \begin{cases} \dfrac{\frac{1}{\mu} e^{-x/\mu}}{P(X \geqslant T)}, & \text{if } x \geqslant T \\ 0, & \text{elsewhere} \end{cases}$$

Since $P(X \geqslant T) = e^{-T/\mu}$, we get, after simplifying,

$$f_X(x \mid X \geqslant T) = \begin{cases} \dfrac{1}{\mu} e^{-\frac{1}{\mu}(x-T)}, & \text{if } x \geqslant T \\ 0, & \text{elsewhere} \end{cases}$$

Therefore, making the substitution $y = (x - T)/\mu$ in the integral and simplifying,

$$E(X \mid X \geqslant T) = \int_T^{\infty} x \cdot \frac{1}{\mu} e^{-\frac{1}{\mu}(x-T)} dx = \mu + T$$

This is not surprising in view of the *lack of memory property* of the exponential distribution.

EXERCISES–SECTION 2

1. Suppose X is a binomial random variable $B(8, 0.2)$. If μ and σ represent respectively the mean and the standard deviation of X, find $P(\mu + \sigma < X < 3\mu + \sigma)$.
2. Suppose X has the binomial distribution consisting of 20 independent trials. If the variance of X is 3.20, find the probability of getting 3 successes.
3. The life (measured in hours) of an electronic component used in a machine has the following pdf:

$$f(x) = \begin{cases} \dfrac{2(100)^2}{x^3}, & x \geqslant 100 \\ 0, & \text{elsewhere} \end{cases}$$

   (a) Find the expected life of the component.
   (b) If a machine uses 180 components which function independently, find the expected number of components still functioning at the end of 300 hours.
4. A coin with $P(\text{head}) = \frac{1}{3}$ is tossed n times. If a person wins C_1 dollars on any toss that shows up heads and loses C_2 dollars on any toss that shows up tails, find the expected net winnings and the variance of these winnings. *Hint:* If Y represents the net winnings and X the number of tosses, then $Y = (C_1 + C_2)X - C_2 n$.
5. Suppose X has the binomial distribution $B(n, p)$. A random variable Y defined by $Y = X/n$ represents the proportion of successes in n independent Bernoulli trials and is called the *sample proportion*. Show that $E(Y) = p$ and $\text{Var}(Y) = p(1-p)/n$.
6. A machine is known to have a defective rate of producing 1 defective item in ten. If 100 items produced by this machine are inspected, what is the expected proportion of defectives? Also find the variance of the proportion.

7. A random variable has a known pdf f.
   (a) If n independent observations are made from this distribution, find the expected number of observations which are less than c, where c is a real number.
   (b) If independent observations are made, find the expected number of observations needed to obtain a value less than c for the first time.
8. Express $\text{Var}(X)$ in terms of $E(X(X+1))$ and $E(X)$.
9. Suppose X has the negative binomial distribution given by

$$P(X=k)=\binom{k-1}{r-1}p^r(1-p)^{k-r}, \qquad k=r, r+1, \ldots$$

Show that $E(X)=r/p$ and $E(X(X+1))=r(1+r)/p^2$. Use these to show that $\text{Var}(X)=r(1-p)/p^2$.

10. Let X be a Poisson random variable with parameter λ. Show that $E(X(X-1)\ldots(X-n+1))=\lambda^n$ for any positive integer n.

11. If X has a Poisson distribution with parameter λ, show that $E(|X-1|)=\lambda+2e^{-\lambda}-1$. *Hint:* Find $E(|X-1|+1)$.

12. The number of misprints on a page is a Poisson random variable with a standard deviation of 0.1.
   (a) What is the probability that a 400-page volume will have at least two misprints?
   (b) Find the number of misprints that have the maximum probability and find this probability.

13. The number of people who visit a shooting gallery in a one-hour period is a Poisson random variable, with a mean rate of six people per hour. The probability that a person wins a prize is equal to 0.2.
   (a) Find the distribution of the number of people who win prizes in an eight-hour period.
   (b) Find the expected number of people winning prizes in an eight-hour period. *Hint:* See Example 1.14 of Chapter 5.

14. Find $E(5X^2+2X+3)$ if X is–
   (a) binomial, $B(10, 0.4)$
   (b) Poisson with parameter $\lambda=3$
   (c) normal, $N(10, 16)$.

15. If a random variable X has a symmetric distribution about c, show that all odd-order moments about c are zero; that is

$$E[(X-c)^{2n+1}]=0$$

for any nonnegative integer n.

Hint: $E[(X-c)^{2n+1}]=\int_{-\infty}^{c}(x-c)^{2n+1}f(x)\,dx+\int_{c}^{\infty}(x-c)^{2n+1}f(x)\,dx.$

16. If X has a symmetric distribution with mean μ and variance σ^2, both finite, show that $E(X^3)=3\mu\sigma^2+\mu^3$.

17. If X has a uniform distribution with $E(X)=5$ and $\text{Var}(X)=3$, find $P(|X-4|>1.5)$.

18. A dart board consists of three concentric circles of radii 2, 5, and 10 inches. A person wins twenty dollars if he throws the dart inside the circle of radius 2, four dollars if he throws inside the annular region between the circles of radii 2 and 5, and one dollar if he throws in the annular region between the circles of radii 5 and 10. If the privilege of throwing a dart costs three dollars, find the expected winnings. Interpret the result as you would to a layman. (Assume that a person is sure to hit some point on the dart board and that the probability of hitting in any region of the board is proportional to the area of the region.)
19. A random variable X has a uniform distribution over the interval $[0, 1]$. Find:
    (a) $E(X^n)$, for any positive integer n
    (b) $\text{Var}(X^n)$, for any positive integer n
20. Suppose X is uniformly distributed over the interval $(-\pi/2, \pi)$. Find $E(\sin X)$.
21. If X is uniformly distributed over the interval $(0, 1)$, find $E(\ln X)$ and $\text{Var}(\ln X)$.
22. The length of a telephone conversation (measured in hours) has an exponential distribution with the parameter $\lambda = 4$. If the telephone company charges at the rate of $0.5 + 5T$ dollars for a phone call that lasts T hours, find the expected cost of a call. Interpret this answer in practical terms.
23. The weight of food packed in certain containers is a normally distributed random variable with a mean weight of 500 lbs and a standard deviation of 5 lbs. Suppose a container is picked at random. Find the probability that the container weighs–
    (a) more than 510 lbs
    (b) less than 498 lbs
    (c) between 491 and 498 lbs.
24. The amount of annual rainfall in a certain region is known to be a normally distributed random variable with a mean rainfall of 50 inches and a standard deviation of 4 inches. Find the probability that the annual rainfall will exceed 58 inches.
25. Consider the distribution of rainfall described in exercise 24. If the rainfall exceeds 58 inches during a year, it leads to floods.
    (a) What is the probability that during a fifteen-year period there will be: (*i*) no floods, (*ii*) at least one flood, (*iii*) three floods?
    (b) During a fifty-year period, what is the expected number of floods? Interpret this result.
26. The radius of a lead shot produced by a machine is a random variable which has a normal distribution with a mean radius of 2 cm and a standard deviation of 0.3 cm. Find the expected volume of the lead shot. *Hint:* See exercise 16.
27. Suppose X is $N(0, \sigma^2)$. Show that $E(e^X) = e^{E(X^2)/2}$.

Hint: $$\frac{1}{\sigma\sqrt{2\pi}}\int_{-\infty}^{\infty} e^{-(x-\sigma^2)^2/(2\sigma^2)}\,dx = 1$$

28. If a random variable X is normally distributed with mean μ and variance σ^2, show that $E(|X-\mu|) = \sigma\sqrt{2/\pi}$. *Hint:* If f is the pdf of X, then $E(|X-\mu|) = 2\int_{\mu}^{\infty}(x-\mu)f(x)\,dx$. Also, $\int_{0}^{\infty} ye^{-y^2/2}\,dy = 1$.

29. Suppose X is $N(0, 1)$. Use integration by parts to derive a recurrence relation for the moments $E(X^{2r})$, $r \geqslant 0$. Hence show that $E(X^n) = (n-1)(n-3) \ldots 3 \cdot 1$ when n is even.

30. If X has a gamma distribution with parameters λ, p show that $E(X) = p/\lambda$ and $\text{Var}(X) = \lambda/p^2$.

31. A radioactive source is emitting particles at the mean rate of five particles per half hour. Assuming that the number of particles emitted has a Poisson distribution, find the expected length of wait until the source emits eight particles. Also find the variance of the wait. *Hint:* See exercise 23 of Section 2, Chapter 5.

Table 2.1. A Glossary of the More Common Discrete Distributions

The Distribution	Parameters	Probability Function, $p(x)$	Mean	Variance
Bernoulli	$0 \leqslant p \leqslant 1$	$p^x(1-p)^{1-x}, \quad x = 0, 1$	p	$p(1-p)$
Binomial	$n = 1, 2, \ldots$ $0 \leqslant p \leqslant 1$	$\binom{n}{x} p^x(1-p)^{n-x}, \quad x = 0, 1, \ldots, n$	np	$np(1-p)$
Geometric	$0 \leqslant p \leqslant 1$	$p(1-p)^{x-1}, \quad x = 1, 2, \ldots$	$\frac{1}{p}$	$\frac{1-p}{p^2}$
Negative Binomial* (Pascal)	$r = 1, 2, \ldots$ $0 \leqslant p \leqslant 1$	$\binom{r+x-1}{r-1} p^r(1-p)^x, \quad x = 0, 1, \ldots$	$\frac{r}{p}$	$\frac{r(1-p)}{p^2}$
Poisson	$t > 0$ $\lambda > 0$	$\frac{e^{-\lambda t}(\lambda t)^x}{x!}, \quad x = 0, 1, \ldots$	λt	λt
Hypergeometric*	$N = 1, 2, \ldots$ $n = 1, 2, \ldots, N$ $p = 0, \frac{1}{N}, \ldots, 1$	$\frac{\binom{Np}{x}\binom{Nq}{n-x}}{\binom{N}{n}}, \quad x = 0, 1, \ldots, n$	np	$np(1-p)\left(\frac{N-n}{N-1}\right)$

*The computation of the expectation and variance are left to the exercises. (Also, an alternate method of computation will be given for these cases in Chapter 11.)

Table 2.2. A Glossary of the More Common Continuous Distributions

The Distribution	Parameters	Probability Density Function	Mean	Variance
Uniform over the interval $[a, b]$	$-\infty < a < \infty$ $-\infty < b < \infty$ with $a < b$	$f(x) = \begin{cases} \frac{1}{b-a}, & a \leqslant x \leqslant b \\ 0, & \text{elsewhere} \end{cases}$	$\frac{a+b}{2}$	$\frac{(b-a)^2}{12}$
Normal	$-\infty < a < \infty$ $b > 0$	$f(x) = \frac{1}{b\sqrt{2\pi}} e^{-(x-a)^2/2b^2}$, $-\infty < x < \infty$	a	b^2
Exponential	$\lambda > 0$	$f(x) = \begin{cases} \lambda e^{-\lambda x}, & x > 0 \\ 0, & \text{elsewhere} \end{cases}$	$\frac{1}{\lambda}$	$\frac{1}{\lambda^2}$
Gamma*	$\lambda > 0$ $p > 0$	$f(x) = \begin{cases} \frac{\lambda^p}{\Gamma(p)} x^{p-1} e^{-\lambda x}, & x > 0 \\ 0, & \text{elsewhere} \end{cases}$	$\frac{p}{\lambda}$	$\frac{p}{\lambda^2}$

*The computation of the expectation and variance are left to the exercises.

8 Joint and Marginal Distributions

INTRODUCTION

Our first encounter with the notion of a random variable was in Chapter 4. The discussion there centered on one random variable defined on a given sample space. This is often called the *univariate case.* However, there are many instances when we are interested in observing more than one numerical value on each outcome in the sample space. For example, if a person is picked at random, we might be interested in his height, weight, and IQ. As another example, the sample space might consist of points on the surface of the earth, and if s represents one such point then we might be interested in $X(s)$, the latitude of s, and $Y(s)$, the longitude of s.

For a better understanding of the concepts involved, we shall concentrate on the case of two random variables. This is referred to as the *bivariate case.* The more general case involving any finite number of random variables is called the *multivariate case,* but will not be considered here in its full generality. It would involve cumbersome notation without adding to our comprehension. We shall present a glossary of results for the general multivariate case in the last section of Chapter 9.

1. JOINT DISTRIBUTIONS

1.1 The Notion of a Random Vector

Let us initiate our discussion with the following example. Suppose two dice are rolled giving the familiar sample space $S = \{(i, j) \mid i, j = 1, 2, \ldots, 6\}$ with 36 outcomes. Consider two random variables X and Y defined on S in the following way:

X = The sum of the numbers on the dice
Y = The absolute value of the difference of the numbers on the dice

For any sample point $s = (i, j)$, $X(s) = i + j$ and $Y(s) = |i - j|$. The range of X consists of the set of integers $\{2, 3, 4, \ldots, 12\}$, and that of Y the set $\{0, 1, 2, 3, 4, 5\}$. The values of the random variables X and Y could be represented as ordered pairs $(X(s), Y(s))$, thereby getting the set

$$\left\{(x, y) \,\middle|\, \begin{array}{l} x = 2, 3, \ldots, 12 \\ y = 0, 1, 2, 3, 4, 5 \end{array}\right\}$$

In a general setting, suppose S is a sample space and X and Y are two real valued functions defined on S.

$$X : S \to \mathbf{R} \quad \text{and} \quad Y : S \to \mathbf{R}$$

For any point $s \in S$, the values of X and Y can be represented as an ordered pair $(X(s), Y(s))$. In a natural way, this procedure defines a function from S to $\mathbf{R}^2$, the Cartesian plane. Denoting this function by (X, Y), we have

$$(X, Y) : S \to \mathbf{R}^2$$

where, for any $s \in S$, $(X, Y)(s) = (X(s), Y(s))$. This is shown in Figure 1.1.

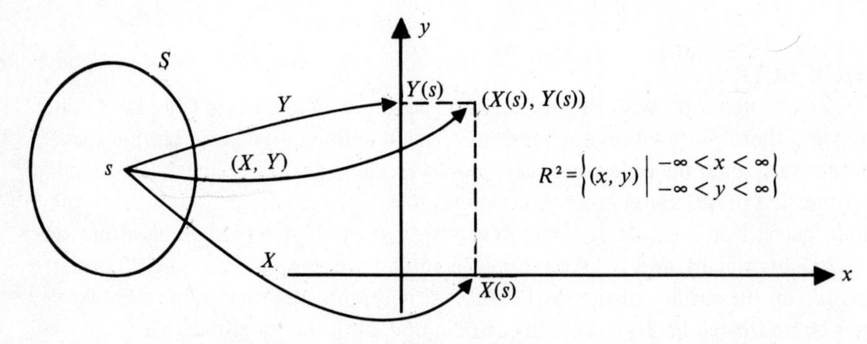

Figure 1.1

So far we have only defined the function (X, Y) which is called a *vector-valued function.* We have not actually given the definition of a *random vector* and this is what we shall do next. For this purpose, suppose $(S, \mathcal{F}, P)$ is a probability space.

A **random vector** (X, Y) is a function $(X, Y) : S \to \mathbf{R}^2$ such that, for any two real numbers u and v, $\{s \mid -\infty < X(s) \leqslant u, -\infty < Y(s) \leqslant v\}$ is a member of $\mathcal{F}$.

Important: In what follows, it is worth noting that the treatment of joint distributions runs for the most part parallel to that of the univariate case. For instance, the definitions of random vectors, joint distribution functions, discrete distributions, and absolutely continuous distributions can all be related to the corresponding definitions in the univariate case.

If X is a random variable, then we know that X induces a probability measure P^X on the subsets of the real line. In like manner, the function (X, Y) induces a probability measure on the (Borel) subsets of the plane. Let us denote this probability measure by $P^{X,Y}$. How is it defined on a (Borel) set E of the plane? Naturally, as in the univariate case, we collect those points of S that go into E. If this set is A, say, then assign to E the probability $P(A)$. (See Figure 1.2.)

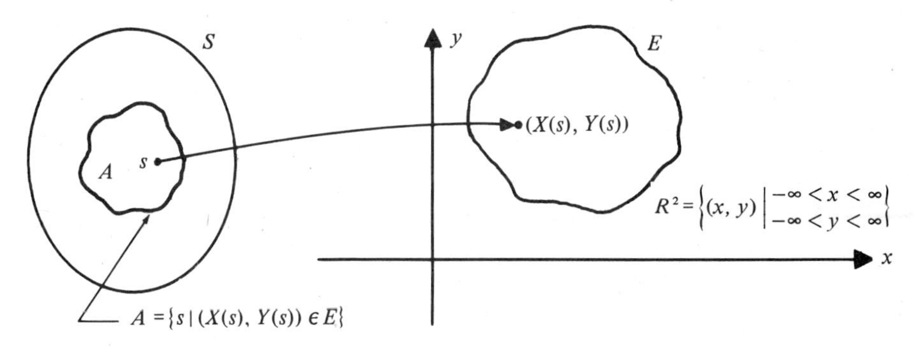

Figure 1.2

In short, we set

$$P^{X,Y}(E) = P(\{s \mid (X(s), Y(s)) \in E\})$$

Following the convention adopted in the univariate case, we shall write $\{(X, Y) \in E\}$ in place of $\{s \mid (X(s), Y(s)) \in E\}$. For example, we write $\{X \leqslant u,\ Y \leqslant v\}$ instead of $\{s \mid X(s) \leqslant u,\ Y(s) \leqslant v\}$, and this represents the set of points in S for which $X(s) \leqslant u$ *and* $Y(s) \leqslant v$.

1.2 The Definition of a Joint Distribution Function

The concept of a joint distribution function is simply an extension of the notion of a distribution function in the univariate case. Whereas we use the infinite interval $(-\infty, u]$ in the univariate case, here we use the infinite rectangle B_{uv} where

$$B_{uv} = \{(x, y) \mid -\infty < x \leqslant u,\ -\infty < y \leqslant v\}$$

B_{uv} consists of all the ordered pairs of real numbers where the first component is less than or equal to u *and* the second component is less than or equal to v, as illustrated by the shaded region in Figure 1.3.

Notice, incidentally, that B_{uv} is the intersection of the two regions of the plane, namely, $\{(x, y) \mid x \leqslant u\}$ (see Figure 1.4(a)) and $\{(x, y) \mid y \leqslant v\}$ (see Figure 1.4(b)).

Now, as we know,

$$\begin{aligned} P^{X,Y}(B_{uv}) &= P(\{s \mid (X(s), Y(s)) \in B_{uv}\}) \\ &= P(\{s \mid -\infty < X(s) \leqslant u,\ -\infty < Y(s) \leqslant v\}) \end{aligned}$$

Or, using the notational convention,

$$P^{X,Y}(B_{uv}) = P(X \leqslant u,\ Y \leqslant v)$$

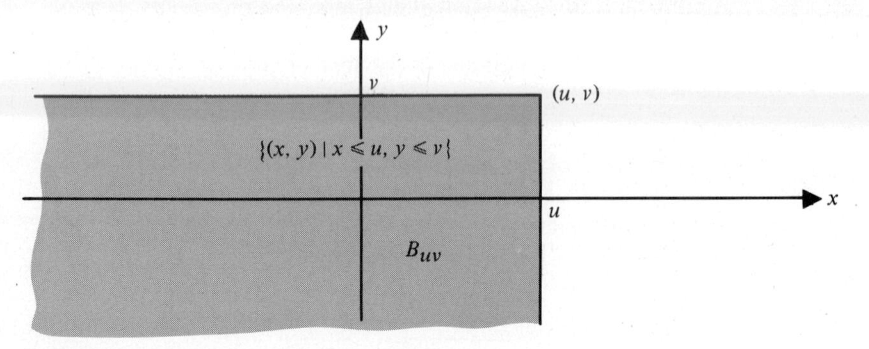

Figure 1.3

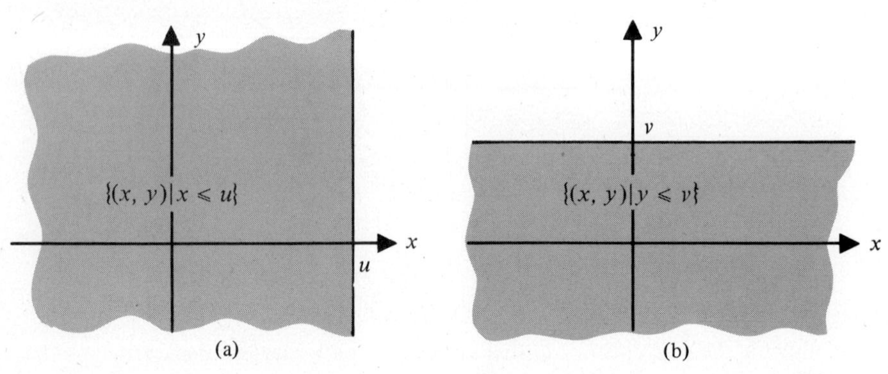

Figure 1.4

As u and v change we get different subsets B_{uv} of the plane and, in the process, a function of two real variables. This function is called the joint distribution function of X and Y. We now give the following definition:

A function F defined by

$$F(u, v) = P(X \leqslant u, \;\; Y \leqslant v)$$

for any real numbers u and v is called the **joint distribution function** of X and Y.

(From now on, in the context of two random variables X, Y, we shall denote the joint distribution function of X and Y by F. The distribution functions of X and Y will be denoted appropriately by using the corresponding subscripts as F_X, F_Y.)

It is tedious to derive the distribution function for a joint distribution, and it is even harder to plot the graph. First of all, two coordinate axes have to be reserved to plot the values of the two variables, and the third axis for plotting the values of F. The following procedure will be adopted frequently: we shall mark off the values of the variables along the two axes in the plane of the paper and *imagine* that the values of F are plotted along the axis perpendicular to the plane of the paper.

Example 1.1. A fair coin is tossed twice. Define X and Y as follows:

X = The number of heads on the first toss
Y = The number of heads on the second toss

Find the joint D.F. of X and Y.

Solution. The sample space has four outcomes and can be written as $S = \{HH, HT, TH, TT\}$. Now, recalling that (X, Y) is defined for any $s \in S$ by the assignment $(X, Y)(s) = (X(s), Y(s))$, we get

$$(X, Y)(HH) = (X(HH), Y(HH)) = (1, 1)$$
$$(X, Y)(HT) = (X(HT), Y(HT)) = (1, 0)$$

and so on.

Therefore, (X, Y) takes four values, namely, $(0, 0)$, $(0, 1)$, $(1, 0)$, and $(1, 1)$, each with probability $\frac{1}{4}$. To obtain the joint D.F., we observe that

$$\{s \mid X(s) \leqslant u,\ Y(s) \leqslant v\} = \begin{cases} \emptyset, & u < 0 \quad \textit{or} \quad v < 0 \\ \{TT\}, & 0 \leqslant u < 1 \quad \textit{and} \quad 0 \leqslant v < 1 \\ \{TT, TH\}, & 0 \leqslant u < 1 \quad \textit{and} \quad v \geqslant 1 \\ \{TT, HT\}, & u \geqslant 1 \quad \textit{and} \quad 0 \leqslant v < 1 \\ S, & u \geqslant 1 \quad \textit{and} \quad v \geqslant 1 \end{cases}$$

For example, in Figure 1.5 we are considering u, v where $u \geqslant 1$ and $0 \leqslant v < 1$. Notice that the only outcomes of S that go into the infinite rectangle are HT, TT.

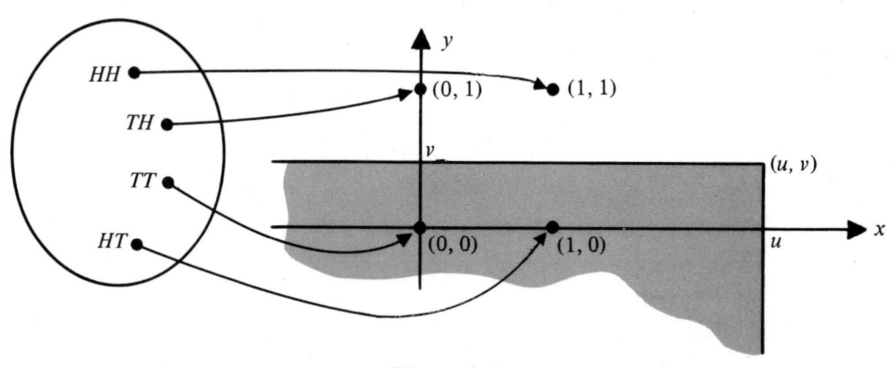

Figure 1.5

Therefore, finding the probabilities, the joint D.F. of X and Y is given by

$$F(u, v) = \begin{cases} 0, & u < 0 \quad \text{or} \quad v < 0 \\ \frac{1}{4}, & 0 \leqslant u < 1 \quad \text{and} \quad 0 \leqslant v < 1 \\ \frac{1}{2}, & 0 \leqslant u < 1 \quad \text{and} \quad v \geqslant 1 \\ \frac{1}{2}, & u \geqslant 1 \quad \text{and} \quad 0 \leqslant v < 1 \\ 1, & u \geqslant 1 \quad \text{and} \quad v \geqslant 1 \end{cases}$$

1.3 Properties of Joint Distribution Functions

A joint distribution function (of two random variables) has the properties given below. The reader should try to appreciate them in conjunction with the properties of a distribution function in the univariate case. It should be mentioned that the proofs in the univariate case were given in terms of the induced measure P^X rather than in terms of P, the probability measure on subsets of S, and it was left to the reader to argue by using the measure P. In what follows, just for a change, the proofs are based on the measure P rather than the induced measure $P^{X,Y}$. This time, the reader might wish to develop for himself the proofs based on the measure $P^{X,Y}$.

(*i*) $0 \leqslant F(u, v) \leqslant 1$ for any pair of real numbers u, v

This follows immediately because $F(u, v)$ represents probability.

(*ii*) **(Monotonicity)** F is a monotone nondecreasing function in each variable separately. That is, for any *fixed u*, $F(u, v)$ is monotone nondecreasing in v and for any *fixed v*, $F(u, v)$ is monotone nondecreasing in u.

This can be proved as follows: Let us fix $u = b$ and let $c < d$.

From Figure 1.6, it can be seen that

$$\{s \mid X(s) \leqslant b,\ Y(s) \leqslant d\} = \{s \mid X(s) \leqslant b,\ Y(s) \leqslant c\} \cup \{s \mid X(s) \leqslant b,\ c < Y(s) \leqslant d\}$$

Since the events on the right-hand side are mutually exclusive, it follows that

$$P(X \leqslant b,\ Y \leqslant d) = P(X \leqslant b,\ Y \leqslant c) + P(X \leqslant b,\ c < Y \leqslant d)$$

That is,

$$F(b, d) = F(b, c) + P(X \leqslant b,\ c < Y \leqslant d)$$

Consequently,

$$P(X \leqslant b,\ c < Y \leqslant d) = F(b, d) - F(b, c)$$

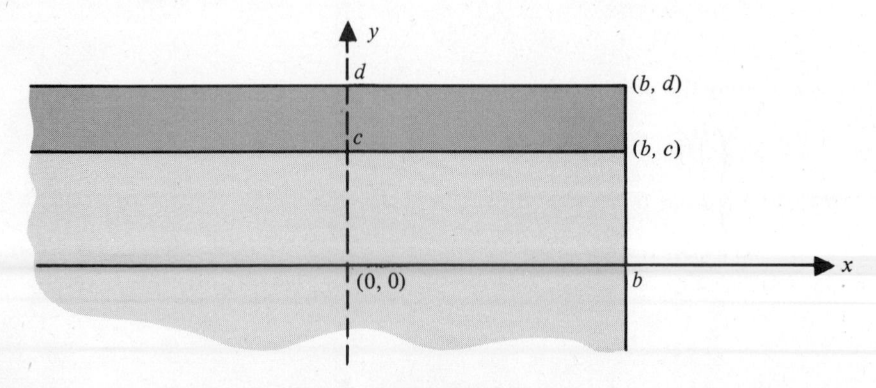

Figure 1.6

With reference to Figure 1.6, the result states that the probability of (X, Y) falling in the region shaded ▬ is equal to the value of the D.F. at (b, d) minus its value at (b, c).

Now $P(X \leqslant b,\ c < Y \leqslant d)$, being a probability, is greater than or equal to zero, and this in turn implies that $F(b, d) - F(b, c) \geqslant 0$; that is, $F(b, d) \geqslant F(b, c)$.

Thus we have shown that for a fixed value of u, say b,

$$F(b, d) \geqslant F(b, c)$$

whenever $d > c$. In other words, for fixed u, $F(u, v)$ is monotone nondecreasing in the other variable.

For fixed v, the monotonicity of $F(u, v)$ with respect to u can be proved similarly.

(*iii*) If a, b, c, d are any real numbers with $a < b$ and $c < d$, then

$$P(a < X \leqslant b,\ c < Y \leqslant d) = F(b, d) - F(b, c) - F(a, d) + F(a, c)$$

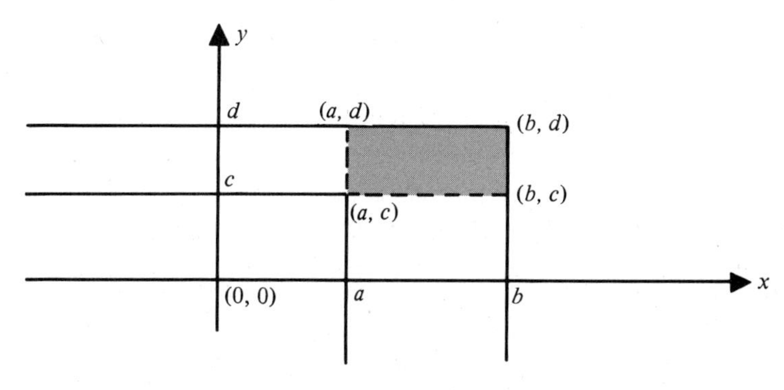

Figure 1.7

To prove the result, we observe from Figure 1.7 that

$$P(a < X \leqslant b,\ c < Y \leqslant d) = P(X \leqslant b,\ c < Y \leqslant d) - P(X \leqslant a,\ c < Y \leqslant d) \quad \text{(why?)}$$

In the proof of property (*ii*) of a D.F., we showed that

$$P(X \leqslant b,\ c < Y \leqslant d) = F(b, d) - F(b, c)$$

and

$$P(X \leqslant a,\ c < Y \leqslant d) = F(a, d) - F(a, c)$$

Substituting, the result follows.

Comment. This result is analogous to $P(a < X \leqslant b) = F(b) - F(a)$ in the univariate case. Its essence is that the probability that the random vector (X, Y) falls in the shaded rectangular region in Figure 1.7 can be found by combining the values of the distribution function at the four corners of the rectangle as indicated.

(*iv*)

$$\lim_{u \to -\infty} F(u, v) = 0 = \lim_{v \to -\infty} F(u, v)$$

Let us prove this. Suppose u_n is a monotone decreasing sequence of real numbers for which $\lim_{n \to \infty} u_n = -\infty$, and let

$$A_n = \{s \mid X(s) \leqslant u_n,\ Y(s) \leqslant v\}$$

Then $\{A_n\}$ is a contracting sequence of sets with

$$\lim_{n \to \infty} A_n = \bigcap_{n=1}^{\infty} A_n = \emptyset \qquad \text{(why?)}$$

Now

$$\begin{aligned} \lim_{u_n \to -\infty} F(u_n, v) &= \lim_{n \to \infty} F(u_n, v) \\ &= \lim_{n \to \infty} P(A_n) \\ &= P(\lim_{n \to \infty} A_n) \qquad \text{(why?)} \\ &= P(\emptyset) = 0 \end{aligned}$$

Similarly, it can be shown that $\lim_{v \to -\infty} F(u, v) = 0$.

(*v*)

$$\lim_{\substack{u \to \infty \\ v \to \infty}} F(u, v) = 1$$

This can be proved as follows: Let $\{u_n\}$ and $\{v_n\}$ be increasing sequences of real numbers for which $\lim_{n \to \infty} u_n = \infty$ and $\lim_{n \to \infty} v_n = \infty$, and let

$$B_n = \{s \mid X(s) \leqslant u_n,\ Y(s) \leqslant v_n\}$$

Then $\{B_n\}$ is an expanding sequence of subsets of S with

$$\lim_{n \to \infty} B_n = \bigcup_{n=1}^{\infty} B_n = S \qquad \text{(why?)}$$

Hence,

$$\begin{aligned} \lim_{\substack{u_n \to \infty \\ v_n \to \infty}} F(u_n, v_n) = \lim_{n \to \infty} F(u_n, v_n) &= \lim_{n \to \infty} P(B_n) \\ &= P(\lim_{n \to \infty} B_n) \qquad \text{(why?)} \\ &= P(S) = 1 \end{aligned}$$

Comment. Notice that we do *not* claim that $\lim_{u\to\infty} F(u, v)$ or $\lim_{v\to\infty} F(u, v)$ is equal to 1.

(*vi*) F is continuous from the right in each of the variables. That is, for any *fixed* u, $F(u, v)$ is continuous from the right in v, and for any *fixed* v, $F(u, v)$ is continuous from the right in u.

The proof is omitted.

The following result is analogous to the result $P(X = a) = F(a) - F(a^-)$ which was proved in the univariate case.

For any two real numbers a, b

$$P(X = a,\ y = b) = F(a, b) - F(a, b^-) - F(a^-, b) + F(a^-, b^-)$$

where $F(a, b^-) = \lim_{v\to b^-} F(a, v)$, and so on.

This can be shown as follows: Let

$$C_n = \left\{ s \;\middle|\; a - \frac{1}{n} < X(s) \leqslant a,\ b - \frac{1}{n} < Y(s) \leqslant b \right\}$$

Then C_n is a contracting sequence of sets and

$$\lim_{n\to\infty} C_n = \bigcap_{n=1}^{\infty} C_n = \{s \mid X(s) = a,\ Y(s) = b\} \qquad \text{(why?)}$$

Therefore,

$$\begin{aligned} P(X = a,\ Y = b) &= P(\lim_{n\to\infty} C_n) \\ &= \lim_{n\to\infty} P(C_n) \qquad \text{(why?)} \\ &= \lim_{n\to\infty} \left[F(a, b) - F\left(a, b - \frac{1}{n}\right) - F\left(a - \frac{1}{n}, b\right) + F\left(a - \frac{1}{n}, b - \frac{1}{n}\right) \right], \\ &\qquad \text{by property } (iii) \\ &= F(a, b) - F(a, b^-) - F(a^-, b) + F(a^-, b^-) \end{aligned}$$

Comment. Suppose F is continuous at (a, b) in one of the variables, say the first variable. Then $F(a, b) = F(a^-, b)$. Also, since F is monotone and probability is nonnegative, $F(a, b^-) = F(a^-, b^-)$. Then, $P(X = a, Y = b) = 0$. *Thus, for any pair of numbers a, b, $P(X = a, Y = b) = 0$ if F is continuous at (a, b) in one of the variables. Consequently, to locate the points (x, y) where $P(X = x, Y = y) > 0$, we consider only those points where F is discontinuous in both variables.*

Example 1.2. Show why the functions F given below do not represent joint distribution functions.

$$(a)\quad F(x, y) = \begin{cases} 1 - e^{-x+y}, & x \geqslant 0 \text{ and } y \geqslant 0 \\ 0, & \text{elsewhere} \end{cases}$$

$$(b)\quad F(x, y) = \begin{cases} 1 - e^{-x-y}, & x \geqslant 0 \text{ and } y \geqslant 0 \\ 0, & \text{elsewhere} \end{cases}$$

$$(c)\quad F(x, y) = \begin{cases} 0, & x + y < 0 \\ 1, & x + y \geqslant 0 \end{cases}$$

Solution

(*a*) For any given real number x such that $x > 0$,

$$\lim_{y \to \infty} F(x, y) = \lim_{y \to \infty} (1 - e^{-x+y}) = -\infty$$

which contradicts property (*i*).

(*b*) In trying to find $P(1 < X \leqslant 2,\ 1 < Y \leqslant 2)$, for instance, we see that

$$\begin{aligned} P(1 < X \leqslant 2,\ 1 < Y \leqslant 2) &= F(2, 2) - F(1, 2) - F(2, 1) + F(1, 1) \\ &= (1 - e^{-4}) - (1 - e^{-3}) - (1 - e^{-3}) + (1 - e^{-2}) \\ &= -e^{-2}(e^{-1} - 1)^2 \end{aligned}$$

which is negative. Therefore, F cannot represent a distribution function.

(*c*) The D.F. is equal to 1 over the shaded region and 0 below the line $x + y = 0$, as shown in Figure 1.8.

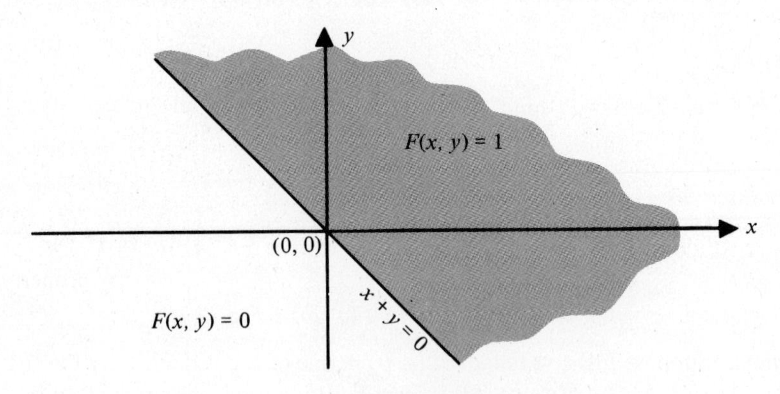

Figure 1.8

Now if we evaluate $P(-1 < X \leqslant 2, -1 < Y \leqslant 3)$, for example, we get

$$\begin{aligned} P(-1 < X \leqslant 2,\ -1 < Y \leqslant 3) &= F(2, 3) - F(2, -1) - F(-1, 3) + F(-1, -1) \\ &= 1 - 1 - 1 + 0 = -1 \end{aligned}$$

which is obviously impossible.

Example 1.3. Suppose the joint distribution of X and Y is given by the following distribution function.

$$F(u, v) = \begin{cases} 0, & u < -2 \quad \text{or} \quad v < -5 \\ \frac{3}{8}, & -2 \leqslant u < 2 \quad \text{and} \quad -5 \leqslant v < 3 \\ \frac{4}{8}, & u \geqslant 2 \quad \text{and} \quad -5 \leqslant v < 3 \\ \frac{4}{8}, & -2 \leqslant u < 2 \quad \text{and} \quad v \geqslant 3 \\ 1, & u \geqslant 2 \quad \text{and} \quad v \geqslant 3 \end{cases}$$

Determine the pairs (x, y) for which $P(X = x, \ Y = y)$ is positive and find these probabilities.

Solution. The pairs (x, y) that we are interested in are the ones where F is discontinuous in both variables. A careful examination will show that these points are $(-2, -5)$, $(-2, 3)$, $(2, -5)$, and $(2, 3)$. (See Figure 1.9.)

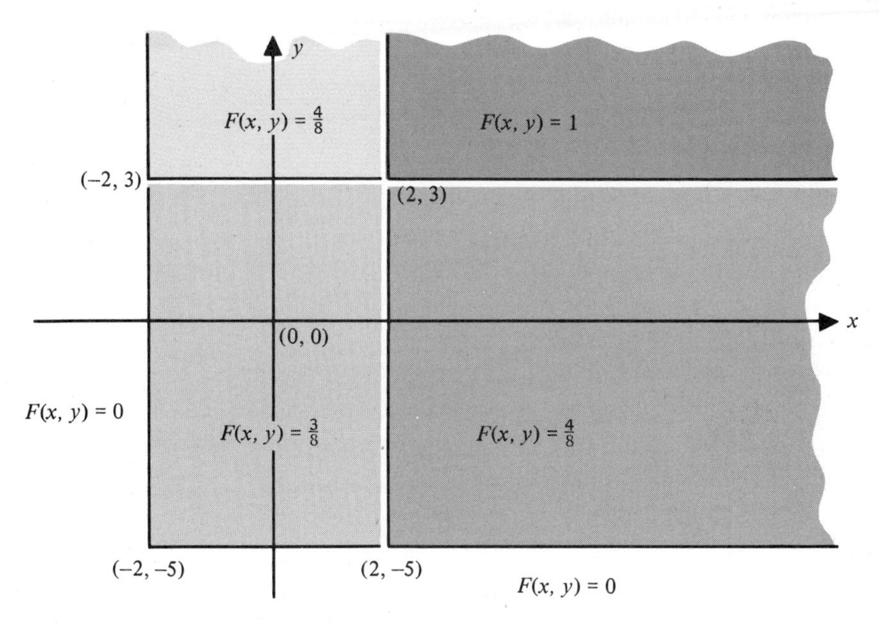

Figure 1.9

We now find

$$P(X = x, \ Y = y) = F(x, y) - F(x^-, y) - F(x, y^-) + F(x^-, y^-)$$

at each of these points. We get

$$\begin{aligned} P(X = -2, \ Y = -5) &= F(-2, -5) - F(-2^-, -5) - F(-2, -5^-) + F(-2^-, -5^-) \\ &= \frac{3}{8} - 0 - 0 + 0 = \frac{3}{8} \\ P(X = -2, \ Y = 3) &= F(-2, 3) - F(-2, 3^-) - F(-2^-, 3) + F(-2^-, 3^-) \\ &= \frac{4}{8} - \frac{3}{8} - 0 + 0 = \frac{1}{8} \end{aligned}$$

$$P(X=2,\ Y=-5) = F(2,-5) - F(2,-5^-) - F(2^-,-5) + F(2^-,-5^-)$$
$$= \frac{4}{8} - 0 - \frac{3}{8} + 0 = \frac{1}{8}$$
$$P(X=2,\ Y=3) = F(2,3) - F(2,3^-) - F(2^-,3) + F(2^-,3^-)$$
$$= 1 - \frac{4}{8} - \frac{4}{8} + \frac{3}{8} = \frac{3}{8}$$

1.4 Classification of Joint Distributions

In classifying joint distributions, the same approach is adopted as in the univariate case; that is, the classification is carried out on the basis of the nature of the joint distribution function. There are two important types of distributions–discrete and (absolutely) continuous. These types of distributions by no means exhaust all the possible cases: it is possible to have bivariate distributions which are absolutely continuous in one variable and discrete in the other variable. Also, cases of singular distributions abound among joint distributions. Only the discrete and the absolutely continuous cases will be treated here.

Joint discrete distributions

For any pair of real numbers x, y, we know that

$$P(X=x,\ Y=y) = F(x,y) - F(x,y^-) - F(x^-,y) + F(x^-,y^-)$$

It can be shown that if X and Y have a joint distribution, then there are at most countably infinite points (x, y) for which $P(X=x,\ Y=y) > 0$. Informally, X and Y are jointly discrete (or have a joint discrete distribution) if the entire probability mass is distributed in lumps at a finite or countably infinite points of the plane. We now give a formal definition.

The joint distribution of X and Y is said to be a **discrete distribution** if there exists a nonnegative function p such that p vanishes everywhere except at a finite or countably infinite points in the plane and

$$p(x,y) = P(X=x,\ Y=y)$$

for any point (x, y).

If the set of points where p might be positive is denoted by $\{(x_i, y_j) \mid i, j = 1, 2, \ldots\}$, then p has the following properties:

$$(i)\quad p(x_i, y_j) \geq 0, \qquad i, j = 1, 2, \ldots$$

and

$$(ii)\quad \sum_{j=1}^{\infty} \sum_{i=1}^{\infty} p(x_i, y_j) = 1$$

A function p with the above properties is called the **joint probability mass function**, or simply, the **joint probability function** of X and Y.

It is often convenient to display the probability function of a bivariate distribution in a rectangular array as shown below:

$x \backslash y$	y_1	y_2	$\cdots$	y_j	$\cdots$
x_1	$p(x_1, y_1)$	$p(x_1, y_2)$	$\cdots$	$p(x_1, y_j)$	$\cdots$
x_2	$p(x_2, y_1)$	$p(x_2, y_2)$	$\cdots$	$p(x_2, y_j)$	$\cdots$
$\vdots$	$\vdots$	$\vdots$		$\vdots$	
x_i	$p(x_i, y_1)$	$p(x_i, y_2)$	$\cdots$	$p(x_i, y_j)$	$\cdots$
$\vdots$	$\vdots$	$\vdots$		$\vdots$	

For any subset B of the plane (where, of course, B is an event), the probability that (X, Y) falls in B is found by adding $p(x, y)$ for all the points (x, y) which lie in B; that is,

$$P((X, Y) \in B) = \sum_{(x_i, y_j) \in B} p(x_i, y_j)$$

For example, in Figure 1.10 the sum of the heights at O_1, O_2, O_3, O_4 and O_5 represents $P((X, Y) \in B)$.

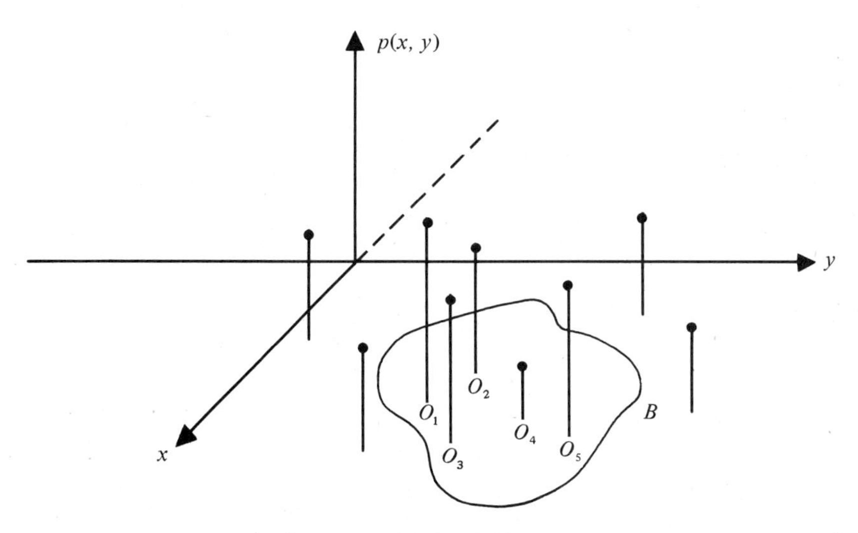

Figure 1.10

In particular, the joint distribution function is obtained from the joint probability function by the relation

$$F(u, v) = P(X \leqslant u, Y \leqslant v) = \sum_{x_i \leqslant u} \sum_{y_j \leqslant v} p(x_i, y_j)$$

where the summation is carried over all the pairs (x_i, y_j) for which $x_i \leqslant u$, $y_j \leqslant v$, and $p(x_i, y_j) > 0$.

In summary, the joint probability function and the joint distribution function of X and Y are connected by the following relations:

$$F(u, v) = \sum_{x_i \leqslant u} \sum_{y_j \leqslant v} p(x_i, y_j)$$

$$P(X = x, \; Y = y) = F(x, y) - F(x, y^-) - F(x^-, y) + F(x^-, y^-)$$

Example 1.4. X and Y have the following joint D.F.:

$$F(u, v) = \begin{cases} 0, & u < -2 \text{ or } v < 0 \\ 0, & -2 \leqslant u < 3 \text{ and } 0 \leqslant v < 5 \\ \frac{3}{5}, & u \geqslant 3 \text{ and } 0 \leqslant v < 5 \\ \frac{2}{5}, & -2 \leqslant u < 3 \text{ and } v \geqslant 5 \\ 1, & u \geqslant 3 \text{ and } v \geqslant 5 \end{cases}$$

Determine that X and Y indeed have a joint discrete distribution and find the joint probability function.

Solution. Figure 1.11 describes the regions of the plane based on the values of F.

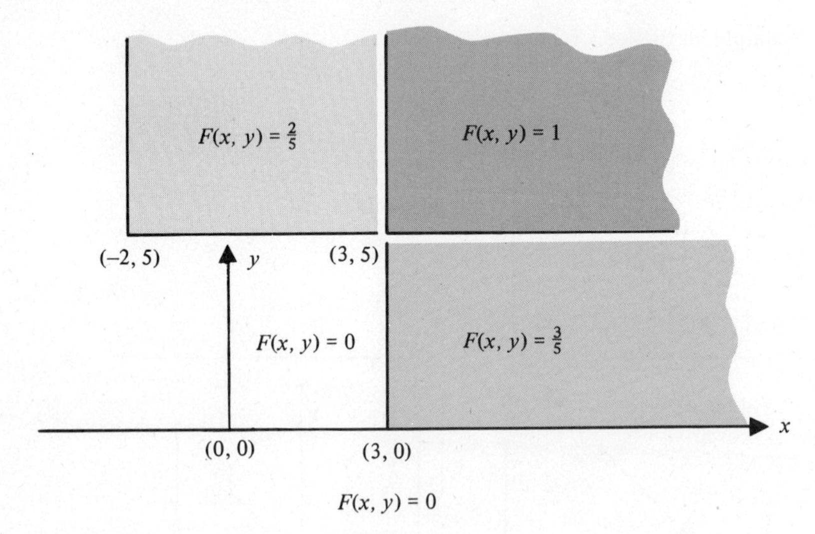

Figure 1.11

As can be seen from the figure, the points where F is discontinuous in both variables are $(-2, 5)$, $(3, 0)$, and $(3, 5)$.

Now,

$$P(X = -2, \; Y = 5) = F(-2, 5) - F(-2^-, 5) - F(-2, 5^-) + F(-2^-, 5^-) = \frac{2}{5} - 0 - 0 + 0 = \frac{2}{5}$$

$$P(X = 3, \; Y = 0) = F(3, 0) - F(3^-, 0) - F(3, 0^-) + F(3^-, 0^-) = \frac{3}{5} - 0 - 0 + 0 = \frac{3}{5}$$

$$P(X=3,\ Y=5) = F(3,5) - F(3^-,5) - F(3,5^-) + F(3^-,5^-)$$
$$= 1 - \frac{2}{5} - \frac{3}{5} + 0 = 0$$

Also observe that $\frac{2}{5} + \frac{3}{5} = 1$. Hence the joint distribution of X and Y is discrete, and the only points which are assumed with positive probability are $(-2, 5)$ and $(3, 0)$.

Example 1.5. The joint probability function of X and Y is given by

$$p(2,3) = \tfrac{1}{12}$$
$$p(2,-5) = \tfrac{5}{12}$$
$$p(-4,3) = \tfrac{3}{12}$$
$$p(-4,-5) = \tfrac{3}{12}$$

(*a*) Plot the probability function.
(*b*) Find the joint D.F.

Solution

(*a*) The probability function is plotted in Figure 1.12.

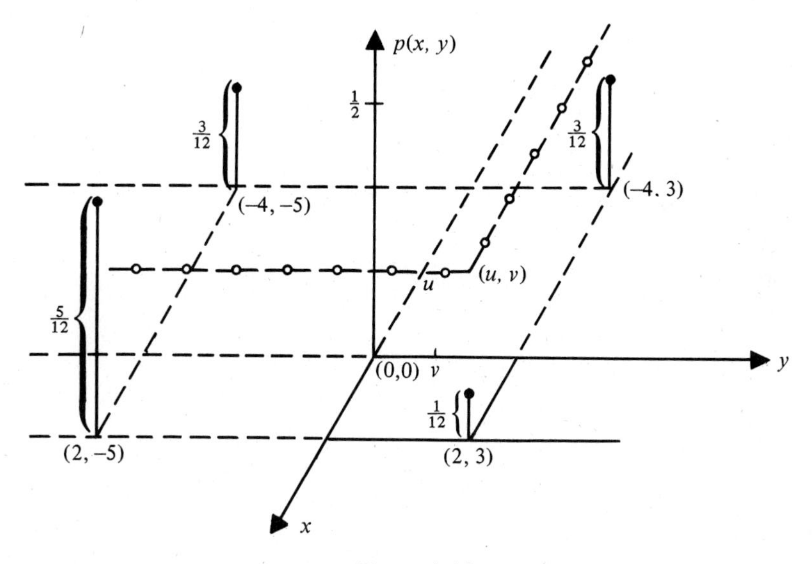

Figure 1.12

(*b*) The probability function can be arranged in a rectangular array as

$x \backslash y$	-5	3
-4	$\frac{3}{12}$	$\frac{3}{12}$
2	$\frac{5}{12}$	$\frac{1}{12}$

It can be seen from the table of the probability function, or from Figure 1.12, that the joint D.F. is given by:

$F(u, v)$

u \ v	$v < -5$	$-5 \leqslant v < 3$	$v \geqslant 3$
$u < -4$	0	0	0
$-4 \leqslant u < 2$	0	$\frac{3}{12}$	$\frac{6}{12}$
$u \geqslant 2$	0	$\frac{8}{12}$	1

For example, from Figure 1.12 (or from the dotted line in the table of the probability function), when u and v satisfy $-4 \leqslant u < 2$ and $-5 \leqslant v < 3$, then $(-4, -5)$ is the only point with positive probability (namely, $\frac{3}{12}$) in the infinite rectangle B_{uv}. Hence, $F(u, v) = \frac{3}{12}$ if $-4 \leqslant u < 2$ and $-5 \leqslant v < 3$. As u and v change we get other values of $F(u, v)$.

The D.F. can also be written as

$$F(u, v) = \begin{cases} 0, & u < -4 \text{ or } v < -5 \\ \frac{3}{12}, & -4 \leqslant u < 2 \text{ and } -5 \leqslant v < 3 \\ \frac{8}{12}, & u \geqslant 2 \text{ and } -5 \leqslant v < 3 \\ \frac{6}{12}, & -4 \leqslant u < 2 \text{ and } v \geqslant 3 \\ 1, & u \geqslant 2 \text{ and } v \geqslant 3 \end{cases}$$

Example 1.6. Suppose X and Y have a joint distribution as given by the following probability function:

x \ y	0	1	3	5
-1	$\frac{1}{12}$	$\frac{1}{6}$	$\frac{1}{12}$	0
3	$\frac{1}{6}$	$\frac{1}{12}$	0	$\frac{1}{12}$
4	0	$\frac{1}{12}$	$\frac{1}{6}$	$\frac{1}{12}$

(*a*) Plot the probability function.
(*b*) Find the joint D.F. of X and Y.
(*c*) Find (*i*) $P(X \leqslant Y)$, (*ii*) $P(X + Y^2 \leqslant 8)$.

Solution.

(*a*) The probability function is plotted in Figure 1.13.
(*b*) The joint D.F. is given as follows:

$F(u, v)$

u \ v	$v < 0$	$0 \leqslant v < 1$	$1 \leqslant v < 3$	$3 \leqslant v < 5$	$v \geqslant 5$
$u < -1$	0	0	0	0	0
$-1 \leqslant u < 3$	0	$\frac{1}{12}$	$\frac{3}{12}$	$\frac{4}{12}$	$\frac{4}{12}$
$3 \leqslant u < 4$	0	$\frac{3}{12}$	$\frac{6}{12}$	$\frac{7}{12}$	$\frac{8}{12}$
$u \geqslant 4$	0	$\frac{3}{12}$	$\frac{7}{12}$	$\frac{10}{12}$	1

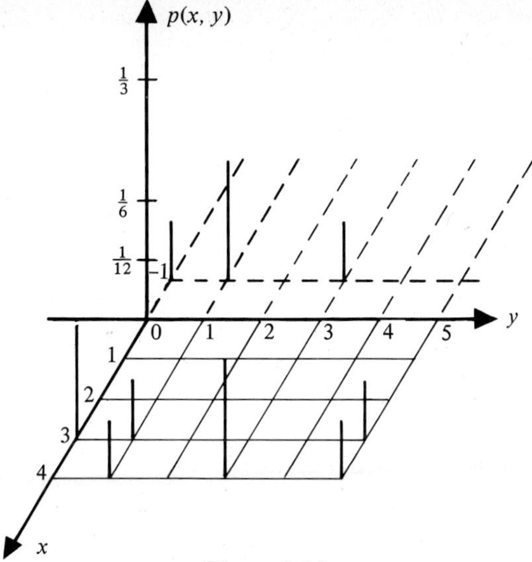

Figure 1.13

Explanation: Suppose, for instance, u and v satisfy $3 \leqslant u < 4$ and $3 \leqslant v < 5$. Then $F(u, v) = \frac{1}{12} + \frac{1}{6} + \frac{1}{6} + \frac{1}{12} + \frac{1}{12} = \frac{7}{12}$. See the dotted lines and the circled quantities in the table below.

x \ y	0	1	3	5
−1	($\frac{1}{12}$)	($\frac{1}{6}$)	($\frac{1}{12}$)	0
3	($\frac{1}{6}$)	($\frac{1}{12}$)	(0)	$\frac{1}{12}$
4	0	$\frac{1}{12}$	$\frac{1}{6}$	$\frac{1}{12}$

(dashed line u below row $x = 3$; dashed line v between columns $y = 3$ and $y = 5$)

(*c*) (*i*) To find $P(X \leqslant Y)$, we pick those pairs (x, y) with $p(x, y) > 0$ for which $x \leqslant y$. Hence,

$$P(X \leqslant Y) = p(-1, 0) + p(-1, 1) + p(-1, 3) + p(-1, 5) + p(3, 3) + p(3, 5) + p(4, 5) = \tfrac{1}{2}$$

(*ii*) Picking the pairs (x, y) for which $x + y^2 \leqslant 8$,

$$P(X + Y^2 \leqslant 8) = p(-1, 0) + p(-1, 1) + p(-1, 3) + p(3, 0) + p(3, 1) + p(4, 0) + p(4, 1) = \tfrac{2}{3}$$

Example 1.7. The joint probability function of X and Y is given by

$$P(X = x,\ Y = y) = K(x^2 + y^2), \qquad x = -1, 0, 1, 3, \quad y = -1, 2, 3$$

(*a*) Find the constant K.
(*b*) Find $P(X/Y \leqslant 0)$.

Solution

(*a*) The sum of all the probabilities has to add up to 1. Therefore,

$$\Sigma K(x^2 + y^2) = 1, \quad \text{that is,} \quad K \Sigma (x^2 + y^2) = 1$$

(Here the summation is carried over all the possible values of x and y.) Hence,

$$K = \frac{1}{\Sigma (x^2 + y^2)} = \frac{1}{89}$$

(*b*)
$$\begin{aligned} P(X/Y \leqslant 0) &= p(-1, 2) + p(-1, 3) + p(0, -1) + p(1, -1) \\ &\quad + p(3, -1) + p(0, 2) + p(0, 3) \\ &= \frac{41}{89} \end{aligned}$$

• *Example 1.8 (Generalization of the hypergeometric distribution)* A box contains M beads; a are white, b black, and $M - a - b$ green. Suppose n beads are picked without replacement. Let

X = The number of white beads in the sample
Y = The number of black beads in the sample

Find the joint distribution of X and Y.

Solution. For any pair of integers i, j, the event $\{X = i, Y = j\}$ signifies that there are i white beads, j black beads, and, consequently, $n - i - j$ green beads. However, note that i cannot be greater than a or n, j cannot be greater than b or n, and $i + j$ cannot be greater than n.

Now there are $\binom{M}{n}$ ways of picking n beads out of M without replacement, and there are $\binom{a}{i}\binom{b}{j}\binom{M-a-b}{n-i-j}$ possibilities where there are i white beads, and j black beads (and consequently, $n - i - j$ green beads). Therefore,

$$P(X = i, \; Y = j) = \frac{\binom{a}{i}\binom{b}{j}\binom{M-a-b}{n-i-j}}{\binom{M}{n}}$$

where $i, j = 0, 1, \ldots, n$ (subject to the above-mentioned restrictions). Actually there is no need to specify the exact range of i and j because, by the convention that $\binom{m}{r} = 0$ if $m < r$ or $r < 0$, the pairs (i, j) that do not satisfy the above restrictions will automatically give zero probabilities.) ∎

The distribution given in the example below is the bivariate analog of the binomial distribution. It is called the *trinomial distribution.* In this case, the experiment consists of n independent and identical trials. On any trial, one of three mutually exclusive and exhaustive possibilities can occur. (In the binomial case, one of the two mutually exclusive possibilities, success or failure, occurs.) For example, a stock could go up, go down, or remain steady. In the present context, let us refer to these three possibilities as success, failure, and draw, with respective probabilities on any trial equal to p, q, and $1 - p - q$.

• *Example 1.9 (The trinomial distribution)* An experiment consists of n independent and identical trials where each trial can result in a success, a failure, or a draw, with respective probabilities equal to p, q, and $1 - p - q$. Let

X = The number of successes in n trials
Y = The number of failures in n trials

Find the joint distribution of X and Y.

Solution. The set of possible values of (X, Y) is

$$\{(i, j) \mid i, j = 0, 1, 2, \ldots, n \text{ and } 0 \leqslant i + j \leqslant n\}$$

We want to find $P(X = i,\ Y = j)$. The probability of getting successes on i *designated* trials and failures on j *designated* trials (consequently, draws on the rest of the trials) is equal to $p^i q^j (1 - p - q)^{n-i-j}$ (why?). Of course, there are $\binom{n}{i}$ ways of picking i trials out of n for successes to occur, and corresponding to any one of these there are $\binom{n-i}{j}$ ways of picking j trials from the remaining $n - i$ trials for failures to occur. Therefore, by the basic counting rule, there are $\binom{n}{i}\binom{n-i}{j}$ ways of picking i trials for successes and j trials for failures. Each has a probability of $p^i q^j (1 - p - q)^{n-i-j}$. Hence

$$P(X = i,\ Y = j) = \binom{n}{i}\binom{n-i}{j} p^i q^j (1 - p - q)^{n-i-j}$$

That is, simplifying,

$$P(X = i,\ Y = j) = \frac{n!}{i!j!(n - i - j)!} p^i q^j (1 - p - q)^{n-i-j}$$

where $i, j = 0, 1, \ldots, n$ with $0 \leqslant i + j \leqslant n$. ■

The distribution above is called a trinomial distribution because the probabilities are given as the terms in the trinomial expansion of $[p + q + (1 - p - q)]^n$:

$$[p + q + (1 - p - q)]^n = \sum_{i=0}^{n} \sum_{j=0}^{n-i} \binom{n}{i}\binom{n-i}{j} p^i q^j (1 - p - q)^{n-i-j}$$

It also follows from this that the sum of the probabilities is equal to 1, since $[p+q+(1-p-q)]^n = 1$.

For future reference, we shall call this distribution the *trinomial distribution of X and Y with n trials and parameters p, q.*

As an illustration of the trinomial distribution, suppose an instructor is in the habit of assigning either an A or a B or an F to a student as his final grade with respective probabilities 0.2, 0.5, 0.3. If the instructor has twenty students in a class, then the probability that he gives 5 A's and 6 B's (and, consequently, 9 F's) is equal to $\frac{20!}{5!6!9!}(0.2)^5(0.5)^6(0.3)^9 = 0.0076$.

Joint absolutely continuous distributions

As in the univariate case, the concept of absolute continuity for joint distributions has to do with being able to recover the joint D.F. by integrating a nonnegative function.

The joint distribution of X and Y is said to be **absolutely continuous** if there exists a nonnegative function f such that, for any real numbers u, v,

$$F(u, v) = \int_{-\infty}^{u}\int_{-\infty}^{v} f(x, y)\, dy\, dx$$

In other words, the joint D.F. can be obtained by integrating a nonnegative function f over $B_{uv} = \{(x, y) \mid -\infty < x \leqslant u, -\infty < y \leqslant v\}$. The function f is called the **joint probability density function** of X and Y. (We shall often abbreviate this as joint pdf.)

Since $F(u, v) = \int_{-\infty}^{u}\int_{-\infty}^{v} f(x, y)\, dy\, dx$, the following consequences are immediate:

(*i*)

$$\int_{-\infty}^{\infty}\int_{-\infty}^{\infty} f(x, y)\, dy\, dx = 1$$

This is because $\int_{-\infty}^{\infty}\int_{-\infty}^{\infty} f(x, y)\, dy\, dx = F(\infty, \infty) = 1$.

(*ii*)

$$P(a < X \leqslant b,\ c < Y \leqslant d) = \int_{a}^{b}\int_{c}^{d} f(x, y)\, dy\, dx$$

This follows, since

$$\begin{aligned} P(a < X \leqslant b,\ c < Y \leqslant d) &= F(b, d) - F(b, c) - F(a, d) + F(a, c) \\ &= \int_{-\infty}^{b}\int_{-\infty}^{d} f(x, y)\, dy\, dx - \int_{-\infty}^{b}\int_{-\infty}^{c} f(x, y)\, dy\, dx \\ &\quad - \int_{-\infty}^{a}\int_{-\infty}^{d} f(x, y)\, dy\, dx + \int_{-\infty}^{a}\int_{-\infty}^{c} f(x, y)\, dy\, dx \\ &= \int_{a}^{b}\int_{c}^{d} f(x, y)\, dy\, dx \end{aligned}$$

Geometrically, since $f(x, y)$ is nonnegative, this implies that $P(a < X \leqslant b, c < Y \leqslant d)$ can be obtained as the volume under the surface $z = f(x, y)$ and above the rectangle $\{(x, y) \mid a < X \leqslant b,\ c < Y \leqslant d\}$. (See Figure 1.14.)

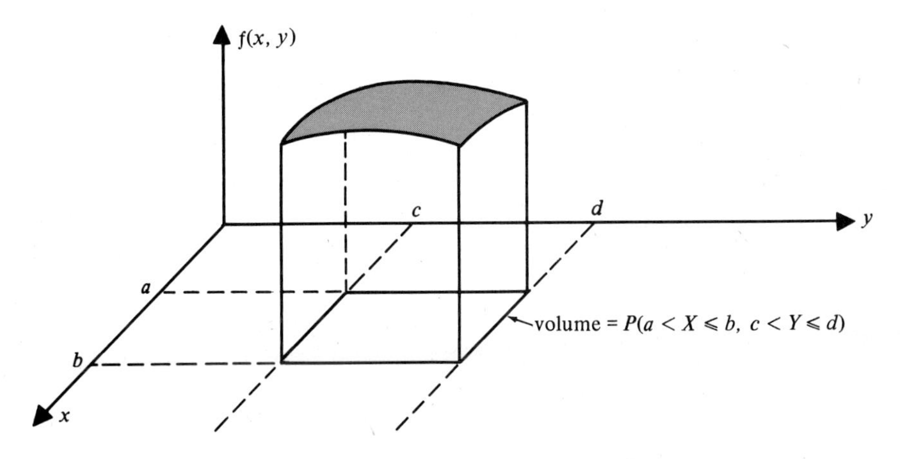

Figure 1.14

As a matter of fact, for any (Borel) subset B of the plane

$$P((X, Y) \in B) = \iint_B f(x, y)\, dy\, dx$$

The proof of this falls beyond the scope of our treatment of the subject.

(*iii*) Since the joint D.F. is given by the integral of the pdf as mentioned, it follows from calculus that the joint pdf can be obtained from the joint D.F. by differentiating as follows:

$$f(x, y) = \frac{\partial^2}{\partial x \partial y} F(x, y)$$

Summarizing, the pdf f has the following properties:

$$f(x, y) \geqslant 0 \qquad \text{for every pair } (x, y)$$

$$\text{and} \quad \int_{-\infty}^{\infty} \int_{-\infty}^{\infty} f(x, y)\, dy\, dx = 1$$

and the joint D.F. and the corresponding pdf are related by the following relations:

$$F(u, v) = \int_{-\infty}^{u} \int_{-\infty}^{v} f(x, y)\, dy\, dx$$

$$\text{and} \quad f(x, y) = \frac{\partial^2}{\partial x \partial y} F(x, y)$$

Comments. (1) Recall from calculus that

$$\frac{\partial^2}{\partial x \partial y} F(x, y) = \lim_{\substack{h \to 0 \\ k \to 0}} \frac{F(x+h, y+k) - F(x, y+k) - F(x+h, y) + F(x, y)}{hk}$$

$$= \lim_{\substack{h \to 0 \\ k \to 0}} \frac{P(x < X \leq x+h,\ y < Y \leq y+k)}{hk}$$

Therefore, the joint pdf represents the limit of the ratio of the amount of probability in a rectangle to the area of the rectangle as the sides of the rectangle shrink to zero. It is due to this fact that a joint pdf reflects how densely the probability mass is spread over the plane.

(2) Also, it follows that if *h and k are small,* then $P(x < X \leq x + h,$ $y < Y \leq y + k)$ is approximately equal to $hk \frac{\partial^2}{\partial x \partial y} F(x, y)$, that is, $hkf(x, y)$.

Example 1.10. The joint distribution of X and Y is said to be uniform over the unit square $\{(x, y) \mid 0 \leq x \leq 1,\ 0 \leq y \leq 1\}$ if the joint pdf is given by

$$f(x, y) = \begin{cases} 1, & 0 \leq x \leq 1,\ 0 \leq y \leq 1 \\ 0, & \text{elsewhere} \end{cases}$$

Find:

(*a*) $P(X + Y \leq 1)$
(*b*) $P(\frac{1}{3} \leq X + Y \leq \frac{3}{2})$
(*c*) $P(X \geq 2Y)$

Solution. The above distribution is analogous to the uniform distribution of a single variable over the unit interval [0, 1]. Figure 1.15 shows the graph of the pdf. Notice that the height at each point in the shaded region of the *xy*-plane is 1 so that the volume of the entire cube is equal to 1.

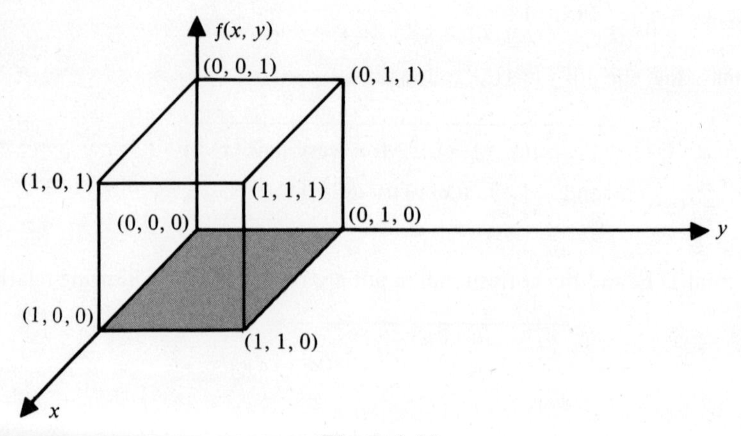

Figure 1.15

(As mentioned earlier, only the xy-plane is drawn in the figure. Imagine that the values of f are plotted along the axis perpendicular to the plane of the paper.)

(*a*) Since $P((X, Y) \in B) = \iint_B f(x, y)\, dy\, dx$ for any Borel set B, we have

$$P(X + Y \leqslant 1) = \iint_{\{(x, y) \mid x + y \leqslant 1\}} f(x, y)\, dy\, dx$$

In other words, $P(X + Y \leqslant 1)$ is obtained by integrating f over the region below the line $x + y = 1$. However, below this line $f(x, y)$ is 0 for any point (x, y) outside the shaded triangle shown in Figure 1.16. Inside the shaded triangle, $f(x, y) = 1$ for any point (x, y).

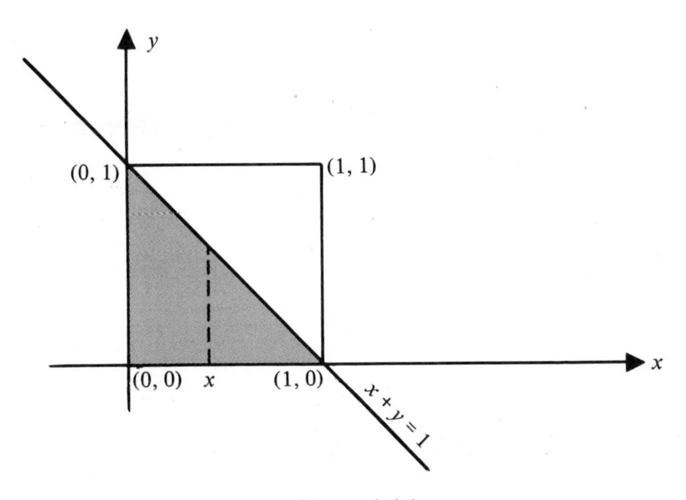

Figure 1.16

Hence, to evaluate the integral we observe that: given any x, y goes from 0 to $1 - x$; then x goes from 0 to 1. Therefore,

$$P(X + Y \leqslant 1) = \int_0^1 \int_0^{1-x} 1\, dy\, dx = \frac{1}{2}$$

As far as this problem is concerned, one can obtain probabilities much more easily without integrating just by realizing the geometric nature of the graph of the pdf. Thus,

$$P(X + Y \leqslant 1) = \text{volume over the shaded triangle} = \frac{1}{2}$$

(*b*) To find $P(\frac{1}{3} \leqslant X + Y \leqslant \frac{3}{2})$, let us first find the area of the shaded region in Figure 1.17.

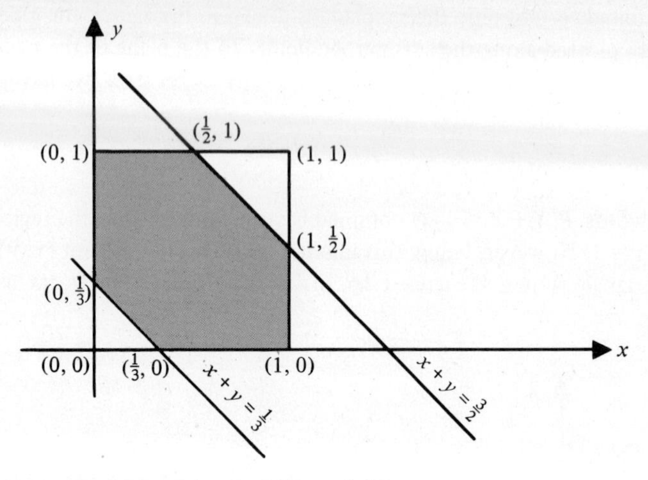

Figure 1.17

The area of the unshaded triangle below the line $x + y = \frac{1}{3}$ is $\frac{1}{2} \cdot \frac{1}{3} \cdot \frac{1}{3} = \frac{1}{18}$; that of the unshaded triangle above the line $x + y = \frac{3}{2}$ is $\frac{1}{2} \cdot \frac{1}{2} \cdot \frac{1}{2} = \frac{1}{8}$. Therefore, the area of the shaded region is $1 - \frac{1}{18} - \frac{1}{8} = \frac{59}{72}$. Consequently, the desired probability, being the volume over the shaded region, is equal to $\frac{59}{72} \cdot 1 = \frac{59}{72}$.

(*c*) We want to find $P(X \geqslant 2Y)$. This is simply the volume over the shaded triangle in Figure 1.18.

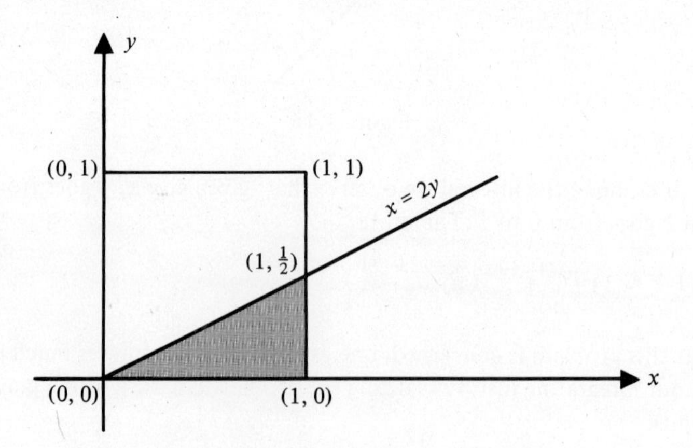

Figure 1.18

Now, the area of the shaded triangle is $\frac{1}{2} \cdot 1 \cdot \frac{1}{2} = \frac{1}{4}$. Therefore,

$$P(X \geqslant 2Y) = \frac{1}{4} \cdot 1 = \frac{1}{4}$$

Alternatively, we could find the probability by integrating as

$$P(X \geqslant 2Y) = \int_0^1 \int_0^{x/2} 1\, dy\, dx = \frac{1}{4}$$

Example 1.11. Suppose a joint pdf is given by

$$f(x, y) = \begin{cases} Kx^2y, & 0 < y < x < 1 \\ 0, & \text{elsewhere} \end{cases}$$

where K is a constant. Find K.

Solution. We know that $\int_{-\infty}^{\infty} \int_{-\infty}^{\infty} f(x, y)\, dy\, dx = 1$.

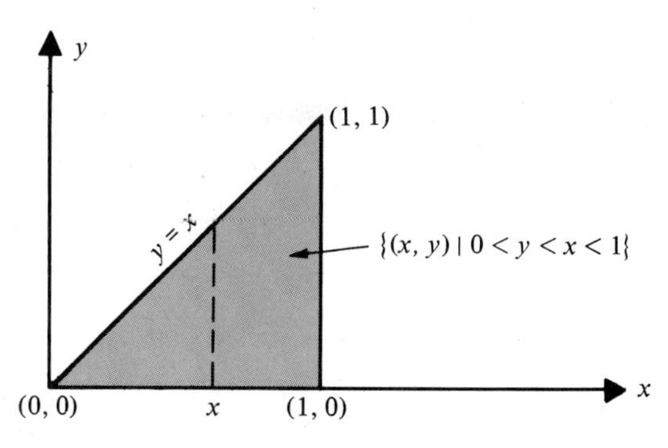

Figure 1.19

$f(x, y) = 0$ for any point (x, y) outside the shaded triangle in Figure 1.19. To find the limits of integration: Given any x, inside the triangle y goes from 0 to x. Of course, x can go from 0 to 1. Therefore we get

$$\int_0^1 \int_0^x Kx^2y\, dy\, dx = 1$$

That is,

$$K \int_0^1 \frac{x^4}{2}\, dx = 1$$

Consequently, after simplifying, $K = 10$.

Example 1.12. Let X and Y have the following joint pdf:

$$f(x, y) = \begin{cases} 2e^{-x-y}, & 0 < x < y < \infty \\ 0, & \text{elsewhere} \end{cases}$$

Find the joint distribution function of X and Y.

Solution. As indicated in Figure 1.20, the pdf is 0 outside the shaded region between the positive y-axis and the line $y = x$. Inside the shaded region, the height at any point (x, y) is $2e^{-x-y}$.

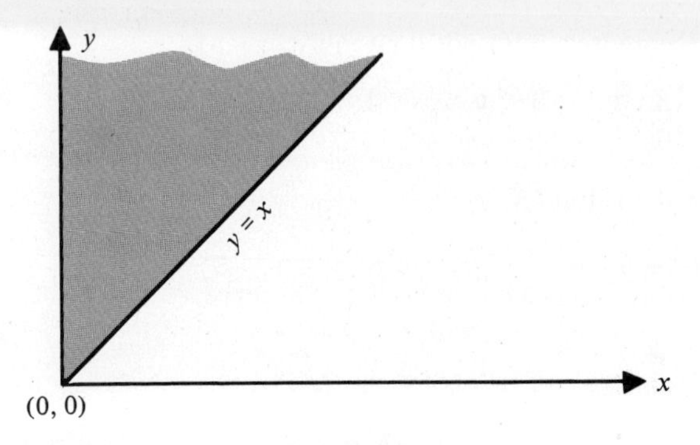

Figure 1.20

To find $F(u, v)$, there are three cases that should be considered, depending upon where the point (u, v) lies. These cases are illustrated in Figures 1.21(a), (b), and (c).

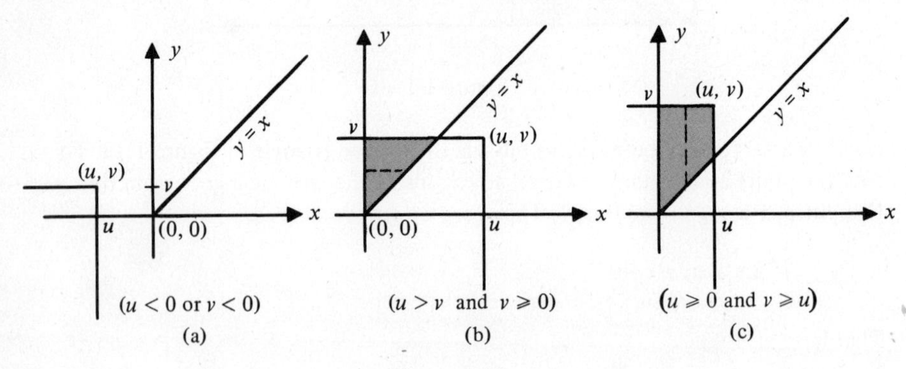

Figure 1.21

It can be seen from these figures that

$$F(u, v) = \begin{cases} 0, & u < 0 \quad \text{or} \quad v < 0 \\ \int_0^v \int_0^y 2e^{-x-y}\,dx\,dy, & u > v \quad \text{and} \quad v \geqslant 0 \\ \int_0^u \int_x^v 2e^{-x-y}\,dy\,dx, & u \geqslant 0 \quad \text{and} \quad v \geqslant u \end{cases}$$

Explanation: (*i*) If $u < 0$ or $v < 0$, then obviously $F(u, v) = 0$. (See Figure 1.21(a).)

(*ii*) If $u > v$ and $v \geqslant 0$, we have the situation shown in Figure 1.21(b). Here we want the integral over the shaded triangle below the line $y = v$. Given any y, x goes from 0 to y; then y goes from 0 to v.

(*iii*) If $u \geqslant 0$ and $v \geqslant u$, we have the situation in Figure 1.21(c). We want the integral over the shaded trapezoidal region below the line $y = v$ and to the left of $x = u$. Given any x, y goes from x to v; then x goes from 0 to u.

Carrying out the integration yields

$$F(u, v) = \begin{cases} 0, & u < 0 \quad \text{or} \quad v < 0 \\ 1 - 2e^{-v} + e^{-2v}, & u > v \quad \text{and} \quad v \geqslant 0 \\ 1 - e^{-2u} - 2e^{-v} + 2e^{-u-v}, & u \geqslant 0 \quad \text{and} \quad v \geqslant u \end{cases}$$

Example 1.13. The joint D.F. of X and Y is given as follows:

$$F(x, y) = \begin{cases} 1 - e^{-x} - e^{-y} + e^{-x-y}, & x \geqslant 0 \quad \text{and} \quad y \geqslant 0 \\ 0, & \text{elsewhere} \end{cases}$$

Find:

(*a*) $P(-2 < X \leqslant 3,\ 1 < Y \leqslant 2)$
(*b*) the joint pdf of X and Y
(*c*) $P(X + Y \geqslant 3)$
(*d*) $P(X \geqslant Y)$

Solution

(*a*) We know that

$$P(-2 < X \leqslant 3,\ 1 < Y \leqslant 2) = F(3, 2) - F(3, 1) - F(-2, 2) + F(-2, 1)$$

Now, $F(-2. 2) = 0$ and $F(-2, 1) = 0$. Therefore,

$$\begin{aligned} P(-2 < X \leqslant 3,\ 1 < Y \leqslant 2) &= F(3, 2) - F(3, 1) \\ &= [1 - e^{-3} - e^{-2} + e^{-5}] - [1 - e^{-3} - e^{-1} + e^{-4}] \\ &= e^{-1}(1 - e^{-1} - e^{-3} + e^{-4}) \end{aligned}$$

(*b*) $f(x, y) = \dfrac{\partial^2}{\partial x \partial y} F(x, y)$. Therefore,

$$f(x, y) = \begin{cases} e^{-x-y}, & x > 0 \quad \text{and} \quad y > 0 \\ 0, & \text{elsewhere} \end{cases}$$

(*c*)
$$\begin{aligned} P(X + Y \geqslant 3) &= 1 - P(X + Y < 3) \\ &= 1 - \iint_{\{(x, y) \mid x + y < 3\}} f(x, y)\, dy\, dx \end{aligned}$$

Now, below the line $x + y = 3$ the pdf is 0 outside the shaded region in Figure 1.22. Given x, y goes from 0 to $3 - x$; then x goes from 0 to 3. Therefore,

$$\begin{aligned} P(X + Y \geqslant 3) &= 1 - \int_0^3 \int_0^{3-x} e^{-x-y}\, dy\, dx \\ &= 1 - \int_0^3 (e^{-x} - e^{-3})\, dx \\ &= 4e^{-3} \end{aligned}$$

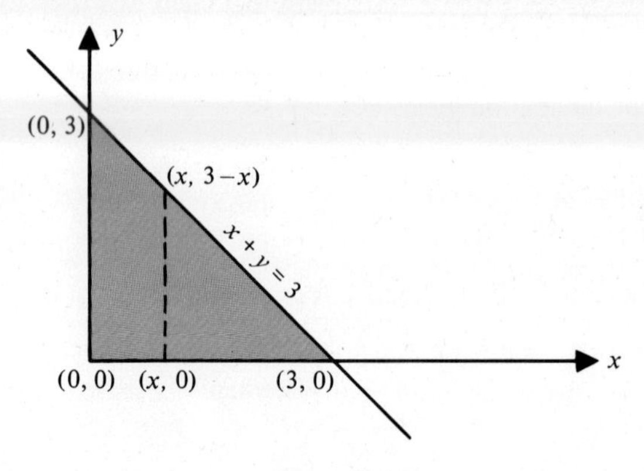

Figure 1.22

(*d*) $P(X \geqslant Y)$ is the volume over the shaded region in Figure 1.23. Given x, y goes from 0 to x; then x goes from 0 to ∞. Therefore,

$$P(X \geqslant Y) = \int_0^\infty \int_0^x e^{-x-y}\,dy\,dx = \int_0^\infty (e^{-x} - e^{-2x})\,dx = \frac{1}{2}$$

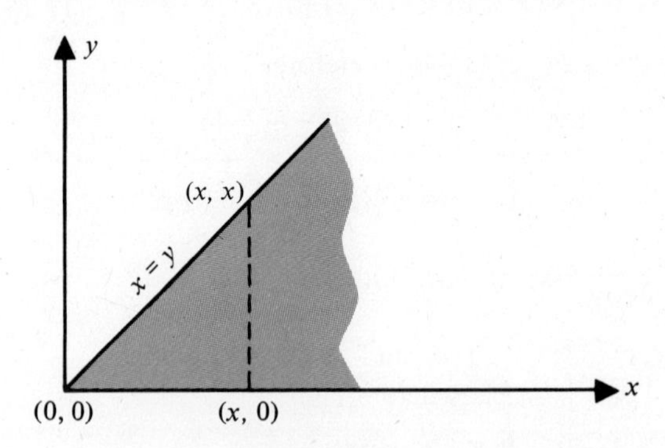

Figure 1.23

EXERCISES–SECTION 1

1. A fair coin is tossed three times in succession. Let X denote the number of heads, and Y the number of heads minus the number of tails. Describe the joint distribution of X and Y.

2. Suppose two chips are picked at random without replacement from a box containing n chips numbered from 1 to n. Let X represent the larger of the two numbers, and Y the smaller. Describe the joint distribution of X and Y.

3. Two numbers are picked at random in the interval [0, 1]. Suppose X represents the larger number and Y the smaller. Describe the joint distribution of X and Y.
4. Let X be a random variable with a probability distribution given by

x	-2	-1	0	1	3
$p(x)$	0.1	0.3	0.2	0.1	0.3

If $Y = X^2 + 1$, find the joint distribution of X and Y.
5. The probability that a United Nations member votes aye on a proposition is 0.6, that he votes nay is 0.3, and that he abstains is 0.1. If there are 120 members in the world body and if they vote independently (an unrealistic assumption indeed!), find the joint distribution of X, the number of members who vote aye, and Y, the number voting nay.
6. Tom and Jim play a series of games. On any play of the game, either Tom wins the game (with probability p), or Jim wins it (with probability q), or the game ends in a draw (with probability $1 - p - q$). Let X denote the number of games which must be played for Tom to win a game, and Y, the number which must be played for Jim to win a game. Find the joint distribution of X and Y.
7. A box contains six balls numbered $1, 2, \ldots, 6$. A ball is picked at random. If the number on this ball is X, then another ball is picked from the balls numbered $1, 2, \ldots, X$. Denoting the number on the second ball drawn as Y, find the joint distribution of X and Y.
8. Prove that the function F defined by

$$F(x, y) = \begin{cases} 1, & x + 3y \geqslant 1 \\ 0, & x + 3y < 1 \end{cases}$$

does not represent a joint D.F.
9. Suppose F, the joint D.F. of X and Y, is known. In each of the cases illustrated in Figure 1.24, express the probability that the random point (X, Y) falls within the shaded region in terms of F.

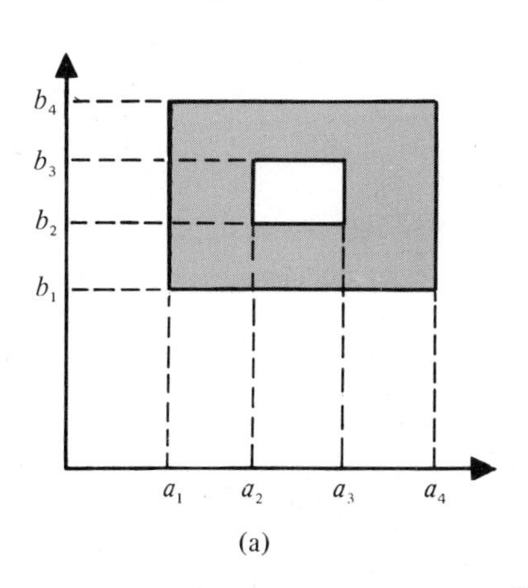

(a)

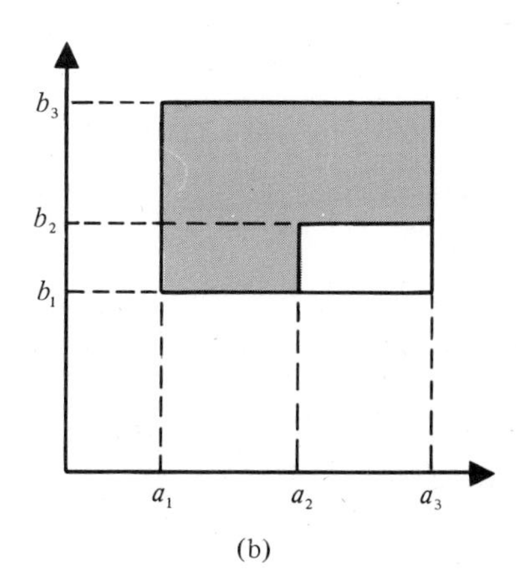

(b)

Figure 1.24

10. If X and Y have a joint distribution F, show that

$$F_X(x) + F_Y(y) - 1 \leqslant F(x, y) \leqslant \sqrt{F_X(x) \cdot F_Y(y)}$$

for all x, y. *Hint:* Consider the regions of the xy-plane.

11. If the distribution function of X and Y is such that $F(x, y) = u(x) \cdot v(y)$ for *every* x, y (u is a function of x only and v is a function of y only), show that

$$\lim_{x \to \infty} u(x) = \frac{1}{\lim_{x \to \infty} v(x)}$$

12. The joint D.F. of X and Y is given by

$$F(x, y) = \begin{cases} 0, & x < 0 \quad \text{or} \quad y < 0 \\ \frac{1}{2}(x^2 y + y^2 x), & 0 \leqslant x < 1 \quad \text{and} \quad 0 \leqslant y < 1 \\ \frac{1}{2}(x^2 + x), & 0 \leqslant x < 1 \quad \text{and} \quad y \geqslant 1 \\ \frac{1}{2}(y^2 + y), & x \geqslant 1 \quad \text{and} \quad 0 \leqslant y < 1 \\ 1, & x \geqslant 1 \quad \text{and} \quad y \geqslant 1 \end{cases}$$

Find: (a) $P(\frac{1}{3} < X < \frac{1}{2}, \frac{1}{3} \leqslant Y < \frac{1}{2})$

(b) $P(\frac{1}{3} < X \leqslant \frac{1}{2}, Y \geqslant \frac{1}{2})$

(c) The joint pdf of X and Y

(d) $P(X \geqslant 2Y)$

(e) $P(X + Y \leqslant 1)$

13. For each of the discrete probability functions given below, obtain the joint distribution function.

(a) $$p(x, y) = \begin{cases} \frac{1}{15}(2x - y + 1), & x = 0, 1, 2, \quad y = 0, 1 \\ 0, & \text{elsewhere} \end{cases}$$

(b) $$p(x, y) = \begin{cases} \frac{1}{39}x(x + y), & x = 2, 3, \quad y = -1, 0, 1 \\ 0, & \text{elsewhere} \end{cases}$$

(c) $$p(x, y) = \begin{cases} \dfrac{x}{12}, & x = 1, 2, 3, \quad y = 1, 2 \\ 0, & \text{elsewhere} \end{cases}$$

14. For each of the joint pdf's given below, obtain the joint distribution function.

(a) $$f(x, y) = \begin{cases} 4xy, & 0 < x < 1, \quad 0 < y < 1 \\ 0, & \text{elsewhere} \end{cases}$$

(b) $$f(x, y) = \begin{cases} 12y, & 0 < y < 2x < 1 \\ 0, & \text{elsewhere} \end{cases}$$

(c) $$f(x, y) = \begin{cases} 24y(1 - x), & 0 \leqslant y \leqslant x \leqslant 1 \\ 0, & \text{elsewhere} \end{cases}$$

15. Show why the function p given below cannot represent a joint probability function for any choice of c.

$$p(x, y) = \begin{cases} cx(2x - y), & x = 0, 1, 2, \quad y = 0, 3 \\ 0, & \text{elsewhere} \end{cases}$$

16. Show why the function f defined by

$$f(x, y) = \begin{cases} cx(3x - y), & 0 < x < 2, \ -x < y < 4x \\ 0, & \text{elsewhere} \end{cases}$$

cannot represent a pdf for any choice of c.

17. The functions p defined below represent joint probability functions for appropriate choice of the constant c. Determine c.

(a) $$p(x, y) = \begin{cases} c(x^2 - y), & x = 0, 2, \ y = -2, -1 \\ 0, & \text{elsewhere} \end{cases}$$

(b) $$p(x, y) = \begin{cases} c|x - y|, & x = -2, 0, 2, \ y = -2, 3 \\ 0, & \text{elsewhere} \end{cases}$$

18. Determine for what constant c the functions given below will represent pdf's.

(a) $$f(x, y) = \begin{cases} c(x + 3y), & 0 \leqslant x \leqslant 1, \ 0 \leqslant y \leqslant 1 \\ 0, & \text{elsewhere} \end{cases}$$

(b) $$f(x, y) = \begin{cases} c(x^2 + y^2), & 0 \leqslant x \leqslant 1, \ 0 \leqslant y \leqslant 1 \\ 0, & \text{elsewhere} \end{cases}$$

(c) $$f(x, y) = \begin{cases} cxy, & 0 \leqslant x \leqslant y, \ 0 \leqslant y \leqslant 2 \\ 0, & \text{elsewhere} \end{cases}$$

(d) $$f(x, y) = \begin{cases} cx(x - y), & 0 < x < 1, \ -x < y < x \\ 0, & \text{elsewhere} \end{cases}$$

(e) $$f(x, y) = \begin{cases} ce^x, & y < 0, \ x < 2y \\ 0, & \text{elsewhere} \end{cases}$$

(f) $$f(x, y) = \begin{cases} c(1 - x^2 - y^2), & x^2 + y^2 \leqslant 1 \\ 0, & \text{elsewhere} \end{cases}$$

19. Two random variables X and Y have a joint discrete distribution with probability function $p(x, y)$ as described in the following table:

$x \backslash y$	-3	0	2	4
-4	$\frac{1}{12}$	0	$\frac{2}{12}$	$\frac{1}{12}$
3	0	$\frac{1}{12}$	0	$\frac{3}{12}$
5	$\frac{2}{12}$	0	$\frac{1}{12}$	$\frac{1}{12}$

Find:

(a) $P(X \geqslant Y)$ (b) $P(X + Y < 0)$
(c) $P(X^2 > Y^2)$ (d) $P(XY > 0)$

20. Suppose the joint pdf of X and Y is given by

$$f(x, y) = \begin{cases} e^{-x-y}, & x > 0, \ y > 0 \\ 0, & \text{elsewhere} \end{cases}$$

Find:

(a) $P(X + Y \leqslant 2)$ (b) $P(X > 2Y)$

21. If the joint pdf of X and Y is given by

$$f(x, y) = \begin{cases} 24y(1-x), & 0<x<1,\ 0<y<x \\ 0, & \text{elsewhere} \end{cases}$$

find:

(a) $P(X = Y)$
(b) $P(X < 2Y)$
(c) $P(\frac{1}{3} < X < \frac{1}{2}, Y \leqslant \frac{1}{2})$
(d) $P(\frac{1}{3} < X < \frac{1}{2})$
(e) $P(Y \leqslant \frac{1}{2})$
(f) $P(\{\frac{1}{3} < X < \frac{1}{2}\} \cup \{Y \leqslant \frac{1}{2}\})$

22. If the joint distribution of X and Y is described by the pdf

$$f(x, y) = \begin{cases} \dfrac{2}{\pi}(1 - x^2 - y^2), & x^2 + y^2 \leqslant 1 \\ 0, & \text{elsewhere} \end{cases}$$

find $P(X^2 + Y^2 \leqslant u)$ for any real number u. *Hint:* Use polar coordinates.

23. Suppose X and Y have a joint absolutely continuous distribution with the following pdf:

$$f(x, y) = \begin{cases} \dfrac{1}{2\pi} e^{-(x^2+y^2)/2}, & -\infty < x < \infty,\ -\infty < y < \infty \\ 0, & \text{elsewhere} \end{cases}$$

Find $P(X^2 + Y^2 \leqslant u)$ for any real number u. What can you say about the distribution of $Z = X^2 + Y^2$? *Hint:* Use polar coordinates.

2. MARGINAL DISTRIBUTIONS

2.1 A General Discussion

As before, suppose $(S, \mathcal{F}, P)$ is a probability space, and X and Y are two real-valued functions defined on S. Several questions are pertinent:

Question 1. Suppose (X, Y) is a random vector. What can be said about X and Y individually? Are they random variables? If they are, can we find their distributions if the joint distribution of X and Y is known?

Question 2. Conversely, suppose X and Y are random variables. Then, is (X, Y) a random vector? If it is, can we find the joint distribution when the individual distributions of X and Y are known?

The answer to Question 1 is contained in the following:

If (X, Y) is a bivariate random vector, then X and Y are each random variables. Furthermore, the distributions of X and Y are given by

$$F_X(u) = F(u, \infty) = \lim_{v\to\infty} F(u, v)$$

$$F_Y(v) = F(\infty, v) = \lim_{u\to\infty} F(u, v)$$

Let us prove this. Since (X, Y) is a bivariate random vector, by definition,

$$\{s \mid X(s) \leqslant u,\ \ Y(s) \leqslant v\} \in \mathcal{F}$$

for *any* pair of real numbers u, v. In particular, then

$$\{s \mid X(s) \leqslant u,\ \ Y(s) < \infty\} \in \mathcal{F}$$

However,

$$\begin{aligned}\{s \mid X(s) \leqslant u,\ \ Y(s) < \infty\} &= \{s \mid X(s) \leqslant u\} \cap \{s \mid Y(s) < \infty\} \\ &= \{s \mid X(s) \leqslant u\} \cap S, \qquad \text{since } Y \text{ is real valued} \\ &= \{s \mid X(s) \leqslant u\}\end{aligned}$$

Hence, if (X, Y) is a random vector, then $\{s \mid X(s) \leqslant u\} \in \mathcal{F}$ for every real number u and, consequently, X is a random variable. A similar argument shows that Y is a random variable.

Incidentally, the equality

$$\{s \mid X(s) \leqslant u,\ \ Y(s) < \infty\} = \{s \mid X(s) \leqslant u\}$$

yields

$$F(u, \infty) = F_X(u)$$

Similarly,

$$\{s \mid X(s) < \infty,\ \ Y(s) \leqslant v\} = \{s \mid Y(s) \leqslant v\}$$

gives

$$F(\infty, v) = F_Y(v)$$

It follows from the above discussion that the distributions of X and Y can be obtained from the knowledge of their joint distribution. The individual distributions of X and Y are called their **marginal distributions**. Thus F_X, F_Y are called respectively the marginal distribution functions of X and Y. There is nothing wrong if the qualifier "marginal" is omitted because, after all, these are the distributions of X and Y in the usual sense.

Example 2.1. Consider the joint distribution function of X and Y given by

$$F(u, v) = \begin{cases} 0, & u < 0 \quad \text{or} \quad v < 0 \\ 1 - 2e^{-v} + e^{-2v}, & u > v \quad \text{and} \quad v \geqslant 0 \\ 1 - e^{-2u} + 2e^{-(u+v)} - 2e^{-v}, & u \geqslant 0 \quad \text{and} \quad v \geqslant u \end{cases}$$

Find the distributions of X and Y.

Solution

(*a*) To give the probability distribution of X, we shall obtain the D.F. of X, by using the relation $F_X(u) = \lim\limits_{v \to \infty} F(u, v)$.

Now, if $u < 0$, $F(u, v) = 0$ for every v. On the other hand, if $u \geqslant 0$, since we let $v \to \infty$, after some stage v will be greater than u, and the functional form

$$F(u, v) = 1 - e^{-2u} + 2e^{-(u+v)} - 2e^{-v}$$

is appropriate. Therefore,

$$F_X(u) = \lim_{v\to\infty} F(u, v) = \begin{cases} 0, & u < 0 \\ \lim\limits_{v\to\infty} (1 - e^{-2u} + 2e^{-(u+v)} - 2e^{-v}), & u \geqslant 0 \end{cases}$$

$$= \begin{cases} 0, & u < 0 \\ 1 - e^{-2u}, & u \geqslant 0 \end{cases}$$

(*b*) To find the D.F. of Y, we note that

$$F(u, v) = \begin{cases} 0, & \text{for every } u \text{ if } v < 0 \\ 1 - 2e^{-v} + e^{-2v}, & \text{for every } u > v \text{ if } v \geqslant 0 \end{cases}$$

It ıollows, therefore, that

$$F_Y(v) = \lim_{u\to\infty} F(u, v) = \begin{cases} 0, & v < 0 \\ \lim\limits_{u\to\infty} (1 - 2e^{-v} + e^{-2v}), & v \geqslant 0 \end{cases}$$

$$= \begin{cases} 0, & v < 0 \\ 1 - 2e^{-v} + e^{-2v}, & v \geqslant 0 \end{cases}$$

Example 2.2. Suppose the joint distribution function of X and Y is given by

$$F(x, y) = \begin{cases} 0, & x < 0 \quad \text{or} \quad y < 0 \\ \dfrac{Kxy}{(1 + 2x)(1 + 3y)}, & x \geqslant 0 \quad \text{and} \quad y \geqslant 0 \end{cases}$$

where K is a constant.

(*a*) Determine the constant K.

(*b*) Find the (marginal) distributions of X and Y.

Solution

(*a*) We know that $\lim\limits_{\substack{x\to\infty \\ y\to\infty}} F(x, y) = 1$. But,

$$\lim_{\substack{x\to\infty \\ y\to\infty}} F(x, y) = \lim_{\substack{x\to\infty \\ y\to\infty}} \frac{Kxy}{(1 + 2x)(1 + 3y)} = \lim_{\substack{x\to\infty \\ y\to\infty}} \frac{K}{\left(\frac{1}{x} + 2\right)\left(\frac{1}{y} + 3\right)} = \frac{K}{6}$$

Hence,

$$K = 6 \quad \text{and} \quad F(x, y) = \begin{cases} 0, & x < 0 \quad \text{or} \quad y < 0 \\ \dfrac{6xy}{(1 + 2x)(1 + 3y)}, & x \geqslant 0 \quad \text{and} \quad y \geqslant 0 \end{cases}$$

(*b*) The D.F. of X is given by

$$F_X(x) = \lim_{y\to\infty} F(x, y) = \begin{cases} 0, & x < 0 \\ \lim_{y\to\infty} \dfrac{6xy}{(1 + 2x)(1 + 3y)}, & x \geqslant 0 \end{cases}$$

Therefore,

$$F_X(x) = \begin{cases} 0, & x < 0 \\ \dfrac{2x}{1 + 2x}, & x \geqslant 0 \end{cases}$$

In a similar manner, we can find the distribution function of Y as

$$F_Y(y) = \begin{cases} 0, & y < 0 \\ \dfrac{3y}{1 + 3y}, & y \geqslant 0 \end{cases}$$

■

Having answered Question 1 in the affirmative, we shall now address ourselves to Question 2. We have the following result as our answer.

If X and Y are random variables, then indeed (X, Y) is a random vector. However, the knowledge of the distribution of X and that of Y is *not* sufficient to determine their joint distribution.

Let us assume that X and Y are random variables. To prove that (X, Y) is a random vector, we observe that since X and Y are random variables,

$$\{s \mid X(s) \leqslant u\} \in \mathcal{F} \quad \text{and} \quad \{s \mid Y(s) \leqslant v\} \in \mathcal{F}$$

for any real numbers u, v. However, $\mathcal{F}$ is closed with respect to intersection. Therefore,

$$\{s \mid X(s) \leqslant u\} \cap \{s \mid Y(s) \leqslant v\} \in \mathcal{F}$$

But

$$\{s \mid X(s) \leqslant u\} \cap \{s \mid Y(s) \leqslant v\} = \{s \mid X(s) \leqslant u,\ Y(s) \leqslant v\}$$

Consequently,

$$\{s \mid X(s) \leqslant u,\ Y(s) \leqslant v\} \in \mathcal{F}$$

for any real numbers u, v.

By the definition of a random vector, it therefore follows that (X, Y) is a random vector.

To see that the knowledge of the marginal distributions of the random variables is not sufficient to determine their joint distribution, consider the following two *distinct* joint distribution functions $F^{(1)}$ and $F^{(2)}$:

$$F^{(1)}(x, y) = \begin{cases} 0, & x < 0 \quad \text{or} \quad y < 0 \\ \frac{6}{7}\left(xy + \frac{x^3y^2}{6}\right), & 0 \leqslant x < 1 \quad \text{and} \quad 0 \leqslant y < 1 \\ \frac{6}{7}\left(x + \frac{x^3}{6}\right), & 0 \leqslant x < 1 \quad \text{and} \quad y \geqslant 1 \\ \frac{6}{7}\left(y + \frac{y^2}{6}\right), & x \geqslant 1 \quad \text{and} \quad 0 \leqslant y < 1 \\ 1, & x \geqslant 1 \quad \text{and} \quad y \geqslant 1 \end{cases}$$

and

$$F^{(2)}(x, y) = \begin{cases} 0, & x < 0 \quad \text{or} \quad y < 0 \\ \frac{36}{49}\left(x + \frac{x^3}{6}\right)\left(y + \frac{y^2}{6}\right), & 0 \leqslant x < 1 \quad \text{and} \quad 0 \leqslant y < 1 \\ \frac{6}{7}\left(x + \frac{x^3}{6}\right), & 0 \leqslant x < 1 \quad \text{and} \quad y \geqslant 1 \\ \frac{6}{7}\left(y + \frac{y^2}{6}\right), & x \geqslant 1 \quad \text{and} \quad 0 \leqslant y < 1 \\ 1, & x \geqslant 1 \quad \text{and} \quad y \geqslant 1 \end{cases}$$

As can be easily verified, they both give rise to the same marginal distribution functions, namely,

$$F_X(x) = \begin{cases} 0, & x < 0 \\ \frac{6}{7}\left(x + \frac{x^3}{6}\right), & 0 \leqslant x < 1 \\ 1, & x \geqslant 1 \end{cases}$$

and

$$F_Y(y) = \begin{cases} 0, & y < 0 \\ \frac{6}{7}\left(y + \frac{y^2}{6}\right), & 0 \leqslant y < 1 \\ 1, & y \geqslant 1 \end{cases}$$

The inference to be drawn from our discussion is that, whereas the marginal distributions are uniquely determined from the joint distribution, the converse is not true.

Now that we know that the joint distribution completely specifies the marginal distributions, we shall take up the discrete and the absolutely continuous cases separately. In the discrete case, the goal will be to obtain the probability functions of the individual random variables from the joint probability function of X and Y. In the absolutely continuous case, the goal will be to get the individual probability density functions from the joint probability density function.

2.2 The Discrete Case

Suppose X and Y have a joint discrete distribution with the joint probability function given by

$$P(X = x_i,\ Y = y_j) = p(x_i, y_j), \qquad i, j = 1, 2, \ldots$$

Then

$$P(X = x_i) = \sum_{j=1}^{\infty} p(x_i, y_j), \qquad i = 1, 2, \ldots$$

$$P(Y = y_j) = \sum_{i=1}^{\infty} p(x_i, y_j), \qquad j = 1, 2, \ldots$$

To prove this, we note (from Section 2.1) that $F_X(x) = F(x, \infty)$. Therefore,

$$F_X(x) = F(x, \infty) = \sum_{x_i \leqslant x} \sum_{y_j < \infty} p(x_i, y_j)$$

Now, recall from Chapter 4 that

$$F_X(x) = \sum_{x_i \leqslant x} p_X(x_i)$$

where p_X is the probability function of X. A comparison of the above two expressions for $F_X(x)$ shows, after proper identification, that

$$p_X(x_i) = \sum_{y_j < \infty} p(x_i, y_j) = \sum_{j=1}^{\infty} p(x_i, y_j)$$

Similarly, it can be shown that $p_Y(y_j) = \sum_{i=1}^{\infty} p(x_i, y_j)$.

An alternate way of showing that

$$P(X = x_i) = \sum_{j=1}^{\infty} p(x_i, y_j)$$

is the following: The event $\{X = x_i\}$ occurs in conjunction with the values of Y and

$$\{X = x_i\} = \{X = x_i,\ Y = y_1\} \cup \{X = x_i,\ Y = y_2\} \cup \ldots$$

Therefore,

$$P(X = x_i) = P\left(\bigcup_{j=1}^{\infty} \{X = x_i,\ Y = y_j\}\right)$$

Since the events $\{X = x_i,\ Y = y_j\}$, $j = 1, 2, \ldots$, are mutually exclusive, this gives

$$P(X = x_i) = \sum_{j=1}^{\infty} P(X = x_i,\ Y = y_j) = \sum_{j=1}^{\infty} p(x_i, y_j)$$

In short, the message contained in the above result is: *To find the marginal probability function of a random variable, sum the joint probabilities with respect to the other variable.*

Displaying the joint probability function in a rectangular array, we get the following table:

$x \backslash y$	y_1	y_2	$\cdots$	y_j	$\cdots$	$P(X=x)$
x_1	$p(x_1, y_1)$	$p(x_1, y_2)$	$\cdots$	$p(x_1, y_j)$	$\cdots$	$\sum_{j=1}^{\infty} p(x_1, y_j)$
x_2	$p(x_2, y_1)$	$p(x_2, y_2)$	$\cdots$	$p(x_2, y_j)$	$\cdots$	$\sum_{j=1}^{\infty} p(x_2, y_j)$
$\vdots$	$\vdots$	$\vdots$		$\vdots$		$\vdots$
x_i	$p(x_i, y_1)$	$p(x_i, y_2)$	$\cdots$	$p(x_i, y_j)$	$\cdots$	$\sum_{j=1}^{\infty} p(x_i, y_j)$
$\vdots$	$\vdots$	$\vdots$		$\vdots$		$\vdots$
$P(Y=y)$	$\sum_{i=1}^{\infty} p(x_i, y_1)$	$\sum_{i=1}^{\infty} p(x_i, y_2)$	$\cdots$	$\sum_{i=1}^{\infty} p(x_i, y_j)$	$\cdots$	1

As can be seen, the totals in the vertical and horizontal margins in fact represent, respectively, the probability functions of X and Y. It is because of this feature that the individual distributions of X and Y are often called the *marginal* distributions.

In Section 2.1, we saw that distinct joint distributions can give rise to the same marginal distributions. As another example of this, consider the family of joint probability functions given below (where $0 \leqslant \epsilon/2 \leqslant \frac{1}{18}$ (why?)):

$x \backslash y$	y_1	y_2	y_3	$P(X=x)$
x_1	$\frac{1}{9} - \epsilon$	$\frac{2}{9} + \epsilon$	$\frac{1}{9}$	$\frac{4}{9}$
x_2	$\frac{1}{18} + \frac{\epsilon}{2}$	$\frac{1}{18} - \frac{\epsilon}{2}$	$\frac{3}{18}$	$\frac{5}{18}$
x_3	$\frac{3}{18} + \frac{\epsilon}{2}$	$\frac{2}{18} - \frac{\epsilon}{2}$	0	$\frac{5}{18}$
$P(Y=y)$	$\frac{1}{3}$	$\frac{7}{18}$	$\frac{5}{18}$	

This table describes a family of joint probability functions for different values of ϵ. But no matter what ϵ is (as long as $0 < \epsilon/2 < \frac{1}{18}$), we always get the same marginal probability function of X, namely, $P(X = x_1) = \frac{4}{9}$, $P(X = x_2) = \frac{5}{18}$, and $P(X = x_3) = \frac{5}{18}$, and the same marginal probability function of Y, namely, $P(Y = y_1) = \frac{1}{3}$, $P(Y = y_2) = \frac{7}{18}$, and $P(Y = y_3) = \frac{5}{18}$.

Example 2.3. If X and Y have the joint probability function given by

$$P(X = x,\ Y = y) = \tfrac{1}{42}(x + y^2), \qquad x = 1, 4, \quad y = -1, 0, 1, 3$$

find the marginal distributions of X and Y.

Solution. The probability function of X is obtained by summing the joint probabilities with respect to all the possible values of y. Therefore, for $x = 1, 4$,

$$P(X = x) = \sum_y \tfrac{1}{42}(x + y^2)$$
$$= \tfrac{1}{42}[(x + 1) + (x + 0) + (x + 1) + (x + 9)]$$

thus

$$P(X = x) = \frac{4x + 11}{42}, \qquad x = 1, 4$$

Similarly, the probability function of Y is given by

$$P(Y = y) = \sum_x \tfrac{1}{42}(x + y^2)$$
$$= \tfrac{1}{42}[(1 + y^2) + (4 + y^2)], \qquad y = -1, 0, 1, 3,$$

Hence

$$P(Y = y) = \frac{5 + 2y^2}{42}, \qquad y = -1, 0, 1, 3$$

- *Example 2.4 (The trinomial distribution)* Suppose X and Y have the trinomial distribution with n trials and parameters p, q. (See Example 1.9.) Find the marginal distributions of X and Y.

Solution. We know that the joint probability function of X and Y is

$$P(X = i,\ Y = j) = \frac{n!}{i!j!(n - i - j)!} p^i q^j (1 - p - q)^{n-i-j}$$

where $i = 0, 1, \ldots, n$; $j = 0, 1, \ldots, n - i$. Therefore, for $i = 0, 1, \ldots, n$, we have

$$P(X = i) = \sum_{j=0}^{n-i} P(X = i,\ Y = j)$$
$$= \sum_{j=0}^{n-i} \frac{n!}{i!j!(n - i - j)!} p^i q^j (1 - p - q)^{n-i-j}$$
$$= \frac{n!}{i!} p^i \sum_{j=0}^{n-i} \frac{1}{j!(n - i - j)!} q^j (1 - p - q)^{n-i-j}$$
$$= \frac{n! p^i}{i!(n - i)!} \sum_{j=0}^{n-i} \frac{(n - i)!}{j!(n - i - j)!} q^j (1 - p - q)^{n-i-j}$$
$$= \binom{n}{i} p^i [q + (1 - p - q)]^{n-i} = \binom{n}{i} p^i (1 - p)^{n-i}$$

Thus, *if X and Y have the trinomial distribution with n trials and parameters p, q, then X has the binomial distribution with n trials and the probability of success p.*

Similarly, *Y has the binomial distribution with n trials and the probability of success q.*

2.3 The Absolutely Continuous Case

If X and Y have a joint absolutely continuous distribution with a joint pdf f, then X and Y each have an absolutely continuous distribution and the pdf's of X and Y are given by

$$f_X(x) = \int_{-\infty}^{\infty} f(x, y)\, dy, \qquad -\infty < x < \infty$$

$$f_Y(y) = \int_{-\infty}^{\infty} f(x, y)\, dx, \qquad -\infty < y < \infty$$

To prove this, recall that the notion of absolute continuity involves the idea of being able to obtain the distribution function by appropriately integrating a non-negative function called the probability density function. Now we already know that

$$F_X(u) = \lim_{v \to \infty} F(u, v) = F(u, \infty) = \int_{-\infty}^{u} \int_{-\infty}^{\infty} f(x, y)\, dy\, dx$$

But since $f(x, y)$ is nonnegative, it follows that $\int_{-\infty}^{\infty} f(x, y)\, dy$ is nonnegative. Consequently, the relation

$$F_X(u) = \int_{-\infty}^{u} \left[\int_{-\infty}^{\infty} f(x, y)\, dy \right] dx$$

shows that the distribution function of X can be obtained by integrating the non-negative function $\int_{-\infty}^{\infty} f(x, y)\, dy$. Therefore, by definition, X has an absolutely continuous distribution with the pdf $\int_{-\infty}^{\infty} f(x, y)\, dy$.

A similar argument shows that Y is absolutely continuous with pdf $\int_{-\infty}^{\infty} f(x, y)\, dx$.

Example 2.5. Suppose the joint pdf of X and Y is given by

$$f(x, y) = \begin{cases} \frac{6}{7}(1 + x^2 y), & 0 < x < 1, \ 0 < y < 1 \\ 0, & \text{elsewhere} \end{cases}$$

Find the marginal distribution of X.

Solution. We have $f_X(x) = \int_{-\infty}^{\infty} f(x, y)\, dy$.

If $x \leqslant 0$ or $x \geqslant 1$, then $f(x, y) = 0$ for every y. Therefore,

$$f_X(x) = \int_{-\infty}^{\infty} 0\, dy = 0 \quad \text{if } x \leqslant 0 \text{ or } x \geqslant 1$$

If $0 < x < 1$, as can be seen from Figure 2.1,

$$f(x, y) = \begin{cases} 0, & \text{if } y < 0 \text{ or } y > 1 \\ \frac{6}{7}(1 + x^2 y), & 0 < y < 1 \end{cases}$$

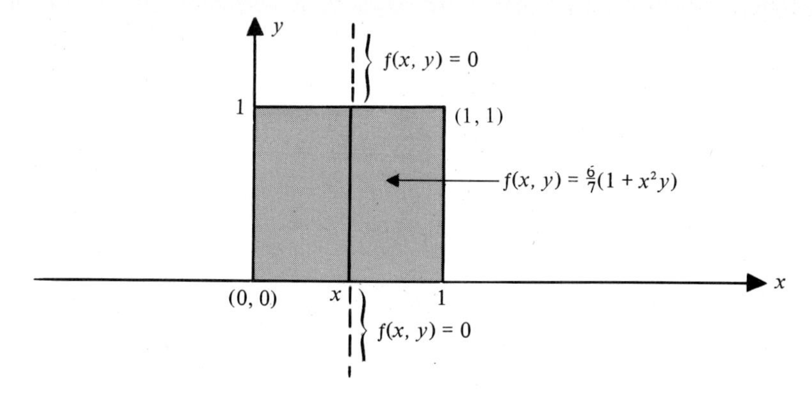

Figure 2.1

Therefore,

$$f_X(x) = \int_{-\infty}^{0} 0\,dy + \int_{0}^{1} \frac{6}{7}(1 + x^2y)\,dy + \int_{1}^{\infty} 0\,dy$$
$$= \frac{6}{7}\left(1 + \frac{x^2}{2}\right) \quad \text{if} \quad 0 < x < 1$$

Combining the above, we get the pdf of X as

$$f_X(x) = \begin{cases} \frac{6}{7}\left(1 + \frac{x^2}{2}\right), & 0 < x < 1 \\ 0, & \text{elsewhere} \end{cases}$$

Example 2.6. Let

$$f(x, y) = \begin{cases} 10x^2y, & 0 \leqslant y \leqslant x \leqslant 1 \\ 0, & \text{elsewhere} \end{cases}$$

(*a*) Find the marginal pdf's of X and Y.
(*b*) Compute $P(Y \leqslant \frac{1}{2})$.

Solution

(*a*) We first find the pdf of X.
If $x < 0$ or $x > 1$, then $f(x, y) = 0$ for every y, and therefore

$$f_X(x) = \int_{-\infty}^{\infty} 0\,dy = 0$$

If $0 \leqslant x \leqslant 1$, it can be seen from Figure 2.2 that

$$f_X(x) = \int_{-\infty}^{0} 0\,dy + \int_{0}^{x} 10x^2y\,dy + \int_{x}^{\infty} 0\,dy = 5x^4$$

Combining these results, we get

$$f_X(x) = \begin{cases} 5x^4, & 0 \leqslant x \leqslant 1 \\ 0, & \text{elsewhere} \end{cases}$$

We can find the pdf of Y similarly, using the relation $f_Y(y) = \int_{-\infty}^{\infty} f(x, y)\,dx$.

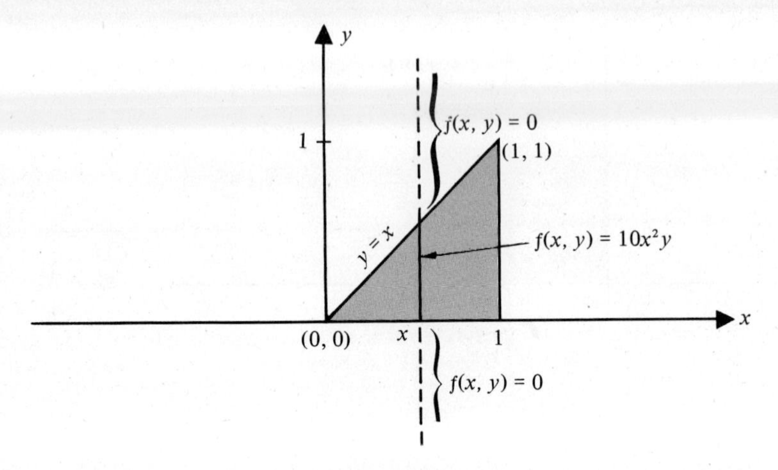

Figure 2.2

If $y < 0$ or $y > 1$, then $f(x, y) = 0$ for every x, so that

$$f_Y(y) = \int_{-\infty}^{\infty} 0 \, dx = 0$$

If $0 \leqslant y \leqslant 1$, it can be seen from Figure 2.3 that

$$f_Y(y) = \int_{-\infty}^{y} 0 \, dx + \int_{y}^{1} 10x^2 y \, dx + \int_{1}^{\infty} 0 \, dx = \tfrac{10}{3} y(1 - y^3)$$

Therefore, combining the above, we get

$$f_Y(y) = \begin{cases} \frac{10}{3} y(1 - y^3), & 0 \leqslant y \leqslant 1 \\ 0, & \text{elsewhere} \end{cases}$$

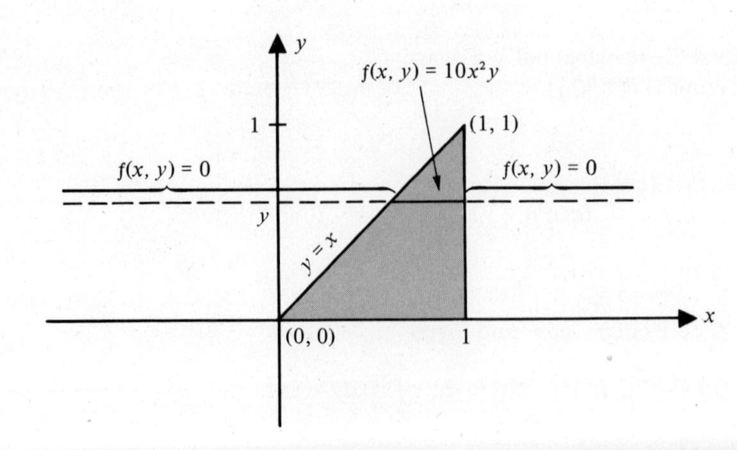

Figure 2.3

(*b*) Since we know the pdf of Y,

$$P(Y \leqslant \tfrac{1}{2}) = \int_{-\infty}^{1/2} f_Y(y)\,dy = \int_0^{1/2} \tfrac{10}{3}y(1-y^3)\,dy = \tfrac{19}{48}$$

Example 2.7. Suppose X and Y have a joint pdf given by

$$f(x, y) = \begin{cases} 5x^2y, & 0 \leqslant x \leqslant 1,\ 0 \leqslant y \leqslant 1,\ x \geqslant y \\ 6x^2, & 0 \leqslant x \leqslant 1,\ 0 \leqslant y \leqslant 1,\ x < y \\ 0, & \text{elsewhere} \end{cases}$$

Find the pdf of X.

Solution. The nature of the joint pdf is illustrated in Figure 2.4.

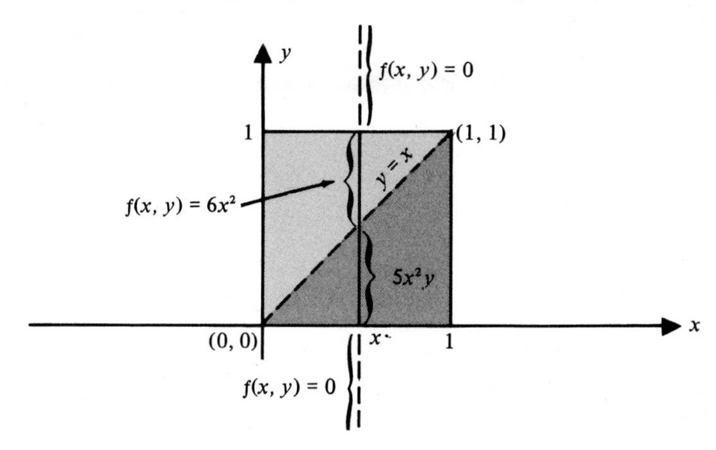

Figure 2.4

If $x < 0$ or $x > 1$, $f(x, y) = 0$ for every y and, consequently,

$$f_X(x) = \int_{-\infty}^{\infty} 0\,dy = 0$$

If $0 \leqslant x \leqslant 1$, it follows from Figure 2.4 that

$$f_X(x) = \int_{-\infty}^{0} 0\,dy + \int_0^x 5x^2y\,dy + \int_x^1 6x^2\,dy + \int_1^{\infty} 0\,dy = \tfrac{5}{2}x^4 - 6x^3 + 6x^2$$

In summary,

$$f_X(x) = \begin{cases} \tfrac{5}{2}x^4 - 6x^3 + 6x^2, & 0 \leqslant x \leqslant 1 \\ 0, & \text{elsewhere} \end{cases}$$

The reader might enjoy showing that

$$f_Y(y) = \begin{cases} \tfrac{5}{3}y + 2y^3 - \tfrac{5}{3}y^4, & 0 \leqslant y \leqslant 1 \\ 0, & \text{elsewhere} \end{cases}$$

- *Example 2.8 (The standard bivariate normal distribution)* Two random variables X and Y are said to have the standard bivariate normal distribution if their joint pdf is given by

$$f(x, y) = \frac{1}{2\pi\sqrt{1-\rho^2}} e^{-(x^2 - 2\rho xy + y^2)/[2(1-\rho^2)]}$$

where $-\infty < x < \infty$, $-\infty < y < \infty$, and $-1 < \rho < 1$. Find the marginal distributions of X and Y.

Solution. The joint pdf describes a surface in the shape of a bell whose bottom rim extends over the entire xy-plane. (See Figure 2.5.)

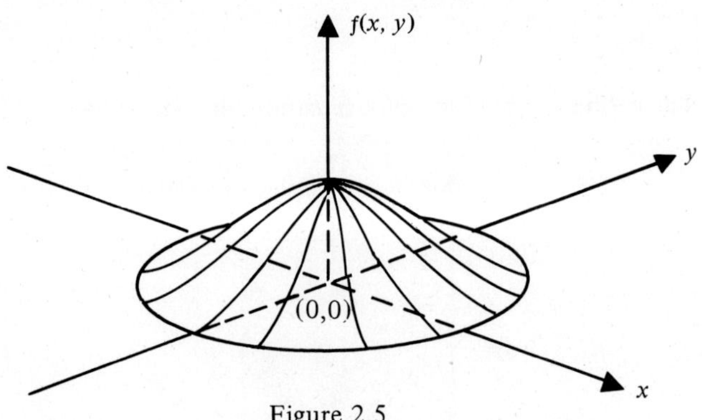

Figure 2.5

Completing the square, we can write

$$x^2 - 2\rho xy + y^2 = x^2 - 2\rho xy + \rho^2 y^2 - \rho^2 y^2 + y^2 = (x - \rho y)^2 + (1 - \rho^2)y^2$$

Hence,

$$f_Y(y) = \int_{-\infty}^{\infty} f(x, y)\, dx = \frac{1}{2\pi\sqrt{1-\rho^2}} e^{-y^2/2} \int_{-\infty}^{\infty} e^{-(x-\rho y)^2/[2(1-\rho^2)]} dx$$

Now, we know from our discussion of the normal distribution in Section 2.2 of Chapter 5 that

$$\int_{-\infty}^{\infty} e^{-(x-a)^2/(2b^2)} dx = b\sqrt{2\pi}$$

Identifying a with ρy and b^2 with $1 - \rho^2$, it follows that

$$\int_{-\infty}^{\infty} e^{-(x-\rho y)^2/[2(1-\rho^2)]} dx = \sqrt{1-\rho^2}\,\sqrt{2\pi}$$

Therefore,

$$f_Y(y) = \frac{1}{2\pi\sqrt{1-\rho^2}} e^{-y^2/2} \sqrt{1-\rho^2}\sqrt{2\pi} = \frac{1}{\sqrt{2\pi}} e^{-y^2/2}, \quad -\infty < y < \infty$$

In conclusion, *if X and Y have the standard bivariate normal distribution, then Y has the standard normal distribution. By symmetry (or directly), it also follows that X has the standard normal distribution.* ■

A random vector (X, Y) is said to have a **uniform bivariate distribution** over a region A if its pdf is given by

$$f(x, y) = \begin{cases} K, & \text{if } (x, y) \in A \\ 0, & \text{elsewhere} \end{cases}$$

where K is a constant which is suitably determined.

Since

$$\int_{-\infty}^{\infty} \int_{-\infty}^{\infty} f(x, y)\, dy\, dx = 1$$

K must be picked so that

$$\iint_A K\, dy\, dx = 1$$

That is,

$$K = \frac{1}{\iint_A dy\, dx} = \frac{1}{\text{the area of } A}$$

Thus, *the constant K is the reciprocal of the area of the base region A.*

Example 2.9. Suppose X and Y have a joint uniform distribution over a circle of radius 1, centered at the origin. Find the marginal distributions of X and Y.

Solution. Since the area of the circle is π, the joint pdf is given as

$$f(x, y) = \begin{cases} \dfrac{1}{\pi}, & \text{if } 0 \leqslant x^2 + y^2 < 1 \\ 0, & \text{elsewhere} \end{cases}$$

If $x \leqslant -1$ or $x \geqslant 1$, then $f(x, y) = 0$ for every y, and consequently $f_X(x) = 0$. If $-1 < x < 1$, it can be seen from Figure 2.6 that

$$f_X(x) = \int_{-\infty}^{-(1-x^2)^{1/2}} 0\, dy + \int_{-(1-x^2)^{1/2}}^{(1-x^2)^{1/2}} \frac{1}{\pi}\, dy + \int_{(1-x^2)^{1/2}}^{\infty} 0\, dy = \frac{2\sqrt{1-x^2}}{\pi}$$

Thus,

$$f_X(x) = \begin{cases} \dfrac{2\sqrt{1-x^2}}{\pi}, & -1 < x < 1 \\ 0, & \text{elsewhere} \end{cases}$$

Similarly, it can be shown that

$$f_Y(y) = \begin{cases} \dfrac{2\sqrt{1-y^2}}{\pi}, & -1 < y < 1 \\ 0, & \text{elsewhere} \end{cases}$$

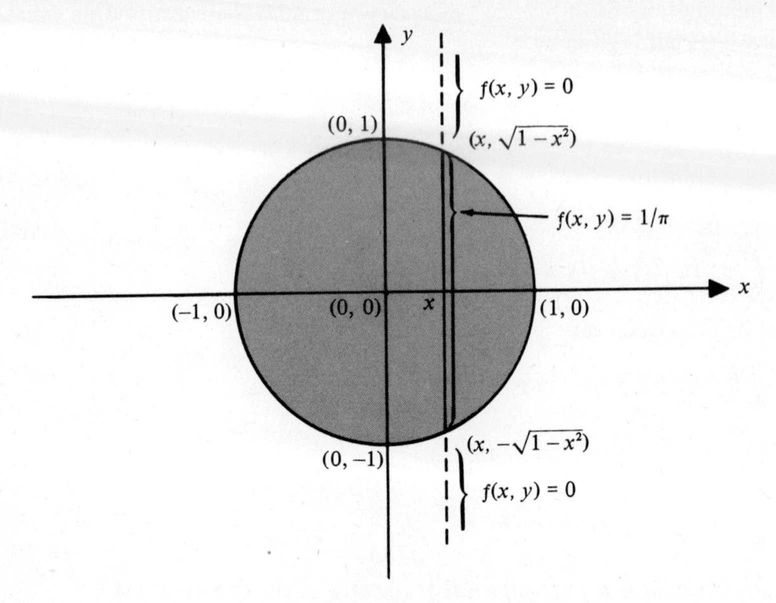

Figure 2.6

*Digression.** We have shown that if X and Y have a joint absolutely continuous distribution, then X and Y have absolutely continuous distributions. The converse of this result is false, in that *if X and Y each have an absolutely continuous distribution, then the joint distribution of X and Y need not be absolutely continuous.* The following example will show this.

Suppose X has an absolutely continuous distribution and that $X = Y$. Specifically let us assume that X is uniformly distributed over the interval $[0, 1]$. (This is not essential, but will be found convenient in the discussion.) Then for any $s \in S$, $X(s) = Y(s)$, and it can be easily seen that (X, Y) has a probability distribution where all the probability mass is distributed uniformly on the line segment L joining $(0, 0)$ to $(1, 1)$. (See Figure 2.7.)

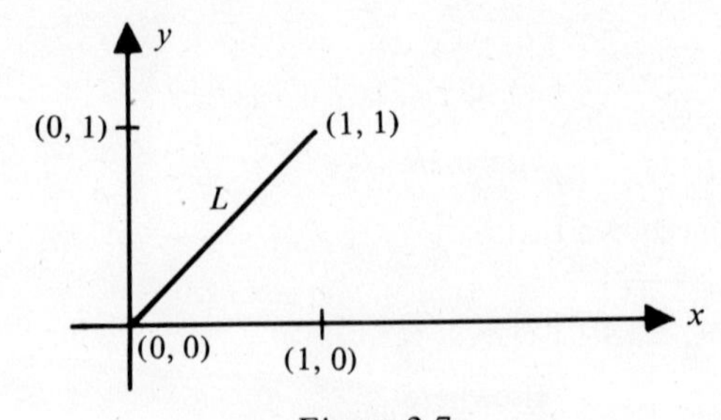

Figure 2.7

*The rest of this section can be safely omitted at first reading. Just take note of the fact that the joint distribution need not be absolutely continuous if the marginal distributions are absolutely continuous.

The random vector (X, Y) is not absolutely continuous. To see this, let us suppose there is a nonnegative function f such that the associated probabilities are obtained by integrating it appropriately.

Now consider the event $(X, Y) \in L$. Since $X = Y$, for every $s \in S$, $X(s) = Y(s)$ and, consequently, for every $s \in S$, $(X(s), Y(s)) \in L$. Therefore,

$$P(\{(X, Y) \in L\}) = 1$$

On the other hand, if (X, Y) is absolutely continuous, we should be able to find $P(\{(X, Y) \in L\})$ by integrating the nonnegative function f as

$$P(\{(X, Y) \in L\}) = \iint_L f(x, y)\, dy\, dx$$

But this is impossible, since the integral on the right-hand side is 0 because the area of L is 0. Hence the probability cannot be obtained by integrating a nonnegative function, thus leading to the conclusion that the distribution of (X, Y) is not absolutely continuous.

Let us carry the above discussion a little further and consider the joint distribution function as sketched in Figure 2.8. We have

$$F(u, v) = P(X \leqslant u,\ Y \leqslant v) = \begin{cases} P(X \leqslant u) & \text{if } u \leqslant v \\ P(Y \leqslant v) & \text{if } v \leqslant u \end{cases}$$

Therefore,

$$F(u, v) = \begin{cases} 0, & u < 0 \text{ or } v < 0 \\ v, & v \leqslant u \text{ and } 0 \leqslant v < 1 \\ u, & u \leqslant v \text{ and } 0 \leqslant u < 1 \\ 1, & u \geqslant 1 \text{ and } v \geqslant 1 \end{cases}$$

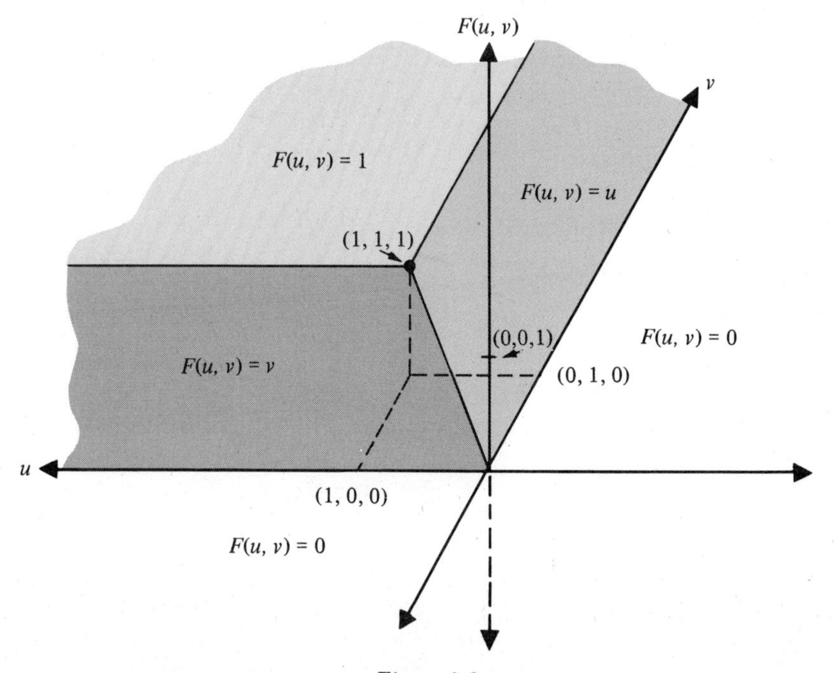

Figure 2.8

The distribution function is *continuous,* but the second partial derivative $\frac{\partial^2}{\partial u\, \partial v} F(u, v)$ is zero everywhere (except where the graph has sharp edges, where the partial derivatives do not exist). The integral of a function which is zero just about everywhere will never give us the distribution function. Hence, there is no joint probability density function. Such joint distributions are said to be singular.

The reason why there is no pdf is because *we have distributed all the probability mass on a region which has zero area, while at the same time there is no probability mass concentrated in lumps at isolated points.* At this stage, recall the discussion of singular distributions in Chapter 4 with reference to a single random variable. We encountered the same situation as above because all the probability mass was distributed on an interval of length zero without lumps at isolated points.

EXERCISES–SECTION 2

1. X and Y have a joint discrete distribution with the entire probability mass distributed at the three points $(1, 4), (2, 4), (2, -1)$, with $p(1, 4) = 0.2, p(2, 4) = 0.5, p(2, -1) = 0.3$. Find the individual probability functions of X and Y.

2. A person is dealt thirteen cards from a deck of bridge cards. Let X represent the number of black cards and Y the number of hearts. Find:
   (a) the distributions of X and Y
   (b) the joint distribution of X and Y
   (c) the marginal distributions of X and Y, using the joint distribution obtained in part (b). Compare this answer with that in part (a).

3. Suppose the joint D.F. of X and Y is given as follows:

$$F(x, y) = \begin{cases} 0, & x < 0 \quad \text{or} \quad y < 0 \\ \frac{1}{2}(1 - \cos y + \sin y), & x \geqslant \pi/2, \quad \text{and} \quad 0 \leqslant y < \pi/2 \\ \frac{1}{2}(1 - \cos x + \sin x), & 0 \leqslant x < \pi/2, \quad \text{and} \quad y \geqslant \pi/2 \\ \frac{1}{2}(\sin x + \sin y - \sin(x + y)), & 0 \leqslant x < \pi/2, \quad \text{and} \quad 0 \leqslant y < \pi/2 \\ 1, & x \geqslant \pi/2, \quad \text{and} \quad y \geqslant \pi/2 \end{cases}$$

Find the marginal distribution functions of X and Y.

4. For the joint distribution function given in exercise 12 of Section 1, obtain:
   (a) F_X and F_Y (b) $P(X \leqslant \frac{1}{2})$ (c) $P(Y > \frac{1}{3})$
   (d) $E(X)$ (e) $\text{Var}(X)$

5. For the joint distribution described in exercise 6 of Section 1, find the marginal distributions.

6. Consider the discrete joint probability functions given below. In each case, determine the marginal probability functions of X and Y.

(a) $p(x, y) = \begin{cases} \frac{1}{15}(2x - y + 1), & x = 0, 1, 2, \ y = 0, 1 \\ 0, & \text{elsewhere} \end{cases}$

(b) $p(x, y) = \begin{cases} \frac{1}{39}x(x + y), & x = 2, 3, \ y = -1, 0, 1 \\ 0, & \text{elsewhere} \end{cases}$

(c) $p(x, y) = \dfrac{2}{n(n+1)}, \quad y = 1, 2, \ldots, x, \ x = 1, 2, \ldots, n$

where n is a positive integer

(d) $p(x, y) = \begin{cases} \dfrac{x}{12}, & x = 1, 2, 3, \ y = 1, 2 \\ 0, & \text{elsewhere} \end{cases}$

7. A number is picked at random in the interval $[0, 0.46)$. Let X and Y be, respectively, the first and second digits in the decimal expansion of the number. Find the joint distribution of X and Y. Also find the marginal distributions.

8. For each of the following joint pdf's of X and Y, obtain the marginal distributions of X and Y.

(a) $f(x, y) = \begin{cases} 4xy, & 0 < x < 1, \ 0 < y < 1 \\ 0, & \text{elsewhere} \end{cases}$

(b) $f(x, y) = \begin{cases} 3x(1 - xy), & 0 \leqslant x \leqslant 1, \ 0 \leqslant y \leqslant 1 \\ 0, & \text{elsewhere} \end{cases}$

(c) $f(x, y) = \begin{cases} 2x(x - y), & 0 < x < 1, \ -x < y < x \\ 0, & \text{elsewhere} \end{cases}$

(d) $f(x, y) = \begin{cases} 12y, & 0 < y < 2x < 1 \\ 0, & \text{elsewhere} \end{cases}$

(e) $f(x, y) = \begin{cases} xe^{-x(1+y)}, & x > 0, \ y > 0 \\ 0, & \text{elsewhere} \end{cases}$

(f) $f(x, y) = \begin{cases} 24y(1 - x), & 0 \leqslant y \leqslant x \leqslant 1 \\ 0, & \text{elsewhere} \end{cases}$

9. Suppose a point (X, Y) is picked at random inside a triangle bounded by the lines $y = 0, y = x$, and $x = 1$. Give the joint distribution of X and Y, and find the marginal distributions.

10. Two random variables X and Y have a joint absolutely continuous distribution with the following joint pdf:

$$f(x, y) = \begin{cases} \frac{6}{5}(x^2 + y), & 0 < x < 1, \ 0 < y < 1 \\ 0, & \text{elsewhere} \end{cases}$$

Find:

(a) $E(X)$ (b) $E(Y)$ (c) $\mathrm{Var}(X)$ (d) $\mathrm{Var}(Y)$

11. Suppose the joint pdf of X and Y is given by

$$f(x, y) = \begin{cases} x + y, & 0 < x < 1,\ 0 < y < 1 \\ 0, & \text{elsewhere} \end{cases}$$

Find:

(a) $P(X \leqslant \frac{1}{2})$ (b) $P(\frac{1}{3} < X \leqslant \frac{1}{2})$ (c) $P(Y > \frac{1}{3})$

12. The amount X (in dollars) that a person earns and the amount Y (in dollars) that he spends during a day are assumed to have the following joint absolutely continuous distribution (an artificial assumption!):

$$f(x, y) = \begin{cases} \frac{1}{25}\left(\frac{20 - x}{x}\right), & 10 < x < 20,\ \frac{x}{2} < y < x \\ 0, & \text{elsewhere} \end{cases}$$

Find:

(a) the distribution of the amount earned
(b) the distribution of the amount spent
(c) the probability that he spends less than two-thirds the amount earned
(d) the expected amount earned
(e) the expected amount spent

13. Suppose X and Y are two random variables with respective pdf's f_X and f_Y and D.F.'s F_X and F_Y. Let $-1 < c < 1$ and define a function g as follows:

$$g(x, y) = f_X(x) f_Y(y) \{1 + c[2F_X(x) - 1] \cdot [2F_Y(y) - 1]\}$$

(a) Show that for any c, g represents a pdf.
(b) Find the marginal pdf's for the joint distribution described by g.
(c) What do you conclude from part (b)?

9 Conditional Distributions and Independent Random Variables

INTRODUCTION

The contents of this chapter are a continuation of the discussion initiated in Chapter 8, where we considered joint distributions of random variables and their marginal distributions. For example, if we are given the joint distribution of heights and weights of adult males, we are now in a position to obtain the distribution of their weights. Of course, we suspect that this distribution will depend on the particular height bracket we are considering. Taller people usually have higher weight, and so on. Thus it is natural for us now to explore how the distribution of one random variable depends on the values of another random variable. This aspect will be considered in Section 1 dealing with conditional distributions.

It is quite possible that the distribution of one random variable is not dependent on the other random variable. For instance, when we consider the joint distribution of arm lengths and waist measurements, we have a strong feeling that the distribution of arm lengths would not depend on the waist measurements, and vice versa. In such cases, the random variables are said to be *independent*. The investigation of such random variables will be taken up in Section 2.

In Section 3, we shall briefly extend the notions developed in the last chapter and the first two sections of this chapter to an arbitrary, but finite, number of random variables.

1. CONDITIONAL DISTRIBUTIONS

1.1 Conditional Distribution Given an Event of Positive Probability

The notion of conditional distribution is closely tied to that of conditional probability. We recall that if A and B are any two events with $P(B) > 0$, then $P(A \mid B) = P(AB)/P(B)$. Suppose we take for the event A the event $\{X \leqslant u\}$

where u is a real number. Then $P(X \leqslant u \mid B)$ is a function of u defined by

$$P(X \leqslant u \mid B) = \frac{P(\{X \leqslant u\} \cap B)}{P(B)}$$

and is called the *conditional distribution function of the random variable X given the event B.*

Example 1.1. Let X and Y have a joint distribution given by the following pdf:

$$f(x, y) = \begin{cases} 10x^2y, & 0 \leqslant y \leqslant x \leqslant 1 \\ 0, & \text{elsewhere} \end{cases}$$

Find:

(*a*) The conditional distribution function of X given $0 < Y \leqslant \frac{1}{2}$
(*b*) $P(\frac{1}{3} < X \leqslant \frac{2}{3} \mid 0 < Y \leqslant \frac{1}{2})$
(*c*) $P(\frac{1}{4} < X < \frac{1}{3} \mid 0 < Y \leqslant \frac{1}{2})$

Solution

(*a*) We want to find $P(X \leqslant u \mid 0 < Y \leqslant \frac{1}{2})$ for any real number u. This is given by

$$\begin{aligned} P(X \leqslant u \mid 0 < Y \leqslant \tfrac{1}{2}) &= \frac{P(X \leqslant u,\ 0 < Y \leqslant \frac{1}{2})}{P(0 < Y \leqslant \frac{1}{2})} \\ &= \tfrac{48}{19} P(X \leqslant u,\ 0 < Y \leqslant \tfrac{1}{2}) \end{aligned}$$

since $P(0 < Y \leqslant \frac{1}{2}) = \frac{19}{48}$, from Example 2.6 in Chapter 8.

There are four cases that need to be considered, depending upon the location of u on the real line. These are $u < 0$, $0 \leqslant u < \frac{1}{2}$, $\frac{1}{2} \leqslant u < 1$, and $u \geqslant 1$. The last three cases are shown in Figures 1.1(a), (b), and (c). It can be seen from these figures that

$$P(X \leqslant u \mid 0 < Y \leqslant \tfrac{1}{2}) = \begin{cases} 0, & u < 0 \\ \frac{48}{19} \int_0^u \int_y^u 10x^2y \, dx \, dy, & 0 \leqslant u < \frac{1}{2} \\ \frac{48}{19} \int_0^{1/2} \int_y^u 10x^2y \, dx \, dy, & \frac{1}{2} \leqslant u < 1 \\ \frac{48}{19} \int_0^{1/2} \int_y^1 10x^2y \, dx \, dy, & u \geqslant 1 \end{cases}$$

Therefore, the conditional distribution function of X given $0 < Y \leqslant \frac{1}{2}$ is obtained as

$$P(X \leqslant u \mid 0 < Y \leqslant \tfrac{1}{2}) = \begin{cases} 0, & u < 0 \\ \frac{48}{19} u^5, & 0 \leqslant u < \frac{1}{2} \\ \frac{1}{19}(20u^3 - 1), & \frac{1}{2} \leqslant u < 1 \\ 1, & u \geqslant 1 \end{cases}$$

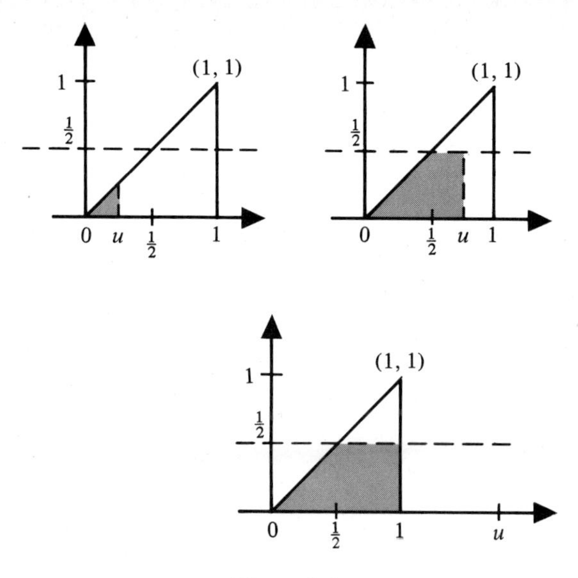

Figure 1.1

(*b*) To find $P(\frac{1}{3} < X \leqslant \frac{2}{3} \mid 0 < Y \leqslant \frac{1}{2})$, let us write $P(X \leqslant u \mid 0 < Y \leqslant \frac{1}{2}) = F(u \mid 0 < Y \leqslant \frac{1}{2})$. Then, recalling that $P(a < Z \leqslant b) = F_Z(b) - F_Z(a)$, we get

$$\begin{aligned} P(\tfrac{1}{3} < X \leqslant \tfrac{2}{3} \mid 0 < Y \leqslant \tfrac{1}{2}) &= F(\tfrac{2}{3} \mid 0 < Y \leqslant \tfrac{1}{2}) - F(\tfrac{1}{3} \mid 0 < Y \leqslant \tfrac{1}{2}) \\ &= \frac{1}{19}\left[20\left(\frac{2}{3}\right)^3 - 1\right] - \frac{48}{19}\left(\frac{1}{3}\right)^5 \\ &= \frac{383}{1539} \end{aligned}$$

(*c*) As in (*b*), we have

$$\begin{aligned} P(\tfrac{1}{4} < X < \tfrac{1}{3} \mid 0 < Y \leqslant \tfrac{1}{2}) &= F(\tfrac{1}{3} \mid 0 < Y \leqslant \tfrac{1}{2}) - F(\tfrac{1}{4} \mid 0 < Y \leqslant \tfrac{1}{2}) \\ &= \frac{48}{19}\left[\left(\frac{1}{3}\right)^5 - \left(\frac{1}{4}\right)^5\right] \end{aligned}$$

1.2 Conditional Distribution Given a Specific Value

Lest the reader be misled by the heading, our goal is to find, specifically, *the conditional distribution of one random variable given that the other random variable has assumed a specific value.* First we will consider the discrete case, then the absolutely continuous case.

The discrete case

Suppose X and Y have a joint discrete distribution where the possible values of X are $x_1, x_2, \ldots$ and those of Y are $y_1, y_2, \ldots$. Then, by directly applying the definition of conditional probability, we can find the probability of the event that X assumes a value x_i given that $Y = y_j$ as

$$P(X = x_i \mid Y = y_j) = \frac{P(X = x_i, Y = y_j)}{P(Y = y_j)} = \frac{p(x_i, y_j)}{p_Y(y_j)}$$

Writing

$$P(X = x_i \mid Y = y_j) = p_{X|Y}(x_i \mid y_j)$$

we get a function $p_{X|Y}(\cdot \mid y_j)$, whose domain consists of all the possible values of X. This function is called the **conditional probability function of X given $Y = y_j$**, and is defined by

$$p_{X|Y}(x_i \mid y_j) = \frac{p(x_i, y_j)}{p_Y(y_j)}, \qquad i = 1, 2, \ldots$$

It is evident that $p_{X|Y}(\cdot \mid y_j)$ indeed represents a probability function, since

$$p_{X|Y}(x_i \mid y_j) = \frac{p(x_i, y_j)}{p_Y(y_j)} \geqslant 0, \qquad \text{for all } x_i$$

and

$$\begin{aligned}\sum_{i=1}^{\infty} p_{X|Y}(x_i \mid y_j) &= \sum_{i=1}^{\infty} \frac{p(x_i, y_j)}{p_Y(y_j)} \\ &= \frac{1}{p_Y(y_j)} \sum_{i=1}^{\infty} p(x_i, y_j) \\ &= \frac{p_Y(y_j)}{p_Y(y_j)} = 1\end{aligned}$$

The **conditional probability function of Y given $X = x_i$** is defined similarly by

$$p_{Y|X}(y_j \mid x_i) = \frac{p(x_i, y_j)}{p_X(x_i)}, \qquad j = 1, 2, \ldots$$

Example 1.2. Suppose X and Y have the joint probability function given by

$$P(X = x,\ Y = y) = \tfrac{1}{42}(x + y^2), \qquad x = 1, 4, \quad y = -1, 0, 1, 3$$

Find:

(a) the conditional distribution of X given $Y = y$ $(y = -1, 0, 1, 3)$
(b) the conditional distribution of Y given $X = x$ $(x = 1, 4)$

Solution. In Example 2.3 of Chapter 8, we found that

$$P(X = x) = \frac{4x + 11}{42}, \qquad x = 1, 4$$

and

$$P(Y = y) = \frac{5 + 2y^2}{42}, \qquad y = -1, 0, 1, 3$$

(a) Let $y \in \{-1, 0, 1, 3\}$. Then

$$\begin{aligned}P(X = x \mid Y = y) &= \frac{P(X = x,\ Y = y)}{P(Y = y)} \\ &= \frac{\frac{1}{42}(x + y^2)}{\frac{1}{42}(5 + 2y^2)} = \frac{x + y^2}{5 + 2y^2}, \qquad x = 1, 4\end{aligned}$$

(*b*) Let $x \in \{1, 4\}$. Then

$$P(Y = y \mid X = x) = \frac{P(X = x,\ Y = y)}{P(X = x)}$$

$$= \frac{\frac{1}{42}(x + y^2)}{\frac{1}{42}(4x + 11)} = \frac{x + y^2}{4x + 11}, \qquad y = -1, 0, 1, 3$$

Example 1.3. If the joint probability function of X and Y is given by

$$P(X = i,\ Y = j) = \frac{1}{2^{i-1} 3^j}, \qquad \begin{array}{l} i = 1, 2, \ldots \\ j = 1, 2, \ldots \end{array}$$

find the conditional distribution of X given $Y = j$.

Solution. Given $Y = j$, X can assume any value $i = 1, 2, \ldots$. Now,

$$P(Y = j) = \sum_{i=1}^{\infty} \frac{1}{2^{i-1} 3^j} = \frac{2}{3^j}, \qquad j = 1, 2, \ldots$$

Therefore,

$$P(X = i \mid Y = j) = \frac{\frac{1}{2^{i-1} 3^j}}{\frac{2}{3^j}}, \qquad i = 1, 2, \ldots$$

That is,

$$P(X = i \mid Y = j) = \frac{1}{2^i}, \qquad i = 1, 2, \ldots$$

Incidentally, observe that

$$P(X = i) = \sum_{j=1}^{\infty} \frac{1}{2^{i-1} 3^j} = \frac{1}{2^i}, \qquad i = 1, 2, \ldots$$

Comparing the conditional distribution of X given $Y = j$ with the marginal distribution of X, we find that they are the same. We shall comment on this later during our discussion of independent random variables.

• *Example 1.4. (The trinomial distribution)* Suppose X and Y have the trinomial distribution with n trials and parameters p, q, namely,

$$P(X = i,\ Y = j) = \frac{n!}{i!\,j!\,(n - i - j)!} p^i q^j (1 - p - q)^{n-i-j}$$

where $i = 0, 1, \ldots, n$; $j = 0, 1, \ldots, n$; and $i + j \leqslant n$. Find the conditional distribution of X given $Y = j$, $0 \leqslant j \leqslant n$.

Solution. Given $Y = j$, X can assume values $0, 1, 2, \ldots, n - j$. Therefore, the conditional distribution of X given $Y = j$ is

$$P(X = i \mid Y = j) = \frac{P(X = i,\ Y = j)}{P(Y = j)}, \quad i = 0, 1, \ldots, n - j$$

Now, we have seen (in Example 2.4 of Chapter 8) that the distribution of Y is binomial with n trials and the probability of success q. Therefore,

$$P(X = i \mid Y = j) = \frac{\dfrac{n!}{i!j!(n-i-j)!} p^i q^j (1-p-q)^{n-i-j}}{\dfrac{n!}{j!(n-j)!} q^j (1-q)^{n-j}}$$

$$= \frac{(n-j)!}{i!(n-i-j)!} \cdot \frac{p^i (1-p-q)^{n-i-j}}{(1-q)^{n-j}}$$

$$= \binom{n-j}{i} \frac{p^i}{(1-q)^i} \, \frac{(1-p-q)^{n-i-j}}{(1-q)^{n-i-j}}$$

$$= \binom{n-j}{i} \left(\frac{p}{1-q}\right)^i \left(1 - \frac{p}{1-q}\right)^{n-j-i}$$

(Notice that $0 \leqslant p/(1-q)$; also since $p + q \leqslant 1$, we have $p \leqslant 1 - q$, and consequently $p/(1-q) \leqslant 1$.) Hence we conclude that *if X and Y have the trinomial distribution with n trials and parameters p, q, then the conditional distribution of X given $Y = j$ is binomial with $n - j$ trials and probability of success equal to $p/(1-q)$.*

The absolutely continuous case

As we know, the definition of conditional probability $P(A \mid B)$ requires that $P(B) > 0$. However, when we consider absolutely continuous distributions, the probability that a random variable assumes a particular value is zero. Thus, for example, since $P(Y = y) = 0$ for any real number y, $P(X \leqslant x \mid Y = y)$ does not exist according to the definition of conditional probability. This is so because we would get

$$P(X \leqslant x \mid Y = y) = \frac{P(X \leqslant x,\ Y = y)}{P(Y = y)} = \frac{0}{0}$$

All the same, it makes good sense to pose a question like "Given that the height of a person is 68 inches, what is the probability that he weighs between 150 lbs and 180 lbs?" It should be borne in mind that, in the discrete case, when we say $P(Y = y) = 0$, we mean that y is *not* one of the possible values of Y. Whereas, in the absolutely continuous case, even if $P(Y = y) = 0$, y might still be a possible value of Y.

A rigorous treatment of the conditional distribution in the absolutely continuous case requires rather sophisticated tools in mathematics and is beyond the scope of this book. The discussion presented below is rather heuristic. We assign meaning to the probability that $X \leqslant x$, given Y assumes a value y, through the following definition:

Let us assume that $f_Y(y) > 0$. The **conditional distribution function of X given $Y = y$** is then defined as

$$F_{X|Y}(x \mid Y = y) = \lim_{h \to 0^+} P(X \leqslant x \mid y \leqslant Y \leqslant y + h)$$

provided the limit exists.

As a result of this definition, we get

$$\begin{aligned} F_{X|Y}(x \mid Y = y) &= \lim_{h \to 0^+} \frac{P(X \leqslant x,\ y \leqslant Y \leqslant y + h)}{P(y \leqslant Y \leqslant y + h)} \\ &= \lim_{h \to 0^+} \frac{F(x, y + h) - F(x, y)}{F_Y(y + h) - F_Y(y)} \end{aligned}$$

(Recall that $P(X \leqslant b,\ c < Y \leqslant d) = F(b, d) - F(b, c)$.) Therefore,

$$\begin{aligned} F_{X|Y}(x \mid Y = y) &= \frac{\lim\limits_{h \to 0^+} \dfrac{F(x, y + h) - F(x, y)}{h}}{\lim\limits_{h \to 0^+} \dfrac{F_Y(y + h) - F_Y(y)}{h}} \\ &= \frac{\dfrac{\partial}{\partial y} F(x, y)}{\dfrac{d}{dy} F_Y(y)} = \frac{\dfrac{\partial}{\partial y} \int_{-\infty}^{y} \int_{-\infty}^{x} f(u, v)\, du\, dv}{f_Y(y)} \\ &= \frac{\int_{-\infty}^{x} f(u, y)\, du}{f_Y(y)} = \int_{-\infty}^{x} \frac{f(u, y)}{f_Y(y)}\, du \end{aligned}$$

In other words, this implies that the conditional distribution function of X given $Y = y$ is obtained by integrating the nonnegative function $f(\cdot, y)/f_Y(y)$; that is,

$$F_{X|Y}(x \mid Y = y) = \int_{-\infty}^{x} \frac{f(u, y)}{f_Y(y)}\, du$$

We therefore conclude that—

(*i*) the conditional distribution of X given $Y = y$ is absolutely continuous, and

(*ii*) the **conditional probability density function of X given $Y = y$**, denoted $f_{X|Y}(\cdot \mid y)$, is given by

$$f_{X|Y}(x \mid y) = \frac{f(x, y)}{f_Y(y)}, \qquad -\infty < x < \infty$$

provided $f_Y(y) > 0$.

We remark that the function $f_{X|Y}(\cdot \mid y)$ as defined above is indeed a probability density function, since

$$f_{X|Y}(x \mid y) = \frac{f(x, y)}{f_Y(y)} \geqslant 0 \qquad \text{for every } x$$

and

$$\int_{-\infty}^{\infty} f_{X|Y}(x \mid y)\, dx = \int_{-\infty}^{\infty} \frac{f(x, y)}{f_Y(y)}\, dx$$
$$= \frac{1}{f_Y(y)} \int_{-\infty}^{\infty} f(x, y)\, dx = \frac{f_Y(y)}{f_Y(y)} = 1$$

The **conditional probability density function of Y given $X = x$** is defined similarly as

$$f_{Y|X}(y \mid x) = \frac{f(x, y)}{f_X(x)}, \qquad -\infty < y < \infty$$

provided $f_X(x) > 0$.

Conditional probability density functions can be used to define conditional probabilities, and results of the following nature are immediate.

$$P(a < X \leqslant b \mid Y = y_0) = \int_a^b f_{X|Y}(x \mid y_0)\, dx = \frac{\int_a^b f(x, y_0)\, dx}{f_Y(y_0)}$$

$$P(c < Y \leqslant d \mid X = x_0) = \int_c^d f_{Y|X}(y \mid x_0)\, dy = \frac{\int_c^d f(x_0, y)\, dy}{f_X(x_0)}$$

We remind the reader once again that $P(a < X \leqslant b \mid Y = y_0)$ in the sense of the definition of conditional probability as $P(a < X \leqslant b,\ Y = y_0)/P(Y = y_0)$ is meaningless. We provide meaning to $P(a < X \leqslant b \mid Y = y_0)$ only through the limit $\lim_{h \to 0^+} P(a < X \leqslant b \mid y_0 \leqslant Y \leqslant y_0 + h)$.

Example 1.5. Suppose X and Y have a joint pdf given by

$$f(x, y) = \begin{cases} 2, & 0 < y < x < 1 \\ 0, & \text{elsewhere} \end{cases}$$

Find:

(*a*) the conditional distribution of X given $Y = y_0$, where $0 < y_0 < 1$
(*b*) the conditional distribution of Y given $X = x_0$, where $0 < x_0 < 1$
(*c*) $P(\frac{1}{3} < Y \leqslant \frac{1}{2} \mid X = \frac{2}{3})$

Solution. First we need the marginal distributions of X and Y since they are essential for finding the conditional pdf's.

It can be easily verified that

$$f_X(x) = \begin{cases} 2x, & 0 < x < 1 \\ 0, & \text{elsewhere} \end{cases}$$

and

$$f_Y(y) = \begin{cases} 2(1-y), & 0 < y < 1 \\ 0, & \text{elsewhere} \end{cases}$$

(*a*) $f_{X|Y}(x \mid y_o)$ is defined if $f_Y(y_o) > 0$, that is, if $0 < y_o < 1$. Now,

$$f(x, y_o) = \begin{cases} 2, & y_o < x < 1 \\ 0, & x \leqslant y_o \text{ or } x \geqslant 1 \end{cases}$$

Therefore,

$$f_{X|Y}(x \mid y_o) = \frac{f(x, y_o)}{f_Y(y_o)} = \begin{cases} \dfrac{2}{2(1-y_o)}, & y_o < x < 1 \\ 0, & \text{elsewhere} \end{cases}$$

(*b*) To find the conditional distribution of Y given $X = x_o$, we must have $f_X(x_o) > 0$. Therefore, suppose $0 < x_o < 1$, so that $f_X(x_o) = 2x_o$. Also notice that

$$f(x_o, y) = \begin{cases} 2, & 0 < y < x_o \\ 0, & y \leqslant 0 \text{ or } y \geqslant x_o \end{cases}$$

Therefore,

$$f_{Y|X}(y \mid x_o) = \begin{cases} \dfrac{2}{2x_o}, & 0 < y < x_o \\ 0, & \text{elsewhere} \end{cases}$$

From (*a*) and (*b*) it follows that the conditional distributions are uniform. The conditional distribution of X given $Y = y_o$ is uniform over the interval $(y_o, 1)$, and that of Y given $X = x_o$ is uniform over $(0, x_o)$.

(*c*) For finding $P(\frac{1}{3} < Y \leqslant \frac{1}{2} \mid X = \frac{2}{3})$, we evaluate $f_{Y|X}(y \mid x_o)$ at $x_o = \frac{2}{3}$. We have

$$f_{Y|X}(y \mid \tfrac{2}{3}) = \begin{cases} \frac{3}{2}, & 0 < y < \frac{2}{3} \\ 0, & \text{elsewhere} \end{cases}$$

Therefore,

$$P(\tfrac{1}{3} < Y \leqslant \tfrac{1}{2} \mid X = \tfrac{2}{3}) = \int_{1/3}^{1/2} \tfrac{3}{2}\, dy = \tfrac{1}{4}$$

Example 1.6. If the joint pdf of X and Y is given by

$$f(x, y) = \begin{cases} e^{-y}, & 0 < x < y < \infty \\ 0, & \text{elsewhere} \end{cases}$$

find–

(*a*) the conditional pdf of Y given $X = x_o$

(*b*) the conditional distribution function of Y given $X = x_o$

(*c*) $P(3 < Y \leqslant 4 \mid X = 2)$.

Solution

(*a*) It can be shown that the marginal pdf of X is

$$f_X(x) = \begin{cases} 0, & x < 0 \\ e^{-x}, & x \geqslant 0 \end{cases}$$

Since $f_{Y|X}(y \mid x_o)$ is defined only when $f_X(x_o) > 0$, we can consider the conditional pdf only if $x_o > 0$. Now,

$$f(x_o, y) = \begin{cases} 0, & y \leqslant x_o \\ e^{-y}, & y > x_o \end{cases}$$

Therefore,

$$f_{Y|X}(y \mid x_o) = \frac{f(x_o, y)}{f_X(x_o)} = \begin{cases} 0, & y \leqslant x_o \\ \dfrac{e^{-y}}{e^{-x_o}}, & y > x_o \end{cases}$$

$$= \begin{cases} 0, & y \leqslant x_o \\ e^{-(y - x_o)}, & y > x_o \end{cases}$$

(*b*) The conditional distribution function of Y given $X = x_o$ is

$$P(Y \leqslant u \mid X = x_o) = \int_{-\infty}^{u} f_{Y|X}(y \mid x_o)\, dy$$

$$= \begin{cases} 0, & u < x_o \\ \int_{x_o}^{u} e^{-(y - x_o)}\, dy, & u \geqslant x_o \end{cases}$$

$$= \begin{cases} 0, & u < x_o \\ 1 - e^{-(u - x_o)}, & u \geqslant x_o \end{cases}$$

(*c*) Since we have already obtained the conditional distribution function in part (*b*), we can find $P(3 < Y \leqslant 4 \mid X = 2)$ as follows:

Writing $F_{Y|X}(u \mid x_o)$ for $P(Y \leqslant u \mid X = x_o)$,

$$\begin{aligned} P(3 < Y \leqslant 4 \mid X = 2) &= F_{Y|X}(4 \mid 2) - F_{Y|X}(3 \mid 2) \qquad \text{(why?)} \\ &= (1 - e^{-(4-2)}) - (1 - e^{-(3-2)}) \\ &= e^{-1} - e^{-2} \end{aligned}$$

Example 1.7. Suppose the joint distribution function of X and Y is given by

$$F(x, y) = \begin{cases} 0, & x < 0 \ \text{ or } \ y < 0 \\ xy(x + y - xy^2), & 0 \leqslant x < 1 \ \text{ and } \ 0 \leqslant y < 1 \\ y(1 + y - y^2), & x \geqslant 1 \ \text{ and } \ 0 \leqslant y < 1 \\ x, & 0 \leqslant x < 1 \ \text{ and } \ y \geqslant 1 \\ 1, & x \geqslant 1 \ \text{ and } \ y \geqslant 1 \end{cases}$$

Find:

(*a*) the conditional pdf of Y given $X = x_o$

(*b*) the conditional pdf of X given $Y = y_o$

(*c*) $P(X > \frac{1}{2} \mid Y = \frac{1}{3})$

Solution. Right away, we get the marginal distribution functions of X and Y as

$$F_X(x) = \lim_{y\to\infty} F(x, y) = \begin{cases} 0, & x < 0 \\ x, & 0 \leqslant x < 1 \\ 1, & x \geqslant 1 \end{cases}$$

and

$$F_Y(y) = \lim_{x\to\infty} F(x, y) = \begin{cases} 0, & y < 0 \\ y(1 + y - y^2), & 0 \leqslant y < 1 \\ 1, & y \geqslant 1 \end{cases}$$

In order to find the conditional pdf's, we require the joint pdf and the marginal pdf's. These are obtained as follows:

$$f(x, y) = \frac{\partial^2}{\partial x\,\partial y} F(x, y) = \begin{cases} 2(x + y - 3xy^2), & 0 < x < 1,\ 0 < y < 1 \\ 0, & \text{elsewhere} \end{cases}$$

$$f_X(x) = \frac{d}{dx} F_X(x) = \begin{cases} 1, & 0 < x < 1 \\ 0, & \text{elsewhere} \end{cases}$$

$$f_Y(y) = \frac{d}{dy} F_Y(y) = \begin{cases} 1 + 2y - 3y^2, & 0 < y < 1 \\ 0, & \text{elsewhere} \end{cases}$$

(a) The conditional pdf of Y given $X = x_o$ is defined provided $0 < x_o < 1$. In that case,

$$f_{Y|X}(y \mid x_o) = \frac{f(x_o, y)}{f_X(x_o)} = \begin{cases} 2(x_o + y - 3x_o y^2), & 0 < y < 1 \\ 0, & \text{elsewhere} \end{cases}$$

(b) The conditional pdf $f_{X|Y}(x \mid y_o)$ is defined only if $0 < y_o < 1$, and we have

$$f_{X|Y}(x \mid y_o) = \frac{f(x, y_o)}{f_Y(y_o)} = \begin{cases} \dfrac{2(x + y_o - 3xy_o^2)}{1 + 2y_o - 3y_o^2}, & 0 < x < 1 \\ 0, & \text{elsewhere} \end{cases}$$

(c) We shall first find $P(X > \frac{1}{2} \mid Y = y_o)$ for any $0 < y_o < 1$.

$$\begin{aligned} P(X > \tfrac{1}{2} \mid Y = y_o) &= 1 - P(X \leqslant \tfrac{1}{2} \mid Y = y_o) \\ &= 1 - \int_{-\infty}^{1/2} f_{X|Y}(x \mid y_o)\, dx \\ &= 1 - \int_0^{1/2} \frac{2(x + y_o - 3xy_o^2)}{1 + 2y_o - 3y_o^2}\, dx \\ &= 1 - \frac{1 + 4y_o - 3y_o^2}{4(1 + 2y_o - 3y_o^2)} \end{aligned}$$

Hence, substituting $y_o = \frac{1}{3}$,

$$P(X > \tfrac{1}{2} \mid Y = \tfrac{1}{3}) = \tfrac{5}{8}$$

■

If the joint pdf of X and Y is given as $f(x, y) = K\,u(x, y)$, where K is a constant, there is actually no need to compute the constant for finding the conditional pdf's. This follows because, for instance,

$$f_{X|Y}(x \mid y) = \frac{f(x, y)}{\int_{-\infty}^{\infty} f(x, y)\,dx} = \frac{K\,u(x, y)}{K\int_{-\infty}^{\infty} u(x, y)\,dx} = \frac{u(x, y)}{\int_{-\infty}^{\infty} u(x, y)\,dx}$$

where K cancels out.

Example 1.8. The joint distribution of X and Y is given by the following joint pdf, where K is a constant.

$$f(x, y) = \begin{cases} K(x^2 + y^2), & 0 \leqslant x^2 + y^2 \leqslant 1 \\ 0, & \text{elsewhere} \end{cases}$$

Find:

(*a*) the conditional pdf of X given $Y = y_0$

(*b*) $P(|X| > \frac{1}{4} \mid Y = \sqrt{\frac{3}{4}})$

Solution

(*a*) First we need the marginal pdf of Y which we shall obtain in terms of K.

$$f_Y(y) = \int_{-\infty}^{\infty} f(x, y)\,dx = \begin{cases} K\int_{-(1-y^2)^{1/2}}^{(1-y^2)^{1/2}} (x^2 + y^2)\,dx, & -1 < y < 1 \\ 0, & \text{elsewhere} \end{cases}$$

$$= \begin{cases} K\left[\frac{2(1-y^2)^{3/2}}{3} + 2y^2\sqrt{1-y^2}\right], & -1 < y < 1 \\ 0, & \text{elsewhere} \end{cases}$$

$$= \begin{cases} K \cdot \frac{2}{3}\sqrt{1-y^2}\,(1 + 2y^2), & -1 < y < 1 \\ 0, & \text{elsewhere} \end{cases}$$

Next, we note that

$$f(x, y_0) = \begin{cases} K(x^2 + y_0^2), & -\sqrt{1-y_0^2} < x < \sqrt{1-y_0^2} \\ 0, & \text{elsewhere} \end{cases}$$

Therefore, for $-1 < y_0 < 1$,

$$f_{X|Y}(x \mid y_0) = \begin{cases} \dfrac{K(x^2 + y_0^2)}{K \cdot \frac{2}{3}\sqrt{1-y_0^2}\,(1 + 2y_0^2)}, & -\sqrt{1-y_0^2} < x < \sqrt{1-y_0^2} \\ 0, & \text{elsewhere} \end{cases}$$

$$= \begin{cases} \dfrac{3(x^2 + y_0^2)}{2\sqrt{1-y_0^2}\,(1 + 2y_0^2)}, & -\sqrt{1-y_0^2} < x < \sqrt{1-y_0^2} \\ 0, & \text{elsewhere} \end{cases}$$

(*b*) Substituting $y_0 = \sqrt{\frac{3}{4}}$ the conditional pdf of X given $Y = \sqrt{\frac{3}{4}}$ is

$$f_{X|Y}(x \mid \sqrt{\tfrac{3}{4}}) = \begin{cases} \frac{6}{5}(x^2 + \frac{3}{4}), & -\frac{1}{2} < x < \frac{1}{2} \\ 0, & \text{elsewhere} \end{cases}$$

Therefore,

$$P(|X| > \tfrac{1}{4} \mid Y = \sqrt{\tfrac{3}{4}}) = 1 - \int_{-1/4}^{1/4} \tfrac{6}{5}(x^2 + \tfrac{3}{4})\, dx = \tfrac{43}{80}$$

• *Example 1.9. (The standard bivariate normal distribution)* Let X and Y have the joint pdf

$$f(x, y) = \frac{1}{2\pi\sqrt{1-\rho^2}} e^{-[1/2(1-\rho^2)](x^2 - 2\rho xy + y^2)}, \quad \begin{array}{l} -\infty < x < \infty \\ -\infty < y < \infty \end{array}$$

where $-1 < \rho < 1$. Find the conditional pdf of Y given $X = x$.

Solution. We have seen in Example 2.8 of Chapter 8 that X has a standard normal distribution. Hence,

$$f_{Y|X}(y \mid x) = \frac{f(x, y)}{f_X(x)} = \frac{\dfrac{1}{2\pi\sqrt{1-\rho^2}} e^{-[2(1-\rho^2)]^{-1}(x^2 - 2\rho xy + y^2)}}{\dfrac{1}{\sqrt{2\pi}} e^{-x^2/2}}$$

$$= \frac{1}{\sqrt{2\pi(1-\rho^2)}} \cdot e^{-[2(1-\rho^2)]^{-1}(x^2 - 2\rho xy + y^2) + x^2/2}$$

$$= \frac{1}{\sqrt{2\pi(1-\rho^2)}} \cdot e^{-[2(1-\rho^2)]^{-1}[x^2 - 2\rho xy + y^2 - (1-\rho^2)x^2]}$$

$$= \frac{1}{\sqrt{2\pi(1-\rho^2)}} \cdot e^{-[2(1-\rho^2)]^{-1}(\rho^2 x^2 - 2\rho xy + y^2)}$$

$$= \frac{1}{\sqrt{2\pi(1-\rho^2)}} \cdot e^{-[2(1-\rho^2)]^{-1}(y - \rho x)^2} \quad -\infty < y < \infty$$

Therefore, from the functional form of the conditional pdf, we recognize that *the conditional distribution of Y, given $X = x$, is normal with mean ρx and variance $1 - \rho^2$. Similarly, the conditional distribution of X, given $Y = y$, is normal with mean ρy and variance $1 - \rho^2$.* ■

It follows immediately from the definition of conditional probability density function that

$$f(x, y) = f_{X|Y}(x \mid y) \cdot f_Y(y) = f_{Y|X}(y \mid x) \cdot f_X(x), \quad \begin{array}{l} -\infty < x < \infty \\ -\infty < y < \infty \end{array}$$

Example 1.10. A nonnegative number X is picked with the probability law given by the following pdf:

$$f_X(x) = \begin{cases} xe^{-x}, & x > 0 \\ 0, & \text{elsewhere} \end{cases}$$

If $X = x$, a number Y is picked with the uniform distribution over the interval $[0, x]$. Find:

(*a*) the joint pdf of X and Y

(*b*) the probability that the sum of the two numbers will be less than c, where c is a positive constant; that is, $P(X + Y \leq c)$

Solution

(*a*) We are given that

$$f_X(x) = \begin{cases} xe^{-x}, & x > 0 \\ 0, & \text{elsewhere} \end{cases}$$

and

$$f_{Y|X}(y \mid x) = \begin{cases} \dfrac{1}{x}, & 0 < y < x \\ 0, & \text{elsewhere} \end{cases}$$

Therefore,

$$f(x, y) = f_{Y|X}(y \mid x) \cdot f_X(x) = \begin{cases} e^{-x}, & 0 < y < x < \infty \\ 0, & \text{elsewhere} \end{cases}$$

(*b*) Recalling that for any Borel set B, $P((X, Y) \in B) = \iint_B f(x, y)\, dx\, dy$

$$P(X + Y \leqslant c) = \iint_{\{(x, y) \mid x + y \leqslant c\}} f(x, y)\, dx\, dy$$

That is, the desired probability is found by integrating the joint pdf over the region below the line $x + y = c$. However, below this line the joint pdf is zero outside the shaded triangle PQR. (See Figure 1.2.)

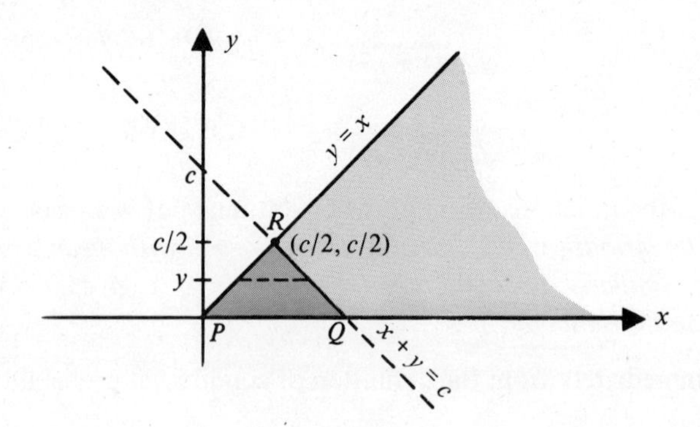

Figure 1.2

Therefore,

$$P(X + Y \leqslant c) = \int_0^{c/2} \int_y^{c-y} e^{-x}\, dx\, dy$$

$$= \int_0^{c/2} (1 - e^{-(c-y)} + e^{-y})\, dy$$

Upon simplifying, this gives

$$P(X+Y \leqslant c) = (1-e^{-c/2})^2$$

EXERCISES–SECTION 1

1. The joint probability function of X and Y is given by

$$p(x, y) = \tfrac{1}{66}(x^2 + 2y), \qquad x = -1, 0, 2, 3, \quad y = 0, 1, 2$$

Show that the conditional probability function of X given $Y = y$ $(y = 0, 1, 2)$ is given by

$$P(X = x \mid Y = y) = \frac{x^2 + 2y}{14 + 8y}, \qquad x = -1, 0, 2, 3$$

2. Suppose X represents the number appearing when a fair die is rolled. If $X = n$, a fair coin is tossed n times. Find the distribution of Y, where Y denotes the number of heads obtained.
3. Suppose two chips are picked one by one without replacement from a box containing n chips numbered $1, 2, \ldots, n$. Let X denote the value on the first chip picked, and Y the value on the second chip. Find:
   (a) the conditional distribution of X given $Y = j$, $j = 1, 2, \ldots, n$
   (b) the conditional distribution of Y given $X = i$, $i = 1, 2, \ldots, n$
   (c) $P(X \geqslant 5 \mid Y = 10)$ if $n = 20$
   (d) $P(4 < X \leqslant 8 \mid Y = 10)$ if $n = 20$
   (e) $P(8 < X < 13 \mid Y = 10)$ if $n = 20$
4. Consider the joint pdf's of X and Y given below. In each case, find the conditional probability densities $f_{X|Y}$ and $f_{Y|X}$.

(a) $$f(x, y) = \begin{cases} 4xy, & 0 < x < 1, \ 0 < y < 1 \\ 0, & \text{elsewhere} \end{cases}$$

(b) $$f(x, y) = \begin{cases} \tfrac{12}{7}(x^2 + xy), & 0 < x < 1, \ 0 < y < 1 \\ 0, & \text{elsewhere} \end{cases}$$

(c) $$f(x, y) = \begin{cases} 12y, & 0 < y < 2x < 1 \\ 0, & \text{elsewhere} \end{cases}$$

(d) $$f(x, y) = \begin{cases} xe^{-x(1+y)}, & x > 0, \ y > 0 \\ 0, & \text{elsewhere} \end{cases}$$

(e) $$f(x, y) = \begin{cases} 24y(1-x), & 0 \leqslant y \leqslant x \leqslant 1 \\ 0, & \text{elsewhere} \end{cases}$$

5. Suppose the joint pdf of X and Y is given by

$$f(x, y) = \begin{cases} \tfrac{2}{3}(x + y), & 1 \leqslant y \leqslant x \leqslant 2 \\ 0, & \text{elsewhere} \end{cases}$$

Compute the following:
   (a) $P(X > \tfrac{3}{2})$ (b) $P(4Y > 3X)$
   (c) $P(Y > \tfrac{4}{3} \mid X < \tfrac{3}{2})$

6. Suppose X and Y have a joint pdf given by

$$f(x, y) = \begin{cases} kxy^2, & 0 < y < x < 1 \\ 0, & \text{elsewhere} \end{cases}$$

Find:

(a) the conditional pdf of Y given $X = x_0$
(b) the conditional pdf of X given $Y = y_0$
(c) $P(\frac{1}{4} < Y < \frac{1}{3} \mid X = \frac{3}{4})$

7. Suppose the conditional distribution of Y given $X = x$ is given by

$$f_{Y|X}(y \mid x) = \begin{cases} \frac{6y}{x^3}(x - y), & 0 \leqslant y \leqslant x \\ 0, & \text{elsewhere} \end{cases}$$

and the marginal distribution of X is given by

$$f_X(x) = \begin{cases} 20x^3(1 - x), & 0 < x < 1 \\ 0, & \text{elsewhere} \end{cases}$$

Find:

(a) the joint pdf of X and Y
(b) $P(\frac{1}{4} < Y < \frac{1}{2} \mid X = \frac{3}{4})$
(c) $P(\frac{1}{4} < Y < \frac{1}{2} \mid X = \frac{1}{3})$
(d) $P(\frac{1}{4} < Y < \frac{1}{2})$

8. If the joint pdf of X and Y is given by

$$f(x, y) = \begin{cases} \frac{e^{-y}}{y}, & 0 < x < y < \infty \\ 0, & \text{elsewhere} \end{cases}$$

find:

(a) the conditional pdf of X given y_0, where $y_0 > 0$
(b) $P(2 < X < 5 \mid Y = 6)$
(c) $P(X > 5 \mid Y = 6)$

9. For the following joint pdf of X and Y given by

$$f(x, y) = \begin{cases} k|x|, & -1 < x < 1, \ -1 < y < 1 \\ 0, & \text{elsewhere} \end{cases}$$

find:

(a) $P(|X| > \frac{1}{2} \mid Y = -\frac{1}{2})$
(b) $P(|X| > \frac{1}{2} \mid Y \leqslant \frac{1}{2})$

10. The joint pdf of X and Y is given by

$$f(x, y) = \begin{cases} ke^x, & y < 0, \ x < 2y \\ 0, & \text{elsewhere} \end{cases}$$

(a) Determine the constant k.
(b) Find the marginal distributions of X and Y.
(c) Find $P(X > -3 \mid -4 < Y < -1)$.

11. A nonnegative number X is picked at random in the interval $[0, 1]$. If the number is x, then another number Y is picked at random in the interval $[0, x]$. Find the probability that the sum of the two numbers is less than $\frac{3}{4}$.

12. Suppose a point (X, Y) is picked at random inside a triangle bounded by the lines $y = 0$, $y = x$, and $x = 1$. Given that the x-coordinate of the point is $\frac{1}{2}$, find the probability that the y-coordinate will be between $\frac{1}{6}$ and $\frac{1}{3}$.

13. Suppose X and Y have a joint uniform distribution over a circle centered at the origin and of radius 1. Find the following probabilities:

(a) $P(-\frac{1}{4} < Y < \frac{1}{4} \mid X = \frac{1}{2}\sqrt{3})$

(b) $P(-\frac{1}{4} < Y < \frac{1}{4} \mid X > 0)$

14. The length X (in feet) of a rectangular region is a random variable with the pdf given by

$$f_X(x) = \begin{cases} \frac{1}{3}(3x^2 - 2x - 1), & 1 < x < 2 \\ 0, & \text{elsewhere} \end{cases}$$

Given that the length is x, the distribution of the breadth Y is given by the following pdf:

$$g(y) = \begin{cases} \dfrac{2(x + y)}{(3x^2 - 2x - 1)}, & 1 < y < x \\ 0, & \text{elsewhere} \end{cases}$$

(a) Find the distribution of the breadth.

(b) Given that the length is 1.5 feet, find the probability that the area is less than 1.8 square feet.

15. Prove the following result, analogous to Bayes' rule discussed in Chapter 3.

$$f_{X|Y}(x \mid y) = \frac{f_X(x) f_{Y|X}(y \mid x)}{\int_{-\infty}^{\infty} f_X(x) f_{Y|X}(y \mid x)\, dx}$$

2. INDEPENDENT RANDOM VARIABLES

The concept of independent random variables is very important in probability theory, as is the concept of independent events. As a matter of fact, the notion of independent random variables is simply an extension of the concept of independent events. There are several equivalent ways of defining the independence of two random variables X and Y (defined on the same probability space, of course), and in the main they convey one simple fact: *any* event associated with the random variable X and *any* event associated with Y are independent.

We begin with the following equivalent definitions:

Definition 1 Two random variables X and Y are said to be **stochastically independent**, or, simply, **independent**, if for any Borel sets B_1 and B_2 of the real line, $P(X \in B_1,\ Y \in B_2) = P(X \in B_1)P(Y \in B_2)$

Equivalently,

Definition 2 Two random variables X and Y are said to be independent if for *every* pair of real numbers x and y the two events $\{s \mid X(s) \leq x\}$ and $\{s \mid Y(s) \leq y\}$ are independent. In other words

$$P(X \leq x,\ Y \leq y) = P(X \leq x) \cdot P(Y \leq y)$$

for every x and y. This criterion, expressed in terms of the distribution functions, states that two random variables are independent if and only if $F(x, y) = F_X(x) \cdot F_Y(y)$ for every pair of real numbers x, y.

That definition 1 implies definition 2 is easy to verify: one has only to take $B_1 = (-\infty, x]$ and $B_2 = (-\infty, y]$. To prove that definition 2 implies definition 1 is much harder, and is omitted.

In Chapter 8 we saw that joint distributions uniquely determine the marginal distributions. Also, we showed through examples that the converse is false, in that it is not possible to determine the joint distribution from knowledge of the marginal distributions. We find now that an important exception to this is provided when the random variables are independent, and that, as a matter of fact, in this case the joint distribution is given by $F(x, y) = F_X(x) \cdot F_Y(y)$ for any real numbers x, y.

We give yet another definition of independence.

Definition 3 Two random variables X and Y are said to be independent if for every choice of real numbers a, b $(a < b)$ and c, d $(c < d)$ the pairs of events $\{a < X \leq b\}, \{c < Y \leq d\}$ are independent; that is,

$$P(a < X \leq b,\ c < Y \leq d) = P(a < X \leq b) P(c < Y \leq d)$$

It is left to the reader to show that definition 2 and definition 3 are equivalent.

Random variables which are not independent are said to be *dependent* random variables.

Example 2.1. The joint D.F. of X and Y is given by

$$F(x, y) = \begin{cases} 0, & x < 0 \quad \text{or} \quad y < 0 \\ 1 - e^{-x} - e^{-y} + e^{-(x+y)}, & x \geq 0 \quad \text{and} \quad y \geq 0 \end{cases}$$

Are X and Y independent?

Solution. The marginal D.F.'s of X and Y are given by

$$F_X(x) = \lim_{y \to \infty} F(x, y) = \begin{cases} 0, & x < 0 \\ 1 - e^{-x}, & x \geq 0 \end{cases}$$

and

$$F_Y(y) = \lim_{x \to \infty} F(x, y) = \begin{cases} 0, & y < 0 \\ 1 - e^{-y}, & y \geq 0 \end{cases}$$

Since $F(x, y) = F_X(x) \cdot F_Y(y)$ for every x, y, the random variables X and Y are independent.

Example 2.2. Suppose two random variables X and Y have the following joint distribution function:

$$F(x, y) = \begin{cases} 0, & x < 0 \quad \text{or} \quad y < 0 \\ 1 - 2e^{-y} + e^{-2y}, & x > y \quad \text{and} \quad y \geqslant 0 \\ 1 - e^{-2x} + 2e^{-(x+y)} - 2e^{-y}, & x \geqslant 0 \quad \text{and} \quad y \geqslant x \end{cases}$$

Show that X and Y are not independent.

Solution. In Example 2.1 of Chapter 8 the marginal distribution functions were found as

$$F_X(x) = \begin{cases} 0, & x < 0 \\ 1 - e^{-2x}, & x \geqslant 0 \end{cases}$$

and

$$F_Y(y) = \begin{cases} 0, & y < 0 \\ 1 - 2e^{-y} + e^{-2y}, & y \geqslant 0 \end{cases}$$

Now, if $x, y > 0$,

$$F_X(x) \cdot F_Y(y) = (1 - e^{-2x}) \cdot (1 - 2e^{-y} + e^{-2y}) \neq F(x, y)$$

Hence X and Y are not independent. ∎

To determine whether two random variables are independent or not, there is actually no need to find the marginal D.F.'s explicitly and then check whether $F(x, y) = F_X(x) \cdot F_Y(y)$. It suffices to check whether the joint D.F. can be factored as a product of two functions (not necessarily D.F.'s), one depending on x only, and the other on y only (see exercise 1). On the basis of this, we could decide immediately that the random variables in Example 2.2 are not independent, without finding the marginal D.F.'s.

Sometimes one can tell if two random variables are not independent pictorially simply by looking at the domain of the definition of the probability distribution. To see this, suppose X and Y have a joint pdf which is positive over the shaded region indicated in Figure 2.1, and zero outside it. (For convenience we are assuming that X and Y have a joint absolutely continuous distribution.)

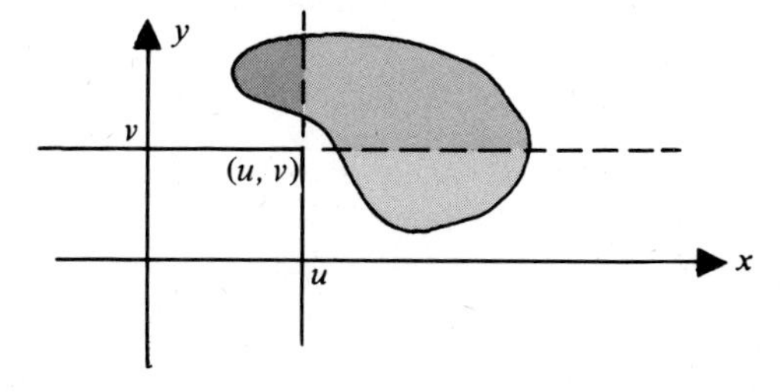

Figure 2.1

Obviously, for the values of u, v that we have picked in the figure, $F(u, v) = 0$.

Now $F_X(u)$ is equal to the volume over the shaded region to the left of the line $x = u$ (shaded ▇) and is therefore *positive*. Also, $F_Y(v)$ is the volume over the shaded region below the line $y = v$ (shaded ▇) and is, again, *positive*. Hence, $F_X(u) \cdot F_Y(v) > 0$ and, consequently, $F_X(u) \cdot F_Y(v)$ is not equal to $F(u, v)$ which, as mentioned, is equal to zero. As a result, X and Y are not independent.

Intuitively, it seems reasonable to expect that if X and Y are independent random variables, then any function of X and any function of Y must be independent also. For instance, X^2 and $\sin Y$ should be independent since, after all, the knowledge of X^2 is only via X and the knowledge of $\sin Y$ is only via Y. In fact, independent random variables do have this property, as the following result shows.

If X and Y are independent random variables, and if h and g are any piecewise-continuous functions, then $h(X)$ and $g(Y)$ are independent random variables.

Let A and B be any two Borel sets of the real line. To show that $h(X)$ and $g(Y)$ are independent, we must establish that $P(h(X) \in A, g(Y) \in B) = P(h(X) \in A) \cdot P(g(Y) \in B)$.

With this in mind, suppose

$$A^* = \{x \in \mathbf{R} \mid h(x) \in A\}$$

and

$$B^* = \{y \in \mathbf{R} \mid g(y) \in B\}$$

Thus, the sets A^* and B^* represent, respectively, the sets of preimages of A and B. It is customary to write $h^{-1}(A)$ for the set A^* and $g^{-1}(B)$ for the set B^*. Since h and g are piecewise continuous functions and A and B are Borel sets, it can be shown that A^* and B^* are also Borel sets. We shall proceed assuming this fact. (In Figure 2.2 we are representing the sets A, B, A^*, B^* as intervals only for convenience; actually these are any Borel sets of the real line.)

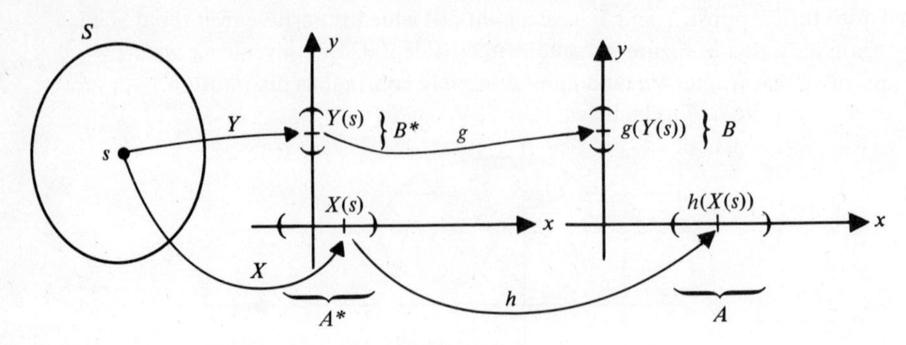

Figure 2.2

We make the following immediate observations: For any $s \in S$,

(*i*) $X(s) \in A^* \Leftrightarrow h(X)(s) \in A$

(Since $X(s)$ is a real number this follows from the fact that $X(s) \in A^* \Leftrightarrow h(X(s)) \in A$, and $h(X(s)) = h(X)(s)$.)

(*ii*) $Y(s) \in B^* \Leftrightarrow g(Y)(s) \in B$

(*iii*) $h(X)(s) \in A$ and $g(Y)(s) \in B \Leftrightarrow X(s) \in A^*$ and $Y(s) \in B^*$

From these observations it follows that

$$\begin{aligned} P(h(X) \in A,\ g(Y) \in B) &= P(X \in A^*,\ Y \in B^*), \qquad \text{from } (iii) \\ &= P(X \in A^*) \cdot P(Y \in B^*), \\ &\qquad \text{since } X \text{ and } Y \text{ are independent} \\ &= P(h(X) \in A) \cdot P(g(Y) \in B) \end{aligned}$$

since, from (*i*), $P(X \in A^*) = P(h(X) \in A)$, and, from (*ii*), $P(Y \in B^*) = P(g(Y) \in B)$.

Hence, by the definition of independence, $h(X)$ and $g(Y)$ are independent random variables.

Comment. We have taken the functions h and g as piecewise continuous. There is no need to be so restrictive. Actually, all that is necessary for the functions h and g is that A^* and B^* be Borel sets whenever A and B are Borel sets. Functions with this property are called ***Borel-measurable functions.*** However, we shall not fuss over this. As far as we are concerned, just about any functions h and g will do.

The three definitions of independence that have been given enable us to provide appropriate criteria in terms of the probability functions for discrete distributions, and in terms of the probability density functions for absolutely continuous distributions. We shall take these up separately.

The discrete case

The independence criterion for discrete distributions boils down to the criterion in terms of the probability functions provided by the following definition:

Definition 4(a) Two random variables X and Y which have a joint discrete distribution are said to be independent if their joint probability function can be written as the product of their individual probability functions; that is

$$P(X = x_i,\ Y = y_j) = P(X = x_i) \cdot P(Y = y_j)$$

for every x_i, y_j.

It is essential to show that this definition is equivalent to one of the three given earlier. We choose here to show that it is equivalent to definition 2. Before proving the equivalence, we shall establish the following result:

If X and Y are independent random variables, then

$$\boxed{F(a, b^-) = F_X(a) \cdot F_Y(b^-)}$$

This is true because

$$\begin{aligned} F(a, b^-) &= \lim_{n\to\infty} F\left(a, b - \frac{1}{n}\right) \\ &= \lim_{n\to\infty} F_X(a) \cdot F_Y\left(b - \frac{1}{n}\right), \qquad \text{since } X \text{ and } Y \text{ are independent} \\ &= F_X(a) \cdot \lim_{n\to\infty} F_Y\left(b - \frac{1}{n}\right) = F_X(a) \cdot F_Y(b^-) \end{aligned}$$

We shall now show that definitions 2 and 4(a) are equivalent.

(*i*) Assume X and Y are independent as per definition 2, that is, $F(x, y) = F_X(x) \cdot F_Y(y)$ for every x, y. Then

$$\begin{aligned} P(X = x_i, Y = y_j) &= F(x_i, y_j) - F(x_i^-, y_j) - F(x_i, y_j^-) + F(x_i^-, y_j^-) \\ &= F_X(x_i)F_Y(y_j) - F_X(x_i^-)F_Y(y_j) - F_X(x_i)F_Y(y_j^-) \\ &\quad + F_X(x_i^-)F_Y(y_j^-) \\ &= [F_X(x_i) - F_X(x_i^-)] \cdot [F_Y(y_j) - F_Y(y_j^-)] \end{aligned}$$

That is,

$$P(X = x_i, Y = y_j) = P(X = x_i) \cdot P(Y = y_j)$$

This implies independence according to definition 4(a).

(*ii*) Conversely, assume independence according to definition 4(a), that is, $P(X = x_i, Y = y_j) = P(X = x_i) \cdot P(Y = y_j)$. Then, by the definition of a joint distribution function,

$$\begin{aligned} F(x, y) &= \sum_{y_j \leqslant y} \sum_{x_i \leqslant x} P(X = x_i, Y = y_j) \\ &= \sum_{y_j \leqslant y} \sum_{x_i \leqslant x} P(X = x_i) \cdot P(Y = y_j), \qquad \text{by assumption} \\ &= \sum_{x_i \leqslant x} P(X = x_i) \sum_{y_j \leqslant y} P(Y = y_j) \\ &= F_X(x) \cdot F_Y(y) \end{aligned}$$

This gives independence according to definition 2.

Hence, from (*i*) and (*ii*), we have the equivalence of the two definitions.

Comment. If X and Y are independent random variables, then, for any x_i,

$$P(X = x_i \mid Y = y_j) = P(X = x_i)$$

because

$$P(X = x_i \mid Y = y_j) = \frac{p(x_i, y_j)}{P(Y = y_j)} = \frac{p_X(x_i)p_Y(y_j)}{p_Y(y_j)} = p_X(x_i)$$

That is, *the conditional distribution of X, given $Y = y_j$, is the same as the distribution of X and hence does not depend on Y.* It is this consequence that might explain the use of the term "independence."

Similarly, if X and Y are independent random variables,

$$P(Y = y_j \mid X = x_i) = P(Y = y_j) = p_Y(y_j)$$

for any y_j.

Example 2.3. If $P(X = i,\ Y = j) = \dfrac{1}{2^{i-1}3^j}$, $i, j = 1, 2, \ldots$, show that X and Y are independent.

Solution. In Example 1.3 we showed that

$$p_X(i) = \frac{1}{2^i}, \qquad i = 1, 2, \ldots$$

and

$$p_Y(j) = \frac{2}{3^j}, \qquad j = 1, 2, \ldots$$

Since $p(i, j) = p_X(i) \cdot p_Y(j)$ for every $i, j = 1, 2, \ldots$, the random variables X and Y are independent.

Example 2.4. Show that the random variables X and Y which have the trinomial distribution with n trials and parameters p, q are not independent.

Solution. We know from Example 2.4 of Chapter 8 that the marginal distribution of X is binomial with n trials and the probability of success p. Also we saw in Example 1.4 that the conditional distribution of X given $Y = j$ is binomial with $n - j$ trials and the probability of success $p/(1 - q)$. Since the conditional distribution of X given $Y = j$ is not the same as the marginal distribution of X, we can conclude that the random variables are not independent.

Example 2.5. Consider a sample space $S = \{s_1, s_2, s_3, s_4\}$ where all the outcomes are equally likely. Let X and Y be two random variables defined on S as shown below

s	$X(s)$	$Y(s)$
s_1	1	1
s_2	0	−1
s_3	−1	0
s_4	0	0

(*a*) Show that X and Y are not independent.
(*b*) What about X^2 and Y^2?

Solution

(*a*) Notice, for example, that

$$P(X = 1) = P(\{s_1\}) = \tfrac{1}{4}, \qquad P(Y = 1) = P(\{s_1\}) = \tfrac{1}{4}$$

and

$$P(X = 1,\ Y = 1) = P(\{s_1\}) = \tfrac{1}{4}$$

Hence,

$$P(X = 1,\ Y = 1) \neq P(X = 1) \cdot P(Y = 1)$$

and, as a result, X and Y are not independent.

(*b*) The distributions of X^2 and Y^2 are given below

u	$P(X^2 = u)$
0	$\frac{1}{2}$
1	$\frac{1}{2}$

v	$P(Y^2 = v)$
0	$\frac{1}{2}$
1	$\frac{1}{2}$

Now

$$\begin{aligned}
&P(X^2 = 0,\ Y^2 = 0) = P(\{s_4\}) = \tfrac{1}{4} = P(X^2 = 0) \cdot P(Y^2 = 0)\\
&P(X^2 = 0,\ Y^2 = 1) = P(\{s_2\}) = \tfrac{1}{4} = P(X^2 = 0) \cdot P(Y^2 = 1)\\
&P(X^2 = 1,\ Y^2 = 0) = P(\{s_3\}) = \tfrac{1}{4} = P(X^2 = 1) \cdot P(Y^2 = 0)\\
\text{and}\quad &P(X^2 = 1,\ Y^2 = 1) = P(\{s_1\}) = \tfrac{1}{4} = P(X^2 = 1) \cdot P(Y^2 = 1)
\end{aligned}$$

Consequently, X^2 and Y^2 are independent.

In conclusion, *even if two random variables are not independent, their functions still can be independent.*

Example 2.6. The following table gives the joint probability function of X and Y.

x \ y	y_1	y_2	y_3	y_4	$P(X = x)$
x_1	$\frac{1}{48}$	$\frac{1}{48}$	$\frac{1}{48}$	$\frac{5}{48}$	$\frac{1}{6}$
x_2	$\frac{1}{48}$	$\frac{1}{48}$	$\frac{1}{96}$	$\frac{11}{96}$	$\frac{1}{6}$
x_3	$\frac{4}{48}$	$\frac{4}{48}$	$\frac{9}{96}$	$\frac{39}{96}$	$\frac{2}{3}$
$P(Y = y)$	$\frac{1}{8}$	$\frac{1}{8}$	$\frac{1}{8}$	$\frac{5}{8}$	

Are X and Y independent random variables?

Solution. For the pairs (x_1, y_1), (x_1, y_2), (x_1, y_3), (x_2, y_1), (x_2, y_2), (x_3, y_1), (x_3, y_2), we see that

$$P(X = x_i,\ Y = y_j) = P(X = x_i) \cdot P(Y = y_j)$$

However, to jump to the conclusion that X and Y are independent on the basis of this evidence would be wrong. We have, for example,

$$P(X = x_3,\ Y = y_4) = \frac{39}{96} \neq \frac{2}{3} \cdot \frac{5}{8} = P(X = x_3) \cdot P(Y = y_4)$$

X and Y are not independent.

The absolutely continuous case

For absolutely continuous distributions, the criterion for the independence of two random variables is provided in terms of the pdf's by the following definition:

Definition 4(b) Two random variables X and Y which have a joint absolutely continuous distribution are said to be independent if their joint pdf can be written as the product of the individual pdf's; that is,

$$\boxed{f(x, y) = f_X(x) \cdot f_Y(y)}$$

for every pair of real numbers x, y.

Of course, it behooves us to show that this definition is equivalent to definition 2 (or definition 1 or definition 3). We shall do this next:

(*i*) Assume independence as per definition 2, that is, $F(x, y) = F_X(x) \cdot F_Y(y)$. Then

$$\begin{aligned} f(x, y) &= \frac{\partial^2}{\partial x\, \partial y} F(x, y), && \text{by definition} \\ &= \frac{\partial^2}{\partial x\, \partial y} F_X(x) \cdot F_Y(y), && \text{by assumption} \\ &= \frac{d}{dx} F_X(x) \cdot \frac{d}{dy} F_Y(y) = f_X(x) \cdot f_Y(y) \end{aligned}$$

Hence follows definition 4(b).

(*ii*) Conversely, assume independence according to definition 4(b), that is, $f(x, y) = f_X(x) \cdot f_Y(y)$. Then

$$\begin{aligned} F(u, v) &= \int_{-\infty}^{u} \int_{-\infty}^{v} f(x, y)\, dy\, dx, && \text{by definition} \\ &= \int_{-\infty}^{u} \int_{-\infty}^{v} f_X(x) f_Y(y)\, dy\, dx, && \text{by assumption} \\ &= \int_{-\infty}^{u} f_X(x)\, dx \int_{-\infty}^{v} f_Y(y)\, dy \\ &= F_X(u) \cdot F_Y(v) \end{aligned}$$

Hence definition 2.

Together, (*i*) and (*ii*) show that the definitions 2 and 4(*b*) imply each other and hence are equivalent.

Comment. If X and Y are independent random variables, then, for any x,

$$f_{X|Y}(x \mid y) = f_X(x)$$

because

$$f_{X|Y}(x \mid y) = \frac{f(x, y)}{f_Y(y)} = \frac{f_X(x) \cdot f_Y(y)}{f_Y(y)} = f_X(x)$$

That is, *the conditional distribution of X given $Y = y$ is the same as the distribution of X.*

Similarly, for any y,

$$f_{Y|X}(y \mid x) = f_Y(y)$$

Example 2.7. If X and Y are two independent random variables with the pdf's

$$f_X(x) = \begin{cases} \dfrac{1}{x^2}, & x \geqslant 1 \\ 0, & \text{elsewhere} \end{cases}$$

and

$$f_Y(y) = \begin{cases} e^{-y}, & y > 0 \\ 0, & \text{elsewhere} \end{cases}$$

find the joint distribution of X and Y.

Solution. We have

$$f(x, y) = f_X(x) \cdot f_Y(y) = \begin{cases} \dfrac{1}{x^2} e^{-y}, & x \geqslant 1, \ y > 0 \\ 0, & \text{elsewhere} \end{cases}$$

Example 2.8. Suppose X and Y are uniformly distributed over a circle with radius 1 centered at the origin. Show that X and Y are not independent.

Solution. It was seen in Example 2.9 of Chapter 8 that

$$f(x, y) = \begin{cases} \dfrac{1}{\pi}, & 0 \leqslant x^2 + y^2 \leqslant 1 \\ 0, & \text{elsewhere} \end{cases}$$

$$f_X(x) = \begin{cases} \dfrac{2\sqrt{1 - x^2}}{\pi}, & -1 < x < 1 \\ 0, & \text{elsewhere} \end{cases}$$

and

$$f_Y(y) = \begin{cases} \dfrac{2\sqrt{1 - y^2}}{\pi}, & -1 < y < 1 \\ 0, & \text{elsewhere} \end{cases}$$

Now, if $-1 < x < 1$ and $-1 < y < 1$,

$$f_X(x) \cdot f_Y(y) = \frac{2\sqrt{1-x^2}}{\pi} \cdot \frac{2\sqrt{1-y^2}}{\pi}$$

which is not equal to $f(x, y)$. Hence X and Y are not independent.

Example 2.9. Show that the random variables X and Y which have the joint pdf

$$f(x, y) = \begin{cases} 2, & 0 < y < x < 1 \\ 0, & \text{elsewhere} \end{cases}$$

are not independent.

Solution. This joint distribution was considered in Example 1.5. The marginal pdf's in this case are

$$f_X(x) = \begin{cases} 2x, & 0 < x < 1 \\ 0, & \text{elsewhere} \end{cases}$$

and

$$f_Y(y) = \begin{cases} 2(1-y), & 0 < y < 1 \\ 0, & \text{elsewhere} \end{cases}$$

Since, for $0 < x < 1$ and $0 < y < 1$, $f_X(x) \cdot f_Y(y) = 2x \cdot 2(1-y) \neq 2 = f(x, y)$, the random variables are not independent.

Alternatively, we can answer the question without actually finding the marginal pdf's, but simply by considering the geometrical picture as follows: The pdf is positive over the shaded region of triangle ABC in Figure 2.3, and zero outside it.

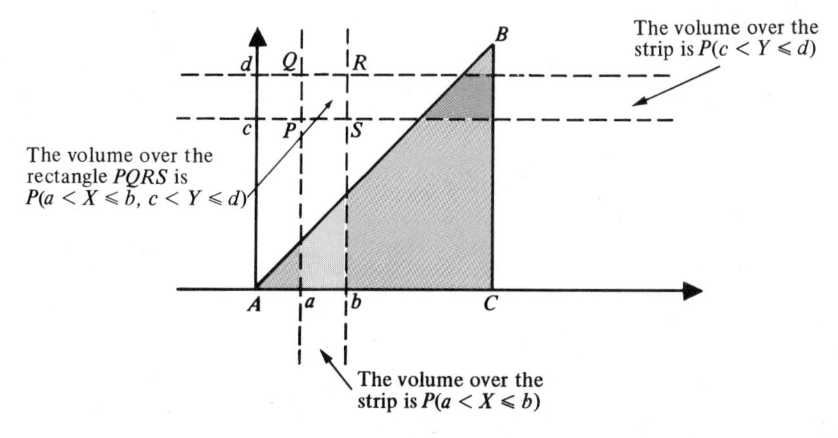

Figure 2.3

From the figure, it can be seen that the volume over the rectangle $PQRS$ is zero, since the joint pdf is zero over this region. However, the volume over the horizontal strip is positive, since this is the volume over the portion of the triangle shaded

Similarly, the volume over the vertical strip, being the volume over the portion of the triangle shaded ▒ is positive. Hence

$$0 = P(a < X \leqslant b,\ c < Y \leqslant d) \neq P(a < X \leqslant b) \cdot P(c < Y \leqslant d) > 0$$

and, consequently, using definition 3, X and Y are not independent. ■

If X and Y have a joint absolutely continuous distribution, then the criterion for independence requires that we be able to express the joint pdf as the product of the marginal pdf's. This would of course necessitate that we first obtain the marginal pdf's. However this is not at all essential. Whether two random variables are independent or not can be determined routinely using the following result. (The essential ingredients of this result are actually contained in exercise 1 at the end of this section.)

Two random variables with an absolutely continuous distribution are independent if their joint pdf at any point (x, y) can be expressed as a product of two nonnegative functions (not necessarily pdf's), one depending on x only, and the other on y only.

To show why this is true, suppose $f(x, y) = u(x) \cdot v(y)$, where u depends only on x and v depends only on y. Then

$$f_X(x) = \int_{-\infty}^{\infty} u(x)v(y)\,dy = u(x)\int_{-\infty}^{\infty} v(y)\,dy$$

and, similarly,

$$f_Y(y) = v(y)\int_{-\infty}^{\infty} u(x)\,dx$$

Next, notice that

$$\begin{aligned}\int_{-\infty}^{\infty} u(x)\,dx \cdot \int_{-\infty}^{\infty} v(y)\,dy &= \int_{-\infty}^{\infty}\int_{-\infty}^{\infty} u(x)v(y)\,dx\,dy \\ &= \int_{-\infty}^{\infty}\int_{-\infty}^{\infty} f(x, y)\,dx\,dy = 1\end{aligned}$$

It follows from these results that

$$\begin{aligned}f(x, y) = u(x) \cdot v(y) &= u(x)\left(\int_{-\infty}^{\infty} v(y)\,dy \cdot \int_{-\infty}^{\infty} u(x)\,dx\right) v(y) \\ &= \left(u(x)\int_{-\infty}^{\infty} v(y)\,dy\right) \cdot \left(v(y)\int_{-\infty}^{\infty} u(x)\,dx\right) \\ &= f_X(x) \cdot f_Y(y)\end{aligned}$$

Hence, by definition 4(b), X and Y are independent.

On the basis of this criterion, we can decide right away, for example, that two random variables with the following joint pdf

$$f(x, y) = \begin{cases} \frac{6}{5}(x + y^2), & 0 < x < 1, \ 0 < y < 1 \\ 0, & \text{elsewhere} \end{cases}$$

are not independent.

We emphasize the fact that $f(x, y)$ must be equal to $u(x) \cdot v(y)$ for *every x and y*. To see the importance of this, consider the following joint pdf of X and Y.

$$f(x, y) = \begin{cases} 8xy, & 0 < y < x < 1 \\ 0, & \text{elsewhere} \end{cases}$$

Just because $8xy$ can be written as, say, $(2x)(4y)$, one might hazard the conclusion that X and Y are independent. This would be erroneous. The simple fact is that we cannot give $u(x)$ and $v(y)$ such that $f(x, y) = u(x) \cdot v(y)$ for *every x and y*. For example, if this were possible then we would be hard put to reconcile the following: $f(\frac{1}{4}, \frac{1}{2}) = 0$, $f(\frac{1}{4}, \frac{1}{6}) = \frac{1}{3}$, $f(\frac{3}{4}, \frac{1}{2}) = 3$. The fact that $f(\frac{1}{4}, \frac{1}{2}) = 0$ would imply that either $u(\frac{1}{4}) = 0$ or $v(\frac{1}{2}) = 0$, whereas $f(\frac{1}{4}, \frac{1}{6}) = \frac{1}{3}$, and $f(\frac{3}{4}, \frac{1}{2}) = 3$ would imply that neither $u(\frac{1}{4})$ nor $v(\frac{1}{2})$ is zero.

Next, consider the following joint pdf of X and Y:

$$f(x, y) = \begin{cases} 4xy, & 0 < x < 1, \ 0 < y < 1 \\ 0, & \text{elsewhere} \end{cases}$$

Then if we take, for example,

$$u(x) = \begin{cases} 4x, & 0 < x < 1 \\ 0, & \text{elsewhere} \end{cases} \quad \text{and} \quad v(y) = \begin{cases} y, & 0 < y < 1 \\ 0, & \text{elsewhere} \end{cases}$$

we see that $f(x, y) = u(x) \cdot v(y)$ for every x and y. Hence X and Y are independent.

Example 2.10. A person lights two fire crackers simultaneously. The amount of time (in seconds) that it takes a cracker to explode is uniformly distributed over the interval $(0, 5)$. Find the probability that the two crackers will explode within 2 seconds of each other.

Solution. Let X represent the time that it takes one cracker to explode and Y the time for the other. By hypothesis, X and Y both have the uniform distribution, and

$$f_X(x) = f_Y(x) = \begin{cases} \frac{1}{5}, & 0 < x < 5 \\ 0, & \text{elsewhere} \end{cases}$$

It seems to be a safe assumption that X and Y are independent. Therefore, their joint pdf is

$$f(x, y) = \begin{cases} \frac{1}{25}, & 0 < x < 5, \ 0 < y < 5 \\ 0, & \text{elsewhere} \end{cases}$$

We are interested in finding $P(|X - Y| \leqslant 2)$. Referring to Figure 2.4, this is simply the probability that a point (X, Y) falls in the unshaded region E of the square, the region bounded by the lines $y = 2 + x$ and $y = -2 + x$.

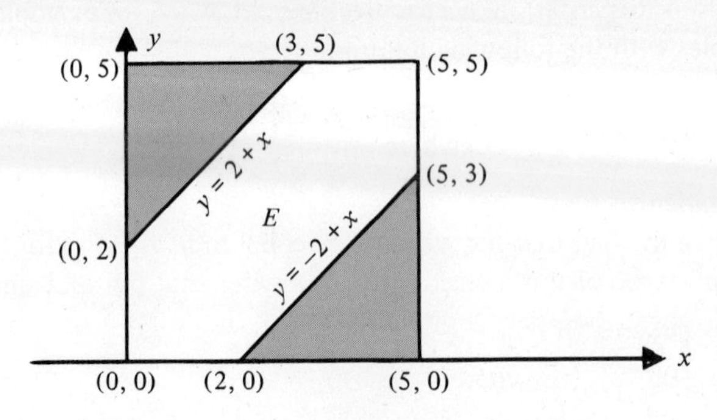

Figure 2.4

Now, the area of the region E is equal to the area of the square minus the area of the two shaded triangles, that is, equal to $25 - 2(\frac{1}{2} \cdot 3 \cdot 3) = 16$. Since the height at each point of the region E is $\frac{1}{25}$, the desired probability is $\frac{16}{25}$.

Example 2.11. Suppose X and Y are independent standard normal variables. Find the joint distribution of X^2 and Y^2.

Solution. Recall that if X and Y are independent, then so are $h(X)$ and $g(Y)$. Hence, X^2 and Y^2 are independent random variables. Therefore, writing f_{X^2, Y^2} for the joint pdf of X^2 and Y^2, we get

$$f_{X^2, Y^2}(u, v) = f_{X^2}(u) \cdot f_{Y^2}(v)$$

for any u, v.

From Example 2.10 in Chapter 6 we know that X^2 and Y^2 each have chi-square distributions with 1 degree of freedom. Therefore,

$$f_{X^2, Y^2}(u, v) = \begin{cases} \dfrac{1}{2\pi\sqrt{uv}} e^{-(u+v)/2}, & u > 0, \ v > 0 \\ 0, & \text{elsewhere} \end{cases}$$

EXERCISES–SECTION 2

1. Show that the criterion for the independence of two random variables amounts to being able to factor the joint D.F. $F(x, y)$ as a product of two functions, one depending on x only, and the other on y only. That is, $F(x, y) = u(x) \cdot v(y)$ for every x, y.
2. If the joint D.F. of X and Y is given by

$$F(x, y) = \begin{cases} 0, & x < 0, \ \text{or} \ y < 0 \\ \dfrac{x^4 y^4}{16}, & 0 \leqslant x < 2 \ \text{and} \ 0 \leqslant y < 2 \\ \dfrac{x^4}{16}, & 0 \leqslant x < 2 \ \text{and} \ y \geqslant 2 \\ \dfrac{y^4}{16}, & x \geqslant 2 \ \text{and} \ 0 \leqslant y < 2 \\ 1, & x \geqslant 2 \ \text{and} \ y \geqslant 2 \end{cases}$$

Show that X and Y are independent, and find–

(a) $P(\frac{1}{2}<X\leqslant 1)$

(b) $P(\frac{1}{2}<Y\leqslant 1)$

(c) $P(\frac{1}{2}<X\leqslant 1,\ \frac{1}{2}<Y\leqslant 1)$.

3. Random variables X and Y have a joint discrete distribution with the probability mass distributed at four points as follows: $P(X=2,\ Y=3)=\frac{1}{3}$, $P(X=2,\ Y=-1)=a$, $P(X=-1,\ Y=3)=b$, $P(X=-1,\ Y=-1)=\frac{1}{6}$. If X and Y are independent, determine a and b.

4. Suppose X and Y are two random variables where Y is a degenerate random variable. Show that X and Y are independent.

5. If X and Y have a joint probability function which is positive at precisely three points, under what conditions will X and Y be independent? Can you generalize this?

6. A number is picked at random from the set of integers $\{1, 2, \ldots, 126\}$. Suppose X represents the remainder when the number is divided by 7, Y the remainder when it is divided by 6, and Z the remainder when divided by 9. Answer the following:

(a) Are X and Y independent?

(b) Are X and Z independent?

(c) Are Y and Z independent?

7. Suppose random variables X and Y are independent, where X is uniformly distributed over the interval $[0, 3]$ and Y has the pdf given by

$$f_Y(y)=\begin{cases}\frac{3}{8}y^2, & 0<y<2\\ 0, & \text{elsewhere}\end{cases}$$

Find the joint D.F. of X and Y.

8. The amount X (in dollars) that Tom earns in a day has the probability function

$$P(X=x)=\frac{10+|x-25|}{50},\qquad x=10, 20, 25$$

and the amount Y (in dollars) that Jane earns in a day has the following probability function:

$$P(Y=y)=\frac{9+|y-16|}{50},\qquad y=9, 12, 13, 16$$

If the amounts that Tom and Jane earn are independent, find the probability that Jane earns more than Tom.

9. The random vector (X, Y) is distributed on two points, $(0, 1)$ and $(2, 3)$, with respective probabilities $\frac{1}{3}$ and $\frac{2}{3}$. Find the joint distribution of two random variables U, V which are independent and have marginal distributions identical with those of X and Y, respectively.

10. Suppose X and Y are independent random variables with $P(a<X\leqslant b)=\frac{1}{3}$ and $P(c<Y\leqslant d)=\frac{3}{4}$. Let $A=\{a<X\leqslant b, -\infty<Y<\infty\}$, $B=\{-\infty<X<\infty, c<Y\leqslant d\}$. Find:

(a) $P(A\cup B)$ (b) $P(A-B)$

11. Consider the shaded regions given in Figure 2.5. In each case, the joint pdf of X and Y is positive over the shaded region and zero outside it. Indicate in which cases the random variables are not independent.

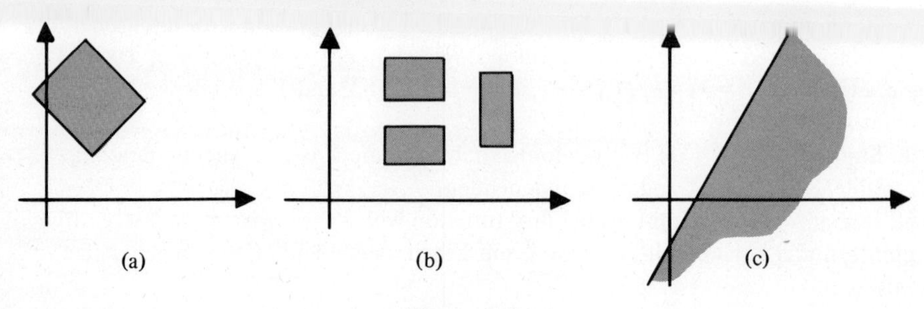

Figure 2.5

Can you say anything categorical about other cases?

12. Suppose X and Y have a joint absolutely continuous distribution with the following pdf:

$$f(x, y) = \begin{cases} 2e^{-(x+y)}, & 0 \leqslant y \leqslant x < \infty \\ 0, & \text{elsewhere} \end{cases}$$

Are X and Y independent? Answer the same question if

$$f(x, y) = \begin{cases} e^{-(x+y)}, & x > 0, \; y > 0 \\ 0, & \text{elsewhere} \end{cases}$$

13. The joint distribution of X and Y is given by the following pdf:

$$f(x, y) = \begin{cases} \frac{1}{4}[1 + xy(x^2 - y^2)], & -1 < x < 1, \; -1 < y < 1 \\ 0, & \text{elsewhere} \end{cases}$$

Obtain a joint distribution of two random variables U and V so that their marginal distributions coincide with those of X and Y.

14. If X and Y are independent random variables with $P(X \geqslant 0) = P(Y \geqslant 0) = 1$:

(a) is it true, for example, that $P(XY \leqslant 6) = P(X \leqslant 2) \cdot P(Y \leqslant 3)$?

(b) is it true that $P(XY \leqslant 6) \geqslant P(X \leqslant 2) \cdot P(Y \leqslant 3)$?

Hint: Sketch the regions in the xy-plane.

15. X and Y are independent random variables, each uniformly distributed over the interval [0, 1]. Find the probability that:

(a) the root of the equation $Xt - Y = 0$ in t is less than $\frac{1}{3}$

(b) the quadratic equation $t^2 + 2Xt + Y = 0$ has two distinct real roots

Hint: In part (b) the roots are distinct if and only if the discriminant $4(X^2 - Y) > 0$.

16. Suppose the height of an adult male (measured in inches) is a normally distributed random variable with a mean height of 68 inches and a standard deviation of 4 inches. If two people (say, Tom and Dick) are picked ar random, assuming independence, find the probability that—

(a) the heights of both are between 64 and 70 inches
(b) Tom is shorter than 70 inches and Dick is taller than 66 inches
(c) one is shorter than 70 inches and the other is taller than 66 inches.

17. A plane is supposed to arrive between 11:00 and 11:30 a.m., and its arrival time is uniformly distributed. A passenger takes a bus which arrives at the airport between 10:50 and 11:20 a.m., with the arrival time, once again, uniformly distributed. If the plane waits for 10 minutes after its arrival, find the probability that the passenger will miss the plane. (Assume that the two arrival times are independent.)

18. A plane leaves from New York for Los Angeles at a random time between 1 p.m. and 6 p.m. Independently, another plane leaves from Los Angeles for New York between 1 p.m. and 6 p.m. at a time Y (in hours) which has the following pdf:

$$f_Y(y) = \begin{cases} \frac{2}{25}(y-1), & 1 < y < 6 \\ 0, & \text{elsewhere} \end{cases}$$

If the flight time either way is three hours, find the probability that the two planes will pass each other en route.

19. If X and Y are independent, absolutely continuous random variables, show that their joint distribution is also absolutely continuous.

3. MORE THAN TWO RANDOM VARIABLES

The discussion of joint distributions has so far been restricted to two random variables. The reason behind this was to maintain simplicity by avoiding cumbersome notation. The ideas that have been developed in the context of two random variables carry over, essentially verbatim, to the general situation where there is an arbitrary (finite) collection of random variables, say $X_1, X_2, \ldots, X_n$. It should be borne in mind that these random variables are defined on the same probability space. We present below a capsule outline, skirting the details. (In the following, we shall denote the joint distribution of $X_1, X_2, \ldots, X_n$ without any subscripts. All other distributions will be represented appropriately by using subscripts.)

3.1 The Joint Distribution Function

For any real numbers $u_1, u_2, \ldots, u_n$, the **joint distribution function** of $X_1, X_2, \ldots, X_n$ is defined by

$$F(u_1, u_2, \ldots, u_n) = P(X_1 \leqslant u_1, X_2 \leqslant u_2, \ldots, X_n \leqslant u_n)$$

Properties of distribution functions

(*i*)
$$0 \leqslant F(u_1, u_2, \ldots, u_n) \leqslant 1$$

for any real numbers $u_1, u_2, \ldots, u_n$.

(*ii*) (**Monotonicity**) A distribution function is a nondecreasing function in each variable separately.

(*iii*) If $a_1, a_2, \ldots, a_n$, $b_1, b_2, \ldots, b_n$ are real numbers for which $a_1 < b_1$, $a_2 < b_2, \ldots, a_n < b_n$, then

$$\begin{aligned} P(a_1 < X_1 \leqslant b_1,\ a_2 < X_2 \leqslant b_2, \ldots,\ a_n < X_n \leqslant b_n) \\ = F(b_1, b_2, \ldots, b_n) \\ - [F(a_1, b_2, \ldots, b_n) + \ldots + F(b_1, b_2, \ldots, b_{n-1}, a_n)] \\ + (\text{all terms with 2 } a\text{'s and } n-2\ b\text{'s}) \\ - (\text{all terms with 3 } a\text{'s and } n-3\ b\text{'s}) \\ \vdots \qquad \vdots \qquad \vdots \\ + (-1)^n F(a_1, a_2, \ldots, a_n) \end{aligned}$$

(*iv*) $$\lim_{u_i \to -\infty} F(u_1, u_2, \ldots, u_{i-1}, u_i, u_{i+1}, \ldots, u_n)$$
$$= F(u_1, u_2, \ldots, u_{i-1}, -\infty, u_{i+1}, \ldots, u_n) = 0 \text{ for any } i.$$

(*v*) $$\lim_{\substack{u_1 \to \infty \\ u_2 \to \infty \\ \vdots \\ u_n \to \infty}} F(u_1, u_2, \ldots, u_n) = F(\infty, \infty, \ldots, \infty) = 1$$

(*vi*) A distribution function is continuous from the right in each variable separately.

3.2 The Discrete Case

The random variables $X_1, X_2, \ldots, X_n$ are said to have a **joint discrete distribution** if there exists a nonnegative function p such that p vanishes everywhere except at a finite or countably infinite points in the n-dimensional space $\mathbf{R}^n$ and

$$p(x_1, x_2, \ldots, x_n) = P(X_1 = x_1,\ X_2 = x_2, \ldots,\ X_n = x_n)$$

The function p is called the *joint probability function.*

We have:

(*i*) $p(x_1, x_2, \ldots, x_n) \geqslant 0$

(*ii*) $\sum_{x_n} \ldots \sum_{x_1} p(x_1, x_2, \ldots, x_n) = 1$

(*iii*) (*Marginal distributions*) Suppose $X_{i_1}, X_{i_2}, \ldots, X_{i_k}$ is a subcollection of random variables from the random variables $X_1, X_2, \ldots, X_n$. The joint distribution of $X_{i_1}, X_{i_2}, \ldots, X_{i_k}$ is obtained by summing the joint probability function with respect to the values of the remaining variables.

For example, if X, Y, Z, and W are four random variables, then

$$P(Z = z) = \sum_w \sum_y \sum_x P(X = x,\ Y = y,\ Z = z,\ W = w)$$

$$P(X = x,\ W = w) = \sum_z \sum_y P(X = x,\ Y = y,\ Z = z,\ W = w)$$

and so on.

(*iv*) (*The conditional distribution*) We shall define this by considering four random variables X, Y, Z, W which have a joint distribution. The conditional distribution of X and Y given $Z = z_o$, $W = w_o$ is defined by

$$p_{X,Y|Z,W}(x, y \mid z_o, w_o) = \frac{p_{X,Y,Z,W}(x, y, z_o, w_o)}{p_{Z,W}(z_o, w_o)}$$

The conditional distribution of X, Y, and W given $Z = z_o$ is defined by

$$p_{X,Y,W|Z}(x, y, w \mid z_o) = \frac{p_{X,Y,Z,W}(x, y, z_o, w)}{p_Z(z_o)}$$

And so on.

(*v*) (*Mutual independence*) The random variables $X_1, X_2, \ldots, X_n$ are said to be *mutually independent* if

$$P(X_1 = x_1,\ X_2 = x_2, \ldots,\ X_n = x_n) = P(X_1 = x_1)P(X_2 = x_2) \ldots P(X_n = x_n)$$

for every $x_1, x_2, \ldots, x_n$.

3.3 The Absolutely Continuous Case

The random variables $X_1, X_2, \ldots, X_n$ are said to have a **joint absolutely continuous distribution** if there exists a nonnegative function f such that for any real numbers $u_1, u_2, \ldots, u_n$

$$F(u_1, u_2, \ldots, u_n) = \int_{-\infty}^{u_n} \cdots \int_{-\infty}^{u_1} f(x_1, x_2, \ldots, x_n)\, dx_1 \ldots dx_n$$

The function f is called the *joint probability density function* of $X_1, X_2, \ldots, X_n$.
We have:

(*i*) $f(x_1, x_2, \ldots, x_n) = \dfrac{\partial^n}{\partial x_1 \partial x_2 \ldots \partial x_n} F(x_1, x_2, \ldots, x_n)$

(*ii*) $f(x_1, x_2, \ldots, x_n) \geqslant 0$

(*iii*) $\int_{-\infty}^{\infty} \cdots \int_{-\infty}^{\infty} f(x_1, x_2, \ldots, x_n)\, dx_1 dx_2 \ldots dx_n = 1$

(*iv*) (*Marginal distributions*) If $X_{i_1}, X_{i_2}, \ldots, X_{i_k}$ is a subcollection of random variables from $X_1, X_2, \ldots, X_n$, their joint pdf is obtained by integrating $f(x_1, x_2, \ldots, x_n)$ with respect to the remaining variables.

For example, if X, Y, Z, and W are four random variables which have a joint absolutely continuous distribution, then

$$f_Z(z) = \int_{-\infty}^{\infty}\int_{-\infty}^{\infty}\int_{-\infty}^{\infty} f_{X,Y,Z,W}(x, y, z, w)\, dx\, dy\, dw$$

$$f_{X,W}(x, w) = \int_{-\infty}^{\infty}\int_{-\infty}^{\infty} f_{X,Y,Z,W}(x, y, z, w)\, dy\, dz$$

and so on.

(*v*) (*The conditional distribution*) We shall define this by considering the random variables X, Y, Z, W which have a joint distribution. The conditional (joint) pdf of X and Y given $Z = z_o$, $W = w_o$ is defined by

$$f_{X,Y|Z,W}(x, y \mid z_o, w_o) = \frac{f_{X,Y,Z,W}(x, y, z_o, w_o)}{f_{Z,W}(z_o, w_o)}$$

The conditional (joint) pdf of X, Y, and W given $Z = z_o$ is defined by

$$f_{X,Y,W|Z}(x, y, w \mid z_o) = \frac{f_{X,Y,Z,W}(x, y, z_o, w)}{f_Z(z_o)}$$

and so on.

(*vi*) (*Mutual independence*) If $X_1, X_2, \ldots, X_n$ have a joint absolutely continuous distribution, they are said to be *mutually independent* if

$$f(x_1, x_2, \ldots, x_n) = f_{X_1}(x_1) f_{X_2}(x_2) \ldots f_{X_n}(x_n)$$

for every $x_1, x_2, \ldots, x_n$.

In the context of mutually independent random variables, the word "mutually" is often omitted.

Example 3.1. Suppose X, Y, and Z have the following joint pdf:

$$f(x, y, z) = \begin{cases} 12x^2yz, & \text{if } 0<x<1,\ 0<y<1,\ 0<z<1 \\ 0, & \text{elsewhere} \end{cases}$$

Find:

(*a*) the pdf's of X, Y, and Z
(*b*) the joint pdf of X and Y
(*c*) the conditional pdf of X and Y given $Z = z_o$, $0<z<1$
(*d*) $P(X \leqslant Y)$
(*e*) $P(X<Y<Z)$

Solution

(*a*) The pdf of X is obtained by integrating, as follows:

$$f_X(x) = \int_{-\infty}^{\infty}\int_{-\infty}^{\infty} f(x, y, z)\, dy\, dz = \begin{cases} \int_0^1\int_0^1 12x^2yz\, dy\, dz, & 0<x<1 \\ 0, & \text{elsewhere} \end{cases}$$

$$= \begin{cases} 3x^2, & 0<x<1 \\ 0, & \text{elsewhere} \end{cases}$$

Similarly,

$$f_Y(y) = \begin{cases} \int_0^1\int_0^1 12x^2yz\, dx\, dz, & 0<y<1 \\ 0, & \text{elsewhere} \end{cases}$$

$$= \begin{cases} 2y, & 0<y<1 \\ 0, & \text{elsewhere} \end{cases}$$

and

$$f_Z(z) = \begin{cases} 2z, & 0 < z < 1 \\ 0, & \text{elsewhere} \end{cases}$$

Notice, incidentally, that $f(x, y, z) = f_X(x) f_Y(y) f_Z(z)$ for every x, y, z, so that X, Y, and Z are mutually independent random variables.

(*b*) The joint pdf of X and Y is given by

$$f_{X,Y}(x, y) = \int_{-\infty}^{\infty} f(x, y, z)\, dz = \begin{cases} \int_0^1 12x^2yz\, dz, & 0 < x < 1,\ 0 < y < 1 \\ 0, & \text{elsewhere} \end{cases}$$

$$= \begin{cases} 6x^2y, & 0 < x < 1,\ 0 < y < 1 \\ 0, & \text{elsewhere} \end{cases}$$

(*c*) The conditional pdf of X and Y given $Z = z_0$, $0 < z_0 < 1$, is obtained as

$$f_{X,Y|Z}(x, y \mid z_0) = \frac{f(x, y, z_0)}{f_Z(z_0)}$$

$$= \begin{cases} 6x^2y, & 0 < x < 1,\ 0 < y < 1 \\ 0, & \text{elsewhere} \end{cases}$$

The answer in (*b*) is the same as in (*c*). Is this surprising?

(*d*) To find $P(X \leqslant Y)$, we use the joint pdf of X and Y obtained in (*b*).

$$P(X \leqslant Y) = \int_0^1 \int_x^1 6x^2y\, dy\, dx = \frac{2}{5}$$

(*e*) We have

$$P(X < Y < Z) = \iiint_{\{(x, y, z) | x < y < z\}} f(x, y, z)\, dx\, dy\, dz$$

$$= \int_0^1 \int_0^z \int_0^y 12x^2yz\, dx\, dy\, dz$$

$$= \frac{4}{35}$$

■

Having discussed the general nature of the distributions of several random variables, we now give a result which deserves special mention:

If $X_1, X_2, \ldots, X_n$ are (mutually) independent random variables, then any subset of these random variables is independent.

The reader will get the idea of the proof in the general case if we show in particular that $X_1, X_2, \ldots, X_k$ (where $k < n$) are independent whenever $X_1, X_2, \ldots, X_k, X_{k+1}, \ldots, X_n$ are. We observe first that, for any $u_1, u_2, \ldots, u_k$,

$$\{X_1 \leqslant u_1, \ldots, X_k \leqslant u_k, X_{k+1} < \infty, \ldots, X_n < \infty\}$$
$$= \{X_1 \leqslant u_1\} \cap \{X_2 \leqslant u_2\} \cap \ldots \cap \{X_k \leqslant u_k\} \cap \{X_{k+1} < \infty\} \cap \ldots \cap \{X_n < \infty\}$$
$$= \{X_1 \leqslant u_1, \ldots, X_k \leqslant u_k\}$$

since $\{X_i < \infty\} = S$ for any i.

Therefore, we get

$$\begin{aligned}
&P(X_1 \leqslant u_1, X_2 \leqslant u_2, \ldots, X_k \leqslant u_k) \\
&\quad = P(X_1 \leqslant u_1, X_2 \leqslant u_2, \ldots, X_k \leqslant u_k, X_{k+1} < \infty, \ldots, X_n < \infty) \\
&\quad = P(X_1 \leqslant u_1) \cdot P(X_2 \leqslant u_2) \cdot \ldots \cdot P(X_k \leqslant u_k) \cdot P(X_{k+1} < \infty) \cdot \ldots \cdot P(X_n < \infty)
\end{aligned}$$

because $X_1, X_2, \ldots, X_n$ are independent.

Now $P(X_i < \infty) = 1$ for any i. Hence,

$$P(X_1 \leqslant u_1, X_2 \leqslant u_2, \ldots, X_k \leqslant u_k) = P(X_1 \leqslant u_1) \cdot P(X_2 \leqslant u_2) \cdot \ldots \cdot P(X_k \leqslant u_k)$$

Consequently, by the definition of independence, $X_1, X_2, \ldots, X_k$ are mutually independent.

Caution. The converse of the above result is not true. We give below an example of three random variables X, Y, and Z which are independent when taken two at a time, but are not independent when taken all together. The example is due to S. Bernstein.

Example 3.2. Suppose X, Y, and Z have a joint discrete distribution given by

$$\begin{aligned}
&p_{X,Y,Z}(0, 1, 0) = \tfrac{1}{4}, \qquad p_{X,Y,Z}(0, 0, 1) = \tfrac{1}{4} \\
&p_{X,Y,Z}(1, 0, 0) = \tfrac{1}{4}, \qquad p_{X,Y,Z}(1, 1, 1) = \tfrac{1}{4}
\end{aligned}$$

Show that any two of the random variables are independent, while the three taken together are not.

Solution. Let us first find the distributions of each of the random variables, and then the joint distributions of X and Y, of X and Z, and of Y and Z.

We have

$$p_X(0) = p_{X,Y,Z}(0, 1, 0) + p_{X,Y,Z}(0, 0, 1) = \tfrac{1}{2}$$

and

$$p_X(1) = p_{X,Y,Z}(1, 0, 0) + p_{X,Y,Z}(1, 1, 1) = \tfrac{1}{2}$$

(*Note:* We are summing with respect to the values of Y and Z to get the marginal of X; for example, $p_X(0) = \sum_{y,z} p(0, y, z)$.) This gives the distribution of X. Similarly, the distribution of Y is given by

$$p_Y(0) = \tfrac{1}{2} \quad \text{and} \quad p_Y(1) = \tfrac{1}{2}$$

and that of Z by

$$p_Z(0) = \tfrac{1}{2} \quad \text{and} \quad p_Z(1) = \tfrac{1}{2}$$

To find the joint distribution of X and Y we observe that $p_{X,Y}(x, y) = \sum_z p_{X,Y,Z}(x, y, z)$. Therefore,

$$p_{X,Y}(0, 0) = \sum_z p_{X,Y,Z}(0, 0, z) = p_{X,Y,Z}(0, 0, 1) = \tfrac{1}{4}$$

Similarly,

$$p_{X,Y}(0, 1) = \tfrac{1}{4}, \quad p_{X,Y}(1, 0) = \tfrac{1}{4}, \quad p_{X,Y}(1, 1) = \tfrac{1}{4}$$

Since

$$p_{X,Y}(0, 0) = p_X(0) \cdot p_Y(0), \quad p_{X,Y}(0, 1) = p_X(0) \cdot p_Y(1)$$
$$p_{X,Y}(1, 0) = p_X(1) \cdot p_Y(0), \quad p_{X,Y}(1, 1) = p_X(1) \cdot p_Y(1)$$

the random variables X and Y are independent. In a similar way, it can be shown that X and Z are independent and that Y and Z are independent.

However, X, Y, and Z are not independent since, for example, $p_{X,Y,Z}(1, 1, 1) \neq p_X(1) \cdot p_Y(1) \cdot p_Z(1)$ because

$$p_{X,Y,Z}(1, 1, 1) = \tfrac{1}{4} \quad \text{and} \quad p_X(1) \cdot p_Y(1) \cdot p_Z(1) = (\tfrac{1}{2})^3 = \tfrac{1}{8}$$

10 Functions of Several Random Variables

INTRODUCTION

As before, we shall confine our discussion mainly to the bivariate case, where there are two random variables X and Y defined on the same sample space. For any sample point $s \in S$, $X(s)$ and $Y(s)$ represent two real numbers. Often it is of interest to find the distribution of a random variable Z which results from combining the random variables X and Y in some way. For example, suppose a physicist is investigating the kinetic energy of atomic particles whose mass X and velocity Y are random variables. Denoting by S the sample space of all the particles, for any particle $s \in S$, $X(s)$ is its mass and $Y(s)$ is its velocity. The kinetic energy of s is $\frac{1}{2}X(s) \cdot [Y(s)]^2$. Thus, what the physicist is interested in is the random variable $Z = \frac{1}{2}XY^2$, where, for any $s \in S$, $Z(s) = \frac{1}{2}X(s) \cdot [Y(s)]^2$.

As another example, suppose two numbers are picked at random between 0 and 1. Denoting these numbers by X and Y, we might be interested in their product $Z = XY$, or we might be interested in their sum $Z = X + Y$, and so on.

In general, then, suppose $Z = h(X, Y)$ where h is a real-valued function of two real variables. If $s \in S$, then $Z(s) = h(X(s), Y(s))$, a real number. We shall adopt the approach which formed the cornerstone of our discussion of functions of a single random variable in Chapter 6: *In the discrete case, the aim will be to give the probability function of Z; whereas, in the continuous case, it will be to obtain the distribution function.* We will investigate the essential features of this process through various examples.

It should be mentioned here that there is a method called the "method of Jacobians" which is often suitable in the continuous case. However, we shall not discuss this technique. Later in the text, in Chapter 12, we shall also see how, in certain situations, it is possible to find the distribution of $h(X, Y)$ by using what are called the moment generating functions.

1. THE DISCRETE CASE

We shall give various examples to illustrate the essence of the approach used in the discrete case.

Example 1.1. X and Y have a joint distribution with the probability function given below.

$x \backslash y$	1	3	5	7	$P(X = x)$
2	$\frac{1}{24}$	$\frac{1}{12}$	$\frac{1}{12}$	$\frac{1}{24}$	$\frac{1}{4}$
4	0	$\frac{1}{12}$	$\frac{1}{4}$	$\frac{1}{12}$	$\frac{5}{12}$
6	$\frac{1}{24}$	$\frac{1}{6}$	$\frac{1}{24}$	$\frac{1}{12}$	$\frac{8}{24}$
$P(Y = y)$	$\frac{1}{12}$	$\frac{1}{3}$	$\frac{9}{24}$	$\frac{5}{24}$	1

Find the distributions of–

(*a*) $Z = X + Y$

(*b*) $V = \max(X, Y)$

(*c*) $W = \min(X, Y)$.

Solution

(*a*) The probability function of Z is given as

z	3	5	7	9	11	13
$P(Z = z)$	$\frac{1}{24}$	$\frac{1}{12}$	$\frac{1}{12} + \frac{1}{12} + \frac{1}{24} = \frac{5}{24}$	$\frac{1}{24} + \frac{1}{4} + \frac{1}{6} = \frac{11}{24}$	$\frac{1}{12} + \frac{1}{24} = \frac{3}{24}$	$\frac{1}{12}$

For example,

$$\{Z = 7\} = \{X = 4,\ Y = 3\} \cup \{X = 2,\ Y = 5\} \cup \{X = 6,\ Y = 1\}$$

Therefore,

$$\begin{aligned} P(Z = 7) &= P(X = 4,\ Y = 3) + P(X = 2,\ Y = 5) + P(X = 6,\ Y = 1) \\ &= \frac{1}{12} + \frac{1}{12} + \frac{1}{24} \end{aligned}$$

(*b*) The random variable $V = \max(X, Y)$ is defined for any $s \in S$ by $V(s) =$ maximum of $X(s)$ and $Y(s)$. For example, if $X(s) = 6$ and $Y(s) = 3$, then $V(s) = 6$, and so on.

Notice that, for example,

$$\{V = 7\} = \{X = 2,\ Y = 7\} \cup \{X = 4,\ Y = 7\} \cup \{X = 6,\ Y = 7\}$$

Therefore,

$$P(V = 7) = \frac{1}{24} + \frac{1}{12} + \frac{1}{12} = \frac{5}{24}$$

Continuing this argument, the probability function of V is given by

t	2	3	4	5	6	7
$P(V = t)$	$\frac{1}{24}$	$\frac{1}{12}$	$\frac{1}{12}$	$\frac{1}{3}$	$\frac{1}{4}$	$\frac{5}{24}$

(*c*) The random variable $W = \min(X, Y)$ is defined for any $s \in S$ by $W(s) =$ minimum of $X(s)$, $Y(s)$. For example, if $X(s) = 6$ and $Y(s) = 3$, then $W(s) = 3$, and so on.

We have, for example,

$$\{W = 4\} = \{X = 4,\ Y = 5\} \cup \{X = 4,\ Y = 7\}$$

Therefore,

$$P(W = 4) = \frac{1}{4} + \frac{1}{12} = \frac{1}{3}$$

Continuing, the probability function of W is given by

t	1	2	3	4	5	6
$P(W = t)$	$\frac{1}{12}$	$\frac{5}{24}$	$\frac{1}{4}$	$\frac{1}{3}$	$\frac{1}{24}$	$\frac{1}{12}$

- *Example 1.2.* X and Y are independent random variables with X binomial $B(n, p)$ and Y binomial $B(m, p)$. Find the distribution of $Z = X + Y$.

Solution. Since the possible values of X are $0, 1, \ldots, n$ and those of Y are $0, 1, \ldots, m$, the possible values of Z are $0, 1, 2, \ldots, n + m$.

Now

$$\begin{aligned}\{X + Y = k\} &= \{X = 0,\ Y = k\} \cup \{X = 1,\ Y = k - 1\} \cup \ldots \cup \{X = k,\ Y = 0\} \\ &= \bigcup_{i=0}^{k} \{X = i,\ Y = k - i\}\end{aligned}$$

Therefore,

$$\begin{aligned}P(Z = k) &= P(X + Y = k) = P\left(\bigcup_{i=0}^{k} \{X = i,\ Y = k - i\}\right) \\ &= \sum_{i=0}^{k} P(X = i,\ Y = k - i), \qquad \text{since the events are mutually exclusive} \\ &= \sum_{i=0}^{k} P(X = i)P(Y = k - i), \qquad \text{since } X \text{ and } Y \text{ are independent} \\ &= \sum_{i=0}^{k} \binom{n}{i} p^i (1 - p)^{n-i} \binom{m}{k-i} p^{k-i} (1 - p)^{m-(k-i)}\end{aligned}$$

since X is $B(n, p)$ and Y is $B(m, p)$.

Hence,

$$P(Z = k) = \sum_{i=0}^{k} \binom{n}{i}\binom{m}{k-i} p^k(1-p)^{n+m-k}$$

$$= p^k(1-p)^{n+m-k} \sum_{i=0}^{k} \binom{n}{i}\binom{m}{k-i}$$

$$= \binom{n+m}{k} p^k(1-p)^{n+m-k}$$

Thus,

$$P(Z = k) = \binom{n+m}{k} p^k(1-p)^{n+m-k}, \qquad k = 0, 1, \ldots, n+m$$

Consequently, we get the following result:

If X is binomial $B(n, p)$ and Y is binomial $B(m, p)$, and if X and Y are independent, then $X + Y$ is binomial $B(n + m, p)$. ■

Comment. The generalization of the result in Example 1.2 is obvious. If $X_1, X_2, \ldots, X_r$ are r independent random variables where X_i is $B(n_i, p)$, $i = 1, 2, \ldots, r$, then $X_1 + X_2 + \ldots + X_r$ is $B\left(\sum_{i=1}^{r} n_i, p\right)$.

• *Example 1.3.* If X and Y are independent Poisson random variables with parameters λ_1 and λ_2, respectively, find the distribution of $Z = X + Y$.

Solution. The possible values of Z are 0, 1, 2, ... As in Example 1.2, for $k = 0, 1, \ldots,$

$$P(Z = k) = \sum_{i=0}^{k} P(X = i)P(Y = k - i)$$

$$= \sum_{i=0}^{k} \frac{e^{-\lambda_1}\lambda_1^i}{i!} \cdot \frac{e^{-\lambda_2}\lambda_2^{k-i}}{(k-i)!}$$

$$= e^{-(\lambda_1+\lambda_2)} \sum_{i=0}^{k} \frac{1}{i!\,(k-i)!} \lambda_1^i \lambda_2^{k-i}$$

$$= \frac{e^{-(\lambda_1+\lambda_2)}}{k!} \sum_{i=0}^{k} \binom{k}{i} \lambda_1^i \lambda_2^{k-i}$$

$$= e^{-(\lambda_1+\lambda_2)} \frac{(\lambda_1+\lambda_2)^k}{k!}$$

This shows that the probability function of Z is that of a Poisson random variable with parameter $\lambda_1 + \lambda_2$. Hence we have the following result:

If X and Y are independent Poisson random variables with parameters λ_1 and λ_2, respectively, then $Z = X + Y$ is a Poisson random variable with parameter $\lambda_1 + \lambda_2$. ■

Comment. The result in Example 1.3 extends to any finite number of independent random variables as follows: If $X_1, X_2, \ldots, X_r$ are r independent random variables where X_i, $i = 1, 2, \ldots, r$, has a Poisson distribution with parameter λ_i, then $X_1 + X_2 + \ldots + X_r$ has a Poisson distribution with parameter $\lambda_1 + \lambda_2 + \ldots + \lambda_r$.

- *Example 1.4.* If X and Y are independent random variables each having a geometric distribution with the probability of success p, find the distribution of $Z = X + Y$.

Solution. The probability functions of X and Y are given by

$$P(X = k) = P(Y = k) = (1 - p)^{k-1}p, \qquad k = 1, 2, \ldots$$

The possible values of X and Y are $1, 2, \ldots$. Therefore, the possible values of Z are $2, 3, 4, \ldots$. Adopting the argument of Example 1.2, for $j = 2, 3, \ldots$, we get

$$\begin{aligned} P(Z = j) &= P(X + Y = j) \\ &= \sum_{r=1}^{j-1} P(X = r) \cdot P(Y = j - r) \\ &= \sum_{r=1}^{j-1} (1 - p)^{r-1} p \cdot (1 - p)^{(j-r)-1} p \\ &= (j - 1)(1 - p)^{j-2} p^2 = \binom{j-1}{1} p^2 (1 - p)^{j-2} \end{aligned}$$

Thus it follows that *if X and Y are independent random variables each having a geometric distribution with parameter p, then $Z = X + Y$ has the negative binomial distribution and represents the number of trials required for an event to occur twice.* ■

Comment. An obvious generalization of Example 1.4 is the following: If $X_1, X_2, \ldots, X_r$ are r independent random variables each having a geometric distribution with parameter p, then $X_1 + X_2 + \ldots + X_r$ has a negative binomial distribution and represents the number of trials required for an event to occur r times.

EXERCISES—SECTION 1

1. Three chips are picked at random from a bowl containing five chips numbered 2, 4, 5, 6, 8. Let X_i represent the number on the ith draw. Find the distribution of $Z = X_1 + X_2 + X_3$, and, using this distribution, find $E(Z)$ and $\text{Var}(Z)$. Consider the two cases, with replacement and without replacement.
2. Suppose the joint distribution of X and Y is given by

$$P(X = x,\ Y = y) = \tfrac{1}{42}(x + y^2), \qquad x = 1, 4, \quad y = -1, 0, 1, 3$$

Find the distribution of—

(a) $Z = X + Y$ (b) $U = X - Y$ (c) $V = \max(X, Y)$
(d) $W = \min(X, Y)$ (e) $R = XY$ (f) $Z = Y/X$

3. Suppose X and Y are independent random variables which assume nonnegative integral values. Let

$$P(X = i) = a_i \quad \text{and} \quad P(Y = i) = b_i, \qquad i = 0, 1, \ldots$$

If $Z = X + Y$, show that

$$P(Z = n) = \sum_{i=0}^{n} a_i b_{n-i}$$

4. The settings available for the current I (in amperes) are 1, 2, 3 with respective probabilities 0.1, 0.3, 0.6, and for the resistance R (in ohms) are 3, 5 with respective probabilities 0.4, 0.6. If the settings for the current and the resistance are independent, find the distribution of the voltage E in the circuit, given by $E = IR$.
5. The amount X (in dollars) that Tom earns in a day has the probability function

$$P(X = x) = \frac{10 + |x - 25|}{50}, \qquad x = 10, 20, 25$$

and the amount Y (in dollars) that Jane earns in a day has the probability function

$$P(Y = y) = \frac{9 + |y - 16|}{50}, \qquad y = 9, 12, 13, 16$$

If the amounts that Tom and Jane earn are independent, find the distribution of the amount that they earn together.
6. While playing basketball, Tom attempts three baskets, John attempts five, and Dick attempts two. Suppose their attempts are independent and the probability of making a basket is 0.6 for each of them. Find the probability that together they score at least five baskets.
7. The number of bacteria on plates of type A is a Poisson random variable with a mean number of one bacteria per plate, and the number on plates of type B is also a Poisson random variable with a mean number of two bacteria per plate. If a biologist has two plates of type A and two of type B, assuming independence, find the probability that there will be at least four bacteria on the four plates.
8. Let X and Y be independent random variables, each with the Poisson distribution, with respective parameters λ_1 and λ_2. Show that

$$P(X = k \mid X + Y = n) = b\left(k; n, \frac{\lambda_1}{\lambda_1 + \lambda_2}\right), \qquad k = 0, 1, 2, \ldots, n$$

That is, the conditional distribution of X given $X + Y$ is binomial with parameter $\lambda_1/(\lambda_1 + \lambda_2)$.
9. If X and Y are independent, identically distributed random variables with

$$P(X = i) = P(Y = i) = \frac{1}{N + 1}, \qquad i = 0, 1, \ldots, N$$

find the distribution of–

(a) $Z = X + Y$ (b) $V = \max(X, Y)$ (c) $W = \min(X, Y)$

10. Suppose X and Y are independent random variables where each has a geometric distribution with parameter p. Find the distribution of–

(a) $W = \min(X, Y)$ (b) $Z = X - Y$

2. THE CONTINUOUS CASE

In the following, we shall obtain distributions of some important functions of X and Y which occur widely in applied problems. Our interest will center mainly in deriving the distributions of $X + Y$, XY, X/Y, $\max(X, Y)$, and $\min(X, Y)$. (In theory, the approach is similar for any function $h(X, Y)$.) In each case, we shall first derive the distribution of $h(X, Y)$ assuming an arbitrary joint distribution of X and Y. The reader would be well advised not to concentrate on the formulae so derived. Rather, he should be aware of the region over which the integration is carried out. When a specific joint distribution is given and one wants to find the distribution of Z, it is instructive to get the solution directly without appealing to the formulae.

In what follows it will be assumed that X and Y have a joint probability density function f and a joint distribution function F.

2.1 Distribution of the sum, $Z = X + Y$

As mentioned in the introduction to the chapter, our goal will be to obtain the D.F. of Z. We have

$$F_Z(t) = F_{X+Y}(t) = P(X + Y \leqslant t)$$

Now $P(X + Y \leqslant t)$ is the probability that (X, Y) falls in the shaded region in Figure 2.1, below the line $x + y = t$. Hence the probability is given as the integral of f over the shaded region. Therefore,

$$F_{X+Y}(t) = \int_{-\infty}^{\infty}\int_{-\infty}^{t-y} f(x, y)\, dx\, dy, \qquad -\infty < t < \infty$$

In the integral $\int_{-\infty}^{t-y} f(x, y)\, dx$, let $u = x + y$. (Notice, here y is fixed). Then $dx = du$, and if $x = t - y$, then $u = t$. Hence,

$$\int_{-\infty}^{t-y} f(x, y)\, dx = \int_{-\infty}^{t} f(u - y, y)\, du$$

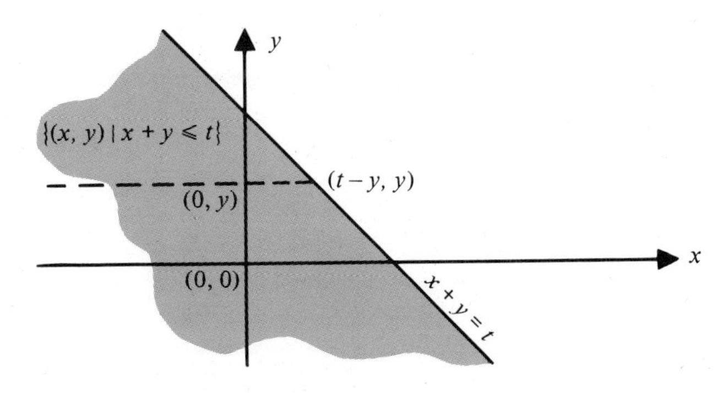

Figure 2.1

Consequently,

$$F_{X+Y}(t) = \int_{-\infty}^{\infty}\left(\int_{-\infty}^{t} f(u-y, y)\, du\right) dy$$
$$= \int_{-\infty}^{t}\left(\int_{-\infty}^{\infty} f(u-y, y)\, dy\right) du$$

interchanging the order of integration.

Therefore, the D.F. of $X + Y$ is given by

$$F_{X+Y}(t) = \int_{-\infty}^{t}\left[\int_{-\infty}^{\infty} f(u-y, y)\, dy\right] du$$

Differentiation gives the pdf of $X + Y$ as

$$f_{X+Y}(t) = \int_{-\infty}^{\infty} f(t-y, y)\, dy, \qquad -\infty < t < \infty$$

In particular, *if X and Y are independent,*

$$f_{X+Y}(t) = \int_{-\infty}^{\infty} f_X(t-y) f_Y(y)\, dy, \qquad -\infty < t < \infty$$

This pdf of $X + Y$ is called the **convolution** of the functions f_X and f_Y.

Example 2.1. Suppose X and Y have a joint pdf given by

$$f(x, y) = \begin{cases} 2e^{-(x+y)}, & 0 \leqslant y \leqslant x < \infty \\ 0, & \text{elsewhere} \end{cases}$$

Find the distribution of $Z = X + Y$.

Solution. From Figure 2.2, we see that the D.F. of Z is given by

$$F_Z(t) = \begin{cases} 0, & t < 0 \\ \int_0^{t/2}\int_y^{t-y} 2e^{-(x+y)} dx\, dy, & t \geqslant 0 \end{cases}$$

Solving the integral gives

$$F_Z(t) = \begin{cases} 0, & t < 0 \\ 1 - e^{-t}(1+t), & t \geqslant 0 \end{cases}$$

Therefore, the pdf of Z is

$$f_Z(t) = \begin{cases} te^{-t}, & t > 0 \\ 0, & \text{elsewhere} \end{cases}$$

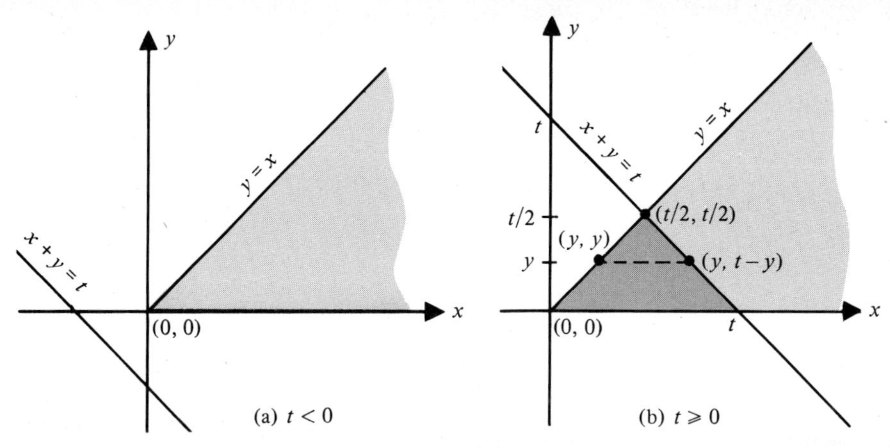

Figure 2.2

Example 2.2. The joint pdf of X and Y is given by

$$f(x, y) = \begin{cases} \dfrac{2y}{x^2}, & 0 \leqslant y \leqslant 1, \quad x \geqslant 1 \\ 0, & \text{elsewhere} \end{cases}$$

Find the distribution of $Z = X + Y$.

Solution. We have three situations to consider, namely, $t < 1$, $1 \leqslant t < 2$, and $t \leqslant 2$, and these are shown respectively in Figures 2.3(a), (b), (c).

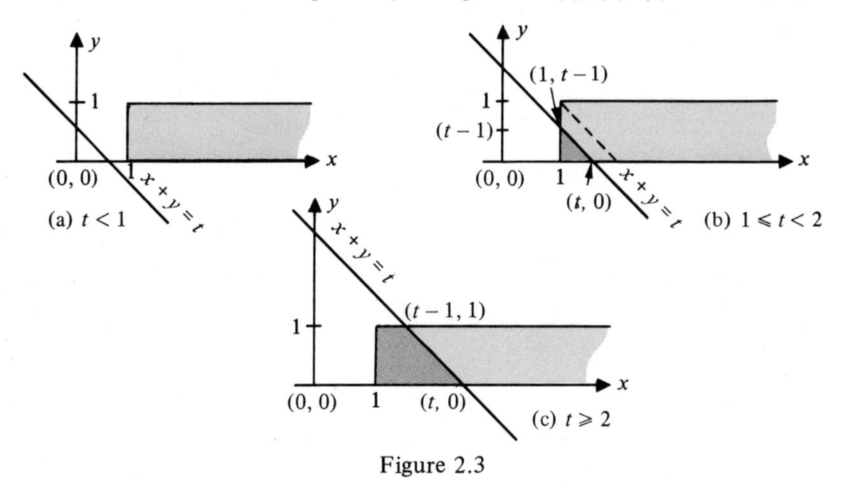

Figure 2.3

From the figures, it follows that the distribution function of Z is

$$F_Z(t) = \begin{cases} 0, & t < 1 \\ \displaystyle\int_0^{t-1}\int_1^{t-y} \frac{2y}{x^2}\,dx\,dy, & 1 \leqslant t < 2 \\ \displaystyle\int_0^{1}\int_1^{t-y} \frac{2y}{x^2}\,dx\,dy, & t \geqslant 2 \end{cases}$$

Calculation of the above integrals gives

$$F_Z(t) = \begin{cases} 0, & t<1 \\ t^2 - 1 - 2t \ln t, & 1 \leqslant t < 2 \\ 2\left[\frac{3}{2} - t \ln \frac{t}{t-1}\right], & t \geqslant 2 \end{cases}$$

Hence,

$$f_Z(t) = \begin{cases} 2(t - 1 - \ln t), & 1 \leqslant t < 2 \\ 2\left[\frac{1}{t-1} - \ln \frac{t}{t-1}\right], & t \geqslant 2 \\ 0, & \text{elsewhere} \end{cases}$$

2.2 Distribution of the product $U = X \cdot Y$

The D.F. of U, $F_U(t) = P(XY \leqslant t)$, is given by integrating the joint pdf over the plane consisting of the points (x, y) for which $xy \leqslant t$. In both cases, whether $t < 0$ (Figure 2.4(a)) or $t \geqslant 0$ (Figure 2.4(b)), this probability is equal to the integral over the shaded region B to the left of the y-axis *plus* the integral over the shaded region A to the right of the y-axis. Hence

$$F_U(t) = \int_{-\infty}^{0} \int_{t/x}^{\infty} f(x, y)\, dy\, dx + \int_{0}^{\infty} \int_{-\infty}^{t/x} f(x, y)\, dy\, dx, \qquad -\infty < t < \infty$$

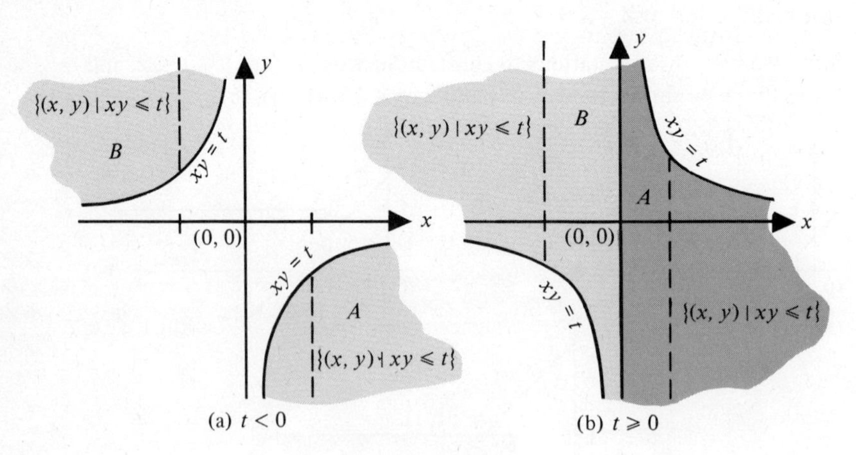

Figure 2.4

Now if we let $u = xy$ (notice, x is fixed), then

$$\int_{t/x}^{\infty} f(x, y)\, dy = \int_{t}^{\infty} f\left(x, \frac{u}{x}\right) \frac{du}{x} \quad \text{and} \quad \int_{-\infty}^{t/x} f(x, y)\, dy = \int_{-\infty}^{t} f\left(x, \frac{u}{x}\right) \frac{du}{x}$$

Therefore,

$$F_U(t) = \int_{-\infty}^{0} \int_{t}^{\infty} f\left(x, \frac{u}{x}\right) \frac{du}{x}\, dx + \int_{0}^{\infty} \int_{-\infty}^{t} f\left(x, \frac{u}{x}\right) \frac{du}{x}\, dx$$

and, interchanging the order of integration,

$$F_U(t) = \int_t^{\infty} \int_{-\infty}^{0} f\left(x, \frac{u}{x}\right) \frac{dx}{x}\, du + \int_{-\infty}^{t} \int_0^{\infty} f\left(x, \frac{u}{x}\right) \frac{dx}{x}\, du$$

Differentiating this gives the pdf of XY as

$$f_U(t) = -\int_{-\infty}^{0} f\left(x, \frac{t}{x}\right) \frac{dx}{x} + \int_0^{\infty} f\left(x, \frac{t}{x}\right) \frac{dx}{x}$$

This can be written more compactly to give

$$f_{XY}(t) = \int_{-\infty}^{\infty} f\left(x, \frac{t}{x}\right) \frac{dx}{|x|}, \qquad -\infty < t < \infty$$

Due to symmetry, we also have

$$f_{XY}(t) = \int_{-\infty}^{\infty} f\left(\frac{t}{y}, y\right) \frac{dy}{|y|}, \qquad -\infty < t < \infty$$

In particular, *if X and Y are independent,*

$$f_{XY}(t) = \int_{-\infty}^{\infty} f_X(x) f_Y\left(\frac{t}{x}\right) \frac{dx}{|x|} = \int_{-\infty}^{\infty} f_X\left(\frac{t}{y}\right) f_Y(y) \frac{dy}{|y|}, \qquad -\infty < t < \infty$$

Example 2.3. If the joint pdf of X and Y is given by

$$f(x, y) = \begin{cases} \dfrac{2y}{x^2}, & 0 \leqslant y \leqslant 1, \ x \geqslant 1 \\ 0, & \text{elsewhere} \end{cases}$$

find the distribution of $U = XY$.

Solution. The three cases that need to be considered, namely, $t < 0$, $0 \leqslant t < 1$, and $t \geqslant 1$, are shown in Figures 2.5(a), (b), and (c), respectively.

From the three figures, we see that

$$P(XY \leqslant t) = \begin{cases} 0, & t < 0 \\ \displaystyle\int_0^t \int_1^{t/y} \frac{2y}{x^2}\, dx\, dy, & 0 \leqslant t < 1 \\ \displaystyle\int_0^1 \int_1^{t/y} \frac{2y}{x^2}\, dx\, dy, & t \geqslant 1 \end{cases}$$

Calculating the above integral yields

$$F_U(t) = \begin{cases} 0, & t < 0 \\ \dfrac{t^2}{3}, & 0 \leqslant t < 1 \\ 1 - \dfrac{2}{3t}, & t \geqslant 1 \end{cases}$$

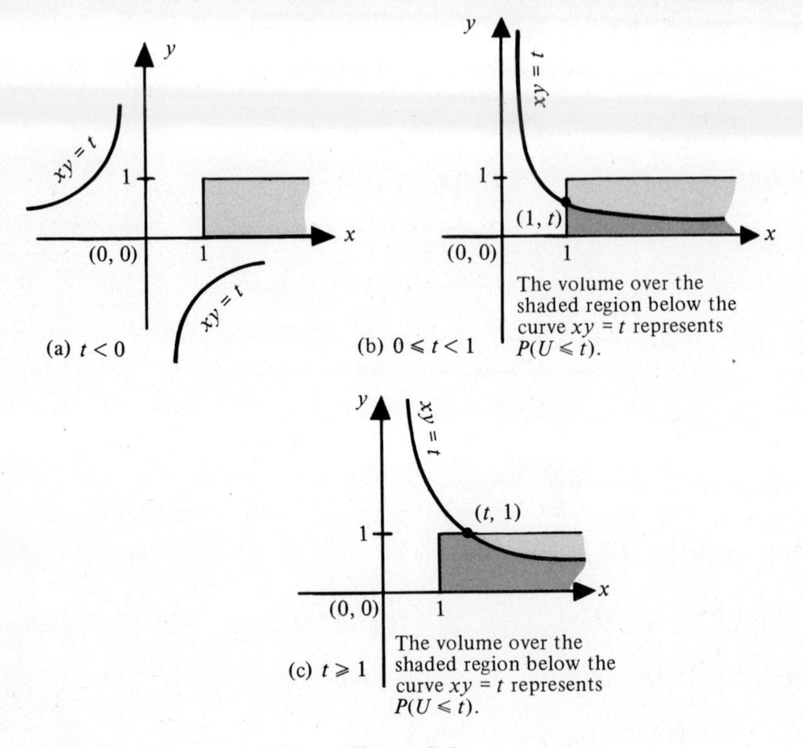

Figure 2.5

By differentiating we get the pdf of XY as

$$f_U(t) = \begin{cases} \dfrac{2t}{3}, & 0 < t < 1 \\ \dfrac{2}{3t^2}, & t \geqslant 1 \\ 0, & \text{elsewhere} \end{cases}$$

2.3 Distribution of the quotient, $Q = X/Y$

In both cases, whether $t < 0$ or $t \geqslant 0$, the D.F. of Q is given as the sum of the integrals of the joint pdf over the regions B and A in Figure 2.6. Therefore,

$$F_{X/Y}(t) = \int_0^\infty \int_{-\infty}^{ty} f(x, y)\, dx\, dy + \int_{-\infty}^0 \int_{ty}^\infty f(x, y)\, dx\, dy, \qquad -\infty < t < \infty$$

Letting $u = x/y$, where y is fixed, we get

$$\int_{-\infty}^{ty} f(x, y)\, dx = \int_{-\infty}^{t} f(uy, y) y\, du \quad \text{and} \quad \int_{ty}^{\infty} f(x, y)\, dx = \int_{t}^{\infty} f(uy, y) y\, du$$

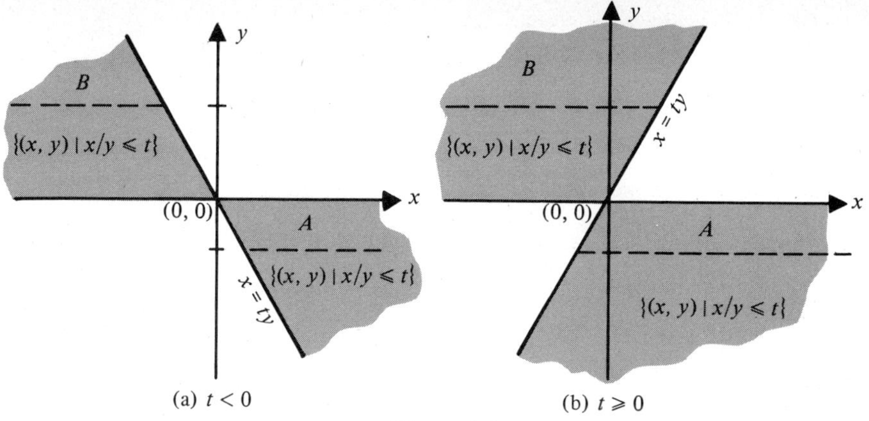

Figure 2.6

Therefore,

$$F_{X/Y}(t) = \int_0^\infty \int_{-\infty}^t f(uy, y)y\, du\, dy + \int_{-\infty}^0 \int_t^\infty f(uy, y)y\, du\, dy$$

and, interchanging the order of integration,

$$F_{X/Y}(t) = \int_{-\infty}^t \int_0^\infty f(uy, y)y\, dy\, du + \int_t^\infty \int_{-\infty}^0 f(uy, y)y\, dy\, du$$

If we differentiate, we get the pdf of X/Y as

$$f_{X/Y}(t) = \int_0^\infty f(ty, y)y\, dy - \int_{-\infty}^0 f(ty, y)y\, dy, \qquad -\infty < t < \infty$$

Rewriting, it follows that

$$f_{X/Y}(t) = \int_{-\infty}^\infty |y| f(ty, y)\, dy, \qquad -\infty < t < \infty$$

In particular, *if X and Y are independent,*

$$f_{X/Y}(t) = \int_{-\infty}^\infty |y| f_X(ty) \cdot f_Y(y)\, dy, \qquad -\infty < t < \infty$$

Example 2.4. If the joint distribution of X and Y is given by

$$f(x, y) = \begin{cases} 2e^{-(x+y)}, & 0 \leq y \leq x < \infty \\ 0, & \text{elsewhere} \end{cases}$$

find the distribution of X/Y.

Solution. (*Note:* f is positive in the part of the plane between the lines $y = x$ and the x-axis, and is zero outside this region.)

The D.F. of X/Y is given as

$$F_{X/Y}(t) = \begin{cases} 0, & t \leq 1 \\ 2\int_0^\infty \int_y^{ty} e^{-(x+y)}\, dx\, dy, & t > 1 \end{cases}$$

(The case $t > 1$ is shown in Figure 2.7.)

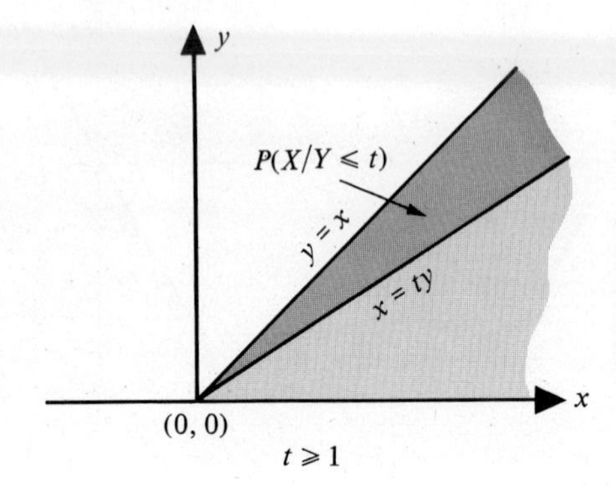

Figure 2.7

Therefore,

$$F_{X/Y}(t) = \begin{cases} 0, & t \leqslant 1 \\ 2\int_0^\infty [e^{-2y} - e^{-(t+1)y}]\,dy, & t > 1 \end{cases}$$

$$= \begin{cases} 0, & t \leqslant 1 \\ 1 - \dfrac{2}{1+t}, & t > 1 \end{cases}$$

Hence

$$f_{X/Y}(t) = \begin{cases} \dfrac{2}{(1+t)^2}, & t > 1 \\ 0, & \text{elsewhere} \end{cases}$$

Example 2.5. Let X and Y have the following joint pdf:

$$f(x, y) = \begin{cases} \dfrac{y}{x^2}, & 0 \leqslant y \leqslant 1, \ |x| \geqslant 1 \\ 0, & \text{elsewhere} \end{cases}$$

Find the distribution of X/Y.

Solution. The joint pdf of X and Y is positive over the shaded region shown in Figure 2.8, and is zero outside it. Also, the integral of the pdf over the shaded region to the left of the y-axis is equal to the integral over the shaded region to the right of the y-axis (why?), and hence equal to $\frac{1}{2}$.

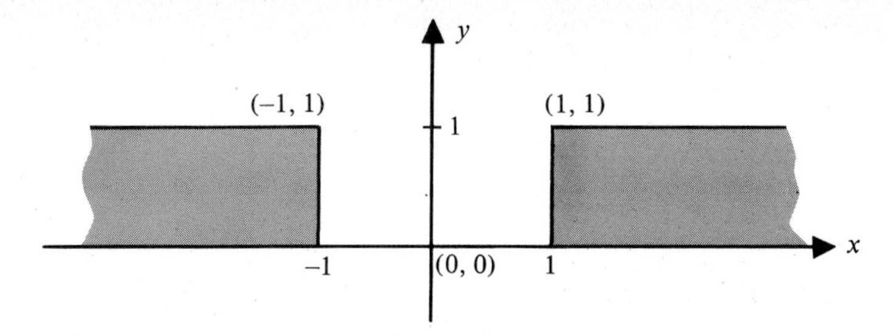

Figure 2.8

To find the D.F. of X/Y, consider the situations as shown in Figures 2.9(a), (b), and (c), corresponding to $t \leqslant -1$, $-1 < t < 1$, and $t \geqslant 1$, respectively.

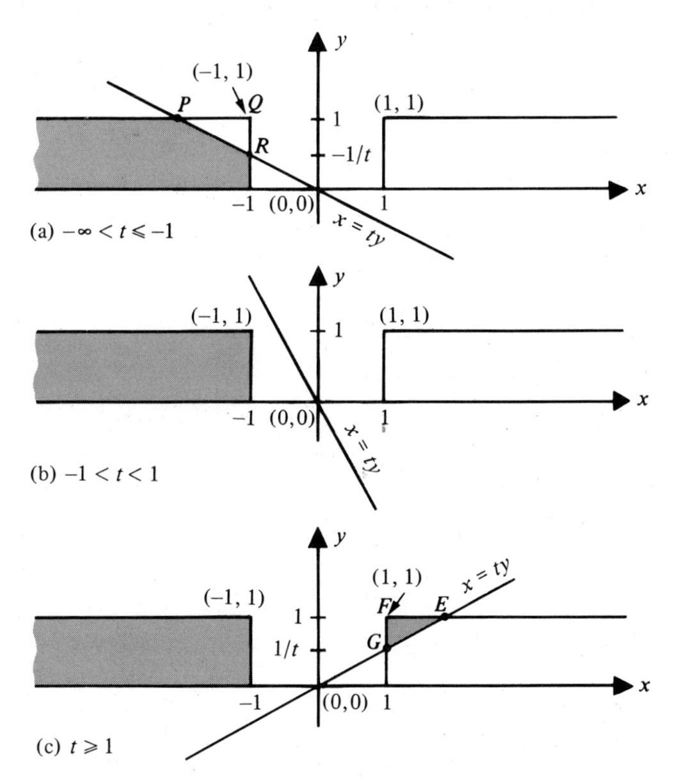

Figure 2.9

From the figures, we see that the D.F. of X/Y can be obtained as follows:

$$(i) \quad F_{X/Y}(t) = \frac{1}{2} - \int_{-1/t}^{1} \int_{ty}^{-1} \frac{y}{x^2}\, dx\, dy, \quad -\infty < t \leqslant -1$$

(This is obtained from Figure 2.9(a) as $\frac{1}{2}$ minus the integral over the unshaded triangle PQR). Further,

$$(ii) \quad F_{X/Y}(t) = \frac{1}{2}, \qquad -1 < t < 1$$

(This can be seen from Figure 2.9(b)). Finally,

$$(iii) \quad F_{X/Y}(t) = \frac{1}{2} + \int_{1/t}^{1} \int_{1}^{ty} \frac{y}{x^2}\, dx\, dy, \qquad t \geqslant 1$$

(This is found from Figure 2.9(c) as $\frac{1}{2}$ plus the integral over the shaded triangle EFG.)

Hence, computing the integrals in (*i*), (*ii*), (*iii*),

$$F_{X/Y}(t) = \begin{cases} -\dfrac{1}{t} - \dfrac{1}{2t^2}, & -\infty < t \leqslant -1 \\ \dfrac{1}{2}, & -1 < t < 1 \\ 1 - \dfrac{1}{t} + \dfrac{1}{2t^2}, & t \geqslant 1 \end{cases}$$

Differentiating, the pdf of X/Y is

$$f_{X/Y}(t) = \begin{cases} \dfrac{1}{t^2} + \dfrac{1}{t^3}, & t \leqslant -1 \\ \dfrac{1}{t^2} - \dfrac{1}{t^3}, & t \geqslant 1 \\ 0, & \text{elsewhere} \end{cases}$$

2.4 Distribution of the maximum, $V = \max(X, Y)$

In this case, a straightforward and intuitively suggestive argument is the following:

For any real number t, $\max(X, Y) \leqslant t$ if and only if *each* of the random variables X, Y is less than or equal to t; that is, if and only if $X \leqslant t$ and $Y \leqslant t$. Hence

$$P(\max(X, Y) \leqslant t) = P(X \leqslant t,\ Y \leqslant t)$$

In other words,

$$F_V(t) = P(X \leqslant t,\ Y \leqslant t) = F(t, t), \qquad -\infty < t < \infty$$

Alternatively, we can derive $F_V(t)$ by adopting our standard argument, noting that $F_V(t)$ is the integral over the shaded region in Figure 2.10.

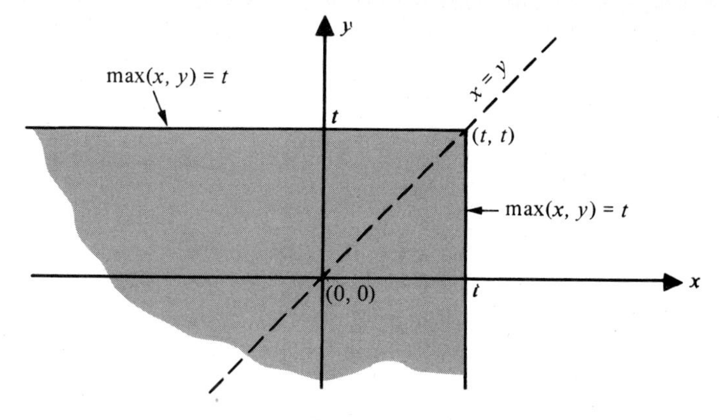

Figure 2.10

Notice that: (*i*) if $x \geqslant y$, then $\max(x, y) = x$, and $\max(x, y) = t$ implies $x = t$; (*ii*) if $x < y$, then $\max(x, y) = y$, and $\max(x, y) = t$ implies $y = t$.

If X and Y are independent, we get

$$F_V(t) = F_X(t)F_Y(t), \qquad -\infty < t < \infty$$

and differentiation with respect to t gives the pdf of $\max(X, Y)$ as

$$f_V(t) = F_X(t)f_Y(t) + f_X(t)F_Y(t), \qquad -\infty < t < \infty$$

In general, if $X_1, X_2, \ldots, X_n$ are n random variables and if we define $V = \max(X_1, X_2, \ldots, X_n)$, then

$$F_V(t) = F_{X_1, X_2, \ldots, X_n}(t, t, \ldots, t), \qquad -\infty < t < \infty$$

Furthermore, *if the random variables are independent,*

$$F_V(t) = F_{X_1}(t)F_{X_2}(t) \ldots F_{X_n}(t), \qquad -\infty < t < \infty$$

Example 2.6. Suppose X and Y have a joint pdf given by

$$f(x, y) = \begin{cases} 2(x + y - 3xy^2), & 0 < x < 1, \ 0 < y < 1 \\ 0, & \text{elsewhere} \end{cases}$$

Find the distribution of $V = \max(X, Y)$

Solution. The joint pdf is zero outside the shaded square *PABC* in Figure 2.11.

We get the D.F. of V in an obvious way as

$$F_V(t) = \begin{cases} 0, & t < 0 \\ \int_0^t \int_0^t 2(x + y - 3xy^2)\, dx\, dy, & 0 \leqslant t < 1 \\ 1, & t \geqslant 1 \end{cases}$$

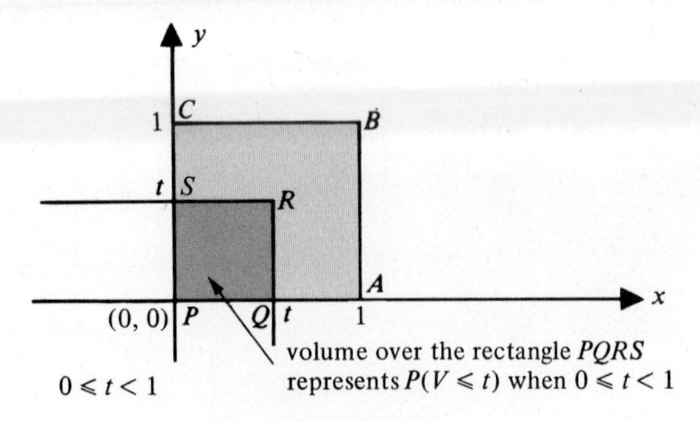

Figure 2.11

(The situation $0 \leqslant t < 1$ is shown in Figure 2.11.)

Upon calculation of the integral, this yields

$$F_V(t) = \begin{cases} 0, & t < 0 \\ t^3\left(\frac{3}{2} - \frac{t^2}{2}\right), & 0 \leqslant t < 1 \\ 1, & t \geqslant 1 \end{cases}$$

The pdf of V is

$$f_V(t) = \begin{cases} \frac{t^2}{2}(9 - 5t^2), & 0 < t < 1 \\ 0, & \text{elsewhere} \end{cases}$$

Example 2.7. If X and Y have the joint pdf given by

$$f(x, y) = \begin{cases} ye^{-xy}, & 0 < x < \infty, \ 0 < y < 1 \\ 0, & \text{elsewhere} \end{cases}$$

find the distribution of $V = \max(X, Y)$.

Solution. $F_V(t)$ is the volume over rectangle $ABCD$ in Figure 2.12 where Figure 2.12(a) considers the case $0 < t < 1$ and Figure 2.12(b) the case $t \geqslant 1$. Of course, $F_V(t) = 0$ if $t < 0$.

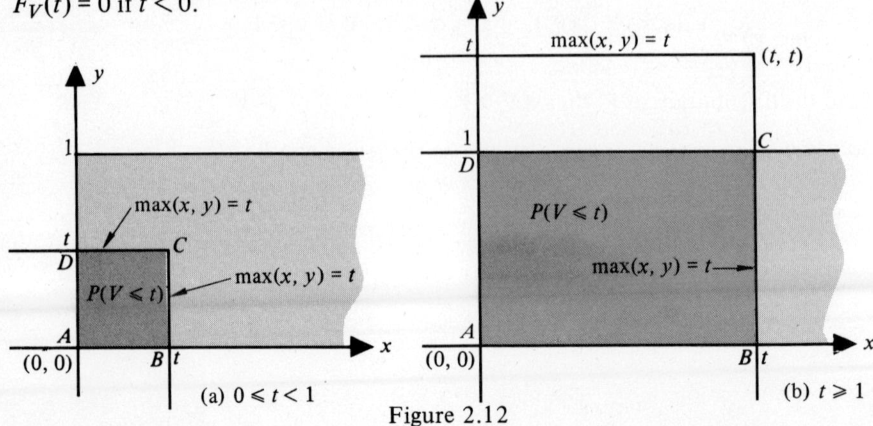

Figure 2.12

Hence it follows that

$$F_V(t) = \begin{cases} 0, & t \leqslant 0 \\ \int_0^t \int_0^t ye^{-xy}\,dx\,dy, & 0 < t < 1 \\ \int_0^1 \int_0^t ye^{-xy}\,dx\,dy, & t \geqslant 1 \end{cases}$$

$$= \begin{cases} 0, & t \leqslant 0 \\ t + \dfrac{1}{t}(e^{-t^2} - 1), & 0 < t < 1 \\ 1 - \dfrac{1}{t}(1 - e^{-t}), & t \geqslant 1 \end{cases}$$

From this, the pdf can be obtained by differentiation.

2.5 Distribution of the minimum, $W = \min(X, Y)$

Arguing as for the distribution of the maximum, it can be easily seen that $P(\min(X, Y) \leqslant t)$ is equal to the integral over the shaded region in Figure 2.13. However, we shall adopt the following suggestive argument. We have

$$F_W(t) = P(\min(X, Y) \leqslant t) = 1 - P(\min(X, Y) > t)$$

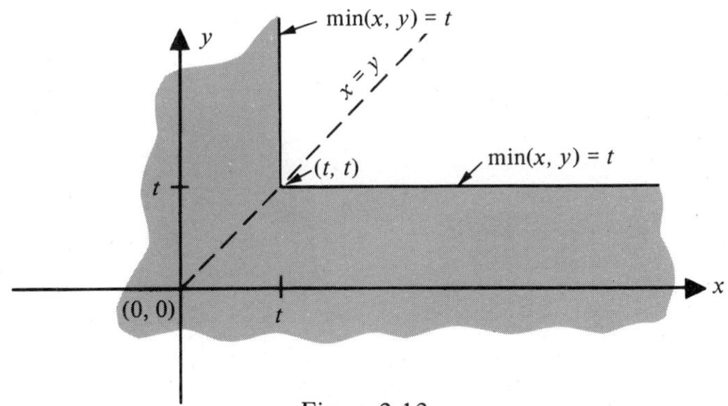

Figure 2.13

But $\min(X, Y) > t$ if and only if $X > t$ *and* $Y > t$. Therefore,

$$F_W(t) = P(\min(X, Y) \leqslant t) = 1 - P(X > t,\ Y > t), \qquad -\infty < t < \infty$$

(Observe that $P(X > t,\ Y > t)$ is the integral over the unshaded region in Figure 2.13.)

If X and Y are independent, this yields

$$F_W(t) = 1 - P(X > t) \cdot P(Y > t) = 1 - (1 - F_X(t))(1 - F_Y(t)), \qquad -\infty < t < \infty$$

In general, if $X_1, X_2, \ldots, X_n$ are n random variables and if we define $W = \min(X_1, X_2, \ldots, X_n)$, we have

$$F_W(t) = 1 - P(X_1 > t,\ X_2 > t, \ldots, X_n > t), \qquad -\infty < t < \infty$$

If, moreover, the random variables are independent,

$$\begin{aligned} F_W(t) &= 1 - P(X_1 > t)P(X_2 > t) \ldots P(X_n > t) \\ &= 1 - (1 - F_{X_1}(t))(1 - F_{X_2}(t)) \ldots (1 - F_{X_n}(t)) \end{aligned}$$

Example 2.8. Suppose X and Y have the following joint pdf:

$$f(x, y) = \begin{cases} 2(x + y - 3xy^2), & 0 < x < 1 \text{ and } 0 < y < 1 \\ 0, & \text{elsewhere} \end{cases}$$

Find the distribution of $W = \min(X, Y)$.

Solution. The D.F. of W is given as

$$F_W(t) = \begin{cases} 0, & t < 0 \\ 1 - \int_t^1 \int_t^1 2(x + y - 3xy^2)\, dx\, dy, & 0 \leq t < 1 \\ 1, & t \geq 1 \end{cases}$$

(If $0 \leq t < 1$, from Figure 2.14 it can be seen that $P(W \leq t)$ is 1 minus the integral over the unshaded rectangle $PQRS$.)

Simplification gives

$$F_W(t) = \begin{cases} 0, & t < 0 \\ 2t + t^2 - 3t^3 + t^5, & 0 \leq t < 1 \\ 1, & t \geq 1 \end{cases}$$

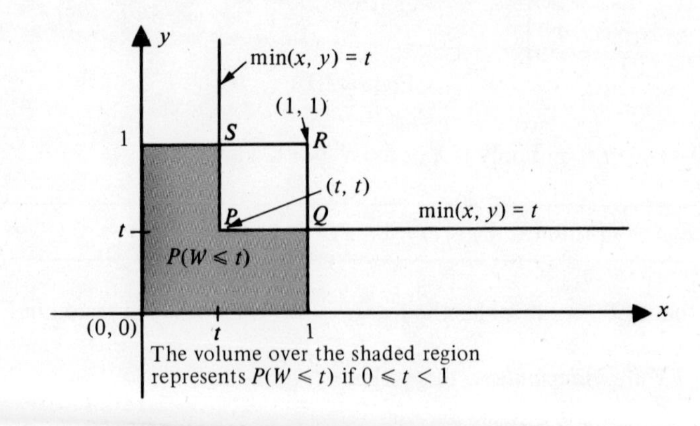

Figure 2.14

Therefore, the pdf of W is

$$f_W(t) = \begin{cases} 2 + 2t - 9t^2 + 5t^4, & 0 < t < 1 \\ 0, & \text{elsewhere} \end{cases}$$

Example 2.9. The joint pdf of X and Y is given by

$$f(x, y) = \begin{cases} ye^{-xy}, & 0 < x < \infty, \ 0 < y < 1 \\ 0, & \text{elsewhere} \end{cases}$$

Find the distribution of $W = \min(X, Y)$.

Solution. (The pdf is zero outside the strip limited by $y = 1$, $y = 0$, and $x \geqslant 0$ in Figure 2.15.)

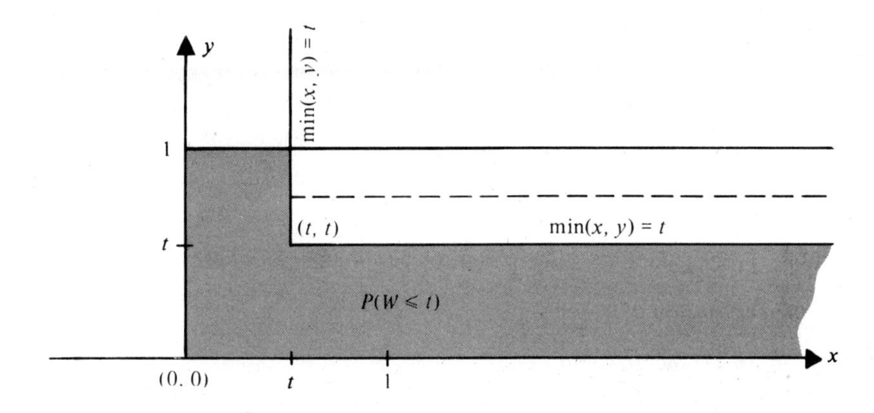

Figure 2.15

The D.F. of W is given as

$$F_W(t) = \begin{cases} 0, & t \leqslant 0 \\ 1 - \int_t^1 \int_t^\infty ye^{-xy}\,dx\,dy, & 0 < t < 1 \\ 1, & t \geqslant 1 \end{cases}$$

(See Figure 2.15 for the case $0 < t < 1$.) After simplifying, this yields

$$F_W(t) = \begin{cases} 0, & t \leqslant 0 \\ 1 + \frac{1}{t}(e^{-t} - e^{-t^2}), & 0 < t < 1 \\ 1, & t \geqslant 1 \end{cases}$$

Hence the pdf of W is

$$f_W(t) = \begin{cases} \frac{e^{-t^2}}{t^2}(2t^2 + 1) - \frac{e^{-t}}{t^2}(t + 1), & 0 < t < 1 \\ 0, & \text{elsewhere} \end{cases}$$

Comment. There are some situations where one can easily identify the random variables $V = \max(X, Y)$, $W = \min(X, Y)$. To be specific, consider the distribution given by the pdf

$$f(x, y) = \begin{cases} 2e^{-(x+y)}, & 0 \leqslant y < x < \infty \\ 0, & \text{elsewhere} \end{cases}$$

In this case, $\max(X, Y) = X$ and $\min(X, Y) = Y$, in view of the fact that $f(x, y)$ is positive for $0 \leqslant y < x < \infty$. Hence, finding the distributions of $\max(X, Y)$ and $\min(X, Y)$ amounts to finding the marginal distributions of X and Y respectively.

The random variables $\max(X_1, X_2, \ldots, X_n)$ and $\min(X_1, X_2, \ldots, X_n)$ are referred to as the **extremes.** Their distributions are very important in reliability and renewal theory. To realize the importance of the distribution of these random variables, consider the following situation: Suppose a machine runs on n electronic components. Let X_i denote the life of the ith component. If the machine breaks down as soon as one component goes bad, then we would be interested in the distribution of $\min(X_1, X_2, \ldots, X_n)$. On the other hand, if the machine breaks down when the last component goes bad, then our interest would lie in finding the distribution of $\max(X_1, X_2, \ldots, X_n)$.

EXERCISES–SECTION 2

Note: Since computation of the integrals in many of the following problems can be quite involved, it will be sufficient to just set the integrals up.

1. If X and Y are independent random variables with the pdf's

$$f_X(x) = \begin{cases} 1, & 1 < x < 2 \\ 0, & \text{elsewhere} \end{cases}$$

$$f_Y(y) = \begin{cases} e^{-(y-1)}, & y > 1 \\ 0, & \text{elsewhere} \end{cases}$$

find the distribution functions of–
(a) $X + Y$ (b) XY (c) X/Y (d) $\max(X, Y)$

2. Suppose X and Y are independent, identically distributed random variables, each having the following pdf:

$$f_X(u) = f_Y(u) = \begin{cases} \dfrac{100}{u^2}, & u \geqslant 100 \\ 0, & \text{elsewhere} \end{cases}$$

Find the distribution functions of–
(a) $X + Y$ (b) XY (c) X/Y (d) $\max(X, Y)$
(e) $\min(X, Y)$

3. Find the distribution functions of
(a) $X + Y$ (b) XY (c) $\max(X, Y)$
if X and Y are independent with the following pdf's:

$$f_X(x) = \begin{cases} 2x, & 0 < x < 1 \\ 0, & \text{elsewhere} \end{cases}$$

and

$$f_Y(y) = \begin{cases} 2(1-y), & 0 < y < 1 \\ 0, & \text{elsewhere} \end{cases}$$

4. Find the distribution functions of
   (a) $X + Y$ (b) XY (c) X/Y
if X and Y are independent random variables with the following pdf's:

$$f_X(x) = \begin{cases} 2x, & 0 < x < 1 \\ 0, & \text{elsewhere} \end{cases}$$

$$f_Y(y) = \begin{cases} \dfrac{100}{y^2}, & y \geqslant 100 \\ 0, & \text{elsewhere} \end{cases}$$

5. The joint pdf of X and Y is given by

$$f(x, y) = \begin{cases} \dfrac{k}{(1+x+y)^3}, & x > 0,\ y > 0 \\ 0, & \text{elsewhere} \end{cases}$$

where k is a constant. Evaluate k and obtain the distribution function of $Z = X + Y$.
6. The random variables X and Y have the following joint pdf:

$$f(x, y) = \begin{cases} \frac{1}{4}[1 + xy(x^2 - y^2)], & -1 \leqslant x \leqslant 1, \quad -1 \leqslant y \leqslant 1 \\ 0, & \text{elsewhere} \end{cases}$$

Find the distribution of $Z = X + Y$.
7. Find the pdf of $Z = XY$ if the joint pdf of X and Y is given by

$$f(x, y) = \begin{cases} xe^{-x(1+y)}, & x > 0,\ y > 0 \\ 0, & \text{elsewhere} \end{cases}$$

8. Find the distribution function of $V = \max(X, Y)$ if the joint pdf of X and Y is given by

$$f(x, y) = \begin{cases} xe^{-x(1+y)}, & x > 0,\ y > 0 \\ 0, & \text{elsewhere} \end{cases}$$

3. MISCELLANEOUS EXAMPLES

We present the following miscellaneous examples in the continuous case as a separate section in view of the importance of the results contained therein.

Example 3.1. Let X and Y be independent and uniformly distributed over the interval $[0, 1]$. Find the distributions of the following random variables:
(*a*) $X + Y$ (*b*) XY (*c*) X/Y (*d*) $\max(X, Y)$

Solution. The joint distribution of X and Y is given by $f(x, y) = f_X(x) \cdot f_Y(y)$, since X and Y are independent. Therefore, the joint distribution is uniform over the rectangle $\{(x, y) \mid 0 \leqslant x \leqslant 1,\ 0 \leqslant y \leqslant 1\}$, since

$$f(x, y) = \begin{cases} 1, & 0 \leqslant x \leqslant 1,\ 0 \leqslant y \leqslant 1 \\ 0, & \text{elsewhere} \end{cases}$$

(*a*) *The distribution of* $X + Y$.

Since $f(x, y) = 1$ for every point (x, y) in the square, the integral over any region in this rectangle is simply the area of the region.

Hence, appealing to the geometry of the figure,

$$F_{X+Y}(t) = P(X + Y \leqslant t) = \begin{cases} 0, & t < 0 \\ \dfrac{t^2}{2}, & 0 \leqslant t < 1 \\ 1 - \dfrac{(2-t)^2}{2}, & 1 \leqslant t < 2 \\ 1, & t \geqslant 2 \end{cases}$$

The cases $0 \leqslant t < 1$ and $1 \leqslant t < 2$ are shown in Figures 3.1(a) and (b), respectively. For example, it follows from Figure 3.1(b) that, for $1 \leqslant t \leqslant 2$,

$$P(X + Y \leqslant t) = 1 - (\text{area of the unshaded triangle } PQR) = 1 - \frac{(2-t)^2}{2}$$

The pdf of $X + Y$ is therefore given as

$$f_{X+Y}(t) = \begin{cases} t, & 0 \leqslant t < 1 \\ 2 - t, & 1 \leqslant t \leqslant 2 \\ 0, & \text{elsewhere} \end{cases}$$

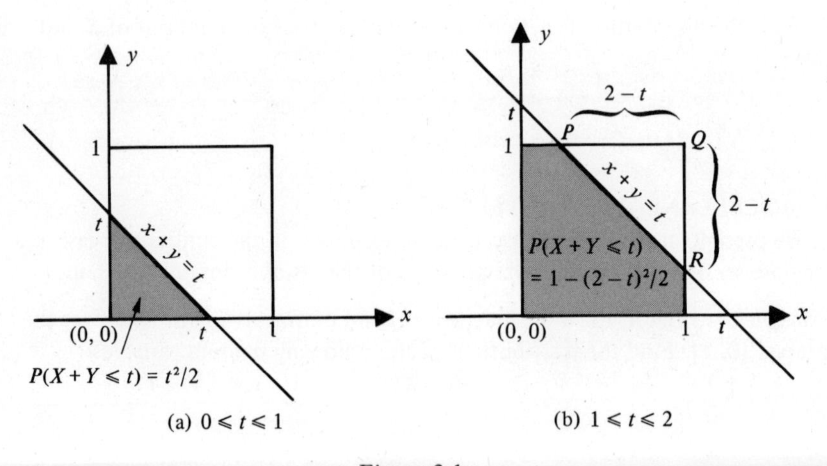

Figure 3.1

The graph of the pdf of $X + Y$ is given in Figure 3.2. Because of its shape, the distribution is called the **triangular distribution.**

In conclusion, *if X and Y are independent and uniformly distributed random variables over the interval $[0, 1]$, then $X + Y$ has the* ***triangular*** *distribution as given above.*

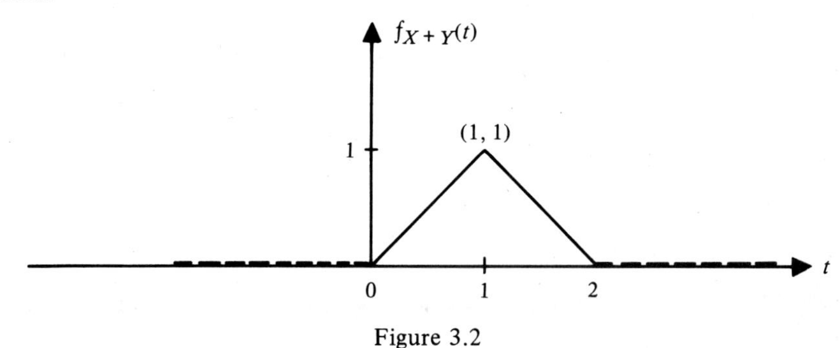

Figure 3.2

(*b*) *The distribution of XY.*

To find the D.F., the case $0 \leqslant t < 1$ is indicated in Figure 3.3. We see that:

$$F_{XY}(t) = \begin{cases} 0, & t < 0 \\ 1 - \begin{pmatrix} \text{area of the unshaded} \\ \text{region of the rectangle} \end{pmatrix}, & 0 \leqslant t < 1 \\ 1, & t \geqslant 1 \end{cases}$$

$$= \begin{cases} 0, & t < 0 \\ 1 - \int_t^1 \int_{t/y}^1 dx\, dy, & 0 \leqslant t < 1 \\ 1, & t \geqslant 1 \end{cases}$$

$$= \begin{cases} 0, & t < 0 \\ t - t \ln t, & 0 \leqslant t < 1 \\ 1, & t \geqslant 1 \end{cases}$$

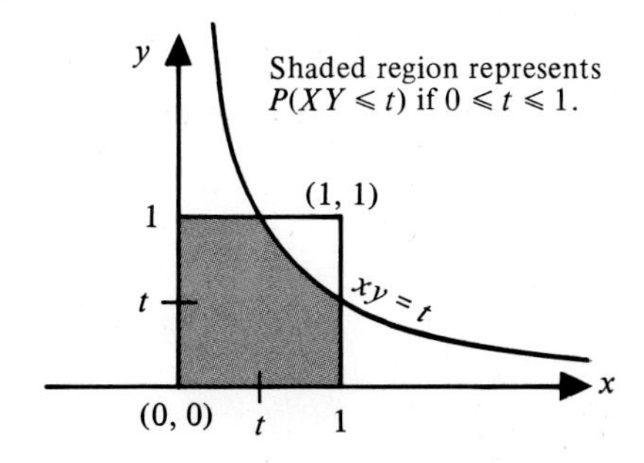

Figure 3.3

Therefore,

$$f_{XY}(t) = \begin{cases} -\ln t, & 0 < t < 1 \\ 0, & \text{elsewhere} \end{cases}$$

(c) *The distribution of X/Y.*

From Figures 3.4(a) and (b), it can be seen that

$$F_{X/Y}(t) = \begin{cases} 0, & t < 0 \\ \dfrac{t}{2}, & 0 \leqslant t < 1 \\ 1 - \dfrac{1}{2t}, & t \geqslant 1 \end{cases}$$

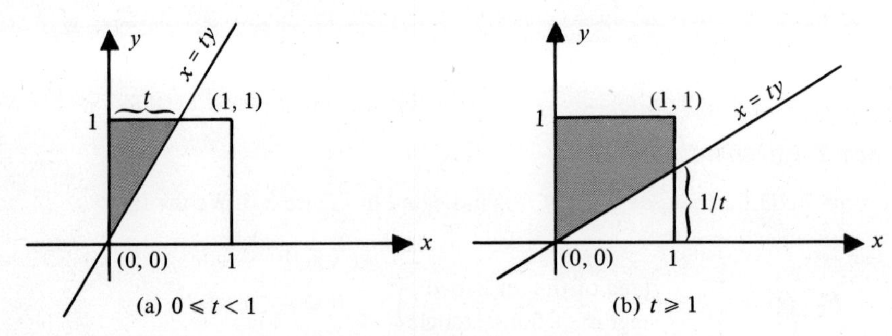

Figure 3.4

(Observe, for example, that for $t \geqslant 1$ (Figure 3.4(b)) the area of the unshaded triangle is $1/(2t)$ and consequently $F_{X/Y}(t) = 1 - [1/(2t)]$.)

Therefore,

$$f_{X/Y}(t) = \begin{cases} \dfrac{1}{2}, & 0 \leqslant t < 1 \\ \dfrac{1}{2t^2}, & t \geqslant 1 \\ 0, & \text{elsewhere} \end{cases}$$

(d) *The distribution of max(X, Y).*

From Figure 3.5 it can be seen that

$$F_{\max(X,\,Y)}(t) = \begin{cases} 0, & t < 0 \\ t^2, & 0 \leqslant t < 1 \\ 1, & t \geqslant 1 \end{cases}$$

and consequently

$$f_{\max(X,\,Y)}(t) = \begin{cases} 2t, & 0 < t < 1 \\ 0, & \text{elsewhere} \end{cases}$$

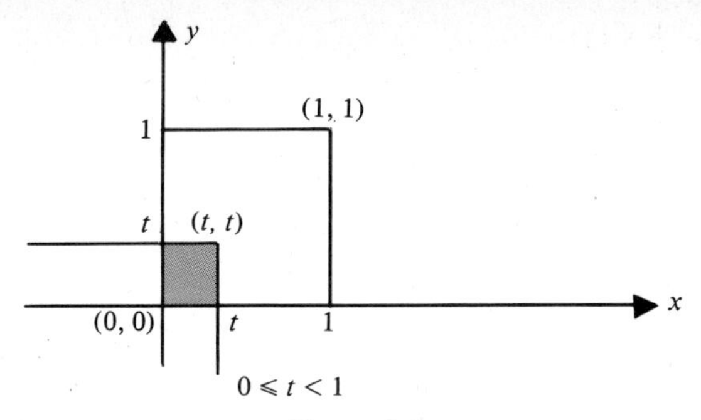

Figure 3.5

The D.F. of max(X, Y) can be obtained alternatively by observing that

$$\begin{aligned} P(\max(X, Y) \leqslant t) &= P(X \leqslant t, Y \leqslant t) \\ &= P(X \leqslant t) \cdot P(Y \leqslant t), \qquad \text{since } X \text{ and } Y \text{ are independent} \\ &= \begin{cases} 0, & t < 0 \\ t^2, & 0 \leqslant t < 1 \\ 1, & t \geqslant 1 \end{cases} \end{aligned}$$

Example 3.2. Suppose X and Y are independent random variables, each with the same exponential distribution with the pdf

$$f(x) = \begin{cases} e^{-x}, & x > 0 \\ 0, & \text{elsewhere} \end{cases}$$

Find the distribution of the following random variables:

(*a*) $X + Y$ (*b*) $X - Y$ (*c*) X/Y

(*d*) max(X, Y) (*e*) min(X, Y)

Solution. Since X and Y are independent, they have a joint distribution with the following pdf:

$$f(x, y) = f_X(x) \cdot f_Y(y) = \begin{cases} e^{-(x+y)}, & x > 0, \ y > 0 \\ 0, & \text{elsewhere} \end{cases}$$

(Note that the pdf is zero everywhere except in the first quadrant.)

The reader is invited to fill in the details in the following:

(*a*) *The distribution of* $X + Y$.

$$F_{X+Y}(t) = \begin{cases} 0, & t < 0 \\ \int_0^t \int_0^{t-x} e^{-(x+y)}\, dy\, dx, & t \geqslant 0 \end{cases}$$

$$= \begin{cases} 0, & t < 0 \\ 1 - e^{-t} - te^{-t}, & t \geqslant 0 \end{cases}$$

(See Figure 3.6.)

Differentiating, the pdf of $X + Y$ is

$$f_{X+Y}(t) = \begin{cases} te^{-t}, & t > 0 \\ 0, & \text{elsewhere} \end{cases}$$

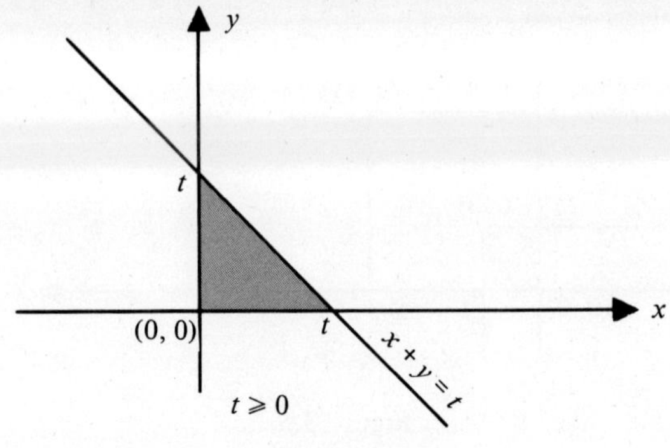

Figure 3.6

Thus, *the distribution of $X + Y$ is gamma with the parameters $\lambda = 1$ and $p = 2$.*

(*b*) *The distribution of $X - Y$.*

In the general case, $F_{X-Y}(t) = P(X - Y \leqslant t)$ will be represented by the integral over the region *above* the line $x - y = t$. In our example, since the joint pdf is zero everywhere except in the first quadrant, it can be seen from Figures 3.7(a) and (b) that the D.F. of $X - Y$ is given by

$$F_{X-Y}(t) = \begin{cases} \int_{-t}^{\infty}\int_{0}^{t+y} e^{-x-y}\,dx\,dy, & t < 0 \\ \int_{0}^{\infty}\int_{0}^{t+y} e^{-x-y}\,dx\,dy, & t \geqslant 0 \end{cases}$$

$$= \begin{cases} \dfrac{e^{t}}{2}, & t < 0 \\ 1 - \dfrac{e^{-t}}{2}, & t \geqslant 0 \end{cases}$$

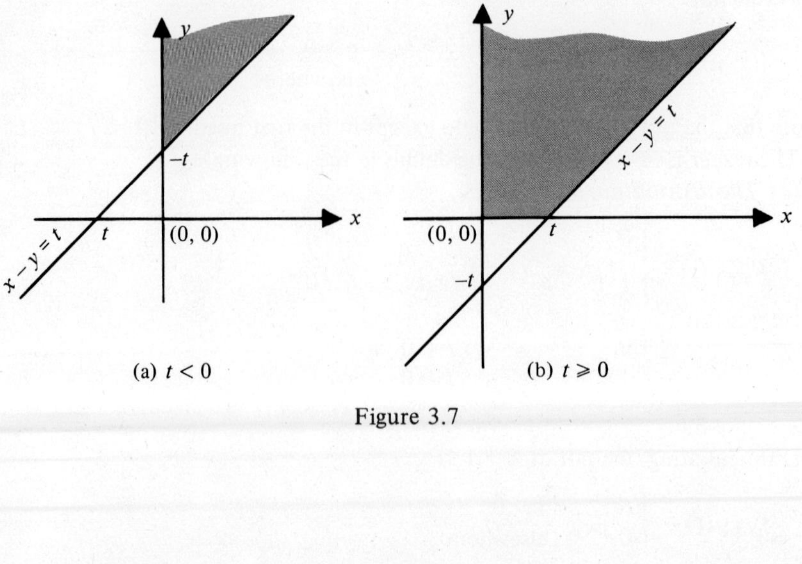

(a) $t < 0$

(b) $t \geqslant 0$

Figure 3.7

Therefore, the pdf of $X-Y$ is

$$f_{X-Y}(t)=\begin{cases}\frac{1}{2}e^{t}, & t<0\\ \frac{1}{2}e^{-t}, & t\geqslant 0\end{cases}$$

That is,

$$f_{X-Y}(t)=\tfrac{1}{2}e^{-|t|},\qquad -\infty<t<\infty$$

In other words, *the distribution of $X-Y$ is a Laplace distribution with the parameters $a=0$ and $b=1$.*

(*c*) *The distribution of X/Y.*

We have

$$F_{X/Y}(t)=\begin{cases}0, & t\leqslant 0\\ \int_0^{\infty}\int_0^{ty} e^{-x-y}\,dx\,dy, & t>0\end{cases}$$

(See Figure 3.8 for the case $t>0$.) Then

$$F_{X/Y}(t)=\begin{cases}0, & t\leqslant 0\\ \int_0^{\infty} e^{-y}[1-e^{-ty}]\,dy, & t>0\end{cases}$$

$$=\begin{cases}0, & t\leqslant 0\\ 1-\dfrac{1}{1+t}, & t>0\end{cases}$$

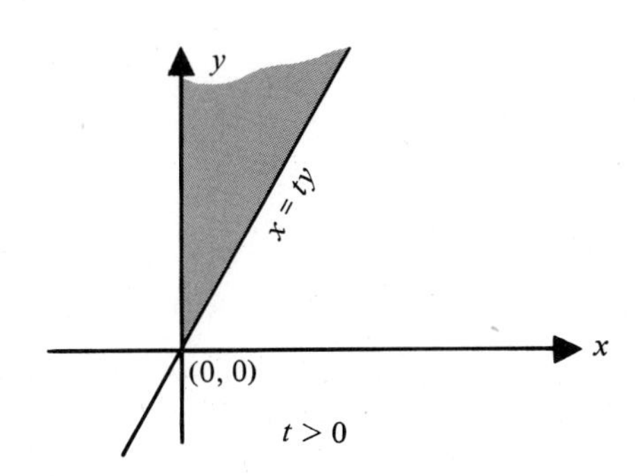

Figure 3.8

Therefore, the pdf of X/Y is

$$f_{X/Y}(t)=\begin{cases}\dfrac{1}{(1+t)^2}, & t>0\\ 0, & \text{elsewhere}\end{cases}$$

(*d*) *The distribution of $\max(X, Y)$.*

We have $P(\max(X,Y)\leqslant t)=P(X\leqslant t)\cdot P(Y\leqslant t)$

since X and Y are independent. Now

$$P(X \leqslant t) = P(Y \leqslant t) = \begin{cases} 0, & t < 0 \\ 1 - e^{-t}, & t \geqslant 0 \end{cases}$$

Therefore,

$$P(\max(X, Y) \leqslant t) = \begin{cases} 0, & t < 0 \\ (1 - e^{-t})(1 - e^{-t}), & t \geqslant 0 \end{cases}$$

And, consequently,

$$f_{\max(X, Y)}(t) = \begin{cases} 2e^{-t}(1 - e^{-t}), & t \geqslant 0 \\ 0, & \text{elsewhere} \end{cases}$$

(e) *The distribution of min(X, Y).*

We have $P(\min(X, Y) \leqslant t) = 1 - P(X > t) \cdot P(Y > t)$

since X and Y are independent. Now

$$P(X > t) = \begin{cases} e^{-t}, & t > 0 \\ 1, & t \leqslant 0 \end{cases}$$

Therefore,

$$P(\min(X, Y) \leqslant t) = \begin{cases} 1 - e^{-2t}, & t > 0 \\ 0, & t \leqslant 0 \end{cases}$$

Hence

$$f_{\min(X, Y)}(t) = \begin{cases} 2e^{-2t}, & t > 0 \\ 0, & \text{elsewhere} \end{cases}$$

• *Example 3.3.* Suppose X and Y have the joint bivariate normal distribution given by

$$f(x, y) = \frac{1}{2\pi\sigma_1\sigma_2\sqrt{1 - \rho^2}} \cdot \exp\left\{-\frac{1}{2(1 - \rho^2)}\left(\frac{x^2}{\sigma_1^2} - 2\rho\frac{xy}{\sigma_1\sigma_2} + \frac{y^2}{\sigma_2^2}\right)\right\},$$
$$-\infty < x < \infty, \quad -\infty < y < \infty$$

Find the distribution of the quotient, X/Y.

Solution. As we have seen, the distribution of the quotient is given by

$$f_{X/Y}(t) = \int_0^{\infty} f(ty, y)y\,dy - \int_{-\infty}^{0} f(ty, y)y\,dy$$

Hence

$$f_{X/Y}(t) = \frac{1}{2\pi\sigma_1\sigma_2\sqrt{1-\rho^2}}\left[\int_0^\infty y\left(\exp\left\{-\frac{1}{2(1-\rho^2)}\left(\frac{t^2y^2}{\sigma_1^2} - 2\rho\frac{ty^2}{\sigma_1\sigma_2} + \frac{y^2}{\sigma_2^2}\right)\right\}\right)dy\right.$$
$$\left. - \int_{-\infty}^0 y\left(\exp\left\{-\frac{1}{2(1-\rho^2)}\left(\frac{t^2y^2}{\sigma_1^2} - 2\rho\frac{ty^2}{\sigma_1\sigma_2} + \frac{y^2}{\sigma_2^2}\right)\right\}\right)dy\right]$$
$$= \frac{1}{\pi\sigma_1\sigma_2\sqrt{1-\rho^2}}\int_0^\infty y\left(\exp\left\{-\frac{y^2}{2(1-\rho^2)}\left(\frac{t^2}{\sigma_1^2} - \frac{2\rho t}{\sigma_1\sigma_2} + \frac{1}{\sigma_2^2}\right)\right\}\right)dy$$

To evaluate the above integral, let

$$u = \frac{y^2}{2(1-\rho^2)}\left(\frac{t^2}{\sigma_1^2} - \frac{2\rho t}{\sigma_1\sigma_2} + \frac{1}{\sigma_2^2}\right)$$

After simplifying, we get

$$f_{X/Y}(t) = \frac{\sigma_1\sigma_2\sqrt{1-\rho^2}}{\pi(\sigma_2^2t^2 - 2\rho\sigma_1\sigma_2 t + \sigma_1^2)}\int_0^\infty e^{-u}\,du$$

Hence

$$f_{X/Y}(t) = \frac{\sigma_1\sigma_2\sqrt{1-\rho^2}}{\pi(\sigma_2^2t^2 - 2\rho\sigma_1\sigma_2 t + \sigma_1^2)}, \qquad -\infty < t < \infty$$

■

Comment. In Example 3.3 if, in particular, we set $\rho = 0$, the distribution of X/Y has the pdf

$$f_{X/Y}(t) = \frac{\sigma_1\sigma_2}{\pi(\sigma_1^2 + \sigma_2^2t^2)} = \frac{1}{\pi}\cdot\frac{\sigma_1/\sigma_2}{[(\sigma_1/\sigma_2)^2 + t^2]}, \qquad -\infty < t < \infty$$

which, as will be recalled from Section 2.4 of Chapter 5, is the Cauchy distribution. In other words, *if X is $N(0, \sigma_1^2)$, Y is $N(0, \sigma_2^2)$, and if X and Y are independent, then X/Y has a Cauchy distribution with parameters $a = \sigma_1/\sigma_2$ and $b = 0$.*

The Student's *t*-distribution

Consider two independent random variables X and Y, where X has the standard normal distribution and Y the chi-square distribution with n degrees of freedom. We are going to find below the distribution of a new random variable T where

$$T = \frac{X}{\sqrt{Y/n}}$$

Letting $Z = \sqrt{Y/n}$, we can write T as $T = X/Z$. As can be easily verified, the distribution of $\sqrt{Y/n}$ is given as

$$f_Z(z) = \begin{cases} \dfrac{n^{n/2}}{2^{(n/2)-1}\Gamma\left(\dfrac{n}{2}\right)} z^{n-1}e^{-nz^2/2}, & z > 0 \\ 0, & \text{elsewhere} \end{cases}$$

Hence, the joint pdf of X and Z is given as

$$f(x, z) = f_X(x) \cdot f_Z(z) = \begin{cases} \dfrac{1}{\sqrt{2\pi}} e^{-x^2/2} \dfrac{n^{n/2}}{2^{(n/2)-1}\Gamma\left(\dfrac{n}{2}\right)} z^{n-1} e^{-nz^2/2}, & z > 0 \\ 0, & z \leqslant 0 \end{cases}$$

since X and Z are independent random variables.

Now, T being the quotient X/Z, it follows from section 2.3 that

$$f_T(t) = f_{X/Z}(t) = \int_0^\infty z f(tz, z)\, dz - \int_{-\infty}^0 z f(tz, z)\, dz, \qquad -\infty < t < \infty$$

$$= \int_0^\infty z f(tz, z)\, dz, \qquad \text{since } f(x, z) = 0 \text{ if } z \text{ is negative}$$

$$= \int_0^\infty z \frac{1}{\sqrt{2\pi}} e^{-(tz)^2/2} \frac{n^{n/2}}{2^{(n/2)-1}\Gamma\left(\dfrac{n}{2}\right)} z^{n-1} e^{-nz^2/2}\, dz$$

$$= \frac{n^{n/2}}{\sqrt{2\pi} \cdot 2^{(n/2)-1}\Gamma\left(\dfrac{n}{2}\right)} \int_0^\infty z^n e^{-(1/2)(n+t^2)z^2}\, dz$$

To evaluate the integral, let

$$u = \frac{1}{2}(n + t^2)z^2$$

Then $du = (n + t^2)z\, dz$ which, after some simplification, gives

$$dz = \frac{du}{\sqrt{2(n + t^2)u}}$$

Hence

$$f_T(t) = \frac{n^{n/2}}{\sqrt{2\pi}\, 2^{(n/2)-1}\Gamma\left(\dfrac{n}{2}\right)} \int_0^\infty \left(\frac{2u}{n + t^2}\right)^{n/2} e^{-u} \frac{du}{\sqrt{2(n + t^2)u}}$$

$$= \frac{n^{n/2}}{\sqrt{\pi}\,\Gamma\left(\dfrac{n}{2}\right)} \int_0^\infty \frac{u^{(n/2)-(1/2)}}{(n + t^2)^{(n+1)/2}} e^{-u}\, du$$

$$= \frac{n^{n/2}}{\sqrt{\pi}\,\Gamma\left(\dfrac{n}{2}\right)(n + t^2)^{(n+1)/2}} \int_0^\infty u^{[(n+1)/2]-1} e^{-u}\, du$$

$$= \frac{n^{n/2}\,\Gamma\left(\dfrac{n+1}{2}\right)}{\sqrt{\pi}\,\Gamma\left(\dfrac{n}{2}\right)(n + t^2)^{(n+1)/2}}$$

$$= \frac{\Gamma\left(\dfrac{n+1}{2}\right)}{\sqrt{n\pi}\,\Gamma\left(\dfrac{n}{2}\right)\left(1 + \dfrac{t^2}{n}\right)^{(n+1)/2}}$$

In conclusion, the distribution of the random variable $T = X/\sqrt{Y/n}$, where X and Y are independent random variables as defined above, is given as

$$f_T(t) = \frac{\Gamma\left(\frac{n+1}{2}\right)}{\sqrt{n\pi}\,\Gamma\left(\frac{n}{2}\right)\left(1 + \frac{t^2}{n}\right)^{(n+1)/2}}, \qquad -\infty < t < \infty$$

This is called the **Student's t-distribution with n degrees of freedom.** It has very important applications in statistics.

Distribution of the range

Suppose $X_1, X_2, \ldots, X_n$ are n random variables. Then the **range** R is defined as

$$R = V - W$$

where $V = \max(X_1, X_2, \ldots, X_n)$ and $W = \min(X_1, X_2, \ldots, X_n)$.

In what follows, the distribution of R is derived assuming that the random variables $X_1, X_2, \ldots, X_n$ are independent, continuous, and have identical distributions. Specifically, we shall assume that $X_1, X_2, \ldots, X_n$ have a common D.F. F_X and a common pdf f_X. We shall obtain the distribution of R in two steps.

Step 1. The joint distribution of V and W. By definition,

$$F_{V,W}(s, t) = P(V \leqslant s,\ W \leqslant t)$$

We have to consider two cases, namely, (*i*) $s \leqslant t$, and (*ii*) $s > t$.

(*i*) If $s \leqslant t$, it is intuitively obvious that $V \leqslant s$ and $W \leqslant t$ if and only if $V \leqslant s$. (A set-theoretic argument would go as follows: Obviously, $\{V \leqslant s,\ W \leqslant t\} \subseteq \{V \leqslant s\}$. On the other hand, $\{V \leqslant s\} \subseteq \{V \leqslant s,\ W \leqslant t\}$ because, if $V \leqslant s$, then certainly $W \leqslant s$; and since $s \leqslant t$, we naturally have $W \leqslant t$. Consequently $\{V \leqslant s,\ W \leqslant t\} = \{V \leqslant s\}$.) Therefore,

$$F_{V,W}(s, t) = P(V \leqslant s) = [F_X(s)]^n \qquad \text{if} \quad s \leqslant t$$

(*ii*) If $s > t$, then $F_{V,W}(s, t)$ represents the probability that each of the random variables $X_1, X_2, \ldots, X_n$ is $\leqslant s$, and *not* all of the random variables take values in the interval $(t, s]$.

Let

$$A = \{X_1 \leqslant s, X_2 \leqslant s, \ldots, X_n \leqslant s\}$$

and

$$B = \{t < X_1 \leqslant s, \ldots, t < X_n \leqslant s\}$$

(*Note:* $P(A) = [F_X(s)]^n$ and $P(B) = [F_X(s) - F_X(t)]^n$.)

Then

$$F_{V,W}(s,t) = P(AB') = P(A) - P(AB) = P(A) - P(B), \qquad \text{since } B \subset A$$
$$= [F_X(s)]^n - [F_X(s) - F_X(t)]^n$$

In summary, *if* $V = max(X_1, X_2, \ldots, X_n)$ *and* $W = min(X_1, X_2, \ldots, X_n)$ *where* $X_1, X_2, \ldots, X_n$ *are continuous, independent, and identically distributed random variables with the distribution function* F_X, *then the joint D.F. of V and W is given by*

$$F_{V,W}(s,t) = \begin{cases} [F_X(s)]^n, & s \leqslant t \\ [F_X(s)]^n - [F_X(s) - F_X(t)]^n, & s > t \end{cases}$$

The joint pdf $f_{V,W}(s,t) = \dfrac{\partial^2}{\partial s \partial t} F_{V,W}(s,t)$ is therefore given by

$$f_{V,W}(s,t) = \begin{cases} 0, & s \leqslant t \\ n(n-1)[F_X(s) - F_X(t)]^{n-2} f_X(s) f_X(t), & s > t \end{cases}$$

The joint pdf of V and W is sketched in Figure 3.9 below. It is zero above the line $s = t$ and nonzero below the line (the shaded region in the figure).

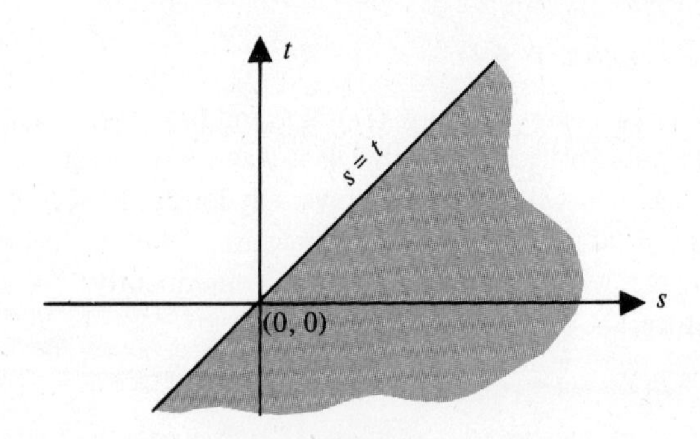

Figure 3.9

Step 2. Distribution of the range R. Having obtained the joint distribution of V and W, it is now a simple matter to find the distribution of the range $R = V - W$.

Since $P(R \leqslant r)$ is given by the integral over the shaded region above the line $s - t = r$, we see from Figures 3.10(a) and (b) that

$$F_R(r) = P(R \leqslant r) = \begin{cases} 0, & r < 0 \\ \int_{-\infty}^{\infty} \int_{t}^{t+r} n(n-1)[F_X(s) - F_X(t)]^{n-2} f_X(s) f_X(t)\, ds\, dt, & r \geqslant 0 \end{cases}$$

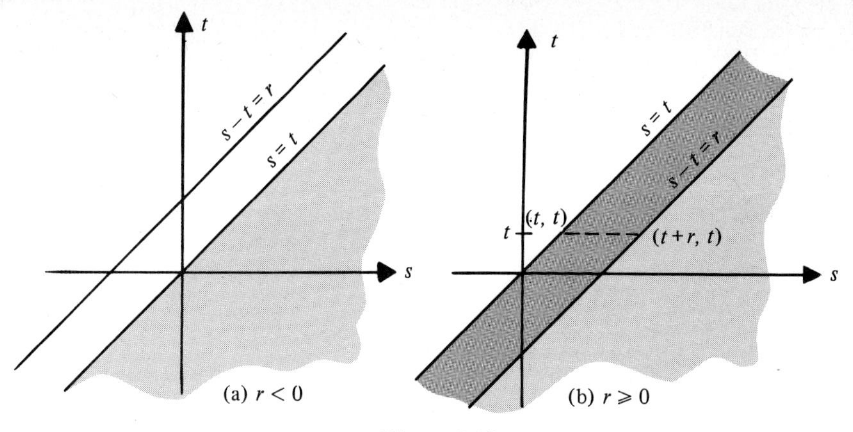

Figure 3.10

Now, making the substitution $F_X(s) - F_X(t) = u$, and noting that $dF_X(s) = f_X(s)\,ds$, it follows that

$$\int_t^{t+r} (n-1)[F_X(s) - F_X(t)]^{n-2} f_X(s)\,ds = [F_X(t+r) - F_X(t)]^{n-1}$$

Consequently,

$$F_R(r) = P(R \leqslant r) = \begin{cases} 0, & r < 0 \\ n\int_{-\infty}^{\infty} [F_X(t+r) - F_X(t)]^{n-1} f_X(t)\,dt, & r \geqslant 0 \end{cases}$$

In conclusion, the D.F. of the range R of n independent, continuous, and identically distributed random variables with a common D.F. F_X is given by

$$F_R(r) = \begin{cases} 0, & r < 0 \\ n\int_{-\infty}^{\infty} [F_X(t+r) - F_X(t)]^{n-1} f_X(t)\,dt, & r \geqslant 0 \end{cases}$$

Example 3.4. Suppose $X_1, X_2, \ldots, X_n$ are n independent random variables where each has the uniform distribution over the interval [0, 1]. Find the distribution of the range.

Solution. Here

$$f_X(x) = \begin{cases} 1, & 0 \leqslant x \leqslant 1 \\ 0, & \text{elsewhere} \end{cases} \quad \text{and} \quad F_X(x) = \begin{cases} 0, & x < 0 \\ x, & 0 \leqslant x < 1 \\ 1, & x \geqslant 1 \end{cases}$$

Now, as we have seen, the distribution function of the range is given by

$$F_R(r) = \begin{cases} 0, & r < 0 \\ n\int_{-\infty}^{\infty} [F_X(t+r) - F_X(t)]^{n-1} f_X(t)\,dt, & r \geqslant 0 \end{cases}$$

Therefore, substituting for $f_X(t)$,

$$F_R(r) = \begin{cases} 0, & r < 0 \\ n \int_0^1 [F_X(t+r) - F_X(t)]^{n-1}\, dt, & r \geqslant 0 \end{cases}$$

If $0 \leqslant r < 1$, it follows that

$$\begin{aligned} n \int_0^1 [F_X(t+r) - F_X(t)]^{n-1}\, dt &= n \left\{ \int_0^{1-r} [F_X(t+r) - F_X(t)]^{n-1}\, dt \right. \\ &\quad \left. + \int_{1-r}^1 [F_X(t+r) - F_X(t)]^{n-1}\, dt \right\} \\ &= n \left\{ \int_0^{1-r} [(t+r) - t]^{n-1}\, dt + \int_{1-r}^1 (1-t)^{n-1}\, dt \right\} \\ &= n \left\{ r^{n-1}(1-r) + \frac{r^n}{n} \right\} \end{aligned}$$

(If $0 \leqslant t < 1 - r$, then $0 \leqslant t + r \leqslant 1$, and consequently $F_X(t+r) = t + r$; if $1 - r \leqslant t$, then $1 \leqslant t + r$, and consequently $F_X(t+r) = 1$.)

If $r \geqslant 1$, it follows that

$$n \int_0^1 [F_X(t+r) - F_X(t)]^{n-1}\, dt = n \int_0^1 [1 - t]^{n-1}\, dt = 1$$

(Since $r \geqslant 1$ and $0 \leqslant t \leqslant 1$, we have $t + r \geqslant 1$. Hence $F_X(t+r) = 1$ and $F_X(t) = t$.)

In summary,

$$F_R(r) = \begin{cases} 0, & r < 0 \\ n\left\{ r^{n-1}(1-r) + \dfrac{r^n}{n} \right\}, & 0 \leqslant r < 1 \\ 1, & r \geqslant 1 \end{cases}$$

Differentiation gives the pdf of R as

$$f_R(r) = \begin{cases} n(n-1)r^{n-2}(1-r), & 0 \leqslant r \leqslant 1 \\ 0, & \text{elsewhere} \end{cases}$$

EXERCISES–SECTION 3

1. Suppose X and Y are independent and identically distributed random variables each having an exponential distribution with parameter λ. Find the distribution of–
   (a) $X + Y$ (b) X/Y
   (c) $\max(X, Y)$ (d) $\min(X, Y)$
2. Let $0 < a < b$. Two numbers are picked independently, one at random in the interval $[a, b]$, and the other at random in the interval $[-b, -a]$. If X represents the number in $[a, b]$, and Y the number in $[-b, -a]$, find the distribution of–
   (a) the sum of the numbers, $X + Y$
   (b) the product of the numbers, XY.
3. In exercise 2, having obtained the distributions of $X + Y$ and XY, find $E(X + Y)$ and $E(XY)$.
4. If X and Y are independent random variables each having an exponential distribution with parameter λ, find the distribution of $Z = X - Y$.

5. Suppose the current I in a circuit is a random variable with the following pdf:

$$f_I(x) = \begin{cases} \frac{3}{2}x(2-x), & 0 \leqslant x \leqslant 1 \\ 0, & \text{elsewhere} \end{cases}$$

Also, suppose the resistance R is a random variable with the pdf

$$f_R(u) = \begin{cases} 2u, & 0 \leqslant u \leqslant 1 \\ 0, & \text{elsewhere} \end{cases}$$

Find the distribution of the voltage E given by $E = IR$.

6. The sides X and Y of a rectangular region are given to be independent random variables with the following pdf's:

$$f_X(x) = \begin{cases} \frac{1}{2}(3-x), & 1 \leqslant x \leqslant 3 \\ 0, & \text{elsewhere} \end{cases}$$

and

$$f_Y(y) = \begin{cases} \frac{3}{7}x^2, & 1 \leqslant x \leqslant 2 \\ 0, & \text{elsewhere} \end{cases}$$

Find:

(a) the distribution of the area of the region
(b) the expected area

7. The length X of a rectangular region is a random variable with the pdf given by

$$f_X(x) = \begin{cases} \frac{1}{3}(3x^2 - 2x - 1), & 1 < x < 2 \\ 0, & \text{elsewhere} \end{cases}$$

Given that the length is x, the distribution of the breadth Y is given by the following pdf

$$g(y) = \begin{cases} \dfrac{2(x+y)}{(3x^2 - 2x - 1)}, & 1 < y < x \\ 0, & \text{elsewhere} \end{cases}$$

Find the distribution of–

(a) the perimeter
(b) the area.

8. If X and Y are independent, normally distributed random variables, each $N(0, \sigma^2)$, find the distribution of $U = \sqrt{X^2 + Y^2}$.

9. The joint pdf of X and Y is given by

$$f(x, y) = \begin{cases} \dfrac{2}{\pi}(1 - x^2 - y^2), & x^2 + y^2 \leqslant 1 \\ 0, & \text{elsewhere} \end{cases}$$

Find the distribution of $Z = \sqrt{X^2 + Y^2}$.

10. A person fires at a circular target of radius r hitting any point inside the circle at random. Find the distribution of R which represents the distance from the point of impact to the center of the circle.

11. Let $X_1, X_2, \ldots, X_n$ be independent random variables each distributed uniformly over $[0, 1]$. If $V = \max(X_1, X_2, \ldots, X_n)$ and $W = \min(X_1, X_2, \ldots, X_n)$, find the joint distribution of V and W.

12. A person lights ten firecrackers simultaneously. The amount of time (in seconds) that it takes a cracker to explode is uniformly distributed over the interval $(0, 10)$. What is the distribution of the time between the first explosion and the last?

13. If the failure time of light bulbs is known to be an exponential random variable with a mean life of five hours, find how many bulbs should be installed in a hall to ensure with a probability of at least 0.95 that the last bulb will not burn out before eight hours.

14. Suppose $X_1, X_2, \ldots, X_n$ are independent random variables, each having an exponential distribution with parameter $\lambda > 0$. Show that the distribution of the range R is given by the following pdf:

$$f_R(r) = \begin{cases} (n-1)\lambda e^{-\lambda r}(1 - e^{-\lambda r})^{n-2}, & r \geqslant 0 \\ 0, & \text{elsewhere} \end{cases}$$

15. A hall is lighted by means of four giant-sized bulbs, each with a time to failure (in hours) given by the exponential distribution with mean time of 5 hours. Find the probability that—

(a) the last bulb will burn out before 6 hours
(b) the last bulb will burn out between 4 and 6 hours
(c) the first bulb will burn out after 6 hours
(d) the first bulb will burn out before 6 hours
(e) the first and the last bulb will burn out within one hour of each other.

11 Expectation–Several Random Variables

INTRODUCTION

The concept of expectation was introduced in Chapter 7. The treatment presented there involved a single random variable. In this chapter, we shall be concerned basically with two topics. In Section 1, we shall treat the general concept of the expectation of a function of several random variables, specifically, two random variables; included in this treatment will be the notions of covariance and of the coefficient of correlation, which provide a quantitative characterization of the interrelation between random variables. In Section 2, we shall take up conditional expectation.

1. EXPECTATION OF A FUNCTION OF SEVERAL RANDOM VARIABLES

1.1 The Definition

Let X and Y be two random variables with a given joint distribution, and let h be a real-valued function of two real variables ($h : \mathbf{R}^2 \to \mathbf{R}$). Our immediate goal is to show how we go about computing the expected value of $h(X, Y)$. If the distribution of $Z = h(X, Y)$ is known, then of course we can find the expected value of Z using the basic definition given in Chapter 7:

$$E(Z) = \begin{cases} \int_{-\infty}^{\infty} z f_Z(z)\, dz, & \text{in the continuous case} \\ \sum_{z_i} z_i P(Z = z_i), & \text{in the discrete case} \end{cases}$$

using the *single* integral in the continuous case, and *single* summation in the discrete case. However, the task of finding the distribution of Z can be quite an involved one, as we have seen in Chapter 10. The following equivalent definition, using the joint distribution of X and Y, avoids this.

The expectation of $Z = h(X, Y)$ is defined by

$$E(Z) = E(h(X, Y)) = \begin{cases} \int_{-\infty}^{\infty}\int_{-\infty}^{\infty} h(x, y)f(x, y)\,dy\,dx, & \text{in the continuous case} \\ \sum_{x_i}\sum_{y_j} h(x_i, y_j)P(X = x_i, Y = y_j), & \text{in the discrete case} \end{cases}$$

If this latter definition is used, there is no need to obtain the distribution of Z, but the computation of expectation involves a *double* integral, or a *double* summation, as the case may be. We shall not bother to verify the equivalence of the two definitions. (The above formula generalizes in an obvious way to situations involving more than two random variables.)

Note. In what follows the proofs of the assertions will be given assuming continuous joint distributions. Proofs in the discrete case follow analogously by replacing integration by summation.

Example 1.1. Suppose X and Y have an absolutely continuous joint distribution. If $Z = X + Y$ show that

$$\int_{-\infty}^{\infty} z f_Z(z)\,dz = \int_{-\infty}^{\infty}\int_{-\infty}^{\infty} (x + y)f(x, y)\,dy\,dx$$

thus establishing the equivalence of the two definitions of expectation, at least in the case $h(X, Y) = X + Y$.

Solution. In Chapter 10 we saw that the distribution of Z is given by

$$f_Z(z) = \int_{-\infty}^{\infty} f(x, z - x)\,dx, \qquad -\infty < z < \infty$$

Therefore,

$$\begin{aligned} E(Z) &= \int_{-\infty}^{\infty} z f_Z(z)\,dz \\ &= \int_{-\infty}^{\infty} z\left[\int_{-\infty}^{\infty} f(x, z - x)\,dx\right] dz \\ &= \int_{-\infty}^{\infty}\int_{-\infty}^{\infty} z f(x, z - x)\,dx\,dz \\ &= \int_{-\infty}^{\infty}\int_{-\infty}^{\infty} (x + y)f(x, y)\,dx\,dy \end{aligned}$$

letting $y = z - x$.

Example 1.2. Suppose X and Y are independent random variables, each with the uniform distribution over the interval $[0, 1]$. Find $E(Z)$ where $Z = \max(X, Y)$.

Solution. We shall find $E(Z)$ using two methods.

Method 1: In Example 3.1(*d*) of Chapter 10 we found the distribution of Z as

$$f_Z(t) = \begin{cases} 2t, & 0 \leqslant t \leqslant 1 \\ 0, & \text{elsewhere} \end{cases}$$

Therefore,

$$E(Z) = \int_{-\infty}^{\infty} t f_Z(t)\, dt = \int_0^1 t \cdot 2t\, dt = \frac{2}{3}$$

Method 2:

$$E(Z) = \int_{-\infty}^{\infty}\int_{-\infty}^{\infty} \max(x, y) f(x, y)\, dy\, dx$$

$$= \int_0^1 \int_0^1 \max(x, y)(1)\, dy\, dx$$

since $f(x, y) = 1$ if $0 \leqslant x \leqslant 1$ and $0 \leqslant y \leqslant 1$, and $f(x, y) = 0$ elsewhere. Hence

$$E(Z) = \int_0^1 \left[\int_0^x \max(x, y)\, dy + \int_x^1 \max(x, y)\, dy\right] dx$$

Now, if $0 < y < x$, then $\max(x, y) = x$, and if $x < y < 1$, then $\max(x, y) = y$. Consequently,

$$E(Z) = \int_0^1 \left[\int_0^x x\, dy + \int_x^1 y\, dy\right] dx$$

$$= \int_0^1 \left[x^2 + \left(\frac{1}{2} - \frac{x^2}{2}\right)\right] dx = \frac{2}{3}$$

Example 1.3. X and Y have the following joint probability function:

$$P(X = x, Y = y) = \begin{cases} \frac{1}{42}(x + y^2), & x = 1, 4, \quad y = -1, 0, 1, 3 \\ 0, & \text{elsewhere} \end{cases}$$

Find (*a*) $E(Y^2/X)$, (*b*) $E(XY)$.

Solution

(*a*)

$$E(Y^2/X) = \sum_x \sum_y \frac{y^2}{x} \cdot P(X = x, Y = y)$$

$$= \sum_x \sum_y \frac{y^2}{x} \frac{(x + y^2)}{42}$$

$$= \frac{1}{42} \sum_x \sum_y \left(y^2 + \frac{y^4}{x}\right)$$

$$= \frac{1}{42} \sum_x \left[\left((-1)^2 + \frac{(-1)^4}{x}\right) + \left(0^2 + \frac{0^4}{x}\right) + \left(1^2 + \frac{1^4}{x}\right) + \left(3^2 + \frac{3^4}{x}\right)\right]$$

$$= \frac{1}{42} \sum_x \left(11 + \frac{83}{x}\right)$$

$$= \frac{1}{42}\left[\left(11 + \frac{83}{1}\right) + \left(11 + \frac{83}{4}\right)\right] = \frac{503}{168}$$

(*b*)

$$E(XY) = \sum_x \sum_y xy \cdot \frac{1}{42}(x + y^2)$$

$$= \frac{1}{42} \sum_x \sum_y (x^2y + xy^3)$$

$$= \frac{1}{42} \sum_x [(x^2(-1) + x(-1)^3) + (x^2 \cdot 0 + x \cdot 0^3) + (x^2 \cdot 1 + x \cdot 1^3) + (x^2 \cdot 3 + x \cdot 3^3)]$$

$$= \frac{1}{42} \sum_x (3x^2 + 27x)$$

$$= \frac{1}{42}[(3 \cdot 1^2 + 27 \cdot 1) + (3 \cdot 4^2 + 27 \cdot 4)] = \frac{93}{21}$$

Example 1.4. A fair coin is tossed three times. Let X represent the number of heads on the first two tosses, and Y the number of heads on the last two tosses. Find (*a*) $E(XY)$, (*b*) $E(X + Y)$.

Solution. Let us first obtain the joint probability function of X and Y. The values of the random variables for each sample point are listed in the table below.

Sample point	*Value of X*	*Value of Y*	*Probability*
HHH	2	2	$\frac{1}{8}$
HHT	2	1	$\frac{1}{8}$
HTH	1	1	$\frac{1}{8}$
THH	1	2	$\frac{1}{8}$
HTT	1	0	$\frac{1}{8}$
THT	1	1	$\frac{1}{8}$
TTH	0	1	$\frac{1}{8}$
TTT	0	0	$\frac{1}{8}$

The joint probability function of X and Y is therefore as displayed in the following table:

x \ y	2	1	0
2	$\frac{1}{8}$	$\frac{1}{8}$	0
1	$\frac{1}{8}$	$\frac{2}{8}$	$\frac{1}{8}$
0	0	$\frac{1}{8}$	$\frac{1}{8}$

(*a*)

$$E(XY) = \sum_x \sum_y xy \cdot P(X = x,\ Y = y)$$

$$= 2 \cdot 2\left(\frac{1}{8}\right) + 2 \cdot 1\left(\frac{1}{8}\right) + 1 \cdot 2\left(\frac{1}{8}\right) + 1 \cdot 1\left(\frac{2}{8}\right) + 1 \cdot 0\left(\frac{1}{8}\right) + 0 \cdot 1\left(\frac{1}{8}\right) + 0 \cdot 0\left(\frac{1}{8}\right)$$

$$= \frac{5}{4}$$

(*b*)

$$E(X+Y) = \sum_x \sum_y (x+y)P(X=x,\ Y=y)$$

$$= (2+2)\frac{1}{8} + (2+1)\frac{1}{8} + (1+2)\frac{1}{8} + (1+1)\frac{2}{8} + (1+0)\frac{1}{8}$$

$$+ (0+1)\frac{1}{8} + (0+0)\frac{1}{8}$$

$$= 2$$

• *Example 1.5 (The trinomial distribution)* Suppose X and Y have a joint trinomial distribution with n trials and parameters p and q, where $0<q<1, 0<p<1$, and $0<p+q<1$. Find $E(XY)$.

Solution. The joint probability function is

$$P(X=i,\ Y=j) = \frac{n!}{i!\,j!\,(n-i-j)!}\, p^i q^j (1-p-q)^{n-i-j}$$

where $i = 0, 1, \ldots, n;\ j = 0, 1, \ldots, n;\ 0 \leqslant i+j \leqslant n.$

Therefore,

$$E(XY) = \sum_{i=0}^{n} \sum_{j=0}^{n-i} ij \cdot \frac{n!}{i!\,j!\,(n-i-j)!}\, p^i q^j (1-p-q)^{n-i-j}$$

$$= \sum_{i=1}^{n-1} \sum_{j=1}^{n-i} \frac{n!}{(i-1)!\,(j-1)!\,(n-i-j)!}\, p^i q^j (1-p-q)^{n-i-j}$$

$$= n(n-1)pq \sum_{i=1}^{n-1} \sum_{j=1}^{n-i} \frac{(n-2)!}{(i-1)!\,(j-1)!\,(n-i-j)!}\, p^{i-1} q^{j-1} (1-p-q)^{n-i-j}$$

Letting $j-1=t,\ i-1=s$, we get

$$E(XY) = n(n-1)pq \sum_{s=0}^{n-2} \sum_{t=0}^{n-2-s} \frac{(n-2)!}{s!\,t!\,(n-2-t-s)!}\, p^s q^t (1-p-q)^{n-2-t-s}$$

$$= n(n-1)pq[p+q+(1-p-q)]^{n-2}$$

$$= n(n-1)pq$$

Hence $E(XY) = n(n-1)pq$.

Example 1.6. The joint pdf of X and Y is given by

$$f(x,y) = \begin{cases} 2x(x-y), & 0<x<1 \text{ and } -x<y<x \\ 0, & \text{elsewhere} \end{cases}$$

Find the variance of X and the variance of Y.

Solution. To find the variances of X and Y, we shall need $E(X)$, $E(X^2)$, $E(Y)$, and $E(Y^2)$. Let us therefore find $E(X^n Y^k)$ where n and k are nonnegative integers. We get

$$E(X^n Y^k) = \int_{-\infty}^{\infty}\int_{-\infty}^{\infty} x^n y^k f(x, y)\, dy\, dx$$

$$= \int_0^1 \int_{-x}^{x} x^n y^k \cdot 2x(x - y)\, dy\, dx$$

$$= 2\int_0^1 x^{n+1} \int_{-x}^{x} (xy^k - y^{k+1})\, dy\, dx$$

$$= 2\int_0^1 x^{n+1} \cdot x^{k+2}\left[\frac{1}{k+1} - \frac{1}{k+2} - \frac{(-1)^{k+1}}{k+1} + \frac{(-1)^{k+2}}{k+2}\right] dx$$

$$= \frac{2}{n+k+4}\left[\frac{1}{(k+1)}(1 - (-1)^{k+1}) - \frac{1}{k+2}(1 - (-1)^{k+2})\right]$$

Assigning appropriate values to n and k, this gives $E(X) = \frac{4}{5}$ (setting $n = 1, k = 0$), $E(X^2) = \frac{4}{6}$ (setting $n = 2, k = 0$), $E(Y) = -\frac{4}{15}$ (setting $n = 0, k = 1$), and $E(Y^2) = \frac{2}{9}$ (setting $n = 0, k = 2$).

Consequently,

$$\operatorname{Var}(X) = E(X^2) - [E(X)]^2 = \frac{2}{75}$$

and

$$\operatorname{Var}(Y) = E(Y^2) - [E(Y)]^2 = \frac{34}{225}$$

Example 1.7. Suppose X and Y have a continuous distribution with the joint pdf

$$f(x, y) = \begin{cases} (x + y), & \text{if } 0 \leqslant x \leqslant 1 \text{ and } 0 \leqslant y \leqslant 1 \\ 0, & \text{elsewhere} \end{cases}$$

Find $E(\min(X, Y))$.

Solution

$$E(\min(X, Y)) = \int_{-\infty}^{\infty}\int_{-\infty}^{\infty} \min(x, y) f(x, y)\, dy\, dx$$

$$= \int_0^1\int_0^1 \min(x, y)(x + y)\, dy\, dx \qquad \text{(why?)}$$

$$= \int_0^1 \left[\int_0^x \min(x, y)(x + y)\, dy + \int_x^1 \min(x, y)(x + y)\, dy\right] dx$$

$$= \int_0^1 \left[\int_0^x y(x + y)\, dy + \int_x^1 x(x + y)\, dy\right] dx \qquad \text{(why?)}$$

Computation of the integrals yields

$$E(\min(X, Y)) = \int_0^1 \left(x^2 + \frac{x}{2} - \frac{2x^3}{3}\right) dx = \frac{5}{12}$$

1.2 The Basic Properties of Expectation

We shall establish below the following two main results: (1) the expected value of the sum of two random variables is equal to the sum of their expected values; (2) the expected value of the product of two random variables is equal to the product of their expected values *provided the random variables are independent* (the converse of this result is not true).

The Expectation of a Linear Combination If h_1 and h_2 are real-valued functions of two real variables, and if a and b are any constants, then

$$E(ah_1(X, Y) + bh_2(X, Y)) = aE(h_1(X, Y)) + bE(h_2(X, Y))$$

The result follows since by the definition of expectation

$$E(ah_1(X, Y) + bh_2(X, Y)) = \int_{-\infty}^{\infty}\int_{-\infty}^{\infty} [ah_1(x, y) + bh_2(x, y)]f(x, y)\, dy\, dx$$

$$= a\int_{-\infty}^{\infty}\int_{-\infty}^{\infty} h_1(x, y)f(x, y)\, dy\, dx + b\int_{-\infty}^{\infty}\int_{-\infty}^{\infty} h_2(x, y)f(x, y)\, dy\, dx$$

$$= aE(h_1(X, Y)) + bE(h_2(X, Y))$$

The expression $ah_1(X, Y) + bh_2(X, Y)$ is called a *linear combination* of the random variables $h_1(X, Y), h_2(X, Y)$, and $aE(h_1(X, Y)) + bE(h_2(X, Y))$ is called a linear combination of the real numbers $E(h_1(X, Y)), E(h_2(X, Y))$. The above result then states that the expected value of a linear combination of random variables is equal to the linear combination of their expected values. This property of the expectation operation is referred to as the **linear property**.

Some particular cases

(*i*) Setting $h_1(X, Y) = X$ and $h_2(X, Y) \equiv 1$ gives

$$E(aX + b) = aE(X) + b, \qquad \text{for any constants } a, b$$

For example, it now follows that

$$E\left(\frac{X - \mu_X}{\sigma_X}\right) = 0$$

(*ii*) Setting $h_1(X, Y) = X$, $h_2(X, Y) = Y$, and $a = b = 1$ gives

$$E(X + Y) = E(X) + E(Y)$$

That is, *the expected value of the sum of two random variables is equal to the sum of their expected values.*

Comment. In proving that $E(X + Y) = E(X) + E(Y)$, we have *not* made any assumption regarding the dependence or independence of X and Y. The result holds irrespective of such considerations.

The foregoing result regarding the expectation of a linear combination generalizes in an obvious way to the case of n random variables $X_1, X_2, \ldots, X_n$. We have

$$E\left(\sum_{i=1}^{n} a_i X_i\right) = \sum_{i=1}^{n} a_i E(X_i)$$

for any set of constants $a_1, a_2, \ldots, a_n$.

In particular, setting $a_1 = a_2 = \ldots = a_n = 1/n$ yields

$$E\left(\frac{1}{n}\sum_{i=1}^{n} X_i\right) = \frac{1}{n}\sum_{i=1}^{n} E(X_i)$$

and, furthermore, *if each random variable has the same expectation,* say μ, then

$$E\left(\frac{1}{n}\sum_{i=1}^{n} X_i\right) = \mu$$

Writing

$$\bar{X} = \frac{1}{n}\sum_{i=1}^{n} X_i$$

we shall refer to $\bar{X}$ as the **sample mean.** What we have established is that, *if each random variable has the same mean μ, then the sample mean $\bar{X}$ also has the mean μ; that is, $E(\bar{X}) = \mu$.*

The Expectation of a Product If X and Y are independent random variables, then, for any real-valued functions h and g of a real variable,

$$E(h(X)g(Y)) = E(h(X))E(g(Y))$$

This can be verified as follows:

$$\begin{aligned} E(h(X)g(Y)) &= \int_{-\infty}^{\infty}\int_{-\infty}^{\infty} h(x)g(y)f(x, y)\,dy\,dx \\ &= \int_{-\infty}^{\infty}\int_{-\infty}^{\infty} h(x)g(y)f_X(x)f_Y(y)\,dy\,dx, \quad \text{since } X \text{ and } Y \text{ are independent} \\ &= \int_{-\infty}^{\infty} h(x)f_X(x)\,dx \int_{-\infty}^{\infty} g(y)f_Y(y)\,dy \\ &= E(h(X))E(g(Y)) \end{aligned}$$

In particular, if we set $h(X) = X$ and $g(Y) = Y$ in the above result, we get

$$E(XY) = E(X)E(Y) \quad \text{if } X \text{ and } Y \text{ are independent}$$

That is, *the expected value of the product of two random variables is equal to the product of their expected values if the random variables are independent.* It is important to bear in mind that the random variables have been assumed to be independent. The following example shows that $E(XY) \neq E(X)E(Y)$ in general, if the condition of independence is relaxed.

Example 1.8. Suppose X and Y have a joint distribution given by the following pdf:

$$f(x, y) = \begin{cases} 2e^{-(x+y)}, & 0 \leqslant y \leqslant x \leqslant \infty \\ 0, & \text{elsewhere} \end{cases}$$

Find $E(X)$, $E(Y)$, and $E(XY)$ and show that $E(XY) \neq E(X)E(Y)$.

Solution. Since $f(x, y)$ cannot be factored as $u(x) \cdot v(y)$ for every x, y, the random variables are not independent. Now,

$$E(X) = \int_0^\infty \int_0^x x \cdot 2e^{-(x+y)}\, dy\, dx$$

$$= \int_0^\infty x \cdot 2e^{-x}(1 - e^{-x})\, dx = \frac{3}{2}$$

$$E(Y) = \int_0^\infty \int_y^\infty y \cdot 2e^{-(x+y)}\, dx\, dy$$

$$= \int_0^\infty y \cdot 2e^{-2y}\, dy = \frac{1}{2}$$

and

$$E(XY) = \int_0^\infty \int_0^x xy \cdot 2e^{-(x+y)}\, dy\, dx$$

$$= 2 \int_0^\infty xe^{-x}[-xe^{-x} - e^{-x} + 1]\, dx = 1$$

Hence

$$E(XY) \neq E(X)E(Y)$$ ■

Caution. We have shown that if X and Y are independent, then $E(XY) = E(X)E(Y)$. *The converse of this assertion is, however, false, in that it does not follow that X and Y are independent just because $E(XY) = E(X)E(Y)$.* The truth of this statement is brought out in the next two examples, Examples 1.9 and 1.10.

Example 1.9. Suppose X has the following probability function:

x	$P(X = x)$
-1	$\frac{1}{4}$
0	$\frac{1}{2}$
1	$\frac{1}{4}$

Let $Y = X^2$. Show that $E(XY) = E(X)E(Y)$ even though X and Y are not independent.

Solution. It follows immediately that the probability function of Y is given by

$$P(Y=1)=\tfrac{1}{2}, \quad P(Y=0)=\tfrac{1}{2}$$

Certainly, X and Y are not independent, since, for example,

$$P(X=0,\ Y=1)=0 \neq P(X=0)\cdot P(Y=1)$$

(Note that $P(X=0,\ Y=1)=0$ because if $X=0$ we must have $Y=0$; Y cannot take the value 1.)

Now,

$$E(X)=0, \quad E(Y)=\tfrac{1}{2}$$

and

$$E(XY)=E(X^3)=(-1)^3\cdot\frac{1}{4}+0^3\cdot\frac{1}{2}+1^3\cdot\frac{1}{4}=0$$

Thus we see that $E(X)E(Y)=0=E(XY)$ even though X and Y are dependent r.v.'s.

Example 1.10. Suppose X and Y have a joint probability function with a probability mass of $\frac{1}{4}$ at each of the points $(0, d)$, $(c, 2d)$, $(2c, d)$, $(c, 0)$, where c and d are any positive numbers. Show that–

(*a*) X and Y are not independent

(*b*) $E(XY)=E(X)E(Y)$.

Solution. Displaying the probability function in tabular form, we have

x \ y	0	d	$2d$	$P(X=x)$
0	0	$\frac{1}{4}$	0	$\frac{1}{4}$
c	$\frac{1}{4}$	0	$\frac{1}{4}$	$\frac{1}{2}$
$2c$	0	$\frac{1}{4}$	0	$\frac{1}{4}$
$P(Y=y)$	$\frac{1}{4}$	$\frac{1}{2}$	$\frac{1}{4}$	

(*a*) The random variables X and Y are not independent since, for example, $P(X=0,\ Y=0)=0$, whereas $P(X=0)P(Y=0)=\frac{1}{4}\cdot\frac{1}{4}=\frac{1}{16}$.

(*b*)

$$E(X)=0\left(\frac{1}{4}\right)+c\left(\frac{1}{2}\right)+2c\left(\frac{1}{4}\right)=c$$

$$E(Y)=0\left(\frac{1}{4}\right)+d\left(\frac{1}{2}\right)+2d\left(\frac{1}{4}\right)=d$$

and $$E(XY)=c\cdot 0\left(\frac{1}{4}\right)+0\cdot d\left(\frac{1}{4}\right)+2c\cdot d\left(\frac{1}{4}\right)+c\cdot 2d\left(\frac{1}{4}\right)=dc$$

Therefore,

$$E(XY)=E(X)E(Y)$$

1.3 The Covariance and the Correlation Coefficient

Two constants which provide a measure of relationship between random variables in the theory of joint distributions are the *covariance* and the *correlation coefficient.*

If X and Y are two random variables then their **covariance**, denoted by $\text{Cov}(X, Y)$, is defined as

$$\text{Cov}(X, Y) = E[(X - \mu_X)(Y - \mu_Y)]$$

where $\mu_X = E(X)$ and $\mu_Y = E(Y)$.

In the above definition, if in particular we take $X = Y$, then the formula yields

$$\text{Cov}(X, X) = E[(X - \mu_X)^2] = \text{Var}(X)$$

In other words, *the covariance of a random variable with itself is its variance.*

The following version of the formula is often convenient for computing the covariance:

$$\text{Cov}(X, Y) = E(XY) - E(X)E(Y)$$

Using the properties of the expectation operation, this can be proved as follows.

$$\begin{aligned}\text{Cov}(X, Y) &= E[(X - \mu_X)(Y - \mu_Y)] \\ &= E(XY - \mu_X Y - \mu_Y X + \mu_X \mu_Y) \\ &= E(XY) - \mu_X E(Y) - \mu_Y E(X) + \mu_X \mu_Y \qquad \text{(why?)} \\ &= E(XY) - E(X)E(Y)\end{aligned}$$

Comments. (1) If X and Y are *independent,* then we know that $E(XY) = E(X)E(Y)$ and, consequently, $\text{Cov}(X, Y) = E(XY) - E(X)E(Y) = 0$. The converse of this is, of course, false, as Examples 1.9 and 1.10 show. Hence, *if two r.v.'s are independent, then it follows that $\text{Cov}(X, Y) = 0$.* However, *if $\text{Cov}(X, Y) = 0$, it is erroneous to conclude that X and Y are independent.*

(2) Notice that $x^2 \pm 2xy + y^2 = (x \pm y)^2 \geqslant 0$. Therefore, $|xy| \leqslant (x^2 + y^2)/2$. Hence

$$\begin{aligned}\int_{-\infty}^{\infty}\int_{-\infty}^{\infty} |xy|\, f(x, y)\, dy\, dx &\leqslant \int_{-\infty}^{\infty}\int_{-\infty}^{\infty} \frac{x^2 + y^2}{2} f(x, y)\, dy\, dx \\ &= \frac{1}{2}\left[\int_{-\infty}^{\infty}\int_{-\infty}^{\infty} x^2 f(x, y)\, dy\, dx + \int_{-\infty}^{\infty}\int_{-\infty}^{\infty} y^2 f(x, y)\, dy\, dx\right] \\ &= \frac{1}{2}[E(X^2) + E(Y^2)]\end{aligned}$$

Consequently, $E(XY)$ exists if $E(X^2) < \infty$ and $E(Y^2) < \infty$. Therefore, the definition of covariance is meaningful if $E(X^2)$ and $E(Y^2)$ are finite.

We shall next define the other constant, namely, the correlation coefficient.

The **correlation coefficient** between X and Y, denoted by $\rho(X, Y)$, is defined by

$$\rho(X, Y) = \frac{\mathrm{Cov}(X, Y)}{\sqrt{\mathrm{Var}(X)\mathrm{Var}(Y)}} = \frac{E(XY) - E(X)E(Y)}{\sqrt{\mathrm{Var}(X)\mathrm{Var}(Y)}}$$

provided, of course, neither variance equals zero.

Random variables for which $\rho(X, Y) = 0$ are said to be *uncorrelated.* It follows immediately that $\rho(X, Y) = 0$ if X and Y are independent since in this case $\mathrm{Cov}(X, Y) = 0$. It should be obvious that the converse is false; that is, $\rho(X, Y) = 0$ need not imply, in general, that X and Y are independent. (Once again, Examples 1.9 and 1.10.) In summary, *independent random variables are uncorrelated; however, uncorrelated random variables need not be independent.*

Example 1.11. A box contains six beads of which three are red, two white, and one green. Two beads are picked at random without replacement. If X represents the number of red beads, and Y the number of white beads, find the correlation coefficient of X and Y.

Solution. We know that

$$P(X = i,\ Y = j) = \frac{\binom{3}{i}\binom{2}{j}\binom{1}{2-i-j}}{\binom{6}{2}}$$

where $i = 0, 1, 2;\ j = 0, 1, 2$ with the understanding that $\binom{k}{r} = 0$ if $r > k$ or $r < 0$.

Displaying the joint probability function in tabular form yields

$x \backslash y$	0	1	2	$P(X = x)$
0	0	$\frac{2}{15}$	$\frac{1}{15}$	$\frac{3}{15}$
1	$\frac{3}{15}$	$\frac{6}{15}$	0	$\frac{9}{15}$
2	$\frac{3}{15}$	0	0	$\frac{3}{15}$
$P(Y = y)$	$\frac{6}{15}$	$\frac{8}{15}$	$\frac{1}{15}$	

Therefore,

$$E(X) = 0\left(\frac{3}{15}\right) + 1\left(\frac{9}{15}\right) + 2\left(\frac{3}{15}\right) = 1$$

$$E(X^2) = 0^2\left(\frac{3}{15}\right) + 1^2\left(\frac{9}{15}\right) + 2^2\left(\frac{3}{15}\right) = \frac{21}{15}$$

$$E(Y) = 0\left(\frac{6}{15}\right) + 1\left(\frac{8}{15}\right) + 2\left(\frac{1}{15}\right) = \frac{10}{15}$$

$$E(Y^2) = 0^2\left(\frac{6}{15}\right) + 1^2\left(\frac{8}{15}\right) + 2^2\left(\frac{1}{15}\right) = \frac{12}{15}$$

and $E(XY) = 0 \cdot 1\left(\frac{2}{15}\right) + 0 \cdot 2\left(\frac{1}{15}\right) + 1 \cdot 0\left(\frac{3}{15}\right) + 1 \cdot 1\left(\frac{6}{15}\right) + 2 \cdot 0\left(\frac{3}{15}\right) = \frac{6}{15}$

As a result,

$$\text{Var}(X) = \frac{21}{15} - 1^2 = \frac{6}{15}$$

$$\text{Var}(Y) = \frac{12}{15} - \left(\frac{10}{15}\right)^2 = \frac{16}{45}$$

and $\text{Cov}(X, Y) = \frac{6}{15} - 1\left(\frac{10}{15}\right) = -\frac{4}{15}$

Hence

$$\rho(X, Y) = \frac{-\frac{4}{15}}{\sqrt{\frac{6}{15} \cdot \frac{16}{45}}} = -\frac{1}{\sqrt{2}} = -0.707$$

• *Example 1.12* (*The trinomial distribution*) Suppose X and Y have a joint trinomial distribution with n trials and parameters p and q, where $0 < q < 1$, $0 < p < 1$, and $0 < p + q < 1$. Find (*a*) $\text{Cov}(X, Y)$, (*b*) $\rho(X, Y)$.

Solution. We are already familiar with the fact that X has the binomial distribution with n trials and the probability of success p, and Y has the binomial distribution with n trials and the probability of success q.

Therefore,

$$E(X) = np, \qquad \text{Var}(X) = np(1 - p)$$

and $E(Y) = nq, \qquad \text{Var}(Y) = nq(1 - q)$

Also, in Example 1.5 in this section, we saw that

$$E(XY) = n(n - 1)pq$$

Hence:

(*a*)

$$\begin{aligned}\text{Cov}(X, Y) &= E(XY) - E(X)E(Y) \\ &= n(n-1)pq - np \cdot nq \\ &= -npq\end{aligned}$$

Incidentally, observe that the covariance of X and Y is negative. Does this seem intuitively reasonable?

(*b*)

$$\rho(X, Y) = \frac{\text{Cov}(X, Y)}{\sqrt{\text{Var}(X) \cdot \text{Var}(Y)}}$$

$$= \frac{-npq}{\sqrt{np(1-p) \cdot nq(1-q)}}$$

$$= -\sqrt{\frac{pq}{(1-p)(1-q)}}$$

In summary, for the *trinomial distribution with n trials and parameters p and q*,

$$\text{Cov}(X, Y) = -npq$$

$$\text{and} \quad \rho(X, Y) = -\sqrt{\frac{pq}{(1-p)(1-q)}}$$

Example 1.13. Suppose X and Y have a joint distribution with the joint pdf given by

$$f(x, y) = \begin{cases} 10x^2 y, & 0 \leqslant y \leqslant x \leqslant 1 \\ 0, & \text{elsewhere} \end{cases}$$

Find:

(*a*) $\text{Var}(X)$ and $\text{Var}(Y)$
(*b*) $\text{Cov}(X, Y)$
(*c*) $\rho(X, Y)$

Solution. For any nonnegative integers n, k, we have

$$E(X^n Y^k) = \int_{-\infty}^{\infty}\int_{-\infty}^{\infty} x^n y^k f(x, y)\, dy\, dx$$

$$= \int_0^1 \int_0^x x^n y^k \cdot 10x^2 y\, dy\, dx$$

$$= \frac{10}{(k+2)(n+k+5)}$$

Therefore, assigning appropriate values to n and k,

$$E(X) = \frac{5}{6}, \quad E(Y) = \frac{5}{9}, \quad E(X^2) = \frac{5}{7}, \quad E(Y^2) = \frac{5}{14} \quad \text{and} \quad E(XY) = \frac{10}{21}$$

Hence:

(*a*)

$$\text{Var}(X) = \frac{5}{7} - \left(\frac{5}{6}\right)^2 = \frac{5}{252}$$

$$\text{Var}(Y) = \frac{5}{14} - \left(\frac{5}{9}\right)^2 = \frac{55}{1134}$$

(*b*)

$$\text{Cov}(X, Y) = E(XY) - E(X)E(Y) = \frac{5}{378}$$

(*c*)

$$\rho(X, Y) = \frac{\text{Cov}(X, Y)}{\sqrt{\text{Var}(X) \cdot \text{Var}(Y)}} = \frac{\frac{5}{378}}{\sqrt{\frac{5}{252} \cdot \frac{55}{1134}}} = \sqrt{\frac{2}{11}}$$

Example 1.4. A number X is picked in the interval (0, 1) with the probability law described by the following pdf:

$$f_X(x) = \begin{cases} 2x, & 0 < x < 1 \\ 0, & \text{elsewhere} \end{cases}$$

If $X = x$, a number Y is picked with the uniform distribution over the interval $(0, x)$. Find the correlation coefficient $\rho(X, Y)$.

Solution. We are given that the conditional distribution of Y given $X = x$ is uniform over $(0, x)$, that is,

$$f_{Y|X}(y \mid x) = \frac{1}{x}, \quad 0 < y < x$$

Therefore,

$$f(x, y) = f_{Y|X}(y \mid x) \cdot f_X(x) = \begin{cases} 2, & 0 < y < x < 1 \\ 0, & \text{elsewhere} \end{cases}$$

Now, for any nonnegative integers n, k,

$$\begin{aligned} E(X^n Y^k) &= \int_0^1 \int_0^x x^n y^k \cdot 2 \, dy \, dx \\ &= 2 \int_0^1 x^n \cdot \frac{x^{k+1}}{k+1} \, dx \\ &= \frac{2}{(k+1)(n+k+2)} \end{aligned}$$

Therefore, in particular,

$$E(X) = \frac{2}{3}, \quad E(Y) = \frac{1}{3}, \quad E(X^2) = \frac{1}{2}, \quad E(Y^2) = \frac{1}{6} \text{ and } E(XY) = \frac{1}{4}$$

Consequently, $\text{Var}(X) = \frac{1}{18}$, $\text{Var}(Y) = \frac{1}{18}$, and

$$\text{Cov}(X, Y) = \frac{1}{4} - \frac{2}{3} \cdot \frac{1}{3} = \frac{1}{36}$$

Hence

$$\rho(X, Y) = \frac{\frac{1}{36}}{\sqrt{\frac{1}{18} \cdot \frac{1}{18}}} = 0.5$$

- *Example 1.15* (*The standard bivariate normal distribution*) If X and Y have the standard bivariate normal distribution, find the correlation coefficient between X and Y.

Solution. We have seen in Example 2.8 of Chapter 8 that both X and Y have a standard normal distribution. Therefore,

$$E(X) = 0, \quad E(Y) = 0, \quad \text{Var}(X) = 1 \text{ and } \text{Var}(Y) = 1$$

and, consequently, the correlation coefficient is equal to $E(XY)$. Now,

$$\begin{aligned}
E(XY) &= \frac{1}{2\pi\sqrt{1-\rho^2}} \int_{-\infty}^{\infty}\int_{-\infty}^{\infty} xy \left[\exp\left\{-\frac{1}{2(1-\rho^2)}(x^2 - 2\rho xy + y^2)\right\}\right] dy\, dx \\
&= \frac{1}{2\pi\sqrt{1-\rho^2}} \int_{-\infty}^{\infty} x \left[\exp\left\{-\frac{x^2}{2(1-\rho^2)}\right\}\right]\left(\int_{-\infty}^{\infty} y \left[\exp\left\{-\frac{1}{2(1-\rho^2)}(y-\rho x)^2 \right.\right.\right. \\
&\qquad \left.\left.\left. + \frac{\rho^2 x^2}{2(1-\rho^2)}\right\}\right] dy\right) dx \\
&= \frac{1}{2\pi\sqrt{1-\rho^2}} \int_{-\infty}^{\infty} xe^{-x^2/2}\left(\int_{-\infty}^{\infty} y \left[\exp\left\{-\frac{1}{2(1-\rho^2)}(y-\rho x)^2\right\}\right] dy\right) dx
\end{aligned}$$

Recalling that

$$\frac{1}{b\sqrt{2\pi}} \int_{-\infty}^{\infty} ue^{-(u-a)^2/2b^2}\, du = a$$

we see that

$$\frac{1}{\sqrt{2\pi}\sqrt{1-\rho^2}} \int_{-\infty}^{\infty} y \left[\exp\left\{-\frac{1}{2(1-\rho^2)}(y-\rho x)^2\right\}\right] dy = \rho x$$

Therefore,

$$\begin{aligned}
E(XY) &= \frac{1}{\sqrt{2\pi}} \int_{-\infty}^{\infty} \rho x \cdot xe^{-x^2/2}\, dx \\
&= \frac{\rho}{\sqrt{2\pi}} \int_{-\infty}^{\infty} x^2 e^{-x^2/2}\, dx \\
&= \rho
\end{aligned}$$

That is, *the correlation coefficient between X and Y is precisely the constant ρ in the joint probability density function of the standard bivariate normal distribution.* ■

Important! The reader is aware that from $\rho(X, Y) = 0$ we cannot infer the independence of X and Y. However, there is an important exception to this. *If X and Y have a bivariate normal distribution, then the fact that the correlation coefficient is zero implies that the random variables are independent.*

This follows because, putting $\rho = 0$ in the joint pdf, we get

$$f(x, y) = \frac{1}{2\pi} e^{-(x^2+y^2)/2} = f_X(x) \cdot f_Y(y)$$

Hence X and Y are independent.

1.4 The Variance of a Linear Combination

We showed earlier that the expectation of a linear combination of random variables is equal to the linear combination of their expectations. The reader is put on guard that, in general, the variance of a linear combination is *not* equal to the linear combination of their variances. We have the following result:

If X and Y are two random variables then

$$\text{Var}(aX + bY) = a^2\text{Var}(X) + b^2\text{Var}(Y) + 2ab\,\text{Cov}(X, Y)$$

for any real numbers a, b.

To prove this, we note that, by the definition of the variance of a random variable,

$$\text{Var}(aX + bY) = E[(aX + bY - E(aX + bY))^2]$$

(Recalling the definition given in Chapter 7.)

Now,

$$E(aX + bY) = aE(X) + bE(Y) = a\mu_X + b\mu_Y$$

Therefore,

$$\begin{aligned}
\text{Var}(aX + bY) &= E[(aX + bY - a\mu_X - b\mu_Y)^2] \\
&= E[(a(X - \mu_X) + b(Y - \mu_Y))^2] \\
&= E[a^2(X - \mu_X)^2 + b^2(Y - \mu_Y)^2 + 2ab(X - \mu_X)(Y - \mu_Y)] \\
&= a^2E[(X - \mu_X)^2] + b^2E[(Y - \mu_Y)^2] \\
&\quad + 2abE[(X - \mu_X)(Y - \mu_Y)] \qquad \text{(why?)} \\
&= a^2\text{Var}(X) + b^2\text{Var}(Y) + 2ab\,\text{Cov}(X, Y)
\end{aligned}$$

Hence the result.

Some particular cases

(*i*) Setting $Y \equiv 1$, we get

$$\text{Var}(aX + b) = a^2\text{Var}(X), \qquad \text{for any real number } a$$

(The reader can fill in the details of the proof.)

For example, it now follows that

$$\text{Var}\left(\frac{X - \mu_X}{\sigma_X}\right) = \text{Var}\left(\frac{X}{\sigma_X} - \frac{\mu_X}{\sigma_X}\right) = \frac{1}{\sigma_X^2}\text{Var}(X) = 1$$

(*ii*) Setting $a = b = 1$, we get

$$\text{Var}(X + Y) = \text{Var}(X) + \text{Var}(Y) + 2\text{Cov}(X, Y)$$

and, setting $a = 1, b = -1$,

$$\text{Var}(X - Y) = \text{Var}(X) + \text{Var}(Y) - 2\text{Cov}(X, Y)$$

(*iii*) If X and Y are *independent*,

$$\text{Var}(aX + bY) = a^2\,\text{Var}(X) + b^2\,\text{Var}(Y) = \text{Var}(aX - bY)$$

since $\text{Cov}(X, Y) = 0$.

The above result regarding the variance of a linear combination generalizes in the following way to the case of n random variables $X_1, X_2, \ldots, X_n$.

$$\text{Var}\left(\sum_{i=1}^{n} a_i X_i\right) = \sum_{i=1}^{n} a_i^2 \text{Var}(X_i) + 2\sum_{i<j} a_i a_j \text{Cov}(X_i, X_j)$$

for any set of constants $a_1, a_2, \ldots, a_n$.

The following corollaries are immediate consequences if $X_1, X_2, \ldots, X_n$ are independent r.v.'s.

(*i*) On account of independence of the random variables $\text{Cov}(X_i, X_j) = 0$ if $i \neq j$, and therefore

$$\text{Var}\left(\sum_{i=1}^{n} a_i X_i\right) = \sum_{i=1}^{n} a_i^2 \text{Var}(X_i)$$

(*ii*) Setting $a_1 = a_2 = \ldots = a_n = 1$, we see right away that, *if $X_1, X_2, \ldots, X_n$ are independent r.v.'s, then the variance of their sum is equal to the sum of their variances.*

(*iii*) If each random variable has the same variance, say σ^2, then setting $a_1 = a_2 = \ldots = a_n = 1/n$ yields

$$\text{Var}(\bar{X}) = \frac{\sigma^2}{n}$$

That is, *if $X_1, X_2, \ldots, X_n$ are independent r.v.'s with the same variance σ^2, then the variance of the sample mean is equal to σ^2/n.* Notice that the variance of the sample mean goes to zero as the sample size n increases.

Example 1.16. A bowl contains four chips numbered 1, 4, 6, 9. If ten chips are picked at random with replacement, find the expected total and the variance of the total.

Solution. Let X_i, $i = 1, 2, \ldots, 10$, represent the number observed on the ith draw. Since the chips are picked with replacement, the random variables $X_1, X_2, \ldots, X_{10}$ are mutually independent. Also,

$$p_{X_i}(1) = p_{X_i}(4) = p_{X_i}(6) = p_{X_i}(9) = \tfrac{1}{4}$$

for $i = 1, 2, \ldots, 10$. Therefore, it follows that

$$E(X_i) = 5 \quad \text{and} \quad E(X_i^2) = 33.5$$

Hence

$$\text{Var}(X_i) = 33.5 - 5^2 = 8.5$$

Consequently, letting T represent the total on the ten chips, $T = X_1 + X_2 + \ldots + X_{10}$, and

$$E(T) = \sum_{i=1}^{10} E(X_i) = 50$$

and, since $X_1, X_2, \ldots, X_{10}$ are independent,

$$\text{Var}(T) = \sum_{i=1}^{10} \text{Var}(X_i) = 85$$

- *Example 1.17* (*Negative binomial distribution*) Suppose a random variable X has the negative binomial distribution representing the number of trials needed to get r successes, where r is a fixed positive integer. If the probability of success on any trial is p, find $E(X)$ and $\text{Var}(X)$.

Solution. Let

Y_1 = the number of trials needed for the first success
Y_2 = the number of trials after the first success needed for the next success

In general, let Y_i = the number of trials after the $(i-1)$th success needed for the next (ith) success, $i = 1, 2, \ldots, r$.

Clearly, $Y_1, Y_2, \ldots, Y_r$ are independent r.v.'s, each having the geometric distribution with

$$E(Y_i) = \frac{1}{p} \quad \text{and} \quad \text{Var}(Y_i) = \frac{1-p}{p^2}, \qquad i = 1, 2, \ldots, r$$

Also, it follows that $X = Y_1 + Y_2 + \ldots + Y_r$. Therefore,

$$E(X) = \sum_{i=1}^{r} E(Y_i) = \frac{r}{p}$$

and, since $Y_1, Y_2, \ldots, Y_r$ are independent,

$$\text{Var}(X) = \sum_{i=1}^{r} \text{Var}(Y_i) = \frac{r(1-p)}{p^2}$$

In summary, therefore, *if X has the negative binomial distribution given by $P(X = r + k) = \binom{r+k-1}{r-1} p^r (1-p)^k$, $k = 0, 1, 2, \ldots$, then*

$$E(X) = \frac{r}{p}$$

and

$$\text{Var}(X) = \frac{r(1-p)}{p^2}$$

1.5 The Method of Indicator Random Variables

For any event A, the indicator r.v. of A was defined in Chapter 4 as one which takes the value 1 at each sample point in A and the value 0 at each sample point in A'. Thus an indicator r.v. assumes only two values, namely, 0 and 1. It is so called because if the value of the random variable is 1, it *indicates* that the event A has occurred, and if the value is 0, it indicates that A has not occurred.

The following identities are immediate and can be proved routinely.

(*i*) $I_{AB} = I_A \cdot I_B$ and, in general,

$$I_{A_1A_2\ldots A_n} = I_{A_1} \cdot I_{A_2} \cdot \ldots \cdot I_{A_n}$$

(*ii*) $I_{A'} = 1 - I_A$

(*iii*) $I_{A \cup B} = I_A + I_B - I_{AB}$ and, in general,

$$I_{A_1 \cup A_2 \cup \ldots \cup A_n} = \sum_{i=1}^{n} I_{A_i} - \sum_{i<j}^{n} I_{A_iA_j} + \ldots + (-1)^{n+1} I_{A_1A_2\ldots A_n}$$

If, in particular, $A_1, A_2, \ldots, A_n$ are *mutually exclusive,* then

$$I_{A_1 \cup A_2 \cup \ldots \cup A_n} = \sum_{i=1}^{n} I_{A_i}$$

(*iv*) $I_{AB'} = I_A - I_{AB}$

Actually, (*iii*) and (*iv*) follow from (*i*) and (*ii*). For example,

$$\begin{aligned} I_{AB'} &= I_A \cdot I_{B'}, \quad \text{by } (i) \\ &= I_A \cdot (1 - I_B), \quad \text{by } (ii) \\ &= I_A - I_A \cdot I_B, \\ &= I_A - I_{AB}, \quad \text{by } (i) \end{aligned}$$

Let us now find $E(I_A)$. Since $I_A(s) = 1$ if and only if $s \in A$, and $I_A(s) = 0$ if and only if $s \in A'$, it follows that

$$P(I_A = 1) = P(A) \quad \text{and} \quad P(I_A = 0) = P(A')$$

Therefore,

$$E(I_A) = 1 \cdot P(A) + 0 \cdot P(A') = P(A)$$

Hence,

$$\boxed{E(I_A) = P(A) \qquad \text{for any event } A}$$

This result shows that *we can regard the probability of an event as the expected value of the corresponding indicator random variable.* In other words, the concept of expectation is an extension of the concept of a probability measure.

This single fact now leads to the various results of the probability measure that are already familiar to us. For example, since $I_{A \cup B} = I_A + I_B - I_{AB}$, using the linear property of expectation it follows that $E(I_{A \cup B}) = E(I_A) + E(I_B) - E(I_{AB})$; that is, $P(A \cup B) = P(A) + P(B) - P(AB)$.

The independence of a collection of events is closely linked to that of the corresponding indicator r.v.'s. As a matter of fact, the following result can be shown to be true. The proof will be left to the exercise set.

If $A_1, A_2, \ldots, A_n$ are n events, then they are independent if and only if the indicator random variables $I_{A_1}, I_{A_2}, \ldots, I_{A_n}$ are independent.

We shall next find $\text{Var}(I_A)$. We immediately have $E(I_A^2) = 1^2 \cdot P(A) + 0^2 \cdot P(A')$ $= P(A)$, so that $\text{Var}(I_A) = E(I_A^2) - [E(I_A)]^2 = P(A) - [P(A)]^2 = P(A)P(A')$.

Hence,

$$\text{Var}(I_A) = P(A) \cdot P(A') \qquad \text{for any event } A$$

The covariance between two indicator random variables can be expressed in terms of the probabilities of the underlying events as follows: Suppose A and B are two events. Then

$$E(I_A \cdot I_B) = E(I_{AB}) = P(AB)$$

and we get

$$\text{Cov}(I_A, I_B) = P(AB) - P(A)P(B), \qquad \text{for any two events } A, B$$

The method of indicator random variables turns out to be a very powerful tool in many instances, as the following examples will illustrate.

Example 1.18 (*The binomial distribution*) Suppose X represents the number of successes in n independent Bernoulli trials, with the probability of success p on each trial. In other words, X is $B(n, p)$. Find $E(X)$ and $\text{Var}(X)$.

Solution. We previously found $E(X)$ and $\text{Var}(X)$ in Chapter 7 using the direct approach which, as will be recalled, involved some tedious algebraic steps. We now give a much simpler approach.

Let A_i represent the event that there is a success on the ith trial, $i = 1, 2, \ldots, n$. Then clearly

$$X = I_{A_1} + I_{A_2} + \ldots + I_{A_n}$$

where $I_{A_1}, I_{A_2}, \ldots, I_{A_n}$ are independent r.v.'s, since the events $A_1, A_2, \ldots, A_n$ are independent.

It follows that

$$E(X) = E(I_{A_1}) + \ldots + E(I_{A_n})$$

$$= \sum_{i=1}^{n} P(A_i) = np$$

and, since $I_{A_1}, I_{A_2}, \ldots, I_{A_n}$ are independent,

$$\begin{aligned}\operatorname{Var}(X) &= \operatorname{Var}(I_{A_1}) + \ldots + \operatorname{Var}(I_{A_n}) \\ &= \sum_{i=1}^{n} P(A_i)P(A_i') = np(1-p)\end{aligned}$$

Thus, $E(X) = np$ and $\operatorname{Var}(X) = np(1-p)$, the results we had obtained previously.

- *Example 1.19* (*The hypergeometric distribution*) Suppose a lot contains M objects, of which D are defective and $M - D$ are nondefective. If n of these objects are drawn *without replacement*, find $E(X)$ and $\operatorname{Var}(X)$, where X represents the number of defective objects in the sample.

Solution. Let A_i be the event that there is a defective object picked on the ith draw, $i = 1, 2, \ldots, n$. Then clearly

$$X = I_{A_1} + I_{A_2} + \ldots + I_{A_n}$$

Therefore,

$$E(X) = \sum_{i=1}^{n} E(I_{A_i}) = \sum_{i=1}^{n} P(A_i) = n\frac{D}{M}$$

since the probability of picking a defective object on any draw is D/M.

To find $\operatorname{Var}(X)$, we note first that $I_{A_1}, I_{A_2}, \ldots, I_{A_n}$ are *not* independent random variables. Therefore,

$$\begin{aligned}\operatorname{Var}(X) &= \sum_{i=1}^{n} \operatorname{Var}(I_{A_i}) + 2\sum_{i<j}^{n} \operatorname{Cov}(I_{A_i}, I_{A_j}) \\ &= \sum_{i=1}^{n} P(A_i)P(A_i') + 2\sum_{i<j}^{n} [P(A_iA_j) - P(A_i)P(A_j)]\end{aligned}$$

Now A_iA_j represents the event that defective objects are picked on the ith and jth draw and, consequently, $P(A_iA_j) = \frac{D}{M} \cdot \frac{D-1}{M-1}$, if $i \neq j$. Substituting for $P(A_i)$ and $P(A_iA_j)$, we get

$$\begin{aligned}\operatorname{Var}(X) &= \sum_{i=1}^{n} \frac{D}{M}\left(1 - \frac{D}{M}\right) + 2\sum_{i<j}^{n} \left[\frac{D}{M} \cdot \frac{D-1}{M-1} - \left(\frac{D}{M}\right)^2\right] \\ &= n\frac{D}{M}\left(1 - \frac{D}{M}\right) + 2 \cdot \binom{n}{2}\left[-\frac{D}{M}\left(1 - \frac{D}{M}\right)\frac{1}{(M-1)}\right]\end{aligned}$$

since there are $\binom{n}{2}$ terms in the summation $\sum_{i<j}^{n}$. Therefore,

$$\begin{aligned}\operatorname{Var}(X) &= n\frac{D}{M}\left(1 - \frac{D}{M}\right)\left[1 - (n-1)\frac{1}{(M-1)}\right] \\ &= n\frac{D}{M}\left(1 - \frac{D}{M}\right)\frac{M-n}{M-1}\end{aligned}$$

Letting $D/M = p$, the proportion of defectives in the original lot, we see that

$$E(X) = np$$

and

$$\text{Var}(X) = np(1-p)\frac{M-n}{M-1}$$

■

Comment. If n objects are picked *with replacement* from a lot containing a proportion p of defectives, then, as we know, the distribution of the number of defectives is binomial, and the expected number of defectives is np and the variance is $np(1-p)$. From Example 1.19, we see that if the sampling is done *without replacement,* then the variance of the number of defectives is $np(1-p)\frac{M-n}{M-1}$. However, if M is very large (relative to n), then the variance of the number of defectives in the hypergeometric case approaches $np(1-p)$, the variance in the binomial case. Intuitively, this stands to reason because, if the lot is extremely large, then whether we sample with replacement or without replacement should not make much difference.

• *Example 1.20* (*Matching problem*) If two decks of n cards, each numbered $1, 2, \ldots, n$, are matched randomly, find the expected number of matches and the variance of the number of matches.

Solution. Let X denote the number of matches. We could go ahead and obtain the distribution of X, and then, by applying the basic definition, find $E(X)$. However, that approach turns out to be very complicated and tedious. The method of indicator r.v.'s proves to be much simpler.

Let A_i, $i = 1, 2, \ldots, n$, be the event that the ith cards in the decks are matched. Then clearly

$$X = I_{A_1} + I_{A_2} + \ldots + I_{A_n}$$

so that

$$E(X) = \sum_{i=1}^{n} P(A_i)$$

and

$$\text{Var}(X) = \sum_{i=1}^{n} P(A_i)P(A_i') + 2\sum_{i<j}^{n} [P(A_iA_j) - P(A_i)P(A_j)]$$

Now, as was shown in Example 3.14 in Chapter 2,

$$P(A_i) = \frac{1}{n}, \quad \text{and} \quad P(A_iA_j) = \frac{1}{n(n-1)} \quad \text{if } i \neq j$$

Therefore,

$$E(X) = n \cdot \frac{1}{n} = 1$$

and

$$\begin{aligned}\mathrm{Var}(X) &= n \cdot \frac{1}{n}\left(1 - \frac{1}{n}\right) + 2\sum_{i<j}^{n}\left[\frac{1}{n(n-1)} - \left(\frac{1}{n}\right)^2\right] \\ &= \left(1 - \frac{1}{n}\right) + 2 \cdot \binom{n}{2}\frac{1}{n^2(n-1)} = 1\end{aligned}$$

• *Example 1.21* (*The trinomial distribution*) Suppose an experiment consists of n independent, identical trials with probability p of success on any trial, probability q of failure, and the probability $1 - p - q$ of a draw. Let X represent the number of successes, and Y the number of failures. Find $E(XY)$.

Solution. The joint distribution of X and Y is trinomial and we have already shown in Example 1.5 that $E(XY) = n(n-1)pq$. Let us now derive the same result by the method of indicators.

Let A_i represent the event that a success occurs on the ith trial, and B_i the event that a failure occurs on the ith trial. Then

$$X = \sum_{i=1}^{n} I_{A_i} \quad \text{and} \quad Y = \sum_{i=1}^{n} I_{B_i}$$

It therefore follows that

$$\begin{aligned}E(XY) &= E\left[\left(\sum_{i=1}^{n} I_{A_i}\right)\left(\sum_{i=1}^{n} I_{B_i}\right)\right] \\ &= E\left[\sum_{i=1}^{n} I_{A_i} \cdot I_{B_i} + \sum_{i\neq j}^{n} I_{A_i} \cdot I_{B_j}\right] \\ &= E\left[\sum_{i=1}^{n} I_{A_iB_i} + \sum_{i\neq j}^{n} I_{A_iB_j}\right] \\ &= \sum_{i=1}^{n} E(I_{A_iB_i}) + \sum_{i\neq j}^{n} E(I_{A_iB_j})\end{aligned}$$

by the linear property of expectation. Hence

$$E(XY) = \sum_{i=1}^{n} P(A_iB_i) + \sum_{i\neq j}^{n} P(A_iB_j)$$

But $P(A_iB_i) = 0$, since we cannot have a success and a failure on the same trial, and $P(A_iB_j) = P(A_i)P(B_j) = pq$ if $i \neq j$, since the trials are independent. Therefore, in conclusion,

$$E(XY) = n(n-1)pq$$

since there are $n(n-1)$ terms in the summation $\sum_{i\neq j}^{n}$.

1.6 Bounds on the Correlation Coefficient

The next topic on the agenda is to set bounds on the correlation coefficient. Towards this we establish a few preliminaries.

(*i*) Suppose X is a random variable with $E(X) = \mu_X$ and $\mathrm{Var}(X) = \sigma_X^2$. Then

$$E\left(\frac{X-\mu_X}{\sigma_X}\right) = 0 \quad \text{and} \quad \mathrm{Var}\left(\frac{X-\mu_X}{\sigma_X}\right) = 1$$

This has already been proved.

The random variable $(X - \mu_X)/\sigma_X$ is called a *standardized random variable.*

(*ii*) Suppose X and Y are two random variables and a_1, a_2, b_1, b_2 are any constants. Then

$$\mathrm{Cov}(a_1 X + b_1, a_2 Y + b_2) = a_1 a_2 \mathrm{Cov}(X, Y)$$

The result can be proved as follows:

$$\begin{aligned} \mathrm{Cov}(a_1 X + b_1, a_2 Y + b_2) &= E[(a_1 X + b_1 - E(a_1 X + b_1)) \\ &\quad \cdot (a_2 Y + b_2 - E(a_2 Y + b_2))] \\ &= E[a_1(X-\mu_X) a_2(Y-\mu_Y)] \qquad \text{(why?)} \\ &= a_1 a_2 E[(X-\mu_X)(Y-\mu_Y)] \\ &= a_1 a_2 \mathrm{Cov}(X, Y) \end{aligned}$$

Observe that the constants b_1 and b_2 do not appear on the right-hand side. This property is referred to as the *invariance of covariance under translation.*

(*iii*) The correlation coefficient between X and Y is equal to the covariance between the corresponding standardized random variables.

To see this, recall that

$$\rho(X, Y) = \frac{E[(X-\mu_X)(Y-\mu_Y)]}{\sigma_X \sigma_Y}$$

That is,

$$\begin{aligned} \rho(X, Y) &= E\left(\frac{X-\mu_X}{\sigma_X} \cdot \frac{Y-\mu_Y}{\sigma_Y}\right) \\ &= E\left(\frac{X-\mu_X}{\sigma_X} \cdot \frac{Y-\mu_Y}{\sigma_Y}\right) - E\left(\frac{X-\mu_X}{\sigma_X}\right) E\left(\frac{Y-\mu_Y}{\sigma_Y}\right), \qquad \text{since } E\left(\frac{X-\mu_X}{\sigma_X}\right) = 0 \\ &= \mathrm{Cov}\left(\frac{X-\mu_X}{\sigma_X}, \frac{Y-\mu_Y}{\sigma_Y}\right) \end{aligned}$$

by definition of covariance.

We now state the result which sets bounds on $\rho(X, Y)$. We have

$$-1 \leqslant \rho(X, Y) \leqslant 1, \qquad \text{for any two random variables } X, Y$$

To prove this, let

$$X^* = \frac{X - \mu_X}{\sigma_X} \quad \text{and} \quad Y^* = \frac{Y - \mu_Y}{\sigma_Y}$$

Now,

$$\begin{aligned} \text{Var}(X^* - Y^*) &= \text{Var}(X^*) + \text{Var}(Y^*) - 2\,\text{Cov}(X^*, Y^*) \\ &= 1 + 1 - 2\rho(X, Y) \\ &= 2 - 2\rho(X, Y) \end{aligned}$$

Since $\text{Var}(X^* - Y^*) \geqslant 0$, this gives $2 - 2\rho(X, Y) \geqslant 0$; that is,

$$\rho(X, Y) \leqslant 1$$

Also,

$$\begin{aligned} \text{Var}(X^* + Y^*) &= \text{Var}(X^*) + \text{Var}(Y^*) + 2\,\text{Cov}(X^*, Y^*) \\ &= 1 + 1 + 2\rho(X, Y) \\ &= 2 + 2\rho(X, Y) \end{aligned}$$

Since $\text{Var}(X^* + Y^*) \geqslant 0$, this gives $2 + 2\rho(X, Y) \geqslant 0$; that is,

$$-1 \leqslant \rho(X, Y)$$

Combining, we get $-1 \leqslant \rho(X, Y) \leqslant 1$.

The cases $\rho(X, Y) = 1$ and $\rho(X, Y) = -1$ have a special significance, as explained in the following results.

Let X be a random variable and suppose that $Y = mX + c$, where $m \neq 0$ and c are constants. Then $\rho(X, Y) = 1$ if $m > 0$ and $\rho(X, Y) = -1$ if $m < 0$.

(In essence the result states that *if Y is a linear function of X, then the correlation is either 1 or −1*. The correlation is 1 if Y increases with X and it is −1 if Y decreases as X increases.)

Let us now prove the result. We see that $\text{Var}(Y) = m^2\sigma_X^2$ so that

$$\sigma_Y = \begin{cases} m\sigma_X & \text{if } m > 0 \\ -m\sigma_X & \text{if } m < 0 \end{cases}$$

Also,

$$\begin{aligned} \text{Cov}(X, Y) &= E[(X - \mu_X)(Y - \mu_Y)] \\ &= E[(X - \mu_X)(mX + c - E(mX + c))] \\ &= E[(X - \mu_X)(m(X - \mu_X))] = m\sigma_X^2 \end{aligned}$$

Hence

$$\rho(X, Y) = \frac{\text{Cov}(X, Y)}{\sigma_X \sigma_Y} = \begin{cases} \dfrac{m\sigma_X^2}{\sigma_X m \sigma_X} & \text{if } m > 0 \\[2ex] \dfrac{m\sigma_X^2}{\sigma_X(-m\sigma_X)} & \text{if } m < 0 \end{cases}$$

$$= \begin{cases} 1 & \text{if } m > 0 \\ -1 & \text{if } m < 0 \end{cases}$$

The converse of the above result is essentially correct, as we shall see next.

Suppose X and Y are random variables with $\rho(X, Y) = 1$ or $\rho(X, Y) = -1$. Then Y is a linear function of X with probability 1. That is, there exist constants $m \neq 0$ and c so that $P(Y = mX + c) = 1$. The constant m is positive if $\rho(X, Y) = 1$ and it is negative if $\rho(X, Y) = -1$.

We shall prove the result when $\rho(X, Y) = 1$ and leave to the reader the proof when $\rho(X, Y) = -1$.

If $\rho(X, Y) = 1$, then

$$\text{Var}(X^* - Y^*) = 2 - 2\rho(X, Y) = 0$$

But we have seen (in Chapter 7) that the variance of a random variable is zero if and only if it assumes a single value with probability one. The value assumed is actually the expected value of the random variable. Now $E(X^* - Y^*) = 0$. Therefore, $\text{Var}(X^* - Y^*) = 0$ implies that

$$P(X^* - Y^* = 0) = 1$$

That is,

$$P\left(\frac{X - \mu_X}{\sigma_X} = \frac{Y - \mu_Y}{\sigma_Y}\right) = 1$$

After simplifying, we get

$$P\left(Y = \frac{\sigma_Y}{\sigma_X} X + \left(\mu_Y - \frac{\sigma_Y}{\sigma_X}\mu_X\right)\right) = 1$$

That is,

$$P(Y = mX + c) = 1$$

if we let

$$m = \frac{\sigma_Y}{\sigma_X} \quad \text{and} \quad c = \mu_Y - \frac{\sigma_Y}{\sigma_X}\mu_X$$

(Note that $m = \sigma_Y/\sigma_X > 0$.)

The significance of the results regarding the correlation coefficient that we have just proved should be emphasized. Recapitulating, the correlation coefficient between two random variables cannot be less than -1 or exceed 1. It is equal to

1 or −1 when, and only when, the functional relation between X and Y is linear—that is, when there exist constants $m \neq 0$ and c for which $P(Y = mX + c) = 1$.

What significance can be attached to $-1 < \rho(X, Y) < 1$? This would imply that there is a positive probability that the relation between X and Y is not linear. That is, for any pair of constants m and c, $P(Y \neq mX + c) > 0$. In other words, no matter what straight line we draw in the xy-plane, there is a *positive* probability that (X, Y) will not be on such a line. The correlation coefficient therefore provides a measure of the linear relationship of X and Y. It is due to this reason that it is sometimes called the *coefficient of linear correlation.*

Example 1.22. We give below three cases of joint distributions of X and Y:

(*a*) X has a standard normal distribution and $Y = 2X + 1$.

(*b*) X has a standard normal distribution and $Y = X^2 + 2X$.

(*c*) The joint pdf of X and Y is

$$f(x, y) = \begin{cases} \frac{1}{\pi} e^{-(x^2 + y^2)/2}, & \text{if } x < 0 \text{ and } y > 0, \text{ or} \\ & \quad x > 0 \text{ and } y < 0 \\ 0, & \text{elsewhere} \end{cases}$$

In each case, comment on the correlation coefficient in the light of the functional relation between X and Y.

Solution

(*a*) The joint distribution of X and Y is singular since all the probability mass is distributed along the line $y = 2x + 1$. Consequently, there exists no joint pdf.

Since the relation between X and Y is linear with $m = 2 > 0$, the correlation coefficient has to be equal to 1. We shall compute it directly anyway.

Since X is a standard normal variable, $E(X) = 0$ and $\text{Var}(X) = E(X^2) = 1$. Next,

$$E(Y) = E(2X + 1) = 2E(X) + 1 = 1$$
$$E(Y^2) = E[(2X + 1)^2] = E(4X^2 + 4X + 1) = 4E(X^2) + 4E(X) + 1 = 5$$

and $$E(XY) = E(X(2X + 1)) = 2E(X^2) + E(X) = 2$$

Hence

$$\rho(X, Y) = \frac{2}{\sqrt{1 \cdot 4}} = 1$$

as was anticipated.

(*b*) The joint distribution of X and Y is singular, since all the probability mass is distributed smoothly on the curve $y = x^2 + 2x$, that is, a region whose area is zero. Hence there exists no joint probability density function.

The relation between X and Y is not linear, so that we already know that $\rho(X, Y)$ should be strictly between −1 and 1. Let us compute the actual correlation coefficient. Since X is $N(0, 1)$, we have

$$E(X) = 0, \quad E(X^2) = 1, \quad E(X^3) = 0, \quad E(X^4) = 3$$

Therefore,

$$E(Y) = E(X^2 + 2X) = E(X^2) + 2E(X) = 1$$
$$E(Y^2) = E[(X^2 + 2X)^2] = E(X^4) + 4E(X^3) + 4E(X^2) = 7$$

and $E(XY) = E(X(X^2 + 2X)) = E(X^3) + 2E(X^2) = 2$

Hence $\mathrm{Var}(X) = 1$, $\mathrm{Var}(Y) = 6$, and $\mathrm{Cov}(X, Y) = 2$. Consequently,

$$\rho(X, Y) = \frac{2}{\sqrt{1 \cdot 6}} = \sqrt{\frac{2}{3}}$$

which is strictly between -1 and 1, as we had anticipated.

(*c*) The joint pdf is positive over the shaded region in Figure 1.1 and is zero outside it. We anticipate here that $-1 < \rho(X, Y) < 1$. Not only that; we anticipate it to be negative (why?).

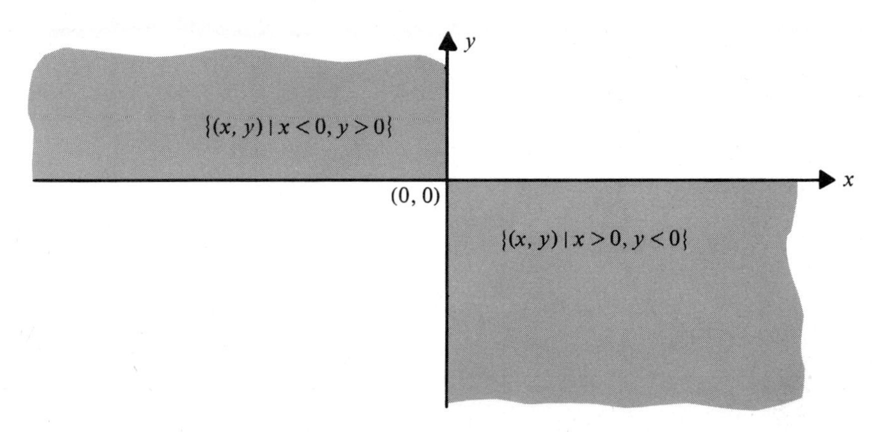

Figure 1.1

It can be easily verified that the marginal distributions of X and Y are standard normal. Hence $E(X) = E(Y) = 0$ and $\mathrm{Var}(X) = \mathrm{Var}(Y) = 1$, so that $\rho(X, Y) = E(XY)$.

Next,

$$E(XY) = \int_{-\infty}^{0}\int_{0}^{\infty} xy \frac{e^{-(x^2+y^2)/2}}{\pi}\, dy\, dx + \int_{0}^{\infty}\int_{-\infty}^{0} xy \frac{e^{-(x^2+y^2)/2}}{\pi}\, dy\, dx$$
$$= \frac{1}{\pi}\left[\int_{-\infty}^{0} xe^{-x^2/2}\, dx \int_{0}^{\infty} ye^{-y^2/2}\, dy + \int_{0}^{\infty} xe^{-x^2/2}\, dx \int_{-\infty}^{0} ye^{-y^2/2}\, dy\right]$$

Letting $u^2/2 = t$, it follows that

$$\int_{0}^{\infty} ue^{-u^2/2}\, du = \int_{0}^{\infty} e^{-t}\, dt = 1$$

Hence

$$E(XY) = \frac{1}{\pi}[(-1)(1) + (1)(-1)] = \frac{-2}{\pi}$$

Thus, finally,

$$\rho(X, Y) = -\frac{2}{\pi}$$

which is between −1 and 0.

EXERCISES–SECTION 1

1. Suppose X and Y are two random variables with a joint distribution. For the continuous case, we defined $E(X)$ in Chapter 7 as

$$E(X) = \int_{-\infty}^{\infty} x f_X(x)\, dx$$

On the other hand, according to the definition given in this chapter,

$$E(X) = \int_{-\infty}^{\infty}\int_{-\infty}^{\infty} x f(x, y)\, dy\, dx$$

How do you reconcile the two definitions?

2. If X and Y have the following joint probability function,

x \ y	0	2	3
−1	0.1	0.2	0.1
0	0.1	0.1	0.2
2	0.1	0.1	0

find–

(a) $E(X)$ (b) $E(X^2Y)$ (c) $E\left(\frac{X}{Y+1}\right)$

3. The amount X (in dollars) that a babysitter earns on a weekend and the amount Y that she spends in the following week have the following joint distribution:

x \ y	1	3	6
2	$\frac{2}{9}$	$\frac{1}{9}$	0
4	$\frac{1}{9}$	$\frac{2}{9}$	$\frac{1}{9}$
7	0	$\frac{1}{9}$	$\frac{1}{9}$

Find:

(a) the expected amount earned
(b) the expected amount spent
(c) the correlation coefficient between the amount earned and the amount spent

4. If X and Y have an absolutely continuous joint distribution with pdf given by

$$f(x, y) = \begin{cases} 24y(1-x), & 0 \leqslant y \leqslant x \leqslant 1 \\ 0, & \text{elsewhere} \end{cases}$$

find:

(a) $E(X)$ (b) $E(X^2Y)$ (c) $E\left(\frac{Y^2}{1-X}\right)$

5. Two points are picked at random and independently inside the interval $[0, a]$. Find the expected distance and the variance of the distance between the points.
6. Suppose the distribution of X and Y is given by the following pdf:

$$f(x, y) = \begin{cases} (x + y), & 0 \leqslant x \leqslant 1 \text{ and } 0 \leqslant y \leqslant 1 \\ 0, & \text{elsewhere} \end{cases}$$

Find $E(\max(X, Y))$.
7. Let $0 < a < b$. Suppose X and Y are uniformly distributed over the intervals $[a, b]$ and $[-b, -a]$, respectively. Find:

(a) $E(X + Y)$
(b) $E(XY)$ if X and Y are independent

(*Comment:* Recall that in exercise 3 of Section 3, Chapter 10, $E(X + Y)$ and $E(XY)$ were found by actually finding the distributions of $X + Y$ and XY. There is no need to go this route!)
8. If X and Y are independent random variables, why is it true that $E(X/Y) = E(X) \cdot E(1/Y)$? What restriction would be required on Y? Use the above result to find $E(X/Y)$ for the random variables X and Y described in exercise 7.
9. Suppose Z is uniformly distributed over the interval $[0, 2\pi]$. Define X and Y as follows:

$$X = \cos Z \quad \text{and} \quad Y = \sin Z$$

Show that–

(a) X and Y are not independent
(b) X and Y are not correlated.

10. If X and Y are two random variables such that each assumes only two values, then show that $\text{Cov}(X, Y) = 0$ implies that X and Y are independent. *Hint:* There is no loss of generality in assuming that X assumes values $0, x$ and Y assumes values $0, y$ (why?).
11. Suppose X has a distribution which is symmetric about 0. Let $Y = X^2$. Show that X and Y are uncorrelated. Are they independent?
12. The dimensions X, Y, Z of a rectangular parallelepiped are known to be independent random variables with the following pdf's:

$$f_X(x) = \begin{cases} \frac{1}{3}, & 1 < x < 4 \\ 0, & \text{elsewhere} \end{cases}$$

$$f_Y(y) = \begin{cases} \frac{1}{2}(3 - y), & 1 < y < 3 \\ 0, & \text{elsewhere} \end{cases}$$

and

$$f_Z(z) = \begin{cases} \frac{3}{124} z^2, & 1 < z < 5 \\ 0, & \text{elsewhere} \end{cases}$$

Find:

(a) the expected surface area
(b) the expected volume of the parallelepiped

13. If X represents the product of the outcomes on six independent tosses of a normal die, find $E(X)$.

14. Suppose n points are picked independently in the interval $[0, 2]$, each governed by the probability law with pdf given by

$$f(x) = \begin{cases} \frac{3}{8}x^2, & 0 \leq x \leq 2 \\ 0, & \text{elsewhere} \end{cases}$$

Find the expected product.

15. The amount of gold X (in ounces) that a prospector digs in a day is a random variable with the following pdf:

$$f(x) = \begin{cases} 6x(1-x), & 0 < x < 1 \\ 0, & \text{elsewhere} \end{cases}$$

If the price of gold Y (in dollars) per ounce is uniformly distributed over the interval $[100, 180]$ and is independent of X, find the expected earnings in a day.

16. Suppose X is a random variable with $0 < \text{Var}(X) < \infty$. If a and b are real numbers with $a \neq 0$, and $Y = aX + b$, show that $E(XY) \neq E(X) \cdot E(Y)$.

17. Suppose X and Y have the following joint discrete distribution:

x \ y	-1	2	4	$P(X = x)$
-6	$\frac{1}{6}$	$\frac{1}{6}$	$\frac{1}{12}$	$\frac{5}{12}$
6	$\frac{1}{3}$	$\frac{1}{12}$	$\frac{1}{6}$	$\frac{7}{12}$
$P(Y = y)$	$\frac{1}{2}$	$\frac{1}{4}$	$\frac{1}{4}$	

Compute the mean and the variance of $Z = 3 + 6X + 2Y$.

18. A fair die with four faces in the shape of a tetrahedron has the faces marked $-4, -1, 2, 7$. Find the expected total and the variance of the total in twenty independent throws of the die.

19. Suppose sixteen observations are picked independently from a distribution with the following pdf:

$$f(x) = \begin{cases} \frac{5}{32}x^4, & 0 < x < 2 \\ 0, & \text{elsewhere} \end{cases}$$

Find the expectation and the variance of the sample mean $\bar{X}$.

20. Suppose X and Y are independent random variables where X is $N(0, \sigma^2)$ and Y is exponential with parameter $\lambda = 1$. Find $E(Z), E(Z^2), E(Z^3)$, and $E(Z^4)$, where $Z = X + Y$.

21. Suppose X and Y are independent random variables with $\text{Var}(X) = 4$, and $\text{Var}(Y) = 9$. Find the correlation coefficient between X and $X + 2Y$.

22. If X and Y are independent random variables with $\text{Var}(2X + 3Y) = 40$ and $\text{Var}(X - 2Y) = 17$, find $\text{Var}(X)$ and $\text{Var}(Y)$.

23. If X and Y are two random variables with $\text{Var}(2X + 3Y) = 67$, $\text{Var}(X - Y) = 3$, and $\text{Var}(X - 2Y) = 8$, find the correlation coefficient $\rho(X, Y)$.

24. If X and Y are standardized random variables with correlation coefficient ρ, find–

(a) $E(X - \rho Y)$ (b) $\text{Var}(X - \rho Y)$ (c) $\text{Cov}(X - \rho Y, Y)$

25. If X and Y are independent random variables, express the following in terms of μ_X, μ_Y, σ_X^2, and σ_Y^2:
(a) $\text{Var}(XY)$ (b) $\text{Cov}(X, XY)$ (c) $\rho(X, XY)$

26. If $X_1, X_2, \ldots, X_n$ are independent random variables with the same variance, find the correlation coefficient between $\sum_{i=1}^{n} X_i$ and X_i for any i.

27. Suppose $X_1, X_2, \ldots, X_l, Y_1, Y_2, \ldots, Y_m, Z_1, Z_2, \ldots, Z_n$ are uncorrelated random variables each having unit variance. Let

$$U = X_1 + X_2 + \ldots + X_l + Y_1 + \ldots + Y_m$$
$$V = X_1 + X_2 + \ldots + X_l + Z_1 + \ldots + Z_n$$

Show that $\rho(U, V) = l/\sqrt{(l+m)(l+n)}$.

28. If A and B are two events, show that–
(a) $I_{AB} = I_A \cdot I_B$
(b) $I_{A'} = 1 - I_A$

29. If $A_1, A_2, \ldots, A_n$ are n events, show that they are independent if and only if the indicator random variables $I_{A_1}, I_{A_2}, \ldots, I_{A_n}$ are independent.

30. Using the fact that $I_{AB} = I_A \cdot I_B$ and $I_{A'} = 1 - I_A$, show that

$$I_{A_1 \cup A_2 \cup \ldots \cup A_n} = \sum_{i=1}^{n} I_{A_i} - \sum_{i<j}^{n} I_{A_i A_j} + \ldots + (-1)^{n+1} I_{A_1 A_2 \ldots A_n}$$

What formula in probability could you derive from this? *Hint:* Use De Morgan's law.

31. Suppose n distinguishable objects are distributed at random into M distinguishable boxes in a way that does not preclude a box from getting more than one object. Use the method of indicators to find the expected number of boxes that are occupied. *Hint:* Find the probability that a given box is occupied.

32. A lake is known to contain 1000 trout and 2000 bass. If a random sample of 100 fish is picked from the lake, find the expected number of trout caught. What is the variance of this number? (Consider both cases, with and without replacement.)

33. A fair die is rolled independently n times. Let X represent the number of 1s obtained, and Y the number of 2s. Find $\text{Cov}(X, Y)$.

34. Let $X_1, X_2, \ldots, X_n$ be independent random variables, each distributed uniformly over $[0, 1]$. If

$$V = \max(X_1, X_2, \ldots, X_n)$$
$$W = \min(X_1, X_2, \ldots, X_n)$$

find–
(a) $E(V^k)$ (b) $E(W^k)$ (c) $E(VW)$ (d) $\rho(V, W)$.

Hint: See exercise 11, Section 3, Chapter 10, and use the fact that

$$\int_0^1 x^{m-1}(1-x)^{n-1}\, dx = \frac{\Gamma(m)\Gamma(n)}{\Gamma(m+n)}$$

2. CONDITIONAL EXPECTATION

2.1 The Definition of Conditional Expectation

Our goal is to define the conditional expectation of a function of two random variables X and Y, given that one of the random variables, say Y, assumes a specific value y. To start with, let us consider the simplest situation, that of the conditional expectation of X given $Y = y$. We shall denote this as $E(X \mid Y = y)$, or simply as $E(X \mid y)$ if there is no room for confusion. Following the same procedure used earlier, we define

$$E(X \mid Y = y) = \begin{cases} \int_{-\infty}^{\infty} x f_{X|Y}(x \mid y)\,dx, & \text{in the continuous case} \\ \sum\limits_{x_i} x_i P(X = x_i \mid Y = y), & \text{in the discrete case} \end{cases}$$

provided $f_Y(y) > 0$ in the continuous case and $P(Y = y) > 0$ in the discrete case.

The reader will see that there is nothing startlingly new about this definition. All we are doing is using the appropriate distribution, namely, the conditional distribution of X given $Y = y$.

$E(X \mid Y = y)$ depends on y and hence is a function of y. The graph of $E(X \mid Y = y)$ as a function of y is called **the regression curve of X on Y.** Similarly, the graph of $E(Y \mid X = x)$ as a function of x is called **the regression of Y on X.**

Comment. If X and Y are independent, then $E(X \mid Y = y) = E(X)$, since in this case $f_{X|Y}(x \mid y) = f_X(x)$

More generally, if h is a real-valued function of two variables, we define

---

$$E(h(X, Y) \mid Y = y) = \begin{cases} \int_{-\infty}^{\infty} h(x, y) f_{X|Y}(x \mid y)\,dx, & \text{in the continuous case} \\ \sum\limits_{x_i} h(x_i, y) P(X = x_i \mid Y = y), & \text{in the discrete case} \end{cases}$$

Example 2.1. If the joint pdf of X and Y is given by

$$f(x, y) = \begin{cases} ye^{-xy}, & 0 < x < \infty,\ 0 < y < 1 \\ 0, & \text{elsewhere} \end{cases}$$

find $E(X^n \mid y_o)$, where n is a nonnegative integer and $0 < y_o < 1$.

Solution. We need the conditional pdf of X given $Y = y_o$ which, for $0 < y_o < 1$, is obtained as

$$f_{X|Y}(x \mid y_o) = \frac{f(x, y_o)}{f_Y(y_o)} = \begin{cases} 0, & x < 0 \\ \dfrac{y_o e^{-xy_o}}{\int_0^{\infty} y_o e^{-xy_o}\,dx}, & x \geq 0 \end{cases}$$

$$= \begin{cases} 0, & x < 0 \\ y_o e^{-xy_o}, & x \geq 0 \end{cases}$$

Therefore, if $0 < y_o < 1$,

$$E(X^n \mid y_o) = \int_0^\infty x^n \cdot y_o e^{-xy_o}\, dx = \frac{1}{y_o^n} \int_0^\infty u^n e^{-u}\, du, \qquad \text{letting } u = xy_o$$

$$= (\Gamma(n+1))/y_o^n$$

In particular, the regression of X on Y is given by

$$E(X \mid y) = \frac{1}{y}, \qquad 0 < y < 1$$

Example 2.2. Suppose the joint pdf of X and Y is given by

$$f(x, y) = \begin{cases} 6xy(2-x-y), & 0 < x < 1,\ 0 < y < 1 \\ 0, & \text{elsewhere} \end{cases}$$

Find the regression of X on Y.

Solution. The conditional pdf of X given $Y = y$ is defined for $0 < y < 1$ and routine calculations give it as

$$f_{X \mid Y}(x \mid y) = \begin{cases} \dfrac{6x(2-x-y)}{4-3y}, & 0 < x < 1 \\ 0, & \text{elsewhere} \end{cases}$$

Therefore, the regression of X on Y is given by

$$E(X \mid y) = \int_{-\infty}^{\infty} x f_{X \mid Y}(x \mid y)\, dx$$

$$= \frac{6}{4-3y} \int_0^1 x \cdot x(2-x-y)\, dx$$

$$= \frac{6}{4-3y} \left[\frac{2x^3}{3} - \frac{x^4}{4} - \frac{yx^3}{3}\right] \Bigg|_{x=0}^{x=1}$$

$$= \frac{1}{2}\left(\frac{5-4y}{4-3y}\right), \qquad 0 < y < 1$$

■

Comment. $E(X \mid Y) = \frac{1}{2}\left(\frac{5-4Y}{4-3Y}\right)$ is a random variable which is a function of the random variable Y. For $Y = y$, its value is $\frac{1}{2}\left(\frac{5-4y}{4-3y}\right)$.

The properties of the expectation operation that were established in Section 1 hold with respect to the conditional expectation. For instance, we have

$$E(ah_1(X, Y) + bh_2(X, Y) \mid y_o) = aE(h_1(X, Y) \mid y_o) + bE(h_2(X, Y) \mid y_o)$$

This follows immediately because

$$\int_{-\infty}^{\infty} (ah_1(x, y_o) + bh_2(x, y_o)) f_{X \mid Y}(x \mid y_o)\, dx = a \int_{-\infty}^{\infty} h_1(x, y_o) f_{X \mid Y}(x \mid y_o)\, dx$$

$$+ b \int_{-\infty}^{\infty} h_2(x, y_o) f_{X \mid Y}(x \mid y_o)\, dx$$

The **conditional variance** *of an r.v. X given $Y = y$* is defined in an obvious way. Denoting it by $\text{Var}(X \mid Y = y)$, or simply as $\text{Var}(X \mid y)$, it is defined as

$$\text{Var}(X \mid Y = y) = E[(X - E(X \mid y))^2 \mid Y = y]$$

$$= \begin{cases} \int_{-\infty}^{\infty} (x - E(X \mid y))^2 f_{X \mid Y}(x \mid y)\, dx, & \text{in the continuous case} \\ \sum_{x_i} (x_i - E(X \mid y))^2 p_{X \mid Y}(x_i \mid y), & \text{in the discrete case} \end{cases}$$

All the properties of the variance operation that were established in Chapter 7 hold, and these can be checked in a routine way. For example, we have

$$\text{Var}(X \mid Y = y) = E(X^2 \mid Y = y) - [E(X \mid Y = y)]^2$$

Example 2.3. Suppose X and Y have the joint probability function given by

$$P(X = x,\ Y = y) = \begin{cases} \frac{1}{42}(x + y^2), & x = 1, 4, \quad y = -1, 0, 1, 3 \\ 0, & \text{elsewhere} \end{cases}$$

Find $\text{Var}(X \mid Y = 0)$.

Solution. In Example 1.2 of Chapter 9 it was shown that

$$P(X = x \mid Y = y) = \frac{x + y^2}{5 + y^2}, \qquad x = 1, 4$$

for $y = -1, 0, 1, 3$. Hence

$$E(X \mid Y = y) = \sum_{x \in \{1,4\}} x \cdot \frac{x + y^2}{5 + y^2} = \frac{1}{5 + y^2}[1(1 + y^2) + 4(4 + y^2)] = \frac{17 + 5y^2}{5 + y^2}$$

$$E(X^2 \mid Y = y) = \sum_{x \in \{1,4\}} x^2 \cdot \frac{x + y^2}{5 + y^2} = \frac{1}{5 + y^2}[1^2(1 + y^2) + 4^2(4 + y^2)] = \frac{65 + 17y^2}{5 + y^2}$$

Therefore,

$$\text{Var}(X \mid Y = y) = \frac{65 + 17y^2}{5 + y^2} - \left(\frac{17 + 5y^2}{5 + y^2}\right)^2$$

Consequently,

$$\text{Var}(X \mid Y = 0) = \frac{65}{5} - \left(\frac{17}{5}\right)^2 = \frac{36}{25}$$

Example 2.4. If the joint pdf of X and Y is given as

$$f(x, y) = \begin{cases} \frac{12}{5}x^2(x + y), & 0 < x < 1, \quad 0 < y < 1 \\ 0, & \text{elsewhere} \end{cases}$$

find $\text{Var}(X \mid Y = y_0)$, where $0 < y_0 < 1$.

Solution. The conditional pdf of X given $Y = y_o$ is

$$f_{X|Y}(x \mid y_o) = \frac{f(x, y_o)}{f_Y(y_o)} = \begin{cases} \dfrac{\frac{12}{5}x^2(x + y_o)}{\int_0^1 \frac{12}{5}x^2(x + y_o)\,dx}, & 0 < x < 1 \\ 0, & \text{elsewhere} \end{cases}$$

$$= \begin{cases} \dfrac{12x^2(x + y_o)}{3 + 4y_o}, & 0 < x < 1 \\ 0, & \text{elsewhere} \end{cases}$$

Therefore, for any positive integer n,

$$E(X^n \mid Y = y_o) = \int_0^1 x^n \cdot \frac{12x^2(x + y_o)}{(3 + 4y_o)}\,dx = \frac{12}{3 + 4y_o}\int_0^1 (x^{n+3} + x^{n+2}y_o)\,dx$$
$$= \frac{12}{3 + 4y_o}\left(\frac{1}{n + 4} + \frac{y_o}{n + 3}\right)$$

Consequently,

$$E(X^2 \mid y_o) = \frac{12}{30} \cdot \frac{5 + 6y_o}{3 + 4y_o} \quad \text{and} \quad E(X \mid y_o) = \frac{12}{20} \cdot \frac{4 + 5y_o}{3 + 4y_o}$$

and

$$\text{Var}(X \mid y_o) = E(X^2 \mid y_o) - [E(X \mid y_o)]^2 = \frac{1}{25}\,\frac{(6 + 20y_o + 15y_o^2)}{(3 + 4y_o)^2}$$

after simplifying.

2.2 The Expected Value of a Random Variable by Conditioning

When there is more than one random variable involved in a discussion, it is often convenient to find the expected value of a random variable by conditioning over the possible values of the other random variables. As remarked earlier, $E(X \mid y)$ is a real number which depends on y. As a matter of fact, it is the value of the random variable $E(X \mid Y)$ at $Y = y$. (See the comment at the end of Example 2.2.) Since $E(X \mid Y)$ is a random variable, it is meaningful to discuss its expected value (if it exists). Since $E(X \mid Y)$ is a function of Y, we would naturally find the expected value of $E(X \mid Y)$, that is, $E(E(X \mid Y))$, by using the distribution of Y.

We shall now show specifically that

$$E(X) = \int_{-\infty}^{\infty} E(X \mid y) f_Y(y)\,dy = E(E(X \mid Y))$$

We have

$$E(X) = \int_{-\infty}^{\infty}\int_{-\infty}^{\infty} x \cdot f(x, y)\,dy\,dx, \qquad \text{by definition}$$
$$= \int_{-\infty}^{\infty}\left[\int_{-\infty}^{\infty} x\frac{f(x, y)}{f_Y(y)}\,dx\right] f_Y(y)\,dy$$

Now,

$$\int_{-\infty}^{\infty} x \frac{f(x, y)}{f_Y(y)}\, dx = \int_{-\infty}^{\infty} x f_{X|Y}(x \mid y)\, dx = E(X \mid y)$$

Therefore,

$$E(X) = \int_{-\infty}^{\infty} E(X \mid y) f_Y(y)\, dy = E(E(X \mid Y))$$

Important. It is essential to bear in mind that in finding $E(X)$ as $E(E(X \mid Y))$ the inside expectation, namely, $E(X \mid Y)$, is taken with respect to the conditional distribution of X given Y, whereas the outside expectation is taken with respect to the distribution of the random variable Y.

Example 2.5. Suppose X_t, the number of phone calls arriving at an exchange during a period of length t, has the Poisson distribution with parameter λt. The probability that an operator answers any given phone call is equal to p, $0 \leqslant p \leqslant 1$. Find the expected number of phone calls answered during a period of length t.

Solution. Letting Y_t denote the number of phone calls answered during a period of length t, it was shown in Example 1.14 of Chapter 5 that the distribution of Y_t is Poisson with parameter $p\lambda t$, so that the expected value of Y_t is $p\lambda t$.

However, we prefer now to find $E(Y_t)$ by conditioning, a method that obviates the need to obtain the distribution of Y_t explicitly. We have

$$E(Y_t) = E(E(Y_t \mid X_t))$$

Now, given that there were X_t calls, the distribution of Y_t is binomial with X_t trials and the probability of success p. Therefore,

$$E(Y_t \mid X_t) = X_t p$$

Consequently,

$$E(Y_t) = E(X_t p) = pE(X_t) = p\lambda t$$

since the distribution of X_t is Poisson with parameter λt.

Example 2.6. A plane has a route which operates between n locations, each 400 miles apart, as shown in Figure 2.1. On any day the plane is equally likely to be found at any one of the n locations and the pilot is likely to be asked to fly to any of the places with equal probability (not precluding that he won't fly at all). Find the expected distance covered in a day.

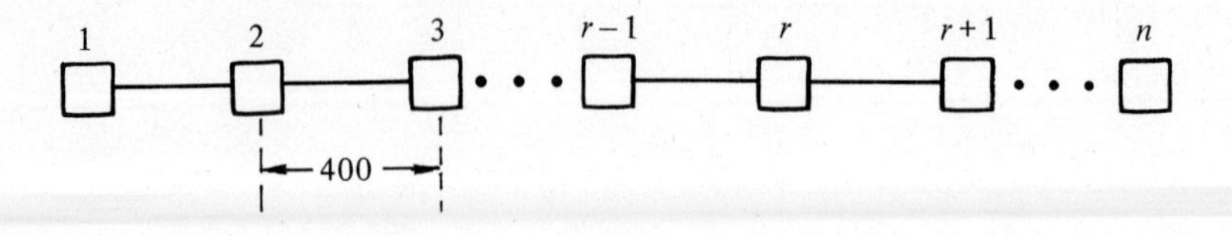

Figure 2.1

Solution. Let X denote the location of the plane, and Y the distance covered. Then

$$E(Y) = \sum_{r=1}^{n} E(Y \mid X = r)P(X = r)$$

If the plane is located at, say, location 4, then Y can assume the values $(4-1)(400), (4-2)(400), (4-3)(400), (4-4)(400), (5-4)(400), \ldots, (n-4)(400)$, each with probability $1/n$. In general, if the plane is at location r, then Y can assume a value y where

$$y = \begin{cases} (r-i)(400), & \text{if } i = 1, 2, \ldots, r \\ (i-r)(400), & \text{if } i = r+1, \ldots, n \end{cases}$$

each with the same probability of $1/n$.

Therefore,

$$\begin{aligned} E(Y \mid X = r) &= \sum_{i=1}^{r} (r-i)(400)\frac{1}{n} + \sum_{i=r+1}^{n} (i-r)(400)\frac{1}{n} \\ &= \frac{400}{n}\left[\frac{r(r-1)}{2} + \frac{(n-r)(n-r+1)}{2}\right] \\ &= \frac{400}{2n}[2r^2 - 2(n+1)r + n(n+1)] \end{aligned}$$

Since $P(X = r) = 1/n$, $r = 1, 2, \ldots, n$,

$$\begin{aligned} E(Y) &= \sum_{r=1}^{n} \frac{400}{2n}[2r^2 - 2(n+1)r + n(n+1)]\frac{1}{n} \\ &= \frac{400}{2n^2}\left[\frac{2n(n+1)(2n+1)}{6} - \frac{2(n+1)n(n+1)}{2} + n^2(n+1)\right] \end{aligned}$$

using the results

$$\sum_{r=1}^{n} r^2 = \frac{n(n+1)(2n+1)}{6} \quad \text{and} \quad \sum_{r=1}^{n} r = \frac{n(n+1)}{2}$$

Upon simplifying, this gives

$$E(Y) = \frac{400(n-1)(n+1)}{3n}$$

Example 2.7. Suppose N, the number of individuals in a population, is a random variable, and that each individual in the population gives rise to a random number of progeny. Find the expected total progeny. (We shall assume that the number Z of progeny to which an individual gives rise is independent of the total number of individuals, N, in the population.)

Solution. Let Z_i represent the number of progeny of the ith individual. Then the total progeny X is given as $X = \sum_{i=1}^{N} Z_i$. Therefore,

$$E(X) = E\left(\sum_{i=1}^{N} Z_i\right) = E\left(E\left(\sum_{i=1}^{N} Z_i \mid N\right)\right)$$

Now, $E\left(\sum_{i=1}^{N} Z_i | N = n\right) = E\left(\sum_{i=1}^{n} Z_i\right)$, since Z_i and N are independent by assumption. Hence $E\left(\sum_{i=1}^{N} Z_i | N = n\right) = nE(Z)$.

Therefore, the random variable $E\left(\sum_{i=1}^{N} Z_i | N\right)$ is $NE(Z)$. As a result,

$$E(X) = E\left(E\left(\sum_{i=1}^{N} Z_i | N\right)\right) = E(NE(Z)) = E(N) \cdot E(Z)$$

2.3 Probabilities by Conditioning

The result that we are about to establish is analogous to the theorem of total probability encountered in Chapter 3.

Suppose $(S, \mathcal{F}, P)$ is the probability space, A any event, and Y some random variable defined on S. Then

$$P(A) = \begin{cases} \int_{-\infty}^{\infty} P(A | Y = y) f_Y(y)\, dy, & \text{if } Y \text{ is continuous} \\ \sum_{y_j} P(A | Y = y_j) P(Y = y_j), & \text{if } Y \text{ is discrete} \end{cases}$$

Let us prove this. If I_A is the indicator random variable of the event A, it was shown in Section 1.5 that $E(I_A) = P(A)$. In an analogous way, it can be shown that

$$E(I_A | Y = y) = P(A | Y = y)$$

In view of this, using the fact that $P(A) = E(I_A) = E(E(I_A | Y))$, it follows that

$$P(A) = \begin{cases} \int_{-\infty}^{\infty} E(I_A | y) f_Y(y)\, dy, & \text{if } Y \text{ is continuous} \\ \sum_{y_j} E(I_A | y_j) P(Y = y_j), & \text{if } Y \text{ is discrete} \end{cases}$$

$$= \begin{cases} \int_{-\infty}^{\infty} P(A | Y = y) f_Y(y)\, dy, & \text{if } Y \text{ is continuous} \\ \sum_{y_j} P(A | Y = y_j) P(Y = y_j), & \text{if } Y \text{ is discrete} \end{cases}$$

Example 2.8. Suppose X and Y are independent, absolutely continuous random variables. Find $P(X < Y)$.

Solution. Using the relation $P(A) = \int_{-\infty}^{\infty} P(A | Y = y) f_Y(y)\, dy$, we get

$$P(X < Y) = \int_{-\infty}^{\infty} P(X < Y | Y = y) f_Y(y)\, dy = \int_{-\infty}^{\infty} P(X < y | Y = y) f_Y(y)\, dy$$

$$= \int_{-\infty}^{\infty} P(X < y) f_Y(y)\, dy, \qquad \text{since } X \text{ and } Y \text{ are independent}$$

$$= \int_{-\infty}^{\infty} F_X(y) f_Y(y)\, dy$$

An alternate method is, of course, to realize that $P(X < Y)$ is the integral over the shaded region shown in Figure 2.2.

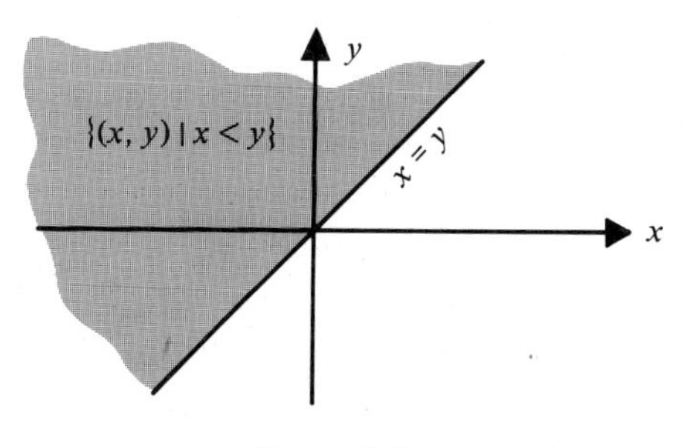

Figure 2.2

■

The method of computing probabilities by conditioning is purely algebraic and there is no need to consider geometric regions of the plane in order to set the limits of integration as we did in Chapter 10. The following two examples, which were considered in Chapter 10, will serve to illustrate the essence of this new approach.

Example 2.9. Suppose X and Y are independent, absolutely continuous random variables. Find the distribution of $Z = X + Y$.

Solution. We are free to condition on any one of the random variables. Conditioning on Y yields

$$\begin{aligned} F_Z(t) &= P(X + Y \leqslant t) \\ &= \int_{-\infty}^{\infty} P(X + Y \leqslant t \mid Y = y) f_Y(y)\, dy \\ &= \int_{-\infty}^{\infty} P(X + y \leqslant t \mid Y = y) f_Y(y)\, dy \\ &= \int_{-\infty}^{\infty} P(X + y \leqslant t) f_Y(y)\, dy, \qquad \text{since } X \text{ and } Y \text{ are independent} \\ &= \int_{-\infty}^{\infty} P(X \leqslant t - y) f_Y(y)\, dy \end{aligned}$$

Hence

$$F_Z(t) = \int_{-\infty}^{\infty} F_X(t - y) f_Y(y)\, dy$$

Example 2.10. Suppose X and Y are independent, absolutely continuous random variables. Find the distribution of $Z = X/Y$.

Solution. Conditioning on Y yields

$$\begin{aligned} F_Z(t) &= P(X/Y \leqslant t) \\ &= \int_{-\infty}^{0} P(X/Y \leqslant t \mid Y = y) f_Y(y)\, dy + \int_{0}^{\infty} P(X/Y \leqslant t \mid Y = y) f_Y(y)\, dy \\ &= \int_{-\infty}^{0} P(X \geqslant yt \mid Y = y) f_Y(y)\, dy + \int_{0}^{\infty} P(X \leqslant yt \mid Y = y) f_Y(y)\, dy \end{aligned}$$

(Recall that if $a \leqslant b$, then $ac \leqslant bc$ if $c > 0$, and $ac \geqslant bc$ if $c < 0$.)

Therefore,

$$F_Z(t) = \int_{-\infty}^{0} P(X \geqslant yt) f_Y(y)\, dy + \int_{0}^{\infty} P(X \leqslant yt) f_Y(y)\, dy$$

since X and Y are independent. Hence

$$F_Z(t) = \int_{-\infty}^{0} [1 - F_X(yt)] f_Y(y)\, dy + \int_{0}^{\infty} F_X(yt) f_Y(y)\, dy$$

Differentiating, this gives the expression that we obtained in Chapter 10 for the distribution of $Z = X/Y$ when X and Y are independent.

Example 2.11 (*Buffon's Needle Problem**) Suppose a floor is marked with parallel lines all at a distance $2a$ apart. A needle of length $2l$ $(l < a)$ is tossed *at random* on the floor. Determine the probability that the needle will intersect one of the lines.

Solution. First of all, we need a more precise formulation of the phrase "at random." To explain this, let the random variable X represent the distance from the center of the needle to the nearest line, and Y the angle between the needle and the direction perpendicular to the given lines (see Figure 2.3).

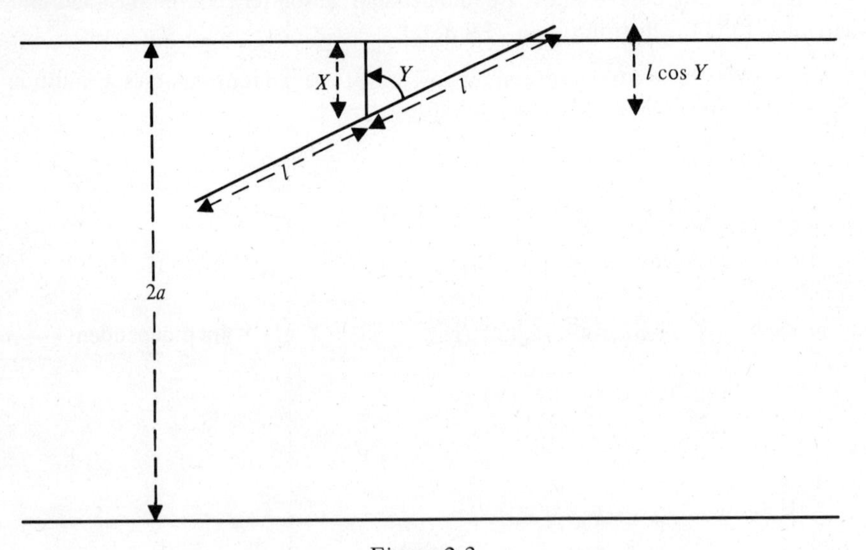

Figure 2.3

The phrase "at random" is meant to connote that the distribution of X is uniform over the interval $[0, a]$, the distribution of Y is uniform over the interval $[-\pi/2, \pi/2]$, and, furthermore, that X and Y are independent. Hence

$$f_X(x) = \begin{cases} \dfrac{1}{a}, & 0 \leqslant x \leqslant a \\ 0, & \text{elsewhere} \end{cases} \qquad f_Y(\theta) = \begin{cases} \dfrac{1}{\pi}, & -\pi/2 \leqslant \theta \leqslant \pi/2 \\ 0, & \text{elsewhere} \end{cases}$$

*Count de Buffon was a French naturalist of the eighteenth century.

and, consequently,

$$f_{X,Y}(x,\theta) = \begin{cases} \dfrac{1}{a\pi}, & \text{if } 0 \leqslant x \leqslant a \text{ and } -\pi/2 \leqslant \theta \leqslant \pi/2 \\ 0, & \text{elsewhere} \end{cases}$$

Now, as can be seen from Figure 2.3, the needle will intersect a line precisely when $X \leqslant l \cos Y$. Hence

$$\begin{aligned} P(\text{the needle intersects a line}) &= P(X \leqslant l \cos Y) \\ &= \int_{-\infty}^{\infty} P(X \leqslant l \cos Y \mid Y = \theta) f_Y(\theta)\, d\theta \\ &= \int_{-\pi/2}^{\pi/2} P(X \leqslant l \cos Y \mid Y = \theta) \frac{d\theta}{\pi} \end{aligned}$$

Since X and Y are independent,

$$\begin{aligned} P(X \leqslant l \cos Y \mid Y = \theta) &= P(X \leqslant l \cos \theta) \\ &= \frac{l \cos \theta}{a} \end{aligned}$$

(Notice that for $-\pi/2 \leqslant \theta \leqslant \pi/2$ we have $0 \leqslant l \cos \theta \leqslant a$.) Consequently,

$$\begin{aligned} P(\text{the needle intersects a line}) &= \int_{-\pi/2}^{\pi/2} \frac{l \cos \theta}{a} \frac{d\theta}{\pi} \\ &= \frac{l}{\pi a} \sin \theta \Big|_{-\pi/2}^{\pi/2} \\ &= \frac{2l}{\pi a} \end{aligned}$$

EXERCISES–SECTION 2

1. Two discrete random variables X and Y have the following joint probability function:

$x \backslash y$	2	3	5
1	$\frac{1}{18}$	$\frac{1}{9}$	$\frac{1}{18}$
2	$\frac{1}{18}$	$\frac{1}{9}$	$\frac{1}{9}$
3	0	$\frac{4}{9}$	$\frac{1}{18}$

Find:

(a) $E(X \mid Y = i)$, $\quad i = 2, 3, 5$

(b) $E(Y \mid X = i)$, $\quad i = 1, 2, 3$

2. The joint probability function of X and Y is given by

$$P(X = x,\ Y = y) = \begin{cases} \dfrac{2}{n(n+1)}, & y = 1, 2, \ldots, x, \quad x = 1, 2, \ldots, n \\ 0, & \text{otherwise} \end{cases}$$

Find:

(a) the regression of Y on X

(b) the regression of X on Y

3. Suppose two chips are picked one by one without replacement from a box containing n chips numbered $1, 2, \ldots, n$. Let X denote the value on the first chip picked and Y the value on the second chip. Find:

(a) $\text{Cov}(X, Y)$

(b) $E(X \mid Y = j), \quad j = 1, 2, \ldots, n$

(c) $E(Y \mid X = i), \quad i = 1, 2, \ldots, n$

4. From a group of fifteen people consisting of four doctors, five lawyers, and six students, five people are picked at random.

(a) Find the expected number of lawyers.

(b) Given that there are two doctors in the sample, find the expected number of lawyers.

5. If the joint pdf of X and Y is given by

$$f(x, y) = \begin{cases} 24y(1-x), & 0 \leqslant y \leqslant x \leqslant 1 \\ 0, & \text{elsewhere} \end{cases}$$

find–

(a) the regression of X on Y

(b) the regression of Y on X.

6. Find the regression of X on Y if the joint pdf of X and Y is given by

$$f(x, y) = \begin{cases} 2[(1-x)(1-y) + xy], & 0 < x < 1, \ 0 < y < 1 \\ 0, & \text{elsewhere} \end{cases}$$

7. If X and Y have a joint pdf given by

$$f(x, y) = \begin{cases} kxy^2, & 0 < y < x < 1 \\ 0, & \text{elsewhere} \end{cases}$$

find–

(a) $E(Y^n \mid X = \frac{3}{4})$

(b) $\text{Var}(Y \mid X = \frac{3}{4})$

8. Two random variables X and Y have the following joint pdf:

$$f(x, y) = \begin{cases} \dfrac{e^{-y}}{y}, & 0 < x < y < \infty \\ 0, & \text{elsewhere} \end{cases}$$

If $y_0 > 0$, find–

(a) $E(X^n \mid y_0)$, where n is a positive integer

(b) $\text{Var}(X \mid y_0)$

9. If X and Y have the joint uniform distribution over the circle with radius 1 and centered at the origin, find $E(X^n \mid Y = y_0)$, where $-1 < y_0 < 1$ and n is any non-negative integer.

10. Suppose X and Y are independent random variables, each exponentially distributed with parameter λ. Find the pdf of $Z = X/(X + Y)$.

11. A point X_1 is picked at random in the interval $[0, 1]$. A second point X_2 is then picked at random in the interval $[0, X_1]$. Show that the distribution of X_2 is identical with that of Y_1Y_2 where Y_1, Y_2 are independent random variables each having the uniform distribution over the interval $[0, 1]$. *Hint:* If $0 < u < 1$,

$$P(X_2 \leqslant u) = \int_{-\infty}^{\infty} P(X_2 \leqslant u \mid X_1 = x) f_{X_1}(x)\, dx = \int_0^u 1\, dx + \int_u^1 \frac{u}{x}\, dx$$

Simplify and compare with Example 3.1(b) of Chapter 10.

12. The number of emergency calls at a hospital on any day is a random variable X with the following distribution:

x	100	150	200	300
$P(X = x)$	0.1	0.3	0.4	0.2

The probability that an emergency is due to a heart attack is 0.05. Find the expected number of calls due to heart attack.

13. A tennis pro gives 8, 10, or 12 lessons during a day with respective probabilities 0.3, 0.5, 0.2. The probability that a lesson is taken by a junior is 0.3, that it is taken by a regular adult student is 0.6, and that it is taken by an infrequent adult visitor is 0.1. If the charges per lesson are 5 dollars for the juniors, 8 dollars for the regular adult student, and 10 dollars for the infrequent adult customer, find the pro's expected earnings in a day.

14. A real number X is picked at random in the interval $[0, 1]$. If $X = x$, a coin with $P(\text{head}) = x$ is tossed n times. Let Y represent the number of heads in n tosses. Find the probability distribution of Y and $E(Y)$.

12 Generating Functions

INTRODUCTION

One of the most powerful tools in probability theory is the technique of generating functions. This mathematical device comes under the broad heading of the theory of transforms in analysis. For our study of the theory of probability, there are two types of generating functions that are of special interest, namely, the moment generating functions and the factorial moment generating functions. There is a complex variable analog of the moment generating function called the *characteristic function* which is used in advanced probability theory. However, we are not equipped to handle this version.

The moment generating function, as the term suggests, is a function which, when used appropriately, generates the moments of a distribution and provides an efficient method of computing these quantities. A factorial moment generating function, used appropriately, generates the factorial moments. It is also called the probability generating function because in some situations it can be used to generate probabilities.

Besides generating moments, generating functions can often be used in studying distributions of functions of random variables, especially sums of independent random variables. The powerful technique of generating functions—specifically, an analog of the probability generating function—was used for the first time, and most successfully, by the great mathematician Euler (1707-1783) to solve a problem in number theory.

1. THE MOMENT GENERATING FUNCTION

1.1 The Definition

Suppose X is an r.v. with a given distribution, and that, for some real number $h > 0, E(e^{sX})$ exists for every value of s in the interval $(-h, h)$. Then the function M defined by

$$M(s) = E(e^{sX})$$

is called the **moment generating function** of X, or of the distribution of X. (The term moment generating function will be abbreviated as mgf.)

Thus,

$$M(s) = \int_{-\infty}^{\infty} e^{sx} f(x)\, dx$$

if X is a continuous random variable, and

$$M(s) = \sum_{x_i} e^{sx_i} P(X = x_i)$$

if X is a discrete random variable.

If two or more random variables are involved in a discussion, we will find it convenient to write the mgf's using the underlying r.v.'s as subscripts; for example, M_X to represent the mgf of X.

Clearly, the expected value of e^{sX} depends on the real number s. If $s = 0$, it always exists and, as a matter of fact, $M(0) = E(e^{0X}) = E(1) = 1$. It should be borne in mind that when we refer to the existence of the mgf M, we are interested in $M(s)$ in some *neighborhood* of $s = 0$, that is, for $s \in (-h, h)$ for some $h > 0$. In short, $M: (-h, h) \to \mathbf{R}$.

• *Example 1.1.* Find the moment generating functions for the following discrete distributions:

(*a*) Bernoulli, with parameter p
(*b*) Binomial, $B(n, p)$
(*c*) Geometric, with parameter p
(*d*) Negative binomial, with parameters r, p
(*e*) Poisson, with parameter λ

Solution

(*a*) (*Bernoulli*) For the Bernoulli distribution, $M(s) = e^{s \cdot 0} P(X = 0) + e^{s \cdot 1} P(X = 1) = (1 - p) + pe^s$. Hence the mgf exists for all s and

$$M(s) = pe^s + 1 - p, \qquad -\infty < s < \infty$$

(*b*) (*Binomial*) For the binomial distribution $B(n, p)$,

$$M(s) = \sum_{k=0}^{n} e^{sk} \binom{n}{k} p^k (1-p)^{n-k}$$
$$= \sum_{k=0}^{n} \binom{n}{k} (pe^s)^k (1-p)^{n-k} = [pe^s + (1-p)]^n$$

Hence, *the mgf for $B(n, p)$ is given by*

$$M(s) = [pe^s + (1-p)]^n, \qquad -\infty < s < \infty$$

(*c*) (*Geometric*) For the geometric distribution

$$M(s) = \sum_{k=1}^{\infty} e^{sk} P(X = k)$$
$$= \sum_{k=1}^{\infty} e^{sk} p(1-p)^{k-1}$$
$$= pe^s \sum_{k=1}^{\infty} [(1-p)e^s]^{k-1}$$
$$= pe^s \sum_{r=0}^{\infty} [(1-p)e^s]^r$$

Now, $\sum_{r=0}^{\infty} [(1-p)e^s]^r$ is a geometric series and converges to $1/[1-(1-p)e^s]$ provided $(1-p)e^s < 1$.

Hence, *the mgf for the geometric distribution with parameter p is given by*

$$M(s) = \frac{pe^s}{1-(1-p)e^s}, \qquad \text{if } (1-p)e^s < 1$$

(*d*) (*Negative Binomial*) For the negative binomial distribution with parameters *r, p*

$$M(s) = \sum_{k=0}^{\infty} e^{s(r+k)} P(X = r + k)$$
$$= \sum_{k=0}^{\infty} e^{s(r+k)} \binom{r+k-1}{r-1} p^r (1-p)^k$$
$$= (pe^s)^r \sum_{k=0}^{\infty} \binom{r+k-1}{r-1} [(1-p)e^s]^k$$

Now, $\sum_{k=0}^{\infty} \binom{r+k-1}{r-1} x^k = (1-x)^{-r}$, provided $|x| < 1$. Therefore, $M(s) = (pe^s)^r [1-(1-p)e^s]^{-r}$, provided $(1-p)e^s < 1$.

In conclusion, *the mgf of the negative binomial distribution with parameters r, p is*

$$M(s) = \left[\frac{pe^s}{1-(1-p)e^s}\right]^r, \qquad \text{if } (1-p)e^s < 1$$

(*e*) (*Poisson*) For the Poisson distribution with parameter λ,

$$
\begin{aligned}
M(s) &= \sum_{k=0}^{\infty} e^{sk} \frac{e^{-\lambda}\lambda^k}{k!} \\
&= e^{-\lambda} \sum_{k=0}^{\infty} \frac{(\lambda e^s)^k}{k!} \\
&= e^{-\lambda} e^{\lambda e^s}, \text{ using the fact } \sum_{k=0}^{\infty} \frac{x^k}{k!} = e^x \\
&= e^{\lambda(e^s - 1)}
\end{aligned}
$$

That is, *the Poisson distribution with parameter λ has a mgf for all s, and*

$$
M(s) = e^{\lambda(e^s - 1)}, \qquad -\infty < s < \infty
$$

• *Example 1.2.* Find the mgf's for the following absolutely continuous distributions:

(*a*) The uniform distribution over the interval $[a, b]$
(*b*) The standard normal distribution
(*c*) The gamma distribution with parameter $\lambda > 0$

Solution

(*a*) (*Uniform distribution*) The mgf is given by

$$
M(s) = \int_a^b e^{sx} \frac{1}{b-a}\, dx = \frac{e^{bs} - e^{as}}{s(b-a)}, \qquad \text{if } s \neq 0
$$

Of course, $M(0) = 1$.

Hence, *if X is uniformly distributed over the interval [a, b], then the mgf exists for all s, and*

$$
M(s) = \begin{cases} \dfrac{e^{bs} - e^{as}}{s(b-a)}, & s \neq 0 \\ 1, & s = 0 \end{cases}
$$

(*b*) (*Standard normal distribution*) For the standard normal distribution,

$$
\begin{aligned}
M(s) &= \int_{-\infty}^{\infty} e^{sx} \frac{1}{\sqrt{2\pi}} e^{-x^2/2}\, dx \\
&= \frac{1}{\sqrt{2\pi}} \int_{-\infty}^{\infty} e^{-(x^2 - 2sx)/2}\, dx \\
&= \frac{1}{\sqrt{2\pi}} \int_{-\infty}^{\infty} e^{[-(x^2 - 2sx + s^2)/2] + s^2/2}\, dx \\
&= e^{s^2/2} \frac{1}{\sqrt{2\pi}} \int_{-\infty}^{\infty} e^{-(x-s)^2/2}\, dx \\
&= e^{s^2/2}
\end{aligned}
$$

since $\frac{1}{\sqrt{2\pi}} \int_{-\infty}^{\infty} e^{-(x-s)^2/2}\, dx$ represents the area under the normal distribution $N(s, 1)$, and hence is equal to 1.

Therefore, *for a standard normal distribution,*

$$M(s) = e^{s^2/2}, \qquad -\infty < s < \infty$$

Comment. For an arbitrary normal distribution $N(\mu, \sigma^2)$, it turns out that $M(s) = e^{\mu s + (\sigma^2 s^2/2)}$ and can be derived by evaluating the integral

$$\int_{-\infty}^{\infty} e^{sx} \frac{1}{\sigma\sqrt{2\pi}} e^{-(x-\mu)^2/(2\sigma^2)}\, dx$$

However, this involves tedious algebraic manipulations, and therefore we will later derive the mgf in this case using an elegant trick.

(c) (*Gamma distribution*) If X has the gamma distribution with parameters $\lambda > 0$ and $p > 0$, then

$$\begin{aligned} M(s) &= \int_0^{\infty} e^{sx} \cdot \frac{\lambda^p}{\Gamma(p)} x^{p-1} e^{-\lambda x}\, dx \\ &= \frac{\lambda^p}{\Gamma(p)} \int_0^{\infty} x^{p-1} e^{-(\lambda - s)x}\, dx \end{aligned}$$

The above integral converges provided $\lambda > s$.

Let $(\lambda - s)x = y$; then $dx = dy/(\lambda - s)$, and we get

$$\begin{aligned} M(s) &= \frac{\lambda^p}{\Gamma(p)} \int_0^{\infty} \left(\frac{y}{\lambda - s}\right)^{p-1} e^{-y} \frac{dy}{(\lambda - s)} \\ &= \left(\frac{\lambda}{\lambda - s}\right)^p \frac{1}{\Gamma(p)} \int_0^{\infty} y^{p-1} e^{-y}\, dy \\ &= \left(\frac{\lambda}{\lambda - s}\right)^p \end{aligned}$$

since $\int_0^{\infty} y^{p-1} e^{-y}\, dy = \Gamma(p)$.

Hence, *if X has the gamma distribution with parameters $\lambda > 0$ and $p > 0$, then*

$$M(s) = \left(\frac{\lambda}{\lambda - s}\right)^p, \qquad \text{if } s < \lambda$$

■

Comments. (1) If we set $p = 1$, then, as we know, the gamma distribution gives the exponential distribution as a special case. Consequently, it follows that *for the exponential distribution with parameter λ, the mgf is given by*

$$M(s) = \frac{\lambda}{\lambda - s}, \qquad \text{if } s < \lambda$$

(2) If we set $\lambda = \frac{1}{2}$ and $r = n/2$ (where n is a positive integer), we get the chi-square distribution with n degrees of freedom as a special case of the gamma distribution. Therefore, it follows that *if X has a chi-square distribution with n degrees of freedom, then the mgf is given by*

$$M(s) = \frac{1}{(1-2s)^{n/2}}, \quad \text{if } s < \tfrac{1}{2}$$

Caution. Not all random variables have mgf's, since the integral or the sum, as the case may be, may not converge absolutely. Examples 1.3 and 1.4 demonstrate this.

Example 1.3. For the probability function given by

$$P(X = k) = \frac{6}{\pi^2 k^2}, \qquad k = 1, 2, \ldots$$

show that the mgf does not exist.

Solution. (Observe that, since the series $\sum_{k=1}^{\infty} (1/k^2)$ converges to $\pi^2/6$, the above does represent an honest probability function.) If the mgf existed, then we would have

$$M(s) = \sum_{k=1}^{\infty} e^{sk} P(X = k) = \sum_{k=1}^{\infty} \frac{6e^{sk}}{\pi^2 k^2}$$

However, using the ratio test, it can be seen that the series diverges if $s > 0$. Hence there is no positive h such that $M(s)$ exists for nonzero s in the interval $(-h, h)$. Hence the mgf does not exist.

Example 1.4. Suppose X has the distribution given by the following pdf:

$$f(x) = \begin{cases} \dfrac{1}{x^2}, & x \geqslant 1 \\ 0, & x < 1 \end{cases}$$

Show that the mgf does not exist.

Solution. In this case, we would get

$$M(s) = \int_{-\infty}^{\infty} e^{sx} f(x)\, dx = \int_{1}^{\infty} \frac{e^{sx}}{x^2}\, dx$$

But, using l'Hospital's rule, we see that if $s > 0$, then $e^{sx}/x^2 \to \infty$ as $x \to \infty$. Consequently, since the integrand diverges, $\int_1^{\infty} (e^{sx}/x^2)\, dx$ does not exist for any $s > 0$. Hence the mgf does not exist.

1.2 How Moments Are Generated

We open this discussion with a word of caution. In what follows, we will have occasion to use the infinite series version of the result $E\left(\sum_{i=1}^{n} a_i X_i\right) = \sum_{i=1}^{n} a_i E(X_i)$.

Also, we will interchange the operations of expectation and differentiation. Whether these operations are justified or not would ordinarily be a moot question, but this aspect of the problem will not concern us here. The conditions justifying such operations are fairly general, and most functions that are of interest to us are well behaved; hence, we can safely carry out these operations.

We shall now see how mgf's can be used to generate the moments of the distribution. Since the Maclaurin's series expansion of e^x is

$$e^x = 1 + x + \frac{x^2}{2!} + \frac{x^3}{3!} + \dots$$

which converges for all x, we can write

$$e^{sX} = \sum_{r=0}^{\infty} \frac{(sX)^r}{r!}$$

If $M(s)$ exists for $s \in (-h, h)$ for some positive number h, we can write

$$M(s) = E(e^{sX}) = E\left(\sum_{r=0}^{\infty} \frac{(sX)^r}{r!} \right)$$

and, interchanging the order of expectation and summation, this yields

$$\boxed{M(s) = \sum_{r=0}^{\infty} E(X^r) \frac{s^r}{r!}}$$

Thus we see that the coefficient of $s^r/r!$ in the power series expansion of $M(s)$ is precisely $E(X^r)$.

On the other hand, the Maclaurin's series for $M(s)$ is

$$M(s) = \sum_{r=0}^{\infty} M^{(r)}(0) \cdot \frac{s^r}{r!}$$

using the standard notation $M^{(r)}(0)$ in place of $\left.\frac{d^r}{ds^r} M(s)\right|_{s=0}$.

Comparing the coefficients of s^r in the two power series representations for $M(s)$, it follows that

$$\boxed{E(X^r) = M^{(r)}(0), \qquad r = 1, 2, \dots}$$

We have arrived at the following very important conclusion: When the mgf is available, it can be used in two ways to find the moments of the distribution: *(i) if the mgf can be expanded as a power series in powers of s, then the coefficient of s^r multiplied by $r!$ gives $E(X^r)$; (ii) if the mgf can be differentiated repeatedly, then the rth order derivative evaluated at 0 is $E(X^r)$.*

That $E(X^r) = M^{(r)}(0)$ can also be shown by formally differentiating $E(e^{sX})$ r times successively and substituting $s = 0$. Assuming that the operations of expectation and differentiation can be interchanged, it can be seen that

$$M^{(r)}(s) = \frac{d^r}{ds^r} E(e^{sX}) = E\left(\frac{d^r}{ds^r} e^{sX}\right) = E(X^r e^{sX})$$

so that setting $s = 0$, this gives

$$M^{(r)}(0) = E(X^r)$$

Example 1.5. In each of the following cases, find $E(X), E(X^2)$ from the mgf's:

(a) X is binomial, $B(n, p)$.

(b) X is Poisson with parameter λ.

Solution

(a) (*Binomial*) In Example 1.1(b) we found the mgf of X as $M(s) = [pe^s + (1-p)]^n$. To obtain $E(X)$ and $E(X^2)$, we evaluate the first and second order derivatives of $M(s)$ at 0. We get

$$E(X) = \frac{d}{ds}M(s)\bigg|_{s=0} = n[pe^s + (1-p)]^{n-1}pe^s\bigg|_{s=0} = np$$

and

$$\begin{aligned} E(X^2) = \frac{d^2}{ds^2}M(s)\bigg|_{s=0} &= n(n-1)[pe^s + (1-p)]^{n-2}p^2e^{2s}\bigg|_{s=0} \\ &\quad + n[pe^s + (1-p)]^{n-1}pe^s\bigg|_{s=0} \\ &= n(n-1)p^2 + np \end{aligned}$$

In this case, the method of expanding as a power series is so unwieldy that it cannot even be contemplated.

(b) (*Poisson*) We have seen in Example 1.1(e) that the mgf is $M(s) = e^{\lambda(e^s-1)}$. Differentiating $M(s)$ repeatedly, this gives

$$E(X) = \frac{d}{ds}M(s)\bigg|_{s=0} = e^{\lambda(e^s-1)}\lambda e^s\bigg|_{s=0} = \lambda$$

and

$$E(X^2) = \frac{d^2}{ds^2}M(s)\bigg|_{s=0} = e^{\lambda(e^s-1)}\lambda e^s + e^{\lambda(e^s-1)}\lambda^2e^{2s}\bigg|_{s=0} = \lambda^2 + \lambda$$

• *Example 1.6.* Suppose X is normally distributed with mean 0 and variance σ^2. Find $E(X^r)$, $r = 0, 1, 2, \ldots$.

Solution. From the comment in Example 1.2 at the end of part (b),

$$\begin{aligned} M(s) = e^{(\sigma^2s^2)/2} &= \sum_{r=0}^{\infty}\left(\frac{\sigma^2s^2}{2}\right)^r \cdot \frac{1}{r!} \\ &= \sum_{r=0}^{\infty}\frac{\sigma^{2r}}{2^r}\frac{s^{2r}}{r!} \\ &= \sum_{r=0}^{\infty}\frac{\sigma^{2r}(2r)!}{2^r r!}\frac{s^{2r}}{(2r)!} \end{aligned}$$

We see that the coefficient of s^{2r+1} is zero for every nonnegative integer r. Hence $E(X^{2r+1}) = 0$. Also, the coefficient of $s^{2r}/(2r)!$ is $(\sigma^{2r}(2r)!)/(2^r r!)$, $r = 0, 1, 2, \ldots$. Hence the moments of X where X is $N(0, \sigma^2)$ are given by

$$E(X^{2r+1}) = 0, \qquad r = 0, 1, 2, \ldots$$

and

$$E(X^{2r}) = \frac{\sigma^{2r}(2r)!}{2^r r!}, \qquad r = 0, 1, \ldots$$

That all the odd order moments are 0 also follows from the fact that a standard normal distribution is symmetric about 0.

Alternatively, the moments can be obtained by differentiating in the following way:

$$\frac{d}{ds}M(s) = \sigma^2 s e^{(\sigma^2 s^2)/2} \quad \text{and} \quad \frac{d^2}{ds^2}M(s) = (\sigma^2 + \sigma^4 s^2)e^{(\sigma^2 s^2)/2}$$

so that

$$E(X) = M^{(1)}(0) = 0 \quad \text{and} \quad E(X^2) = M^{(2)}(0) = \sigma^2$$

However, this approach is not as efficient as the earlier method. For one thing, it does not indicate a pattern to provide an expression for $E(X^r)$ for $r = 0, 1, 2, \ldots$.

1.3 Some Important Results

We now present some results which illustrate the importance of mgf's in probability theory, beyond simply being a device for generating moments.

Suppose an r.v. X has the mgf M_X. Let $Y = aX + b$, where a and b are any real numbers. Then the mgf of Y is given by

$$M_Y(s) = M_{aX+b}(s) = e^{bs}M_X(as)$$

This can be established as follows:

$$\begin{aligned} M_{aX+b}(s) &= E(e^{(aX+b)s}) = E(e^{bs} \cdot e^{asX}) \\ &= e^{bs}E(e^{asX}) \qquad \text{(why?)} \\ &= e^{bs}M_X(as) \end{aligned}$$

In other words, the mgf of an r.v. $aX + b$ is obtained by multiplying the mgf of X evaluated at as by e^{bs}. For instance, if the mgf of X is $(\frac{1}{3} + \frac{2}{3}e^s)^{10}$, then the mgf of $2X + 3$ is $e^{3s} \cdot M_X(2s)$, that is, $e^{3s}(\frac{1}{3} + \frac{2}{3}e^{2s})^{10}$.

• *Example 1.7.* If Y is $N(\mu, \sigma^2)$, find the mgf of Y.

Solution. Let $X = (Y - \mu)/\sigma$. Then from exercise 18 in Chapter 6 the random variable X is $N(0, 1)$. Now $Y = \sigma X + \mu$, and consequently

$$M_Y(s) = M_{\sigma X + \mu}(s) = e^{\mu s}M_X(\sigma s)$$

However, we know from Example 1.2(*b*) that, since X is $N(0, 1)$, $M_X(s) = e^{s^2/2}$, so that $M_X(\sigma s) = e^{(\sigma s)^2/2}$. Hence

$$M_Y(s) = e^{\mu s} e^{(\sigma^2 s^2)/2} = e^{\mu s + (\sigma^2 s^2/2)}$$

In conclusion, *if an r.v. is $N(\mu, \sigma^2)$, then its mgf is given by*

$$M(s) = e^{\mu s + (\sigma^2 s^2/2)}, \quad -\infty < s < \infty$$

It turns out that, in some situations, the mgf's can be used most elegantly to derive distributions of random variables. This is especially true in situations involving linear combinations of independent random variables. The following two results are central to this purpose.

If X and Y are *independent* r.v.'s with mgf's M_X and M_Y, then for any constants a, b,

$$M_{aX+bY}(s) = M_X(as) \cdot M_Y(bs)$$

The result can be verified as follows:

$$\begin{aligned} M_{aX+bY}(s) &= E(e^{s(aX+bY)}) \\ &= E(e^{saX} \cdot e^{sbY}) \end{aligned}$$

Now since X and Y are independent it follows that e^{saX} and e^{sbY} are independent also. Therefore,

$$\begin{aligned} M_{aX+bY}(s) &= E(e^{saX}) \cdot E(e^{sbY}) \\ &= M_X(as) \cdot M_Y(bs) \end{aligned}$$

The above result can be extended to any finite collection of *independent* r.v.'s $X_1, X_2, \ldots, X_n$. Thus, if $Z = a_1 X_1 + \ldots + a_n X_n$, then

$$M_Z(s) = M_{X_1}(a_1 s) M_{X_2}(a_2 s) \ldots M_{X_n}(a_n s)$$

In particular, if $Z = X_1 + X_2 + \ldots + X_n$, then

$$M_{X_1+X_2+\ldots+X_n}(s) = M_{X_1}(s) M_{X_2}(s) \ldots M_{X_n}(s)$$

That is, *the mgf of the sum of a finite number of independent r.v.'s is equal to the product of their mgf's.*

We could have used this result to obtain the mgf of the binomial distribution $B(n, p)$ from the mgf for the Bernoulli distribution. Recall that if X is $B(n, p)$, then X can be written as $X = Y_1 + \ldots + Y_n$ where $Y_1, Y_2, \ldots, Y_n$ are independent Bernoulli r.v.'s, each with parameter p. Since $M_{Y_i}(s) = pe^s + (1-p)$ for $i = 1, 2, \ldots, n$, we get

$$M_X(s) = M_{Y_1}(s)M_{Y_2}(s) \ldots M_{Y_n}(s) = [pe^s + (1-p)]^n$$

Similarly, recalling that if X has the negative binomial distribution with parameters r, p, then X can be written as $X = Y_1 + Y_2 + \ldots + Y_r$, where $Y_1, Y_2, \ldots, Y_r$ are independent r.v.'s, each with the same geometric distribution, we could have found the mgf for the negative binomial distribution from the mgf for the geometric distribution.

A remarkable fact about a mgf is that it is unique when it exists. Given any distribution, there is precisely one mgf that corresponds to it, and, conversely, given any mgf there is precisely one distribution that can give rise to it. The discussion of this difficult problem is beyond the scope of this book. We shall simply state the result and accept it as such.

Uniqueness Theorem If two random variables have the same mgf's, then they have the same distribution, and conversely.

In other words, if X and Y are two r.v.'s, then $M_X(s) = M_Y(s)$ for $s \in (-h, h)$ if and only if $F_X(u) = F_Y(u)$ for all real u. Thus the two concepts, namely, having the same mgf and having the same distribution, are equivalent.

For example, if an r.v. X has the mgf $M(s) = e^{2(e^s - 1)}$, then we can conclude that X has the Poisson distribution with parameter $\lambda = 2$; no other distribution can have such a mgf.

The importance of the uniqueness theorem derives from the following: Suppose we want to find the distribution of $Z = h(X, Y)$ and are in a position to obtain its mgf. If it turns out that this mgf has a recognizable form as the mgf of an r.v. with a distribution function F, then it is safe to conclude that the distribution of Z is given by the distribution function F.

Example 1.8. Suppose X and Y are independent r.v.'s with the following probability functions:

$$P(X = x) = \frac{x}{6}, \quad x = 1, 2, 3$$

and

$$P(Y = y) = \frac{y+2}{10}, \quad y = -1, 2, 3$$

Using the mgf's, find the distributions of—

(*a*) $X + Y$ (*b*) $X - Y$

Solution. The mgf's of X and Y are obtained as

$$M_X(s) = \frac{1}{6}e^s + \frac{2}{6}e^{2s} + \frac{3}{6}e^{3s}$$

and

$$M_Y(s) = \frac{1}{10}e^{-s} + \frac{4}{10}e^{2s} + \frac{1}{2}e^{3s}$$

(*a*) Since X and Y are independent, we know that $M_{X+Y}(s) = M_X(s)M_Y(s)$ which, after simplifying, gives

$$M_{X+Y}(s) = \frac{1}{60}e^{0s} + \frac{2}{60}e^s + \frac{3}{60}e^{2s} + \frac{4}{60}e^{3s} + \frac{13}{60}e^{4s} + \frac{22}{60}e^{5s} + \frac{15}{60}e^{6s}$$

From the form of the mgf, it can be concluded that $X + Y$ assumes the values 0, 1, 2, 3, 4, 5, 6 with respective probabilities $\frac{1}{60}, \frac{2}{60}, \frac{3}{60}, \frac{4}{60}, \frac{13}{60}, \frac{22}{60}, \frac{15}{60}$.

(*b*) We have

$$\begin{aligned} M_{X-Y}(s) &= M_X(s) \cdot M_Y(-s) \\ &= \left(\frac{1}{6}e^s + \frac{2}{6}e^{2s} + \frac{3}{6}e^{3s}\right)\left(\frac{1}{10}e^s + \frac{4}{10}e^{-2s} + \frac{1}{2}e^{-3s}\right) \end{aligned}$$

Multiplying out and simplifying, this yields

$$M_{X-Y}(s) = \frac{1}{12}e^{-2s} + \frac{14}{60}e^{-s} + \frac{23}{60}e^{0s} + \frac{12}{60}e^s + \frac{1}{60}e^{2s} + \frac{2}{60}e^{3s} + \frac{3}{60}e^{4s}$$

From the mgf it follows that the possible values of $X - Y$ are $-2, -1, 0, 1, 2, 3, 4$ with respective probabilities $\frac{1}{12}, \frac{7}{30}, \frac{23}{60}, \frac{1}{5}, \frac{1}{60}, \frac{1}{30}, \frac{1}{20}$. ∎

On the basis of the above example, the reader might feel that there is very little to recommend the mgf technique over the direct approach discussed in Chapter 10. However, in many complicated cases the merits of the mgf approach are unquestionable.

Example 1.9. If X is $N(0, 1)$, show that the distribution of X^2 is chi-square with 1 degree of freedom. (See Example 2.10 in Chapter 6.)

Solution. Let $Y = X^2$. Then the mgf of Y is

$$\begin{aligned} M_Y(s) = E(e^{sX^2}) &= \frac{1}{\sqrt{2\pi}} \int_{-\infty}^{\infty} e^{sx^2} \cdot e^{-x^2/2}\, dx \\ &= \frac{1}{\sqrt{2\pi}} \int_{-\infty}^{\infty} e^{-(1-2s)(x^2/2)}\, dx \end{aligned}$$

Let $(1 - 2s)^{1/2}x = u$. Then

$$\frac{1}{\sqrt{2\pi}} \int_{-\infty}^{\infty} e^{-(1-2s)(x^2/2)}\, dx = \frac{1}{\sqrt{2\pi}} \int_{-\infty}^{\infty} e^{-u^2/2} \frac{du}{(1-2s)^{1/2}} = (1-2s)^{-1/2}$$

Hence

$$M_Y(s) = (1 - 2s)^{-1/2}$$

But this is the mgf of the chi-square distribution with 1 degree of freedom. (See the comment in Example 1.2(*c*).) By the uniqueness theorem, the distribution of Y must be chi-square with 1 degree of freedom.

Example 1.10. Let X be any continuous r.v. having the distribution function F and pdf f. Show that the distribution of $Y = F(X)$ is uniform over the interval $[0, 1]$. (See Example 2.12, Chapter 6.)

Solution. The mgf of Y can be found as

$$M_Y(s) = E(e^{sY}) = E(e^{sF(x)}) = \int_{-\infty}^{\infty} e^{sF(x)} f(x)\, dx$$

Since $dF(x) = f(x)\, dx$ it can be seen that, if $s \neq 0$,

$$\int_{-\infty}^{\infty} e^{sF(x)} f(x)\, dx = \left. \frac{e^{sF(x)}}{s} \right|_{-\infty}^{\infty} = \frac{e^s - 1}{s}$$

because $F(\infty) = 1$ and $F(-\infty) = 0$. Of course, $M_Y(0) = E(e^{0Y}) = 1$. Hence

$$M_Y(s) = \begin{cases} \dfrac{e^s - 1}{s}, & \text{if } s \neq 0 \\ 1, & \text{if } s = 0 \end{cases}$$

But, by Example 1.2(*a*), this is the mgf of the uniform distribution over the interval $[0, 1]$. The result now follows from the uniqueness property.

• *Example 1.11.* Suppose $X_1, X_2, \ldots, X_n$ are independent random variables where X_i is $N(\mu_i, \sigma_i^2)$, $i = 1, 2, \ldots, n$. Find the distribution of $\bar{X} = \sum_{i=1}^{n} X_i/n$.

Solution. Let $Z = \sum_{i=1}^{n} X_i$. Then

$$M_{\bar{X}}(s) = M_{Z/n}(s) = M_Z(s/n)$$

($Z/n = 0 + \frac{1}{n} \cdot Z$ and hence $M_{Z/n}(s) = e^{0s} M_Z(\frac{1}{n} \cdot s)$.)

Since $X_1, X_2, \ldots, X_n$ are independent, and $M_{X_i}(s) = e^{\mu_i s + (\sigma_i^2 s^2/2)}$, $i = 1, 2, \ldots, n$, it follows that

$$\begin{aligned} M_{\bar{X}}(s) = M_Z\left(\frac{s}{n}\right) &= M_{X_1}\left(\frac{s}{n}\right) M_{X_2}\left(\frac{s}{n}\right) \ldots M_{X_n}\left(\frac{s}{n}\right) \\ &= \exp\left\{\left(\sum_{i=1}^{n} \mu_i\right)\frac{s}{n} + \left(\sum_{i=1}^{n} \sigma_i^2\right)\frac{(s/n)^2}{2}\right\} \\ &= \exp\left\{\left(\frac{1}{n}\sum_{i=1}^{n} \mu_i\right)s + \left(\frac{1}{n^2}\sum_{i=1}^{n} \sigma_i^2\right)\frac{s^2}{2}\right\} \end{aligned}$$

As can be recognized, this is the mgf of a random variable which is $N\left(\sum_{i=1}^{n} \mu_i/n, \sum_{i=1}^{n} \sigma_i^2/n^2\right)$. Hence *if $X_1, X_2, \ldots, X_n$ are independent random variables* *where X_i is $N(\mu_i, \sigma_i^2)$, $i = 1, 2, \ldots, n$, then $\bar{X}$ has the distribution* $N\left(\sum_{i=1}^{n} \mu_i/n, \sum_{i=1}^{n} \sigma_i^2/n^2\right)$. In particular, if the random variables are identically distributed with a common mean μ and a common variance σ^2, then $\bar{X}$ is $N(\mu, \sigma^2/n)$.

1.4 Reproductive Properties

We saw in Chapter 10 that if $X_1, X_2, \ldots, X_r$ are independent binomial random variables with parameter p, then $X_1 + X_2 + \ldots + X_r$ is also a binomial random variable. We also found that the same is true of Poisson distribution, in that if $X_1, X_2, \ldots, X_r$ are independent Poisson random variables, then so is $X_1 + X_2 + \ldots + X_r$. This property, called the reproductive property, holds for a number of distributions and is defined as follows:

Suppose $X_1, X_2, \ldots, X_r$ is *any finite* collection of independent random variables belonging to a family of probability distributions. Then the family is said to have the **reproductive property** if the distribution of $X_1 + X_2 + \ldots + X_r$ also belongs to the family.

Among such families are the family of binomial distributions with the same parameter p, and Poisson, normal, and chi-square distributions.

The reproductive property of the binomial and Poisson distributions was established in Chapter 10. In view of the instructive nature of the approach using mgf's, we will prove these results again.

The reproductive property of the binomial distribution with parameter p. Suppose $X_1, X_2, \ldots, X_r$ are independent random variables where X_i is $B(n_i, p)$, $i = 1, 2, \ldots, r$. Then $M_{X_i}(s) = [pe^s + (1-p)]^{n_i}$, $i = 1, 2, \ldots, r$.

Therefore, if $Z = X_1 + X_2 + \ldots + X_r$, we get

$$\begin{aligned} M_Z(s) &= M_{X_1}(s)M_{X_2}(s) \ldots M_{X_r}(s) \\ &= [pe^s + (1-p)]^{n_1 + n_2 + \ldots + n_r} \end{aligned}$$

Since this is the mgf of $B(n_1 + n_2 + \ldots + n_r, p)$, it follows that the distribution of $X_1 + X_2 + \ldots + X_r$ is binomial with $n_1 + n_2 + \ldots + n_r$ trials and probability of success p, thereby exhibiting the reproductive property.

The reproductive property of the Poisson distribution. Suppose X_i, $i = 1, 2, \ldots, r$, are independent Poisson random variables where the parameter of X_i is λ_i. Then $M_{X_i}(s) = e^{\lambda_i(e^s - 1)}$, $i = 1, 2, \ldots, r$.

Letting $Z = X_1 + X_2 + \ldots + X_r$,

$$\begin{aligned} M_Z(s) &= e^{\lambda_1(e^s - 1)} \ldots e^{\lambda_r(e^s - 1)} \\ &= e^{(\lambda_1 + \lambda_2 + \ldots + \lambda_r)(e^s - 1)} \end{aligned}$$

But this is the mgf of a random variable with a Poisson distribution with parameter $\lambda_1 + \lambda_2 + \ldots + \lambda_r$.

The reproductive property of the normal distribution. Suppose $X_1, X_2, \ldots, X_r$ are independent random variables and that X_i is $N(\mu_i, \sigma_i^2)$, $i = 1, 2, \ldots, r$. Then $M_{X_i}(s) = e^{\mu_i s + (\sigma_i^2 s^2/2)}$, $i = 1, 2, \ldots, r$.

Since $X_1, X_2, \ldots, X_r$ are independent, writing $Z = X_1 + X_2 + \ldots + X_r$, we get

$$M_Z(s) = e^{\mu_1 s + (\sigma_1^2 s^2/2)} \cdot e^{\mu_2 s + (\sigma_2^2 s^2/2)} \ldots e^{\mu_r s + (\sigma_r^2 s^2/2)}$$
$$= e^{(\mu_1 + \mu_2 + \ldots + \mu_r)s + [(\sigma_1^2 + \sigma_2^2 + \ldots + \sigma_r^2)s^2/2]}$$

But we recognize this as the mgf of a random variable which is normally distributed with mean $\sum_{i=1}^{r} \mu_i$ and variance $\sum_{i=1}^{r} \sigma_i^2$.

Hence, if $X_1, X_2, \ldots, X_r$ are independent random variables where X_i is $N(\mu_i, \sigma_i^2)$, $i = 1, 2, \ldots, r$, then $X_1 + X_2 + \ldots + X_r$ is $N\left(\sum_{i=1}^{r} \mu_i, \sum_{i=1}^{r} \sigma_i^2\right)$.

Comment. As far as the mean and variance of Z are concerned, these go by the rules $E\left(\sum_{i=1}^{r} X_i\right) = \sum_{i=1}^{r} E(X_i)$, and, since $X_1, X_2, \ldots, X_r$ are independent, $\text{Var}\left(\sum_{i=1}^{r} X_i\right) = \sum_{i=1}^{r} \text{Var}(X_i)$. The important fact that is brought out in the above discussion (hitherto not proved) is that the distribution of $\sum_{i=1}^{r} X_i$ is normal.

The reproductive property of the chi-square distribution. Suppose $X_1, X_2, \ldots, X_r$ are independent random variables where the distribution of X_i is chi-square with n_i degrees of freedom, $i = 1, 2, \ldots, r$. Then from the comment which follows part (*c*) of Example 1.2, $M_{X_i}(s) = (1 - 2s)^{-n_i/2}$, $i = 1, 2, \ldots, r$.

Hence, letting $Z = X_1 + \ldots + X_r$

$$M_Z(s) = (1 - 2s)^{-n_1/2} \ldots (1 - 2s)^{-n_r/2}$$
$$= (1 - 2s)^{-(n_1 + \ldots + n_r)/2}$$

But this is easily recognized as the mgf of a random variable which has the chi-square distribution with $n_1 + n_2 + \ldots + n_r$ degrees of freedom.

Hence, in conclusion, if $X_1, X_2, \ldots, X_r$ are independent random variables where X_i is chi-square with n_i degrees of freedom, $i = 1, 2, \ldots, r$, then $X_1 + \ldots + X_r$ is chi-square with $n_1 + n_2 + \ldots + n_r$ degrees of freedom.

Example 1.12. Suppose ten particles, each of mass 2 grams, are moving independently, each with a velocity (cm/sec) which is normally distributed with mean 0 and variance 9. Find the distribution of the total kinetic energy of all these particles. (The kinetic energy of a particle of mass m grams traveling at a velocity v cm/sec is given as $\frac{1}{2}mv^2$ ergs.)

Solution. Let $V_1, V_2, \ldots, V_{10}$ represent the velocities of the particles. Then the total kinetic energy Z is given by $Z = \sum_{i=1}^{10} \frac{1}{2} \cdot 2V_i^2 = \sum_{i=1}^{10} V_i^2$.

Now V_i is $N(0, 9)$. Therefore, $V_i/3$ is $N(0, 1)$ and consequently $(V_i/3)^2 = V_i^2/9$ has a chi-square distribution with 1 degree of freedom. Since $V_1, V_2, \ldots, V_{10}$ are independent, by the reproductive property of the chi-square distribution it follows that $\sum_{i=1}^{10} V_i^2/9$ has a chi-square distribution with 10 degrees of freedom. Hence the pdf of $U = \sum_{i=1}^{10} V_i^2/9$ is given by

$$f_U(u) = \begin{cases} \dfrac{1}{2^5\Gamma(5)} u^4 e^{-u/2}, & u > 0 \\ 0, & \text{elsewhere} \end{cases}$$

From this it can easily be shown that the distribution of $Z = 9U$ is

$$f_Z(z) = \begin{cases} \dfrac{1}{18^5\Gamma(5)} z^4 e^{-z/18}, & z > 0 \\ 0, & \text{elsewhere} \end{cases}$$

The verification of this is left to the reader.

EXERCISES–SECTION 1

1. Consider a degenerate random variable X with $P(X = c) = 1$. Obtain the mgf of X and $E(X^r)$, $r = 0, 1, 2, \ldots$.
2. Suppose X has the probability function defined by $P(X = 1) = \frac{1}{3}$ and $P(X = -2) = \frac{2}{3}$. For any positive integer n, find $E(X^n)$ in the following two ways:
   (a) By using the basic definition of $E(X^n)$
   (b) By expanding the mgf of X as a power series
3. A random variable X assumes the three values $-2, 3, 4$ with respective probabilities $\frac{1}{6}, \frac{1}{2}, \frac{1}{3}$.
   (a) Find the mgf of X.
   (b) Compute $E(X)$, $E(X^2)$, and $E(X^3)$ by differentiating the mgf.
4. A fair die is rolled repeatedly until a 1 or a 6 shows up. Find the mgf of the number of throws required.
5. If X has a negative binomial distribution with parameters r, p, use the moment generating function of X to find $E(X)$ and $E(X^2)$.
6. If X has the pdf

$$f(x) = \begin{cases} |x|, & -1 < x < 1 \\ 0, & \text{elsewhere} \end{cases}$$

find the mgf of X.

7. If the mgf of X is

$$M(s) = \begin{cases} \dfrac{e^{4s} - e^{-2s}}{6s}, & s \neq 0 \\ 1, & s = 0 \end{cases}$$

determine the distribution of X.

8. For the mgf's given in the following cases, identify the underlying distribution of the random variable.

(a) $M(s) = \frac{1}{3}e^{-2s} + \frac{1}{6} + \frac{1}{12}e^{5s} + \frac{5}{12}e^{8s}$

(b) $M(s) = \frac{1}{3}\sum_{r=0}^{\infty} [1 + (-2)^r 2] \frac{s^r}{r!}$

(c) $M(s) = \left[\frac{1}{4} + \frac{3}{4}\sum_{r=0}^{\infty} \frac{s^r}{r!}\right]^{10}$

(d) $M(s) = \sum_{r=0}^{\infty} \frac{s^{2r}}{r!}$

9. Consider the following mgf's expressed as power series in s:

$$M_X(s) = \sum_{r=0}^{\infty} s^r, \quad |s| < 1$$

$$M_Y(s) = \sum_{r=0}^{\infty} \sum_{i=0}^{r} \frac{2^{r-i}}{i!} s^r, \quad |s| < \tfrac{1}{2}$$

(a) Find $E(X^r)$ and $E(Y^r)$, $r = 0, 1, 2, \ldots$.

(b) How are the random variables X and Y related?

10. If X is uniformly distributed over the interval $[a, b]$, use the mgf of X to show that

$$E(X^r) = \frac{b^{r+1} - a^{r+1}}{(r+1)(b-a)}, \qquad r = 1, 2, \ldots$$

11. Suppose X has a continuous distribution with the following pdf:

$$f(x) = \tfrac{1}{2}e^{-|x|}, \qquad -\infty < x < \infty$$

(a) Obtain the mgf of X.

(b) Using the mgf, find $E(X)$, $E(Xe^{X/2})$, and $\text{Var}(X)$.

12. The mgf of a random variable X is given by $M(s) = (1 - s)^{-3}$, $s < 1$. Use the power series expansion of $M(s)$ to obtain $E(X^r)$ for any nonnegative integer r. *Hint:* Use $1/(1 - s) = \sum_{r=0}^{\infty} s^r$ and differentiate twice.

13. Suppose the mgf of X is given by

$$M(s) = (0.2 + 0.8e^s)^{10}$$

Identify the distribution of X and compute $P(4.3 < X < 7.8)$, using an appropriate table.

14. If the distribution of X is symmetric about c, show that the mgf of X (if it exists) is given by

$$M_X(s) = e^{cs} \int_0^{\infty} (e^{-sy} + e^{sy}) f(c + y)\, dy$$

15. If the distribution of X is symmetric about c, show that $X - c$ and $-X + c$ have the same distribution. (Here you are expected to prove by using the mgf's. See exercise 2 of Chapter 6 for an alternate approach.) *Hint:* Use exercise 14 and the fact that $M_{aX+b}(s) = e^{bs}M_X(as)$.

16. If X has the triangular distribution given by

$$f(x) = \begin{cases} x, & 0 \leqslant x \leqslant 1 \\ 2-x, & 1 < x \leqslant 2 \\ 0, & \text{elsewhere} \end{cases}$$

find its mgf.

17. If X and Y are independent random variables, both uniformly distributed over the interval $[0, 1]$, find the mgf of $X+Y$. On the basis of exercise 16, what can you conclude about the distribution of $X+Y$?

18. If $X_1, X_2, \ldots, X_n$ are independent random variables, each having the same distribution as a random variable X–

(a) express the mgf of $X_1+X_2+\ldots+X_n$ in terms of the mgf of X

(b) use the mgf obtained in (a) to find $E[(X_1+X_2+\ldots+X_n)^2]$ and $E[(X_1+X_2+\ldots+X_n)^3]$ in terms of the moments of X.

19. Suppose X and Y are independent random variables where X is $N(0, \sigma^2)$ and Y is exponential with parameter $\lambda = 1$. Find the mgf of $Z = X+Y$ and expand it in power series to obtain $E(Z)$, $E(Z^2)$, $E(Z^3)$, and $E(Z^4)$. Compare this with exercise 20 in Section 1 of Chapter 11.

20. If X and Y are *correlated* random variables each having a Poisson distribution, show that the distribution of $X+Y$ cannot be Poisson. *Hint:* $E(X+Y) \neq \text{Var}(X+Y)$.

21. Suppose X and Y are independent random variables where X is $N(15, 9)$ and Y is $N(20, 4)$. If $Z = X-2Y$, find $P(-35 < Z < -20)$.

22. Suppose the heights of people are normally distributed with a mean height of 68 inches and a standard deviation of 3 inches. If a sample of nine people is picked at random find $P(67 < \bar{X} < 70)$.

23. The life span of an electronic component is an exponential random variable with an expected life of 30 hours. If five components are available and are to be used consecutively (in series), find the probability that the stock will last more than 140 hours. (Give the answer in the form of an integral.)

24. A random variable X has a normal distribution, $N(10, 4)$. If eight independent observations $X_1, X_2, \ldots, X_8$ are made, find the probability that the sum of the squares of the deviations from 10 will be between 4 and 28; that is, find $P\left(4 < \sum_{i=1}^{8} (X_i - 10)^2 < 28\right)$. Give the answer as an integral.

2. THE FACTORIAL MOMENT GENERATING FUNCTION

As we saw in the last section, a moment generating function generates the raw moments of the distribution. There is another type of generating function, and this one generates the factorial moments. Recall that, in Chapter 7, when finding the variance of X in the binomial and Poisson cases, we found it expedient to obtain the second factorial moment $E(X(X-1))$. So, in some way, we have seen the importance of these moments and a consequent need to generate them efficiently, if possible. The generating function that can be used to generate the factorial moments is called the **factorial moment generating function** (abbreviated fmgf). It is defined as a function g given by

$$g(s) = E(s^X)$$

for those values of s in the neighborhood of 1 for which the expectation exists. Observe that $g(1) = E(1) = 1$.

Repeated differentiation of g with respect to s yields (assuming that differentiation and expectation can be interchanged)

$$g^{(r)}(s) = E(X(X-1)\dots(X-r+1)s^{X-r})$$

If we set $s = 1$, using the standard notation $g^{(r)}(s)\Big|_{s=1} = g^{(r)}(1)$, this gives

$$g^{(r)}(1) = E(X(X-1)\dots(X-r+1))$$

We do not propose to dwell on the topic of factorial moment generating functions in great detail, because the mgf and fmgf of a random variable are closely related to each other, so that the results that we have proved for mgf's carry over to fmgf's, with appropriate changes. (In this connection exercises 3 and 4 at the end of this section should be considered important.) In fact, we have

$$g(s) = E(s^X) = E(e^{(\log_e s)X}) = M(\log_e s)$$

and

$$M(s) = E(e^{sX}) = E((e^s)^X) = g(e^s)$$

Hence the relationship between the mgf and the fmgf of a random variable is given by

$$g(s) = M(\log_e s)$$
$$M(s) = g(e^s)$$

Translated into words, the mgf evaluated at $\log_e s$ gives the fmgf, and the fmgf evaluated at e^s gives the mgf. As special cases, if the mgf involves s only through e^s, then simply replace e^s by s in order to obtain the fmgf from the mgf; if the fmgf involves s only through powers of s, then replace s by e^s to obtain the mgf from the fmgf.

For example, if X is $B(n, p)$ we know that $M(s) = [pe^s + (1-p)]^n$. Hence, in this case, the fmgf is given by $g(s) = [ps + (1-p)]^n$. As another illustration, if the mgf is given as $M(s) = \frac{1}{3}e^{-s} + \frac{1}{6} + \frac{1}{2}e^{2s}$, then the fmgf is given by $g(s) = \frac{1}{3} \cdot s^{-1} + \frac{1}{6} + \frac{1}{2}s^2$.

As an example of obtaining the mgf when the fmgf is given, consider $g(s) = \frac{1}{5}s^{-2} + \frac{1}{10} + \frac{3}{10}s^3 + \frac{2}{5}s^5$. Then $M(s) = \frac{1}{5}e^{-2s} + \frac{1}{10} + \frac{3}{10}e^{3s} + \frac{2}{5}e^{5s}$.

Example 2.1. Using the factorial moment generating function for the binomial distribution $B(n, p)$, find $E(X)$, $E(X(X-1))$, and, in general, $E(X(X-1)\dots(X-r+1))$ for any positive integer r, $1 \leqslant r \leqslant n$.

Solution. The fmgf is given as

$$g(s) = [ps + (1-p)]^n$$

Therefore, it can be seen that

$$g^{(1)}(s) = np\,[ps + (1 - p)]^{n-1}$$
$$g^{(2)}(s) = n(n-1)p^2[ps + (1 - p)]^{n-2}$$

and, in general, for any positive integer r, $1 \leqslant r \leqslant n$

$$g^{(r)}(s) = n(n-1)\ldots(n-r+1)p^r\,[ps + (1-p)]^{n-r}$$

Evaluating these derivatives at $s = 1$, we get

$$E(X) = g^{(1)}(1) = np$$
$$E(X(X-1)) = g^{(2)}(1) = n(n-1)p^2$$

and, in general,

$$E(X(X-1)\ldots(X-r+1)) = g^{(r)}(1) = n(n-1)\ldots(n-r+1)p^r$$ ■

The factorial moment generating functions are of special interest if the random variables are *integer valued,* that is, where the set of possible values is contained in the set $\{0, 1, 2, \ldots\}$. In this case, the fmgf of a random variable X can be used to generate $P(X = r)$, $r = 0, 1, 2, \ldots$. It is because of this that, in discrete stochastic processes, the fmgf is referred to as a **probability generating function** (abbreviated pgf).

Let us see how these probabilities are generated:

Writing $P(X = i) = p(i)$, $i = 0, 1, 2, \ldots$, we get

$$g(s) = E(s^X) = \sum_{i=0}^{\infty} p(i)s^i$$

In this case, since $0 \leqslant p(i) < 1$, a comparison with the geometric series $1 + s + s^2 + \ldots$ shows that $g(s)$ always exists for $|s| < 1$. Successive differentiation gives

$$g^{(r)}(s) = \sum_{i=r}^{\infty} i(i-1)\ldots(i-r+1)s^{i-r}p(i)$$

If s is set equal to zero, all the terms in the series on the right-hand side drop out except the one for which $i = r$. For $i = r$, we get the constant term of the series, namely, $r!p(r)$. Therefore,

$$g^{(r)}(0) = r!p(r)$$

giving the following expression for computing $p(r)$ if X is integer valued:

$$P(X = r) = \frac{g^{(r)}(0)}{r!}, \qquad r = 0, 1, 2, \ldots$$

(*Note:* In view of the fact that $g(0) = P(X = 0)$, we will interpret $g^{(0)}(0)$ as equal to $g(0)$.)

In conclusion, we have the following two ways that a pgf can be used to generate probabilities for *integer-valued* random variables: (*i*) *if the power series expansion of the pgf in powers of s is available, then* $P(X = r)$ *is obtained as the coefficient of* s^r; (*ii*) *if the pgf is differentiated r times, its value at zero divided by r! gives* $P(X = r)$.

The next two examples show how the pgf's can be used effectively to compute probabilities for integer-valued random variables.

Example 2.2. Tom rolls a fair die until an even number shows up, and John rolls the same die until a 1 or a 2 shows up. Find the distribution of the total number of rolls required by Tom and John together if Tom and John perform their experiments independently.

Solution. Let X denote the number of attempts required by Tom and Y the number required by John. Then X and Y are independent random variables where X is geometric with parameter $p = \frac{1}{2}$ and Y is geometric with parameter $p = \frac{1}{3}$. The probability generating functions of X and Y are respectively $(\frac{1}{2}s)/(1 - \frac{1}{2}s)$ and $(\frac{1}{3}s)/(1 - \frac{2}{3}s)$. Since X and Y are independent, we get

$$g_{X+Y}(s) = g_X(s)g_Y(s) = \frac{s^2}{6}\left[\frac{1}{1-\frac{2}{3}s} \cdot \frac{1}{1-\frac{1}{2}s}\right]$$

The expression on the right-hand side can be decomposed into *partial fractions*, and we can write

$$\begin{aligned} g_{X+Y}(s) &= s^2\left[\frac{\frac{2}{3}}{1-\frac{2}{3}s} - \frac{\frac{1}{2}}{1-\frac{1}{2}s}\right] \\ &= s^2\left[\frac{2}{3}\sum_{r=0}^{\infty}(\tfrac{2}{3})^r s^r - \frac{1}{2}\sum_{r=0}^{\infty}(\tfrac{1}{2})^r s^r\right] \\ &= \sum_{r=0}^{\infty}[(\tfrac{2}{3})^{r+1} - (\tfrac{1}{2})^{r+1}]s^{r+2} \end{aligned}$$

Hence

$$\begin{aligned} P(X+Y=r) &= \text{coefficient of } s^r \\ &= (\tfrac{2}{3})^{r-1} - (\tfrac{1}{2})^{r-1}, \qquad r = 2, 3, \ldots \end{aligned}$$

Example 2.3. If the pgf of a random variable X is given by $g(s) = [ps + (1-p)]^n$, find $P(X = 0)$, $P(X = 1)$, $P(X = 2)$, and, in general, $P(X = r)$, $0 \leqslant r \leqslant n$, by differentiating the pgf.

Solution. We recognize that the given pgf is that of a binomial random variable $B(n, p)$, so that $P(X = r) = \binom{n}{r}p^r(1-p)^{n-r}$, $r = 0, 1, \ldots, n$.

However, in the absence of such knowledge, the probabilities can be obtained by differentiating and evaluating the derivatives at zero:

$$P(X = 0) = \frac{g^{(0)}(0)}{0!} = g(0) = (1-p)^n$$

$$P(X = 1) = \frac{g^{(1)}(0)}{1!} = n[ps + (1-p)]^{n-1}p\Big|_{s=0}$$

$$= np(1-p)^{n-1} = \binom{n}{1}p(1-p)^{n-1}$$

$$P(X = 2) = \frac{g^{(2)}(0)}{2!} = \frac{n(n-1)}{2!}[ps + (1-p)]^{n-2}p^2\Big|_{s=0}$$

$$= \binom{n}{2}p^2(1-p)^{n-2}$$

In general, for $r = 0, 1, 2, \ldots, n$, it can be seen that

$$P(X = r) = \frac{g^{(r)}(0)}{r!} = \frac{n(n-1)\ldots(n-r+1)}{r!}[ps + (1-p)]^{n-r}p^r\Big|_{s=0}$$

$$= \binom{n}{r}p^r(1-p)^{n-r}$$

EXERCISES–SECTION 2

1. In the following cases, obtain the factorial moment generating functions from the mgf's.

   (a) X is Poisson with parameter λ.
   (b) X is geometric with parameter p.
   (c) X is negative binomial with parameters r, p.
   (d) X is uniform over the interval $[a, b]$.
   (e) X has the mgf given by

$$M(s) = \tfrac{1}{5}e^{-5s} + \tfrac{3}{10}e^{-3s} + \tfrac{1}{10}e^{-2s} + \tfrac{2}{5}$$

2. Consider the following factorial moment generating functions. In each case, determine the corresponding mgf.

   (a)
$$g(s) = \begin{cases} \dfrac{s^2 - s^1}{\log_e s}, & s \neq 1, \ s > 0 \\ 1, & s = 1 \end{cases}$$

   (b) $g(s) = \dfrac{1}{3s^5} + \dfrac{1}{6s^3} + \dfrac{1}{12}s^3 + \dfrac{5}{12}s^4, \qquad s > 0$

   (c) $g(s) = e^{3(s-1)}$

3. For any random variable X, show that

$$g_{aX+b}(s) = s^b g(s^a)$$

where a and b are any constants.

4. If X and Y are independent random variables, show that

$$g_{aX+bY}(s) = g_X(s^a)g_Y(s^b)$$

for any constants a, b.

5. For any random variable X, show that

$$\text{Var}(X) = g''(1) + g'(1) - [g'(1)]^2$$

6. Suppose an experiment consists of n independent trials where the probability of success on the ith trial is p_i, $i = 1, 2, \ldots, n$. Find the probability generating function of X, the number of successes in n trials.

7. Using the factorial moment generating function, find $E(X)$, $E(X(X-1))$, and $E(X(X-1)(X-2))$ if X is Poisson with parameter λ. Generalize and find $E(X(X-1)\ldots(X-n+1))$ for any positive integer n.

8. Suppose $0 < a < 1$. If the probability generating function of a random variable X is given by

$$g(s) = \frac{2s}{(1+a) + [(1+a)^2 - 4as]^{1/2}}, \qquad |s| \leqslant 1$$

find $E(X)$.

9. If the probability generating function of a random variable X is given by

$$g(s) = \frac{3}{3s^2 - 16s + 16}$$

use the partial fraction decomposition to find the probability function of X.

10. Suppose a random variable X has the following probability generating function:

$$g(s) = \frac{s^3}{6^3}(1 - 3s^6 + 3s^{12} - s^{18}) \sum_{k=0}^{\infty} \binom{k+2}{2} s^k$$

Find:

(a) $P(X = 3)$
(b) $P(X = 5)$
(c) $P(X = 6)$
(d) $P(X = 9)$

11. Suppose X, Y, and Z are independent random variables with the following probability functions:

$$P(X = r) = \frac{2}{3}\left(\frac{1}{3}\right)^{r-1}, \qquad r = 1, 2, \ldots$$

$$P(Y = r) = \frac{1}{2}\left(\frac{1}{2}\right)^{r}, \qquad r = 0, 1, 2, \ldots$$

$$P(Z = r) = \frac{1}{3}\left(\frac{2}{3}\right)^{r}, \qquad r = 0, 1, 2, \ldots$$

Find $g_X(s)$, $g_Y(s)$, and $g_Z(s)$, and use these to find the distribution of $X+Y+Z$. *Hint:* Decompose the generating function into partial fractions.

13 Limit Theorems in Probability

INTRODUCTION

We initiated our study of the theory of probability in Chapter 1 with an intuitive "frequency interpretation" of the probability of an event. This was based on the finding that if an experiment is repeated a large number of times under identical conditions, say n times, and if in these trials an event occurs f times, then it generally happens that, after some initial fluctuations, there is a tendency for the ratio f/n to stabilize around a specific value as n increases. Many gamblers found this to be the case, and any person who has the patience to try will also invariably find this to be true. Implicit in this is the inference that if the experiment is continued indefinitely (an infinite number of times) the relative frequency f/n will approach a certain value. In practice, it is this value that we take to represent the probability of the event. In the same vein, in Chapter 7 we provided an intuitive meaning to the expected value of a random variable as the limiting value of the "ordinary average."

We should, of course, realize that any conjecture that entails an infinite number of trials does not lend itself to experimental verification, since we can never perform an infinite sequence of experiments. We have, so to speak, accepted the interpretation on faith, for the reason that our experience bears it out and as such must have universal validity.

For the inquisitive mind, many questions remain unanswered. For example, does the ratio really have a limit? In what sense do we mean "limit"? Does the mathematical theory of probability contain an assertion that embodies the theoretical counterpart of the empirical interpretation? We shall address ourselves to these aspects in Section 1, and see that the mathematical justification for the empirical interpretation is provided through the *laws of large numbers.*

The most important result to be encountered in Section 2 is the *central limit theorem.* It will be considered in the general framework of the concept of convergence in distribution. From a practical standpoint, the concept of convergence in distribution is useful for computing probabilities by approximation. The central limit theorem is remarkable for its universality. In essence, it establishes that, in certain circumstances, a large number of random factors acting independently coalesce to give a normal distribution.

Together, the laws of large numbers and the central limit theorem constitute perhaps the most important theorems in probability. The pacesetters who contributed a lion's share in the development of this branch of probability theory are the great Russian mathematicians P. L. Chebyshev (1821-1894), A. A. Markov (1856-1922), A. Khintchine (1894-1959), and A. Kolmogorov (1903-   ).

1. LAWS OF LARGE NUMBERS

1.1 Chebyshev's Inequality

We shall begin with the following celebrated inequality due to Chebyshev and named after him. It forms the cornerstone of the discussion in this section.

Chebyshev's Inequality Suppose X is a random variable with finite expectation μ and variance σ^2. Then for any positive real number ϵ,

$$P(|X-\mu| \geqslant \epsilon) \leqslant \frac{\sigma^2}{\epsilon^2}$$

or, equivalently,

$$P(|X-\mu| < \epsilon) \geqslant 1 - \frac{\sigma^2}{\epsilon^2}$$

The inequality holds for discrete as well as continuous random variables. However, we shall prove it only for the continuous case. What the result states is indicated pictorially in Figure 1.1. It asserts that the shaded area can be no greater than σ^2/ϵ^2.

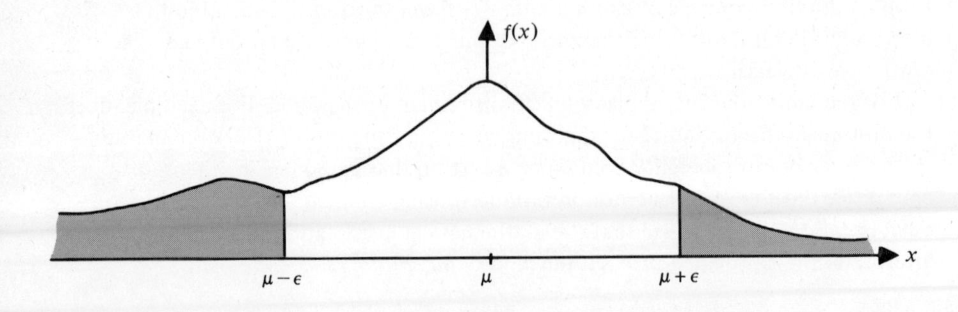

Figure 1.1

To prove the inequality, we have, by the definition,

$$\sigma^2 = \int_{-\infty}^{\infty} (x-\mu)^2 f(x)\,dx$$

Breaking the region $(-\infty, \infty)$ into three parts, as indicated in Figure 1.1,

$$\sigma^2 = \int_{-\infty}^{\mu-\epsilon} (x-\mu)^2 f(x)\,dx + \int_{\mu-\epsilon}^{\mu+\epsilon} (x-\mu)^2 f(x)\,dx + \int_{\mu+\epsilon}^{\infty} (x-\mu)^2 f(x)\,dx$$

Now,

$$\int_{\mu-\epsilon}^{\mu+\epsilon} (x-\mu)^2 f(x)\,dx \geqslant 0$$

Therefore,

$$\sigma^2 \geqslant \int_{-\infty}^{\mu-\epsilon} (x-\mu)^2 f(x)\,dx + \int_{\mu+\epsilon}^{\infty} (x-\mu)^2 f(x)\,dx$$

In both cases, as x increases from $-\infty$ to $\mu-\epsilon$, and as x increases from $\mu+\epsilon$ to ∞, the smallest value of $(x-\mu)^2$ is ϵ^2. Consequently,

$$\sigma^2 \geqslant \int_{-\infty}^{\mu-\epsilon} \epsilon^2 f(x)\,dx + \int_{\mu+\epsilon}^{\infty} \epsilon^2 f(x)\,dx$$

That is,

$$\sigma^2 \geqslant \epsilon^2\left[\int_{-\infty}^{\mu-\epsilon} f(x)\,dx + \int_{\mu+\epsilon}^{\infty} f(x)\,dx\right]$$

$$= \epsilon^2 P(|X-\mu| \geqslant \epsilon)$$

Hence

$$P(|X-\mu| \geqslant \epsilon) \leqslant \frac{\sigma^2}{\epsilon^2}$$

which is the desired result.

Letting $\epsilon = h\sigma$, where $h > 0$, we get an alternate version of Chebyshev's inequality as

$$P(|X-\mu| \geqslant h\sigma) \leqslant \frac{1}{h^2}$$

By assigning different values to h, we can find upper bounds on the probability that X assumes a value outside the interval $(\mu - h\sigma, \mu + h\sigma)$. The result asserts that, *irrespective of the actual distribution of X,* this probability is at most equal to $1/h^2$. For example, no matter what the distribution of X (whether normal, binomial, Poisson, or what have you), the probability that X will differ from the mean μ by more than 2σ (two standard deviations) is at most 0.25, the probability that it will differ by more than four standard deviations is at most 0.0625, and the probability that it will differ by more than ten standard deviations is at most 0.01. Of course,

Chebyshev's inequality is of no value if $0 < h \leqslant 1$, because it does not tell us anything that we don't already know. For instance, Chebyshev's inequality would give $P(|X - \mu| \geqslant \sigma/2) \leqslant 1/(\frac{1}{2})^2 = 4$.

We should realize that Chebyshev's inequality makes a very general, all-encompassing statement regardless of the precise form of the distribution of the random variable, and, consequently, can only provide a very crude bound. We may be able to improve on this if more information is available about the nature of the distribution of the random variable. For example, suppose X is an r.v. with mean μ and variance σ^2. In the absence of Chebyshev's inequality, all that we would be able to assert is that $P(|X - \mu| \geqslant 2\sigma) \leqslant 1$. With the aid of Chebyshev's inequality, we are in a position to make a more substantial statement that the probability is in fact less than $\frac{1}{4}$ (quite an improvement!). And, if X is given to be, say, $N(\mu, \sigma^2)$, then we can do even better and, from the standard normal table, provide the *exact* value of $P(|X - \mu| \geqslant 2\sigma)$ as $2\Phi(-2) = 0.0455$. Nothing beats knowing the exact distribution; but in the absence of it, Chebyshev's inequality often provides a lot of information that we may not have otherwise.

Since $P(|X - \mu| \geqslant \epsilon) \leqslant \sigma^2/\epsilon^2$, we see that $P(|X - \mu| \geqslant \epsilon)$ will be small if the variance σ^2 is small. Thus, Chebyshev's inequality lends precision to the statement that a small variance means that large deviations from the mean are improbable, and that the probability distribution tends to be concentrated around the mean. It thus indicates the sense in which the variance may be used as a measure of the scatter of the distribution about the mean.

Example 1.1. Suppose X is uniformly distributed over the interval $[0, 2]$.

(*a*) Applying Chebyshev's inequality, find an upper bound on the probability $P(X \leqslant 0.2 \text{ or } X \geqslant 1.8)$ and compare it with the exact value.

(*b*) Find an upper bound on $P(X \leqslant 0.3 \text{ or } X \geqslant 1.8)$.

Solution. Since X is uniformly distributed over $[0, 2]$, we know that $E(X) = 1$ and $\text{Var}(X) = (2 - 0)^2/12 = \frac{1}{3}$.

(*a*) We have

$$P(X \leqslant 0.2 \text{ or } X \geqslant 1.8) = P(|X - 1| \geqslant 0.8)$$

and, by Chebyshev's inequality,

$$P(|X - 1| \geqslant 0.8) \leqslant \frac{\text{Var}(X)}{(0.8)^2} = 0.52$$

On the other hand, the exact probability is equal to $(0.4)\frac{1}{2} = 0.2$.

(*b*) In this case,

$$\begin{aligned} P(X \leqslant 0.3 \text{ or } X \geqslant 1.8) &\leqslant P(X \leqslant 0.3 \text{ or } X \geqslant 1.7) \\ &= P(|X - 1| \geqslant 0.7) \\ &\leqslant \frac{\text{Var}(X)}{(0.7)^2}, \quad \text{by Chebyshev's inequality} \\ &= 0.68 \end{aligned}$$

The exact answer is $(0.3)\frac{1}{2} + (0.2)\frac{1}{2} = 0.25$. ■

Before we proceed with more examples, we introduce the following notation which will be used in the rest of this chapter. We also state a result from calculus which we will find extremely useful.

Suppose $X_1, X_2, \ldots$ is a sequence of random variables, that is, a countable collection of random variables. We shall often write X_n, $n \geqslant 1$, to denote such a sequence.

A sequence X_n, $n \geqslant 1$, is said to constitute an independent sequence of random variables if any finite subcollection of these random variables is independent.

Let $S_n = \sum_{i=1}^{n} X_i$. Then S_n is a random variable which represents the sum of a sample of size n and S_n/n represents the sample mean. Whenever convenient, we shall write the sample mean as $\bar{X}_n$, using the subscript n to emphasize the fact that the mean is based on n observations.

We know that if $E(X_i) = \mu_i$ and $\text{Var}(X_i) = \sigma_i^2$, then

$$E(\bar{X}_n) = \frac{\sum_{i=1}^{n} \mu_i}{n}$$

and, if the random variables are independent,

$$\text{Var}(\bar{X}_n) = \frac{\sum_{i=1}^{n} \sigma_i^2}{n^2}$$

In particular, if $\mu_i = \mu$ and $\sigma_i^2 = \sigma^2$ for every i, then

$$E(\bar{X}_n) = \mu \quad \text{and} \quad \text{Var}(\bar{X}_n) = \frac{\sigma^2}{n}$$

The result from calculus that we shall find particularly helpful is the following:

If a is a given real number and c_n is a sequence of real numbers with $\lim_{n\to\infty} c_n = 0$, then

$$\lim_{n\to\infty} \left(1 + \frac{a + c_n}{n}\right)^n = e^a$$

We shall accept this result without proof. As a trivial special case of this result, we have

$$\lim_{n\to\infty} \left(1 + \frac{a}{n}\right)^n = e^a$$

for any real number a.

Example 1.2. Suppose a fair die is rolled thirty times and the number showing on the die noted each time. Use Chebyshev's inequality to find a lower bound on the probability that the total score will be between 90 and 120, both inclusive.

Solution. Let X_i denote the score on the ith roll. Then, as can be easily verified, $E(X_i) = \frac{7}{2}$ and $\mathrm{Var}(X_i) = \frac{35}{12}$, so that $E(S_{30}) = 105$ and $\mathrm{Var}(S_{30}) = 30(\frac{35}{12}) = \frac{175}{2}$. Now

$$\begin{aligned} P(90 \leqslant S_{30} \leqslant 120) &= P(|S_{30} - 105| \leqslant 15) \\ &\geqslant P(|S_{30} - 105| < 15) \\ &\geqslant 1 - \frac{\mathrm{Var}(S_{30})}{15^2} \end{aligned}$$

applying Chebyshev's inequality. Hence

$$P(90 \leqslant S_{30} \leqslant 120) \geqslant 1 - \frac{7}{18} = 0.61$$

Later, in Example 2.5, we shall find an approximate answer to the probability. ■

Chebyshev's inequality is of considerable value to a field statistician in obtaining a cursory estimate of the sample size, as the following example will show.

Example 1.3. Suppose that the probability that a voter will vote for the candidacy of Mr. Smith is p, $0 < p < 1$, and that the voter response is independent. Find how large a sample of voters should be polled in order that, with probability at least as great as 0.95, the sample proportion will lie within 0.1 of the true (unknown) proportion p disposed towards Mr. Smith's candidacy.

Solution. Let n denote the number of voters polled. Let

$$X_i = \begin{cases} 1, & \text{if the voter is disposed in favor of Mr. Smith} \\ 0, & \text{if not disposed in favor} \end{cases}$$

Then $S_n = \sum_{i=1}^{n} X_i$ represents the total number of voters in the sample who are in favor of Mr. Smith and S_n/n represents the sample mean, and is simply the sample proportion. Now, since $E(X_i) = p$ and $\mathrm{Var}(X_i) = p(1-p)$,

$$E\left(\frac{S_n}{n}\right) = \frac{1}{n}E(S_n) = \frac{np}{n} = p$$

and

$$\mathrm{Var}\left(\frac{S_n}{n}\right) = \frac{1}{n^2}\mathrm{Var}(S_n) = \frac{np(1-p)}{n^2} = \frac{p(1-p)}{n}$$

The sample size n is to be determined so as to satisfy

$$P\left(\left|\frac{S_n}{n} - p\right| < 0.1\right) \geqslant 0.95 = 1 - 0.05$$

But by Chebyshev's inequality

$$\begin{aligned} P\left(\left|\frac{S_n}{n} - p\right| < 0.1\right) &\geqslant 1 - \frac{\mathrm{Var}(S_n/n)}{(0.1)^2} \\ &= 1 - \frac{p(1-p)}{n(0.1)^2} \\ &\geqslant 1 - \frac{1}{4(0.1)^2 n} \end{aligned}$$

since, for $0 \leqslant p \leqslant 1$, the maximum value of $p(1-p)$ is $\frac{1}{4}$. Consequently, n will have to be picked such that

$$\frac{1}{4(0.1)^2 n} \leqslant 0.05$$

that is,

$$n \geqslant \frac{1}{4(0.1)^2(0.05)} = 500$$

Hence, if $n \geqslant 500$, we will obtain

$$P\left(\left|\frac{S_n}{n} - p\right| < 0.1\right) \geqslant 0.95$$

1.2 The Laws

A sequence of random variables X_n, $n \geqslant 1$, each with finite mean, is said to obey the **Weak Law of Large Numbers** (abbreviated WLLN) if, for every $\epsilon > 0$,

$$\lim_{n\to\infty} P\left(\left|\frac{S_n}{n} - E\left(\frac{S_n}{n}\right)\right| \geqslant \epsilon\right) = 0$$

or, equivalently,

$$\lim_{n\to\infty} P\left(\left|\frac{S_n}{n} - E\left(\frac{S_n}{n}\right)\right| < \epsilon\right) = 1$$

We now give the following well-known results which state the conditions under which a sequence of random variables obeys the WLLN.

Chebyshev's Theorem If X_n, $n \geqslant 1$, is a sequence of independent random variables, each with finite variance σ^2, then it obeys the WLLN.

To see this, apply Chebyshev's inequality to the random variable S_n/n. Then

$$P\left(\left|\frac{S_n}{n} - E\left(\frac{S_n}{n}\right)\right| \geqslant \epsilon\right) \leqslant \frac{\mathrm{Var}(S_n/n)}{\epsilon^2} = \frac{\sigma^2}{n\epsilon^2}$$

But, since σ^2 is bounded, $\sigma^2/n\epsilon^2 \to 0$ as $n \to \infty$. Hence the result follows.

Markov's Theorem If X_n, $n \geqslant 1$, is a sequence of random variables for which

$$\lim_{n\to\infty} \frac{1}{n^2}\mathrm{Var}\left(\sum_{i=1}^{n} X_i\right) = 0$$

then the sequence obeys the WLLN.

The proof follows, once again, by Chebyshev's inequality, since

$$\operatorname{Var}\left(\frac{S_n}{n}\right) = \frac{1}{n^2}\operatorname{Var}\left(\sum_{i=1}^{n} X_i\right)$$

Note: Chebyshev's theorem is included in Markov's theorem because if $\operatorname{Var}(X_i) = \sigma^2 < \infty$ for every i, and if the random variables are independent, then $(1/n^2)\operatorname{Var}\left(\sum_{i=1}^{n} X_i\right) = \sigma^2/n \to 0$ as $n \to \infty$. Also observe that in Markov's theorem the random variables need not be independent.

Khintchine's Theorem If the random variables $X_1, X_2, \ldots$ are independent, *identically distributed,* and have the same mean μ, then the sequence obeys the WLLN.

The proof of Khintchine's theorem is based on the so-called "method of truncation." It is a rather tricky proof and will not be presented here.

From Khintchine's theorem, it follows that if the random variables are identically distributed, then the condition of Markov's theorem, namely,

$$\lim_{n\to\infty} \frac{1}{n^2}\operatorname{Var}\left(\sum_{i=1}^{n} X_i\right) = 0$$

is not a necessary condition. As a matter of fact, Khintchine's theorem does not impose any restriction on the nature of the variance, and, in that sense, is a considerable improvement over Markov's theorem.

For example, suppose a random variable X has the pdf f given by $f(x) = 2/x^3$ if $x \geqslant 1$, and $f(x) = 0$ elsewhere. Then, as can be easily verified, $E(X) = 2$ and $\operatorname{Var}(X) = \infty$. Let $X_1, X_2, \ldots$ be a sequence of independent random variables each having the same distribution as X. The Markov theorem is in no position to settle the question as to whether the sequence obeys the WLLN or not. But, employing Khintchine's theorem, we can assert that the sequence does obey the WLLN.

Example 1.4. Suppose $X_1, X_2, \ldots$ is a sequence of independent random variables with

$$P(X_i = i^{1/4}) = \tfrac{1}{2} \quad \text{and} \quad P(X_i = -i^{1/4}) = \tfrac{1}{2}$$

Show that the sequence obeys the WLLN.

Solution. For every $i \geqslant 1, E(X_i) = 0$. However, we cannot use Khintchine's theorem, since the random variables are not identically distributed.

Now, $\operatorname{Var}(X_i) = i^{1/2}$, and consequently

$$\frac{1}{n^2}\operatorname{Var}\left(\sum_{i=1}^{n} X_i\right) = \frac{1}{n^2}\sum_{i=1}^{n} i^{1/2}$$

Next, if we consider the Riemann sum, we see that

$$\int_0^1 x^{1/2}dx \approx \sum_{i=1}^{n}\left(\frac{i}{n}\right)^{1/2}\frac{1}{n} = \sum_{i=1}^{n} i^{1/2}\frac{1}{n^{3/2}}$$

That is,

$$\sum_{i=1}^{n} i^{1/2} \approx n^{3/2}\cdot\frac{2}{3}, \quad \text{since} \quad \int_0^1 x^{1/2}dx = \frac{2}{3}$$

As a result,

$$\frac{1}{n^2}\operatorname{Var}\left(\sum_{i=1}^{n} X_i\right) \approx \frac{1}{n^{1/2}}\cdot\frac{2}{3}$$

But the right-hand side goes to 0 as n goes to ∞. Hence, by Markov's theorem, the sequence obeys the WLLN.

There is a much deeper result called the strong law of large numbers (abbreviated SLLN) which applies to a sequence of independent, identically distributed random variables. It involves a rather complicated proof which relies on an inequality (due to Kolmogorov) which generalizes Chebyshev's inequality. We shall enunciate the result without a proof.

Strong Law of Large Numbers (SLLN) Suppose $X_1, X_2, \ldots$ are independent, identically distributed random variables with a finite mean μ. Then

$$P\left(\lim_{n\to\infty}\frac{S_n}{n} = \mu\right) = 1$$

Recall that

$$\left\{\lim_{n\to\infty}\frac{S_n}{n} = \mu\right\}$$

is a convenient way of writing the event

$$\left\{s \,\middle|\, \lim_{n\to\infty}\frac{S_n(s)}{n} = \mu\right\}$$

The SLLN states that the event representing all the points in the sample space for which $\lim_{n\to\infty}(S_n(s)/n) = \mu$ has probability one. The SLLN can be written equivalently as

$$\lim_{n\to\infty} P\left(\left|\frac{S_r}{r} - \mu\right| \geqslant \epsilon \text{ for some } r \geqslant n\right) = 0$$

for any $\epsilon > 0$. (We shall accept this fact without proof.) Now we can see why the SLLN is deeper than the WLLN. Remember that the event

$$\left\{\left|\frac{S_r}{r} - \mu\right| \geqslant \epsilon \text{ for some } r \geqslant n\right\}$$

is the union

$$\bigcup_{r=n}^{\infty}\left\{\left|\frac{S_r}{r}-\mu\right|\geqslant\epsilon\right\}$$

and consequently is a much larger event than $\{|(S_n/n)-\mu|\geqslant\epsilon\}$, and therefore, by the monotone property of the probability measure, has greater probability. Consequently, if a sequence of random variables obeys the SLLN, then it necessarily obeys the WLLN. Examples can be constructed (and this is by no means an easy matter) to show that the converse is false. Under the SLLN, we are not only demanding that S_r/r be close to μ, but also that it remain close to μ as r increases.

In summary: If the sequence obeys the WLLN, we can be almost sure that S_n/n will be close to μ if n is large. Thus, for example, assuming the random variables have a continuous distribution,

$$\int_{\mu-\epsilon}^{\mu+\epsilon} f_{S_n/n}(x)\,dx$$

will be very close to 1; how close to 1, for a given $\epsilon>0$, will depend on how large an n we take. (Here $f_{S_n/n}$ represents the pdf of S_n/n.)

If the sequence obeys the SLLN, then for large n we can be almost sure that S_n/n, $S_{n+1}/(n+1)$, . . . will *all* be close to μ.

There! We have given a meaning in concrete mathematical terms to the intuitive notion that the expected value of a random variable represents the limit of the "ordinary sample average."

Laws of large numbers applied to the Bernoulli sequence

The following discussion is important to us because it serves to explain the frequency interpretation of the probability of an event. In the context of the Bernoulli sequence of random variables, the WLLN is called the **Bernoulli Law of Large Numbers,** and the SLLN, the **Borel Law of Large Numbers.** Both the Bernoulli law and the Borel law were precursors to the WLLN and SLLN in their general formulation.

Let $X_1, X_2, \ldots$ be independent, identical Bernoulli random variables with parameter p, $0\leqslant p\leqslant 1$. Suppose we identify success with the occurrence of an event B. Then, for any $i\geqslant 1$, $X_i=1$ if B occurs and $X_i=0$ if B does not occur. Conceptually, we can regard the underlying sample space S as consisting of infinite sequences and given as

$$S=\{(x_1,x_2,\ldots)\mid x_i=1 \text{ or } 0\}$$

As we know, $E(S_n/n)=p$ and $\mathrm{Var}(S_n/n)=p(1-p)/n$. By the WLLN, it follows that

$$\lim_{n\to\infty} P\left(\left|\frac{S_n}{n}-p\right|\geqslant\epsilon\right)=0$$

for any $\epsilon>0$. Now notice that, in this instance, S_n represents the number of times B occurs in n trials, and S_n/n, the relative frequency of B. Informally, the Bernoulli law therefore states that if the number of trials is large, then the "relative frequency" will be close to p with high probability.

The Borel law allows us to make an even stronger statement than is permitted by the Bernoulli law of large numbers. It asserts that, with probability one, $\lim_{n\to\infty} (S_n/n)$ exists and is equal to p. That is,

$$P\left(\lim_{n\to\infty} \frac{S_n}{n} = p\right) = P\left(\left\{s \,\middle|\, \lim_{n\to\infty} \frac{S_n(s)}{n} = p\right\}\right) = 1$$

Thus, *the Borel law assures that the limit of the relative frequency exists and is equal to p for all sample points* (infinite sequences consisting of one's and zero's), *except perhaps for a set of points whose probability is zero.* It is therefore the Borel law which is at the heart of the intuitive frequency interpretation of the probability of an event.

Let us digress for a moment to discuss the subtlety involved in the phrase "... except perhaps for a set of points whose probability is zero." Suppose a *fair* coin is tossed repeatedly and we write 1 if heads show up on a toss and 0 if tails show up. As mentioned earlier, the sample space S can be written as a collection of infinite sequences of one's and zero's, and for any $s \in S$, $S_n(s)$ represents the number of heads in n tosses. For example, if $s = (1, 0, 0, 1, \ldots)$ then $S_1(s) = 1$, $S_2(s) = 1$, $S_3(s) = 1$, $S_4(s) = 2, \ldots$.

Let C represent the set of points $s \in S$ for which $\lim_{n\to\infty} (S_n(s)/n) = \frac{1}{2}$. That is,

$$C = \left\{s \in S \,\middle|\, \lim_{n\to\infty} \frac{S_n(s)}{n} = \frac{1}{2}\right\}$$

Then C', the complement of C, is given by

$$C' = \left\{s \in S \,\middle|\, \lim_{n\to\infty} \frac{S_n(s)}{n} \neq \frac{1}{2}\right\}$$

Now consider the following points s_1, s_2, s_3 in S:

$$s_1 = (1, 1, 1, \ldots, 1, \ldots)$$
$$s_2 = (0, 1, 1, \ldots, 1, \ldots)$$
$$s_3 = (0, 0, 0, \ldots, 0, \ldots)$$

Then

$$\frac{S_n(s_1)}{n} = 1 \quad \text{for every } n, \text{ so that } \lim_{n\to\infty} \frac{S_n(s_1)}{n} = 1$$

$$\frac{S_n(s_2)}{n} = \frac{n-1}{n} \quad \text{for every } n, \text{ so that } \lim_{n\to\infty} \frac{S_n(s_2)}{n} = 1$$

$$\frac{S_n(s_3)}{n} = \frac{0}{n} \quad \text{for every } n, \text{ so that } \lim_{n\to\infty} \frac{S_n(s_3)}{n} = 0$$

Therefore, s_1, s_2, s_3 all belong to the set C', showing that the set of points s for which $S_n(s)/n$ fails to converge to $\frac{1}{2}$ is by no means an empty set.

All the same, what the Borel law guarantees is that the set of all such points constitutes a negligible event, in that it has zero probability; that is, $P(C') = 0$.

It is the event C, consisting of all s for which $\lim_{n\to\infty} (S_n(s)/n) = \frac{1}{2}$, that has probability one. This is the essence of the Borel law. This should be quite reassuring, because intuitively, deep down, we feel that, for example, a sequence $H, H, H, \ldots$ will almost certainly never occur if the coin is fair. Not only that, any sequence will almost certainly yield a ratio of the number of heads to the number of tosses which will get closer to $\frac{1}{2}$ as we consider more results in the sequence.

EXERCISES–SECTION 1

1. A random variable X is said to be nonnegative if $P(X < 0) = 0$. Suppose X is a nonnegative random variable, and c a positive real number. Show that

$$P(X \geqslant c) \leqslant \frac{E(X)}{c}$$

2. If $X_1, X_2, \ldots, X_n$ are nonnegative random variables with $E(X_i) = \mu_i < \infty$, prove that, for any $c > 0$,

$$P\left(\sum_{i=1}^{n} X_i \geqslant c\right) \leqslant \frac{1}{c}(\mu_1 + \mu_2 + \ldots + \mu_n)$$

Hint: Letting $Y = \sum_{i=1}^{n} X_i$, in the continuous case,

$$E(Y) = \int_0^\infty u f_Y(u)\, du \geqslant \int_c^\infty u f_Y(u)\, du$$

3. If $X_1, X_2, \ldots, X_n$ are independent random variables, each having zero mean and unit variance, show that $P\left(\sum_{i=1}^{n} X_i^2 \geqslant \lambda n\right) \leqslant 1/\lambda$. *Hint:* Letting $Y = \sum_{i=1}^{n} X_i^2$, in the continuous case,

$$E(Y) = \int_0^\infty u f_Y(u)\, du \geqslant \int_{\lambda n}^\infty u f_Y(u)\, du$$

4. (*The Markov Inequality*) Suppose $\alpha \geqslant 0$ and $E(|X|^\alpha)$ exists. Show that

$$P(|X| \geqslant c) \leqslant \frac{1}{c^\alpha} E(|X|^\alpha)$$

for any $c > 0$. (Notice that, with $\alpha = 2$, and replacing $|X|$ by $|X - \mu|$, we get Chebyshev's inequality as a special case.)

5. If X is a random variable for which $E(X) = 0$, $\text{Var}(X) = 1$, and $|X| \leqslant M$, show that, for $0 < \lambda < 1$,

$$P(|X| \geqslant \lambda) \geqslant \frac{1 - \lambda^2}{M^2 - \lambda^2}$$

Hint:
$$\text{Var}(X) = \int_{-\lambda}^{\lambda} x^2 f(x)\, dx + \int_{|x| \geqslant \lambda} x^2 f(x)\, dx$$
$$\leqslant \lambda^2 + (M^2 - \lambda^2) \int_{|x| \geqslant \lambda} f(x)\, dx$$

6. A random variable X has the following probability function:

$$P(X = x) = \tfrac{1}{35}(4 + x), \qquad x \in \{-3, -1, 0, 1, 2, 3, 5\}$$

Compute $P(|X - \mu| \geqslant h\sigma)$ for $h = 1.2, 2, 3$ and compare with the bounds provided by Chebyshev's inequality.

7. Suppose X has an exponential distribution

$$f(x) = \begin{cases} e^{-x}, & x \geqslant 0 \\ 0, & \text{elsewhere} \end{cases}$$

Compute $P(|X - \mu| \geqslant h\sigma)$ and compare it with the bounds provided by Chebyshev's inequality for $h = 1.2, 2, 3$.

8. A random variable X has the gamma distribution given by

$$f(x) = \begin{cases} \dfrac{x^{p-1}}{\Gamma(p)} e^{-x}, & x > 0 \\ 0, & \text{elsewhere} \end{cases}$$

Use Chebyshev's inequality to show that

$$P(0 < X < 2p) > \frac{p-1}{p}$$

9. Suppose X and Y are independent random variables where X is $N(5, 9)$ and Y is $N(10, 4)$. Let $Z = X + 2Y$. Using Chebyshev's inequality, find the lower bound on $P(12 < Z \leqslant 35)$ and compare this with the exact value.

10. Suppose X has a Poisson distribution with parameter λ. Use Chebyshev's inequality to establish the following inequalities:

(a) $P(X \geqslant 2\lambda) \leqslant \dfrac{1}{\lambda}$

(b) $P\left(X \leqslant \dfrac{\lambda}{3}\right) \leqslant \dfrac{9}{4\lambda}$

Hint: (a) $\{|X - \lambda| \geqslant a\} \supset \{X - \lambda \geqslant a\}$
(b) $\{|X - \lambda| \geqslant a\} \supset \{X - \lambda \leqslant -a\}$

for any $a \geqslant 0$. Pick a appropriately.

11. (*Chebyshev's inequality cannot be improved!*) Suppose X has a discrete distribution given by

$$P(X = x) = \begin{cases} \frac{1}{6}, & x = -2 \\ \frac{2}{3}, & x = 0 \\ \frac{1}{6}, & x = 2 \end{cases}$$

Find a real value h such that

$$P(|X - \mu| \geqslant h\sigma) = \frac{1}{h^2}$$

thus showing that the probability attains the bound of Chebyshev's inequality. (Remember that Chebyshev's inequality makes the statement for *every* $h > 0$ and *every* random variable X.)

12. Consider the sequence of independent random variables X_i where

$$P(X_i = 2^i) = 2^{-(2i+1)}, \quad P(X_i = -2^i) = 2^{-(2i+1)}$$

and

$$P(X_i = 0) = 1 - 2^{-2i}$$

Show that the sequence obeys the WLLN.

13. Suppose that $X_1, X_2, \ldots$ is a sequence of mutually independent random variables, with $E(X_i) = \mu$ and $\mathrm{Var}(X_i) = \sigma^2$ for every i. Show that

$$\lim_{n\to\infty} P\left(\frac{X_1 + X_2 + \ldots + X_n}{n} \leqslant t\right) = \begin{cases} 0, & t < \mu \\ 1, & t > \mu \end{cases}$$

Hint: (*i*) Suppose $t = \mu - \epsilon$. Then

$$\left\{\frac{S_n}{n} \leqslant \mu - \epsilon\right\} \subset \left\{\left|\frac{S_n}{n} - \mu\right| \geqslant \epsilon\right\}$$

(*ii*) Suppose $t = \mu + \epsilon$. Then

$$\left\{\frac{S_n}{n} \leqslant \mu + \epsilon\right\} \supset \left\{\left|\frac{S_n}{n} - \mu\right| > \epsilon\right\}'$$

Use Chebyshev's theorem.

2. CONVERGENCE IN DISTRIBUTION

The main result of this section is the central limit theorem. We shall initiate our discussion with the general notion of convergence in distribution.

2.1 The General Notion of Convergence in Distribution

A sequence of random variables X_n, $n \geqslant 1$, is said to **converge in distribution** (or **in law**) to a random variable X if

$$\lim_{n\to\infty} F_{X_n}(u) = F_X(u)$$

at each real number u where F_X is continuous

This is written compactly as $X_n \xrightarrow{d} X$. (The points where F_X is continuous are called the continuity points of F_X).

The importance of the notion of convergence in distribution is to be seen in the following fact: Often one is interested in finding the probabilities associated with the distribution of X_n when n is large. However, the problem of finding the distribution of X_n may be quite complicated and, sometimes, even if the distribution of X_n is available, the actual computations may be quite involved. For example, suppose X_n is the sum of n independent, Bernoulli random variables each with the probability of success p. The distribution of X_n is known for every n as being binomial, $B(n, p)$. But even for moderately small values of n, the computations of $\binom{n}{k}p^k(1-p)^{n-k}$ are messy. Now, if it is known that the sequence $X_1, X_2, \ldots$

converges in distribution to X, then one can use the distribution of X to obtain the approximate probabilities. (Of course, one assumes that the distribution of X is such that the probabilities can be obtained from it easily.) This can be proved for any sequence of r.v.'s as follows:

Suppose $X_1, X_2, \ldots$ converges in distribution to X, that is, F_{X_n} converges to F_X at the continuity points of F_X. Suppose a and b $(a < b)$ are continuity points of F_X. Then, by the definition,

$$\lim_{n\to\infty} F_{X_n}(b) = F_X(b) \quad \text{and} \quad \lim_{n\to\infty} F_{X_n}(a) = F_X(a)$$

Hence

$$\lim_{n\to\infty} P(a < X_n \leqslant b) = \lim_{n\to\infty} [F_{X_n}(b) - F_{X_n}(a)]$$
$$= F_X(b) - F_X(a) = P(a < X \leqslant b)$$

In other words, if n is large, $P(a < X_n \leqslant b)$ can be approximated by $P(a < X \leqslant b)$.

Example 2.1. Suppose for each n, X_n is $N(0, 1/n)$. Show that X_n converges in distribution.

Solution. The D.F. of X_n is given by

$$F_{X_n}(u) = \int_{-\infty}^{u} \frac{\sqrt{n}}{\sqrt{2\pi}} e^{-nx^2/2}\, dx = \Phi(u\sqrt{n})$$

Therefore,

$$\lim_{n\to\infty} F_{X_n}(u) = \begin{cases} \Phi(-\infty) & \text{if } u < 0 \\ \Phi(0) & \text{if } u = 0 \\ \Phi(\infty) & \text{if } u > 0 \end{cases}$$

$$= \begin{cases} 0 & \text{if } u < 0 \\ \frac{1}{2} & \text{if } u = 0 \\ 1 & \text{if } u > 0 \end{cases}$$

Hence, if we consider a random variable X whose D.F. is given by

$$F_X(u) = \begin{cases} 0 & \text{if } u < 0 \\ 1 & \text{if } u \geqslant 0 \end{cases}$$

then $X_n \xrightarrow{d} X$. (Notice that $F_{X_n}(0) = \frac{1}{2}$ for every n and $F_X(0) = 1$, so that $\lim_{n\to\infty} F_{X_n}(0) \neq F_X(0)$. But 0 is not a continuity point of F_X, and for convergence in distribution we do not need convergence at the discontinuity points.)

Example 2.2 Discuss the convergence in distribution of the sequences of random variables X_n, $n \geqslant 1$, in the following cases:

(*a*) For every integer $n \geqslant 1$, the pdf of X_n is given by

$$f_{X_n}(x) = \begin{cases} nx^{n-1}, & 0 < x < 1 \\ 0, & \text{elsewhere} \end{cases}$$

(*b*) For every integer $n \geqslant 1$, the D.F. of X_n is given by

$$F_{X_n}(x) = \begin{cases} 0, & x < n \\ 1, & x \geqslant n \end{cases}$$

Solution

(*a*) The D.F. of X_n can be easily obtained as

$$F_{X_n}(u) = \begin{cases} 0, & u < 0 \\ u^n, & 0 \leqslant u < 1 \\ 1, & u \geqslant 1 \end{cases}$$

As n goes to infinity,

$$\lim_{n\to\infty} F_{X_n}(u) = \begin{cases} 0, & u < 1 \\ 1, & u \geqslant 1 \end{cases}$$

Hence $X_n \xrightarrow{d} X$ where the D.F. of X is

$$F_X(u) = \begin{cases} 0, & u < 1 \\ 1, & u \geqslant 1 \end{cases}$$

(*b*) In this case, $\lim_{n\to\infty} F_{X_n}(u) = 0$ for every real number u. But there is no distribution function which, at its continuity points, will agree with a function which is identically zero. Hence X_n does not converge in distribution. ■

The reader is cautioned that, if $X_n \xrightarrow{d} X$, there is nothing implied about the convergence of the sequence X_n, $n \geqslant 1$, to the random variable X per se.

To see this, suppose X is a standard normal variable. Define a sequence X_n, $n \geqslant 1$, by

$$X_n = (-1)^n X$$

Thus, for any sample point $s \in S$, $X_n(s) = (-1)^n X(s)$. Now, since X is $N(0, 1)$, it follows that for *every* integer $n \geqslant 1$, X_n is $N(0, 1)$. (Recall that if X is $N(\mu, \sigma^2)$, then $aX + b$ is $N(a\mu + b,\ a^2\sigma^2)$. Here $a = (-1)^n$ and $b = 0$.) Therefore, it follows that $X_n \xrightarrow{d} X$.

However, for *every* $s \in S$ the sequence of real numbers $(-1)^n X(s)$, $n \geqslant 1$, diverges. Hence, $P(s \mid X_n(s) \text{ diverges}) = P(X_n \text{ diverges}) = 1$.

In conclusion, although $X_n \xrightarrow{d} X$, the sequence of random variables itself diverges.

Poisson approximation to the binomial

In Chapter 5, we enunciated the postulates under which the random variable representing the number of changes during a given interval of length t has the Poisson distribution. We shall now establish that under certain conditions the binomial probabilities $b(k; n, p)$ can be approximated by the Poisson probabilities if the number of trials n is large.

Suppose p, the probability of success, depends on n in such a way that $np = \lambda$, where λ is a constant which does not depend on n. Then

$$\lim_{n\to\infty} b(k; n, p) = \frac{e^{-\lambda}\lambda^k}{k!}, \qquad k = 0, 1, \ldots$$

Let us first find the limit for $k = 0$. We get

$$\lim_{n\to\infty} b(0; n, p) = \lim_{n\to\infty} (1-p)^n = \lim_{n\to\infty} \left(1 - \frac{\lambda}{n}\right)^n = e^{-\lambda}$$

Next, taking the ratio of the consecutive terms of the binomial probabilities, we have

$$\frac{b(k; n, p)}{b(k-1; n, p)} = \frac{\binom{n}{k} p^k (1-p)^{n-k}}{\binom{n}{k-1} p^{k-1}(1-p)^{n-k+1}} = \frac{(n-k+1)}{k}\frac{p}{1-p}$$

$$= \frac{(n-k+1)}{k} \cdot \frac{\lambda/n}{1-(\lambda/n)}$$

$$= \frac{\lambda}{k}\left[\left(1 - \frac{k-1}{n}\right)\frac{1}{1-(\lambda/n)}\right]$$

Therefore,

$$\lim_{n\to\infty} \frac{b(k; n, p)}{b(k-1; n, p)} = \frac{\lambda}{k}$$

That is,

$$\lim_{n\to\infty} b(k; n, p) = \frac{\lambda}{k} \lim_{n\to\infty} b(k-1; n, p)$$

Proceeding recursively,

$$\lim_{n\to\infty} b(k; n, p) = \frac{\lambda}{k} \cdot \frac{\lambda}{k-1} \lim_{n\to\infty} b(k-2; n, p)$$

$$= \frac{\lambda}{k} \cdot \frac{\lambda}{k-1} \cdot \ldots \cdot \frac{\lambda}{2} \cdot \frac{\lambda}{1} \cdot \lim_{n\to\infty} b(0; n, p)$$

$$= \frac{\lambda^k e^{-\lambda}}{k!}$$

Thus we see that the limiting values are the probabilities associated with a Poisson random variable with parameter λ. (Notice, incidentally, that if X is $B(n, p)$, then $E(X) = np = \lambda$ and $\mathrm{Var}(X) = np(1-p) = \lambda[1 - (\lambda/n)] \to \lambda$ as $n \to \infty$, and λ is indeed the mean and the variance of the Poisson random variable.)

Comments. (1) In the body of the proof above we obtained

$$\frac{b(k;n,p)}{b(k-1;n,p)} = \frac{\lambda}{k}\left[\left(1-\frac{k-1}{n}\right)\frac{1}{1-(\lambda/n)}\right]$$

and we wrote the limit as equal to λ/k. In the limiting process, $1-[(k-1)/n] \to 1$. Now, in a practical situation where we apply the approximation, the value so obtained will provide a good approximation for those k for which $(k-1)/n$ can reasonably be assumed to be close to zero. This will happen when k is considerably smaller than n. A similar reasoning shows that, for a good approximation, λ should be small in comparison with n on account of the term $[1-(\lambda/n)]$ involved in the denominator. In other words, *the Poisson distribution provides a good approximation for the binomial probabilities $b(k; n, p)$ for moderate values of k if $p = \lambda/n$ is small.*

(2) We have indeed shown that the sequence of random variables X_n, $n \geqslant 1$, where X_n is $B(n, p)$ for every n, converges in distribution to a random variable X which has a Poisson distribution. For, suppose u is a continuity point of F_X. Then

$$\begin{aligned}\lim_{n\to\infty} P(X_n \leqslant u) &= \lim_{n\to\infty} \sum_{j\leqslant u} P(X_n = j)\\ &= \sum_{j\leqslant u} \lim_{n\to\infty} P(X_n = j)\\ &= \sum_{j\leqslant u} P(X = j) = F_X(u)\end{aligned}$$

As it turns out, in this case $\lim_{n\to\infty} F_{X_n}(u) = F_X(u)$ even at the discontinuity points of F_X.

Example 2.3

(*a*) If X is $B(30, 0.1)$, find (*i*) $P(X = 2)$ and (*ii*) $P(X = 9)$.
(*b*) If X is $B(30, 0.5)$, find $P(X = 9)$.

Solution

(*a*) (*i*) The exact value of $P(X = 2)$ is $\binom{30}{2}(0.1)^2(0.9)^{28}$ which, using the standard binomial tables, is equal to 0.22766.

Using the Poisson approximation, since $np = 30(0.1) = 3$, $P(X = 2) \approx (e^{-3}3^2)/2!$ $= 0.22404$. The approximation is remarkably close to the exact value.

(*ii*) The exact value of $P(X = 9)$ is equal to $\binom{30}{9}(0.1)^9(0.9)^{21} = 0.00157$, whereas the value given by the Poisson approximation is $(e^{-3}3^9)/9! = 0.0027$. In this case, the approximation is not close because $k(=9)$ is not small in comparison with $n(=30)$.

(*b*) The exact probability is $\binom{30}{9}(\frac{1}{2})^{30} = 0.01332$, and the approximate probability using the Poisson approximation is $(e^{-15}15^9)/9! = 0.03241$. The approximation is rather poor on two counts: the value of p is not close to zero, and $k/n(=\frac{9}{30})$ is not negligible. ■

The key result which is useful in proving convergence in distribution is the following theorem called the continuity theorem. We shall state it without giving the proof, which is quite involved.

Continuity Theorem Suppose X_n, $n \geqslant 1$, is a sequence of random variables where for every n the mgf (moment generating function) M_{X_n} of X_n exists for $s \in (-h, h)$. Also, suppose X is a random variable whose mgf M_X exists for $s \in (-h, h)$. If $\lim_{n\to\infty} M_{X_n}(s) = M_X(s)$ for every $s \in (-h, h)$, then $\lim_{n\to\infty} F_{X_n}(x) = F_X(x)$ at each continuity point x of F_X; that is, $X_n \xrightarrow{d} X$.

As an example of an application of the continuity theorem, suppose for each $n \geqslant 1$, X_n is $B(n, p)$. We have shown earlier that if $np = \lambda$, where λ does not depend on n, then X_n converges in distribution to a random variable which has a Poisson distribution with parameter λ. Let us now view this in the light of the continuity theorem.

The mgf of X_n is given by

$$M_{X_n}(s) = [pe^s + (1-p)]^n, \qquad n = 1, 2, \ldots$$

Under the assumption that $np = \lambda$, where λ does not involve n,

$$\begin{aligned} \lim_{n\to\infty} M_{X_n}(s) &= \lim_{n\to\infty} [1 + p(e^s - 1)]^n \\ &= \lim_{n\to\infty} \left[1 + \frac{\lambda(e^s - 1)}{n}\right]^n \\ &= e^{\lambda(e^s - 1)} \end{aligned}$$

which is the mgf of a Poisson random variable with parameter λ. Hence the conclusion.

Example 2.4. Suppose X_n, $n \geqslant 1$, is a sequence of random variables where for each n the probability function of X_n is given by

$$P\left(X_n = \frac{k}{n}\right) = \frac{1}{n}, \qquad k = 1, 2, \ldots, n$$

Show that $X_n \xrightarrow{d} X$, where X is uniformly distributed over the interval $[0, 1]$.

Solution. X_n assumes n values $1/n, 2/n, \ldots, (n-1)/n, n/n$, each with probability $1/n$, and its mgf is given by

$$M_{X_n}(s) = E(e^{X_n s}) = \frac{1}{n} \sum_{k=1}^{n} e^{(k/n)s}$$

Letting $e^{s/n} = u$,

$$M_{X_n}(s) = \frac{1}{n} \sum_{k=1}^{n} u^k = \frac{u(u^n - 1)}{n(u-1)}$$

Resubstituting,

$$M_{X_n}(s) = \frac{e^{s/n}(e^s - 1)}{n(e^{s/n} - 1)}$$

Now,

$$n(e^{s/n} - 1) = n\left[\frac{s}{n} + \frac{(s/n)^2}{2!} + \ldots\right] = \left[s + \frac{s^2}{2n} + \ldots\right]$$

so that

$$\lim_{n\to\infty} n(e^{s/n} - 1) = s$$

Also,

$$\lim_{n\to\infty} e^{s/n} = e^0 = 1$$

Therefore,

$$\lim_{n\to\infty} M_{X_n}(s) = \frac{e^s - 1}{s}, \quad \text{if } s \neq 0$$

Of course, $M_{X_n}(0) = 1$ for every n and, as a result, $\lim_{n\to\infty} M_{X_n}(0) = 1$. In summary, therefore,

$$\lim_{n\to\infty} M_{X_n}(s) = \begin{cases} \dfrac{e^s - 1}{s}, & \text{if } s \neq 0 \\ 1, & \text{if } s = 0 \end{cases}$$

But the right-hand side represents the mgf of a random variable X which is uniformly distributed over the interval $[0, 1]$. Hence, by the continuity theorem, $X_n \xrightarrow{d} X$.

2.2 Central Limit Theorem

If $X_1, X_2, \ldots, X_n$ represents a sample consisting of n independent observations from a distribution which is $N(\mu, \sigma^2)$, then we know that the distribution of S_n is normal with mean $n\mu$ and variance $n\sigma^2$ for *every* positive integer n. Of course, if the random variables are not normally distributed, then the distribution of S_n need not be normal. However, it is a remarkable fact that if the random variables satisfy some general conditions, and if n is large (but fixed), then the distribution of S_n is close to a normal distribution. Mathematically, what we have in mind is a limiting process as n tends to infinity. Now, since $E(S_n) = n\mu$ and $\text{Var}(S_n) = n\sigma^2$, as n goes to infinity the center of location of the distribution of S_n drifts to infinity and the scatter becomes ever larger. As a result, a limiting distribution of S_n wouldn't give us anything worthwhile because we will end up getting a distribution whose mean is ∞ or $-\infty$ and whose variance is ∞. To accommodate the limiting argument, we make a transformation so that the resulting random variable has a mean and a variance that do not involve n. This is accomplished simply by considering the standardized random variable $Z_n = (S_n - n\mu)/(\sigma\sqrt{n})$. As we know, its mean is zero and its variance is one. Informally, the central limit theorem asserts that the distribution of Z_n approaches the standard normal distribution.

We shall state and prove the central limit theorem for the special case where the random variables are independent and identically distributed with a common mean μ and variance σ^2, and where the mgf for each exists. The central limit theorem actually holds under much weaker conditions. In recent years, research in this area has been directed towards investigating conditions under which the central limit theorem holds for dependent random variables.

The proof given below should be regarded as a plausibility argument; it is in no way a rigorous mathematical presentation, which would call for advanced tools in mathematical analysis.

Central Limit Theorem If $X_1, X_2, \ldots$ are independent, and identically distributed like a random variable X with finite mean μ and finite, *positive* variance σ^2, then

$$\lim_{n\to\infty} P\left(\frac{S_n - n\mu}{\sigma\sqrt{n}} \leqslant t\right) = \frac{1}{\sqrt{2\pi}} \int_{-\infty}^{t} e^{-u^2/2}\, du = \Phi(t)$$

In other words, $(S_n - n\mu)/(\sigma\sqrt{n}) \overset{d}{\to} Z$, where Z is $N(0, 1)$. This is often expressed by stating that $(S_n - n\mu)/(\sigma\sqrt{n})$ is *asymptotically* $N(0, 1)$, or that S_n is *asymptotically* $N(n\mu, \sigma\sqrt{n})$.

To prove the theorem, let

$$Z_n = \frac{S_n - n\mu}{\sigma\sqrt{n}} = \sum_{i=1}^{n} \frac{(X_i - \mu)}{\sigma\sqrt{n}}$$

Since $(X_i - \mu)/(\sigma\sqrt{n})$, $i = 1, 2, \ldots, n$, are independent random variables, and since the mgf of the sum of independent random variables is equal to the product of their mgf's, we can write

$$\begin{aligned} M_{Z_n}(s) &= E\left[\exp\left\{s \cdot \frac{(X_1 - \mu)}{\sigma\sqrt{n}}\right\}\right] E\left[\exp\left\{s \cdot \frac{(X_2 - \mu)}{\sigma\sqrt{n}}\right\}\right] \ldots E\left[\exp\left\{s \cdot \frac{(X_n - \mu)}{\sigma\sqrt{n}}\right\}\right] \\ &= \{E[e^{(s/(\sigma\sqrt{n}))(X-\mu)}]\}^n \end{aligned}$$

since the random variables are identically distributed as X. Hence

$$M_{Z_n}(s) = \left[M_{X-\mu}\left(\frac{s}{\sigma\sqrt{n}}\right)\right]^n$$

Now, we know that

$$M_{X-\mu}(s) = 1 + sE(X - \mu) + \frac{s^2}{2!}E[(X - \mu)^2] + \ldots + \frac{s^r}{r!}E[(X - \mu)^r] + \ldots$$

and, since $E(X - \mu) = 0$ and $E[(X - \mu)^2] = \sigma^2$, we get

$$M_{X-\mu}(s) = 1 + \sigma^2\frac{s^2}{2!} + \ldots + E[(X - \mu)^r]\frac{s^r}{r!} + \ldots$$

Therefore,

$$M_{X-\mu}\left(\frac{s}{\sigma\sqrt{n}}\right) = 1 + \sigma^2\frac{s^2}{2\sigma^2 n} + \text{higher degree terms in } s$$

It turns out that these higher degree terms can be written as

$$\frac{\theta_n s^2}{2n\sigma^2}$$

where $\theta_n \to 0$ as $n \to \infty$. Consequently,

$$M_{Z_n}(s) = \left(1 + \frac{s^2}{2n} + \frac{\theta_n s^2}{2n\sigma^2}\right)^n$$

$$= \left(1 + \frac{\frac{s^2}{2} + \frac{\theta_n}{2} \cdot \frac{s^2}{\sigma^2}}{n}\right)^n$$

Identifying $\frac{\theta_n}{2} \cdot \frac{s^2}{\sigma^2}$ with c_n, and recalling that $\lim_{n\to\infty} \left(1 + \frac{a + c_n}{n}\right)^n = e^a$ if $\lim_{n\to\infty} c_n = 0$, we get

$$\lim_{n\to\infty} M_{Z_n}(s) = e^{s^2/2}$$

But $e^{s^2/2}$ is the mgf of a random variable which is $N(0, 1)$. Hence, by the continuity theorem, $(S_n - n\mu)/(\sigma\sqrt{n}) \overset{d}{\to} Z$, where Z is $N(0, 1)$.

In Figure 2.1(a) we give an illustration of a discrete distribution function approaching a normal distribution function, and in Figure 2.1(b) an illustration in the case of a continuous distribution function.

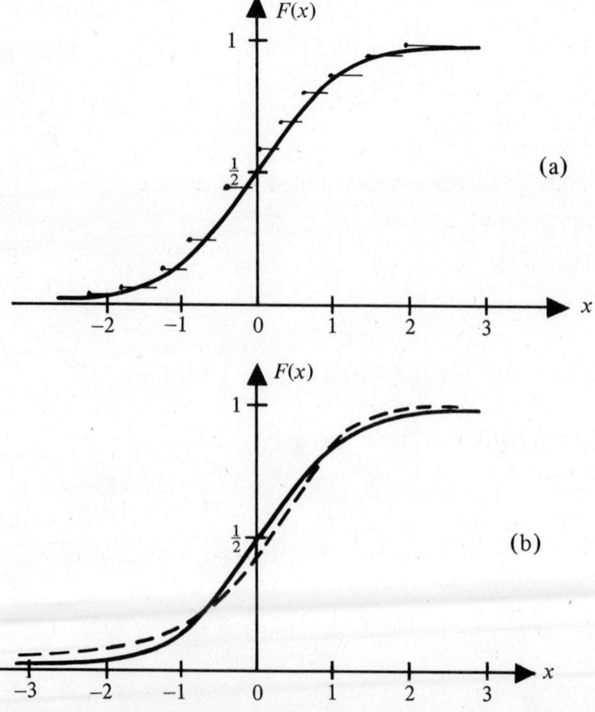

Figure 2.1 The solid curve in (a) and (b) represents the graph of Φ.

The central limit theorem equivalently implies that if a and b $(a < b)$ are any real numbers which do not depend on n, then

$$\lim_{n\to\infty}\left(a < \frac{S_n - n\mu}{\sigma\sqrt{n}} < b\right) = \Phi(b) - \Phi(a)$$

Caution. The statement "S_n is asymptotically $N(\alpha, \beta)$" is *not* meant to suggest that the mean and the variance of S_n converge to α and β. It only states, for instance, that if a and b $(a < b)$ are real numbers which do not depend on n, then

$$P\left(a < \frac{S_n - \alpha}{\beta} < b\right) \approx \Phi(b) - \Phi(a)$$

that is,

$$P(\alpha + a\beta < S_n < \alpha + b\beta) \approx \Phi(b) - \Phi(a)$$

if n is large.

Example 2.5. Suppose a fair die is rolled thirty times. Find the probability (approximately) that the total score will be between 90 and 120, both inclusive.

Solution. In Example 1.2, we found that this probability had to be at least 0.61. We shall now find the approximate value. Since $E(S_{30}) = 105$, and $\text{Var}(S_{30}) = \frac{175}{2}$,

$$\begin{aligned} P(90 \leqslant S_{30} \leqslant 120) &\approx \Phi\left(\frac{120 - 105}{\sqrt{\frac{175}{2}}}\right) - \Phi\left(\frac{90 - 105}{\sqrt{\frac{175}{2}}}\right) \\ &= \Phi(1.6) - \Phi(-1.6) \\ &= 0.8904 \end{aligned}$$

Example 2.6. Suppose a random sample of 48 is picked from a distribution which is uniform over the interval $[0, 2]$. Find the approximate probability that the sample mean is between 0.8 and 1.1.

Solution. In essence, we have 48 independent random variables $X_1, X_2, \ldots, X_{48}$, each of which is uniformly distributed over $[0, 2]$ and consequently has a common mean 1 and common variance $\frac{1}{3}$. Hence $E(\bar{X}_{48}) = 1$ and $\text{Var}(\bar{X}_{48}) = (\frac{1}{3})/48 = \frac{1}{144}$. Therefore,

$$\begin{aligned} P(0.8 < \bar{X}_{48} \leqslant 1.1) &\approx \Phi\left(\frac{1.1 - 1}{\frac{1}{12}}\right) - \Phi\left(\frac{0.8 - 1}{\frac{1}{12}}\right) \\ &= \Phi(1.2) - \Phi(-2.4) \\ &= 0.8767 \end{aligned}$$

Example 2.7. The diameter of a bolt produced by a machine is a random variable with an unknown distribution, but with a known standard deviation of 0.1 inches. Find how many bolts should be tested so that the probability will be approximately 0.95 that the sample mean will differ from the "true mean" by less than 0.02.

Solution. (The phrase "true mean" is a substitute for "the expected diameter.") We want to find n such that

$$P(|\bar{X}_n - \mu| < 0.02) \approx 0.95$$

that is,

$$P\left(\frac{|\bar{X}_n - \mu|}{\sigma/\sqrt{n}} < \frac{0.02}{\sigma/\sqrt{n}}\right) \approx 0.95$$

or, equivalently,

$$\Phi\left(\frac{0.02}{\sigma/\sqrt{n}}\right) - \Phi\left(-\frac{0.02}{\sigma/\sqrt{n}}\right) \approx 0.95$$

or

$$\Phi\left(\frac{0.02}{\sigma/\sqrt{n}}\right) \approx 0.975$$

Therefore, from the standard normal table, n satisfies

$$\frac{0.02}{\sigma/\sqrt{n}} \approx 1.96$$

Since $\sigma = 0.1$, this gives

$$n \approx \left(\frac{1.96 \times 0.1}{0.02}\right)^2 \approx 96$$

■

Comment. In general, the sample size n which will accomplish $P(|\bar{X}_n - \mu| < \delta) \approx 1 - \alpha$ is given by the formula

$$n \approx \left(\frac{z_{1-(\alpha/2)} \cdot \sigma}{\delta}\right)^2$$

where $z_{1-(\alpha/2)}$ is defined by $\Phi(z_{1-(\alpha/2)}) = 1 - (\alpha/2)$. (See Figure 2.2.)

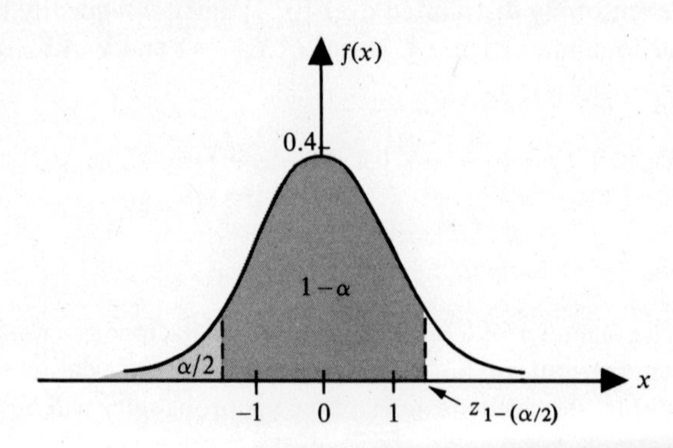

Figure 2.2

Normal approximation to the binomial

Classically, this result is known as the **De Moivre-Laplace theorem** and was proved much before the central limit theorem was established in its wider generality. It turns out to be a special case of the central limit theorem applied to a sequence of independent, identical Bernoulli random variables with the probability of success p.

Let $X_1, X_2, \ldots$ represent independent Bernoulli random variables with parameter p. Then $S_n = \sum_{i=1}^{n} X_i$ has the binomial distribution $B(n, p)$. Since $E(S_n) = np$ and $\text{Var}(S_n) = np(1-p)$, applying the central limit theorem, we see that S_n is asymptotically $N(np, np(1-p))$ and

$$\lim_{n\to\infty} P\left(\frac{S_n - np}{\sqrt{np(1-p)}} \leqslant t\right) = \Phi(t)$$

Thus, for any real number b, if n is large,

$$\begin{aligned} P(S_n \leqslant b) &= P\left(\frac{S_n - np}{\sqrt{np(1-p)}} \leqslant \frac{b - np}{\sqrt{np(1-p)}}\right) \\ &\approx \Phi\left(\frac{b - np}{\sqrt{np(1-p)}}\right) \\ &= \frac{1}{\sqrt{2\pi}} \int_{-\infty}^{(b-np)/\sqrt{np(1-p)}} e^{-u^2/2}\, du \end{aligned}$$

If b is a positive integer, it is best to approximate $P(S_n \leqslant b)$ by $\Phi\left(\frac{b + \frac{1}{2} - np}{\sqrt{np(1-p)}}\right)$ rather than by $\Phi\left(\frac{b - np}{\sqrt{np(1-p)}}\right)$. This is often referred to as a *continuity correction.* A heuristic rationale can be provided as follows:

Since S_n is $B(n, p)$, the exact value of $P(S_n = i)$ is given as

$$P(S_n = i) = \binom{n}{i} p^i (1-p)^{n-i} = \binom{n}{i} p^i (1-p)^{n-i} \cdot 1$$

In other words, $P(S_n = i)$ is *precisely* the area of the shaded rectangle in Figure 2.3(a) whose height is $\binom{n}{i} p^i (1-p)^{n-i}$ and whose base, from $i - \frac{1}{2}$ to $i + \frac{1}{2}$, is of length 1.

Under the normal approximation, this area of the rectangle representing $P(S_n = i)$ is approximated as the shaded region shown in Figure 2.3(b).

It is for this reason that we take $P(S_n = i)$ as approximately equal to the area under the normal distribution $N(np, np(1-p))$ from $i - \frac{1}{2}$ to $i + \frac{1}{2}$, and, as a result, $P(S_n \leqslant b)$ as the area to the left of $b + \frac{1}{2}$ (where b is a positive integer).

The reader will also be able to easily see that, if r and s are positive integers $(0 \leqslant r < s \leqslant n)$, then

$$P(r \leqslant S_n \leqslant s) \approx \Phi\left(\frac{s + \frac{1}{2} - np}{\sqrt{np(1-p)}}\right) - \Phi\left(\frac{r - \frac{1}{2} - np}{\sqrt{np(1-p)}}\right)$$

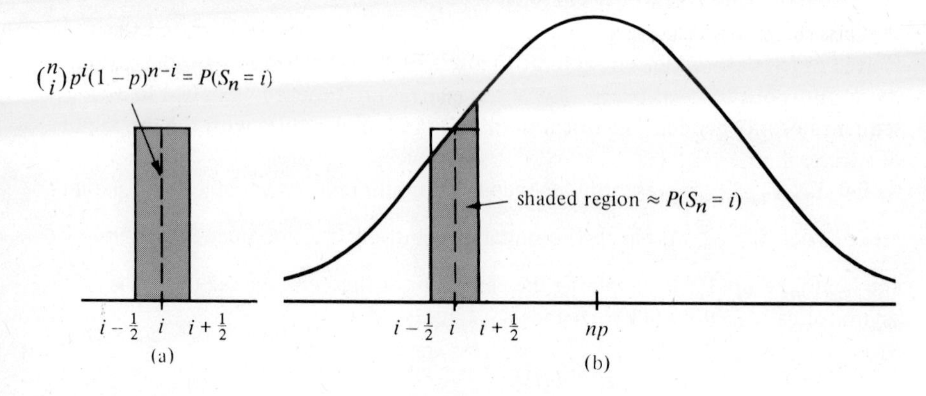

Figure 2.3

$$P(r < S_n < s) \approx \Phi\left(\frac{s - \frac{1}{2} - np}{\sqrt{np(1-p)}}\right) - \Phi\left(\frac{r + \frac{1}{2} - np}{\sqrt{np(1-p)}}\right)$$

and so on. (See Figure 2.4 for $P(r \leqslant S_n \leqslant s)$.)

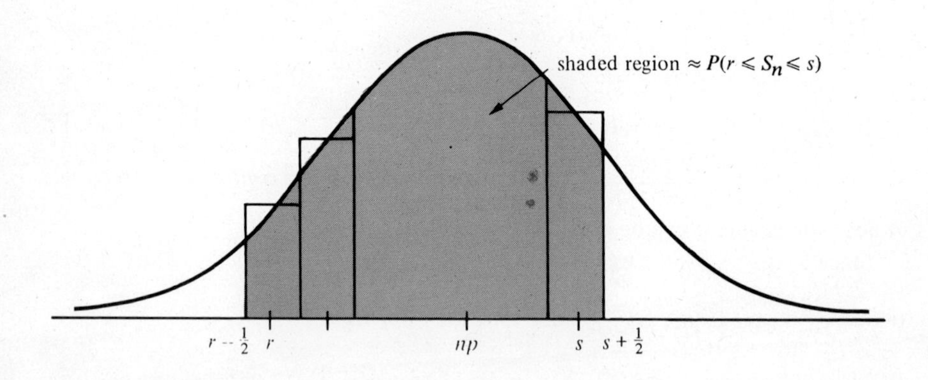

Figure 2.4

It is obvious that the continuity correction will not be important if $\sqrt{np(1-p)}$ is fairly large, since in that case $(\frac{1}{2})/\sqrt{np(1-p)}$ will be negligible.

Comment. We have shown earlier that, if n is large, the binomial probabilities can be approximated by the Poisson. We have now shown that the binomial probabilities can be approximated by the normal distribution. To avoid confusion, we clarify this with the following remark: The Poisson distribution provides a good approximation when n is large and p is small, whereas the normal distribution provides a good approximation when n is large and p is in the neighborhood of $\frac{1}{2}$.

Example 2.8. If X is $B(20, 0.3)$, find the following probabilities:

(*a*) $P(X = 7)$

(*b*) $P(7 < X \leqslant 10)$

Solution

(*a*) The exact answer to $P(X = 7)$ is $\binom{20}{7}(0.3)^7(0.7)^{13}$ which, from standard statistical tables for binomial probabilities, is 0.1643. Let us see how this compares with the approximate answer, where we approximate $P(X = 7)$ by the area from 6.5 to 7.5 under the normal curve with mean $20(0.3) = 6$ and variance $20(0.3)(0.7) = 4.2$. We have, therefore,

$$P(X = 7) \approx \Phi\left(\frac{7.5 - 6.0}{\sqrt{4.2}}\right) - \Phi\left(\frac{6.5 - 6.0}{\sqrt{4.2}}\right)$$

$$= \Phi(0.73) - \Phi(0.24)$$

$$= 0.1725$$

Thus, in this case, even when $n = 20$, the approximation is close.

(*b*) The exact probability, $P(7 < X \leqslant 10)$, is given as

$$P(7 < X \leqslant 10) = \sum_{r=8}^{10} \binom{20}{r} (0.3)^r (0.7)^{20-r}$$

which, from the tables of binomial probabilities, can be shown to be equal to 0.2106.

Using De Moivre's theorem, we can approximate this as the area under the normal curve $N(6, 4.2)$ from 7.5 to 10.5. Hence

$$P(7 < X \leqslant 10) \approx \Phi\left(\frac{10.5 - 6.0}{\sqrt{4.2}}\right) - \Phi\left(\frac{7.5 - 6.0}{\sqrt{4.2}}\right)$$

$$= \Phi(2.2) - \Phi(0.73)$$

$$= 0.2188$$

Again, the approximation is close.

Example 2.9. (*Normal approximation to the Poisson*) The number of phone calls arriving at an exchange during a given period of the day has a Poisson distribution with the mean rate of 100 calls. Find the approximate probability that, during one month (30 days), the total number of calls during the given period of the day will be between 3200 and 2900 (both inclusive).

Solution. From the reproductive property of the Poisson distribution discussed in Chapter 12, we know that if $X_1, X_2, \ldots, X_n$ are independent Poisson random variables, each with parameter λ, then the exact distribution of $Y = \sum_{i=1}^{n} X_i$ is Poisson with parameter $n\lambda$. Therefore, since $n = 30$ and $\lambda = 100$, the exact probability is given by

$$P(2900 \leqslant Y \leqslant 3200) = \sum_{r=2900}^{3200} e^{-3000} \frac{(3000)^r}{r!}$$

(Have fun computing this!)

Employing the central limit theorem, since $E(Y) = n\lambda$ and $\text{Var}(Y) = n\lambda$, we have for any nonnegative integer b,

$$P(Y \leqslant b) \approx \Phi\left(\frac{b + \frac{1}{2} - n\lambda}{\sqrt{n\lambda}}\right)$$

Notice that, as in the binomial case, we are applying the correction and writing $b + \frac{1}{2}$ rather than b. If n is large, this correction is not necessary, because $(\frac{1}{2})/\sqrt{n\lambda}$ will be negligible. It follows that, if a and b are any nonnegative integers, then

$$P(a < Y \leqslant b) \approx \Phi\left(\frac{b + \frac{1}{2} - n\lambda}{\sqrt{n\lambda}}\right) - \Phi\left(\frac{a + \frac{1}{2} - n\lambda}{\sqrt{n\lambda}}\right)$$

In our example, $n\lambda = 3000$, and consequently

$$\begin{aligned} P(2900 \leqslant Y \leqslant 3200) &\approx \Phi\left(\frac{3200 - 3000}{\sqrt{3000}}\right) - \Phi\left(\frac{2900 - 3000}{\sqrt{3000}}\right) \\ &= \Phi(3.65) - \Phi(-1.82) \\ &= 0.9655 \end{aligned}$$

■

Next, we give some miscellaneous examples bearing on the central limit theorem.

Example 2.10. A bowl contains ten chips—five marked 2, three marked 4, and two marked 8. If 100 chips are picked at random with replacement, find the approximate probability that the product of the numbers observed will be—

(*a*) between 2^{160} and 2^{200}
(*b*) greater than 2^{200}.

Solution

(*a*) Letting X_i represent the number observed on the ith draw, X_i takes values 2, 4, and 8 with respective probabilities 0.5, 0.3, and 0.2. Now notice that for any integer i we can write $X_i = 2^{Z_i}$, where Z_i assumes values 1, 2, 3 with respective probabilities 0.5, 0.3, 0.2. We want to find $P(2^{160} < Y < 2^{200})$ where $Y = X_1X_2 \ldots X_{100}$. We see right away that

$$2^{160} < Y < 2^{200} \quad \text{if and only if} \quad 2^{160} < 2^{\sum_{i=1}^{100} Z_i} < 2^{200}$$

that is, if and only if $160 < \sum_{i=1}^{100} Z_i < 200$. Hence

$$P(2^{160} < Y < 2^{200}) = P\left(160 < \sum_{i=1}^{100} Z_i < 200\right)$$

We can now find the approximate probability by applying the central limit theorem to the sequence Z_i, $i \geqslant 1$. It can easily be seen that $E(Z_i) = 1.7$ and $\text{Var}(Z_i) = 0.61$, so that $E\left(\sum_{i=1}^{100} Z_i\right) = 170$ and $\sqrt{\text{Var}\left(\sum_{i=1}^{100} Z_i\right)} = \sqrt{61} = 7.81$.

Therefore,

$$P(2^{160} < Y < 2^{200}) \approx \Phi\left(\frac{200-170}{7.81}\right) - \Phi\left(\frac{160-170}{7.81}\right)$$

$$= \Phi(3.84) - \Phi(-1.28)$$

$$= 0.8997$$

(*b*) It follows from part (*a*) that

$$P(Y > 2^{200}) \approx 1 - \Phi(3.84)$$

which is close to zero.

Example 2.11. A random point $(X_1, X_2, \ldots, X_{100})$ is picked in 100-dimensional space. Assuming that the random variables $X_1, X_2, \ldots, X_{100}$ are independent, each with the uniform distribution over the interval $[-1, 1]$, find the approximate probability that the square of the distance of the point from the origin is less than 40.

Solution. We want to find $P\left(\sum_{i=1}^{100} X_i^2 \leqslant 40\right)$. If we let $X_i^2 = Z_i$, we want to find $P\left(\sum_{i=1}^{100} Z_i \leqslant 40\right)$. Now

$$E(Z_i) = E(X_i^2) = \int_{-1}^{1} x^2 \cdot \frac{1}{2}\, dx = \frac{1}{3}$$

$$E(Z_i^2) = E(X_i^4) = \int_{-1}^{1} x^4 \cdot \frac{1}{2}\, dx = \frac{1}{5}$$

and consequently

$$\text{Var}(Z_i) = \frac{4}{45}$$

Therefore,

$$E\left(\sum_{i=1}^{100} Z_i\right) = \frac{100}{3}$$

and

$$\sqrt{\text{Var}\left(\sum_{i=1}^{100} Z_i\right)} = \sqrt{\frac{400}{45}} = \frac{20}{6.71}$$

Applying the central limit theorem to the sequence Z_i, $i \geqslant 1$, we get

$$P\left(\sum_{i=1}^{100} X_i^2 \leqslant 40\right) \approx \Phi\left(\frac{40-(100/3)}{20/(6.71)}\right)$$

$$= \Phi(2.236) = 0.9873$$

Example 2.12. (*Monte Carlo simulation*) In order to compute the integral $I = \int_0^1 x^3\, dx$, the following simulation method is adopted: n numbers $X_1, X_2, \ldots, X_n$ are picked at random in the interval $[0, 1]$, and the approximate value of the integral is given as $I_n = \frac{1}{n} \sum_{i=1}^{n} X_i^3$. Based on 1000 numbers, find the approximate probability that this computed value will not differ from the true value of the integral by more than 0.01.

Solution. Notice that $I = \int_0^1 x^3 dx$ can be regarded as $E(X^3)$ where X is uniformly distributed over [0, 1]. Let

$$I_n = \frac{1}{n} \sum_{i=1}^{n} X_i^3 = \frac{1}{n} \sum_{i=1}^{n} Z_i, \quad \text{where} \quad Z_i = X_i^3$$

Now,

$$E(Z_i) = E(X_i^3) = \int_0^1 x^3 dx = \frac{1}{4}$$

and

$$E(Z_i^2) = E(X_i^6) = \int_0^1 x^6 dx = \frac{1}{7}$$

Hence

$$\text{Var}(Z_i) = \frac{1}{7} - \left(\frac{1}{4}\right)^2 = \frac{9}{112}$$

Therefore,

$$E(I_n) = \frac{1}{4} = I \quad \text{and} \quad \text{Var}(I_n) = \frac{9}{112n}$$

Since $n = 1000$, we get $\sqrt{\text{Var}(I_n)} = 1/(111.56)$. Hence, applying the central limit theorem,

$$\begin{aligned} P(|I_n - I| < 0.01) &= P\left(\frac{|I_n - I|}{\sqrt{\text{Var}(I_n)}} < \frac{0.01}{\sqrt{\text{Var}(I_n)}}\right) \\ &= P\left(\frac{|I_n - I|}{\sqrt{\text{Var}(I_n)}} < 1.12\right) \\ &\approx \Phi(1.12) - \Phi(-1.12) \\ &= 0.7372 \end{aligned}$$

EXERCISES–SECTION 2

1. Discuss the convergence in distribution for the following sequences of random variables:
   (a) $P(X_n = 1 + \frac{1}{n}) = \frac{1}{3}, \quad P(X_n = \frac{1}{n}) = \frac{2}{3}$
   (b) X_n is uniformly distributed over the interval $[-1/n, 1/n)$.
2. The probability that a person contracts a rare disease is 0.01. If there are 500 people in a town, use the Poisson approximation to find the probability that–
   (a) 3 people contract the disease
   (b) at least 1 person contracts the disease.

3. The exact values of the binomial probabilities $b(k; n, p)$ for $n = 30$ and $p = 0.05$ are given below for different values of k.

k	0	1	2	3	4
$b(k; 30, 0.05)$	0.2146	0.3389	0.2586	0.1271	0.0451

Compare these values with those obtained using the Poisson approximation.

4. Suppose X has the probability function given by $p(1) = 0.2, p(3) = 0.3, p(5) = 0.2, p(8) = 0.3$. Let Y denote the sum of 25 independent observations. Find approximately $P(80 < Y < 120)$.

5. Suppose X has the binomial distribution $B(25, 0.4)$. Use the normal approximation to the binomial to compute the following probabilities:
   (a) $P(X = 10)$ (b) $P(8 < X \leqslant 12)$
   (c) $P(8 \leqslant X \leqslant 12)$ (d) $P(X \geqslant 12)$

6. Suppose X is $B(n, 0.4)$ and we want to find n such that $P(0.35 < X/n < 0.45) \geqslant 0.90$.
   (a) Find how large n should be using Chebyshev's inequality.
   (b) Find the smallest n using the central limit theorem.

7. If $0 < p < 1$, find

$$\lim_{n\to\infty} \sum_{np-2\sqrt{np(1-p)} < r \leqslant np+2\sqrt{np(1-p)}} \binom{n}{r} p^r (1-p)^{n-r}$$

8. Suppose X is $B(28, 0.3)$, Y is $B(10, 0.3)$, and Z is $B(12, 0.3)$. If X, Y, and Z are independent, find approximately $P(X + Y + Z \geqslant 18)$.

9. A new vaccine was tested on 100 persons to determine its effectiveness. If the claim of the drug company is that a random person who is injected with the vaccine will develop immunity with probability 0.8, find the probability that less than 74 people will develop immunity.

10. The probability that a worker working in a factory will meet with an accident during any month is 0.02. If there are 200 workers, find the approximate probability that more than 2 workers will meet with accidents during a month.

11. Suppose the probability that a worker meets with an accident during a one-year period is 0.4. If there are 200 workers in a factory, find the approximate probability that more than 75 workers will have an accident in one year.

12. An electronic component is known to have life length T (measured in hours) which is exponentially distributed with the D.F. given as

$$F_T(t) = \begin{cases} 0 & \text{if } t \leqslant 0 \\ 1 - e^{-2t}, & \text{if } t > 0 \end{cases}$$

If a machine uses 100 such components, which function independently, find the probability that at the end of one hour–
   (a) at least 20 components will be functioning
   (b) between 10 and 15 components (both inclusive) will be functioning.

13. The number of particles emitted by a radioactive sample during a one-hour period is a Poisson random variable with parameter $\lambda = 9$. If 100 such samples are observed during a one-hour period, find the probability that at least 850 particles are emitted.

14. One hundred billiards players are participating in a tournament and each player shoots until he misses a shot. Find the approximate probability that together they will attempt a total of between 800 and 1100 shots, assuming that the probability that a player makes any shot successfully is equal to 0.9.
15. It is known that a student's score on a test is a random variable with unknown mean μ and standard deviation 15 points. If 100 students are taking the test, find the probability that the mean score of these students will differ from their "true mean score" by at most 3 points.
16. Find the approximate probability that a random sample of 24 observations, $X_1, X_2, \ldots, X_{24}$, each having the pdf

$$f(x) = \begin{cases} 4x^3, & 0 < x < 1 \\ 0, & \text{elsewhere} \end{cases}$$

will yield a mean which lies between 0.7 and 0.85.
17. The amount of time (in minutes) that a TV channel devotes to commercials during any half-hour program is a uniform random variable over the interval [5, 8]. What is the probability that a person who watches television from 3 p.m. to 12 midnight will be exposed to–

(a) between 115 and 125 minutes of commercials
(b) over 110 minutes of commercials.

18. A computer is programmed in such a way that the error involved in carrying out any computation is a uniform random variable over the interval $[0, 10^{-8}]$. If 10,000 independent computations are made, find the approximate probability that the total error will be less than $\frac{1}{2}(10^{-4} + 10^{-6})$.
19. Suppose $X_1, X_2, \ldots, X_{100}$ represents a sample of a hundred independent observations from a distribution which is uniform over the open interval (0, 1). Find the probability that their product lies between e^{-110} and e^{-75}.
20. Suppose X has a chi-square distribution with n degrees of freedom. It is claimed that if n is large, then the probability $P(X \leqslant u)$ can be obtained approximately as $\Phi\left(\frac{u-n}{\sqrt{2n}}\right)$. Justify this claim.
21. Let us suppose the velocity (measured in centimeters per second) of a particle with unit mass has a normal distribution with mean 0 and standard deviation 2. Find the probability that the total kinetic energy of fifty such particles will not exceed 60 ergs. (Kinetic emergy of a particle of mass m with velocity v cm/sec is $(mv^2)/2$ ergs.)
22. In order to calculate the integral

$$I = \int_0^{\pi/2} \sin x \, dx$$

the following technique is adopted: Numbers are drawn repeatedly and at random in the interval $[0, \pi/2]$. If X_i represents the number picked on the ith draw, then compute

$$I_n = \frac{\pi}{2n} \sum_{i=1}^{n} \sin X_i$$

and use this as an approximation of I. If 1000 numbers are picked, find the probability that the approximation will differ from the true integral by no more than 0.02.

References

The following is a list of texts, by no means exhaustive, from which the reader might benefit.

[1] Ash, R. *Basic Probability Theory.* New York: Wiley, 1970

[2] Breiman, L. *Probability and Stochastic Processes.* Boston: Houghton Mifflin, 1969

[3] Dwass, M. *Probability.* New York: Benjamin, 1970

[4] Feller, W. *An Introduction to Probability Theory and Its Applications.* 3rd ed. New York: Wiley, 1967

[5] Gangooli, R. & Ylvisaker, D. *Discrete Probability.* New York: Harcourt Brace and World, 1967

[6] Goldberg, S. *Probability: An Introduction.* Englewood Cliffs, N.J.: Prentice-Hall, 1960

[7] Parzen, E. *Modern Probability Theory and Its Applications.* New York: Wiley, 1960

[8] Ross, S. *Introduction to Probability Models.* New York: Academic Press, 1972

[9] Rozanov, Y.A. *Introductory Probability Theory.* Englewood Cliffs, N.J.: Prentice-Hall, 1969

[10] Yeh, R.Z. *Modern Probability Theory.* New York: Harper & Row, 1973

Solutions to Selected Problems

CHAPTER 1

Section 1

1. (a) $\{1, 2, 3, 4, 5, 6, 9, 10, 11, 12\}$ (b) $\{3, 6, 10\}$ (c) $\{1, 4, 11\}$
   (d) $\{2, 5, 9, 12\}$ (e) $\{2, 5, 7, 8, 9, 12\}$
3. Positive integers which are—
   (a) even or odd; that is, all the positive integers
   (b) even or multiples of 3, and which are odd; that is, multiples of 3 which are odd
   (c) even or multiples of 3
   (d) even, or odd and multiples of 3
   (e) even and odd; that is, the empty set
   (f) odd and multiples of 3.
5. (a) True (b) False (c) True (d) False
9. (a) $\{x \mid 0 \leqslant x \leqslant 1 \text{ or } 3 < x \leqslant 4\}$ (b) $\{x \mid 0 \leqslant x < 2\}$ (c) $\{x \mid 1 < x \leqslant 4\}$
   (d) $\{x \mid 0 \leqslant x < 2 \text{ or } 3 < x \leqslant 4\}$ (e) $\{x \mid 0 \leqslant x \leqslant 1\}$ (f) $\{x \mid 0 \leqslant x \leqslant 1\}$
   (g) $\{x \mid 0 \leqslant x < 2 \text{ or } 3 < x \leqslant 4\}$
11. All statements are true.
13. (a) $\{(3, 3)\}$ (b) $\{(1, 3), (2, 3), (3, 3)\}$ (c) $\{(1, 3), (2, 3), (3, 1), (3, 2), (3, 3)\}$
15. The sequence is contracting; $\varnothing$

Section 2

3. $\{d, gd, ggd, gggd, ggggd\}$, where d means defective and g means nondefective.
5. (a) $\{(x, y) \mid x, y = 1, 2, 3, 4\}$ (b) $\{(1, 2), (2, 4), (2, 1), (4, 2)\}$
   (c) $\{(1, 1), (2, 2), (3, 3), (4, 4)\}$

7. (a) $\{b_1b_2, b_1w_1, b_1w_2, b_1w_3, b_2w_1, b_2w_2, b_2w_3, w_1w_2, w_1w_3, w_2w_3, b_1b_2w_1w_2, b_1b_2w_1w_3, b_1w_1w_2w_3, b_2w_1w_2w_3, b_1b_2w_2w_3\}$
   (b) $\{b_1w_1, b_1w_2, b_1w_3, b_2w_1, b_2w_2, b_2w_3, b_1b_2w_1w_2, b_1b_2w_1w_3, b_1b_2w_2w_3\}$
9. (a) $S = \{(x, y) \mid 0 \leqslant x \leqslant 2,\ 9 \leqslant y \leqslant 10\}$
   (b) (i) $\{(x, y) \in S \mid y \leqslant 9.5\}$ (ii) $\{(x, y) \in S \mid y - x < 9.5\}$
11. The student is–
   (a) either a mathematics major, or a history minor, or a senior
   (b) a mathematics major, a history minor, and a senior
   (c) not *both* a mathematics major and a history minor; that is, either not a mathematics major, or not a history minor
   (d) neither a mathematics major nor a senior
   (e) either not a mathematics major, or not a history minor, or not a senior
   (f) not a mathematics major, not a history minor, and not a senior
   (g) a history minor and a senior, but not a mathematics major.
13. (a) At least one fish (bass, trout, salmon) is caught.
   (b) At least one fish of each kind is caught.
   (c) No bass is caught, but at least one fish of each of the other two kinds is caught
   (d) No bass is caught, and at least one fish of either of the other two kinds is caught
15. No; it simply means $A' \subset B$.

Section 3

3. $\{\emptyset, A, B, A', B', A \cup B, A \cup B', B \cup A', A' \cup B', AB, AB', BA', A'B', S\}$

CHAPTER 2

Section 1

1. (a) 0.6 (b) 0 (c) 0.7 (d) 0.4 (e) 1 (f) 0.3
3. (a) 0.6 (b) 0.1
7. (a) 0.50 (b) 0.89 (c) 0.11
10. $k = 0$, 0.0256; $k = 1$, 0.1536; $k = 2$, 0.3456; $k = 3$, 0.3456; $k = 4$, 0.1296
11. (a) $k = 2$ (b) (i) $\frac{1}{4}$ (ii) $\frac{3}{4}$

Section 2

1. (a) $\frac{1}{2}$ (b) $\frac{2}{3}$
3. (a) $\frac{16}{41}$ (b) $\frac{28}{41}$ (c) $\frac{29}{41}$
5. (a) $\frac{1}{26}$ (b) $\frac{5}{26}$ (c) $\frac{21}{26}$ (d) $\frac{6}{26}$ (e) $\frac{3}{26}$

Section 3

1. (a) 26 (b) 12
5. 746496
7. $5 \times 3 \times 3 = 45$
9. (a) $\frac{3}{7}$ (b) $\frac{2}{7}$
11. $\frac{15}{28}$

13. (a) $\frac{1}{11}$ (b) $\frac{17}{33}$

15. (a) $\binom{4}{1}\binom{48}{3}/\binom{52}{4}$ (b) $\binom{12}{1}\binom{40}{3}/\binom{52}{4}$ (c) $\binom{26}{4}/\binom{52}{4}$ (d) $13^4/\binom{52}{4}$
(e) $[\binom{4}{2}\binom{48}{2} + \binom{4}{3}\binom{48}{1} + \binom{4}{4}\binom{48}{0}]/\binom{52}{4}$

17. $\frac{1}{4}$

19. (a) $\binom{4}{3}\binom{4}{2}/\binom{52}{5}$ (b) $\binom{4}{3}\binom{48}{2}/\binom{52}{5}$ (c) $\binom{4}{3}\binom{4}{1}\binom{44}{1}/\binom{52}{5}$
(d) $[\binom{4}{3}\binom{4}{1}\binom{44}{1} + \binom{4}{4}\binom{4}{1}]/\binom{52}{5}$ (e) $[\binom{4}{3}\binom{48}{2} + \binom{4}{4}\binom{48}{1}]/\binom{52}{5}$

21. $\frac{55}{266}$

23. (a) $\dfrac{1}{n(n-1)(n-2)}$ (b) $\dfrac{24}{n(n-1)(n-2)}$

25. $3[\binom{4}{3}\binom{48}{5} - \binom{4}{3}\binom{4}{3}\binom{44}{2}]/\binom{52}{8}$

27. (a) $\dfrac{1}{2^{10}}$ (b) $1 - \dfrac{1}{2^{10}}$ (c) $\dfrac{\binom{10}{r}}{2^{10}}$

29. $\frac{1}{4}$

31. $\binom{20}{10}\binom{20}{10}/\binom{40}{20}$

32. $\dfrac{2(n!n!)}{(2n)!}$

33. $\dfrac{[\binom{4}{1}\binom{12}{3}][\binom{3}{1}\binom{9}{3}][\binom{2}{1}\binom{6}{3}][\binom{1}{1}\binom{3}{3}]}{\binom{16}{4}\binom{12}{4}\binom{8}{4}\binom{4}{4}}$

CHAPTER 3

Section 1

1. (a) $\frac{2}{5}$ (b) $\frac{4}{19}$ (c) $\frac{45}{67}$ (d) $\frac{3}{5}$
3. (a) 0.8 (b) 0.5 (c) 0.4
5. $\frac{3}{8}$
6. (a) False (b) True (c) False (d) True
11. $\frac{1}{4}$
13. $\frac{1}{3}$
15. 0.84
17. (a) $\dbinom{4}{i}\dbinom{4}{j}\dbinom{44}{13-i-j}\Big/\dbinom{52}{13}$ (b) $\dbinom{4}{j}\dbinom{44}{13-i-j}\Big/\dbinom{48}{13-i}$
19. $1/\binom{n}{k}$
20. $\dfrac{(M-W)_{(i-1)}}{(M)_{(i-1)}} \cdot \dfrac{W}{M-(i-1)}, \quad i = 1, 2, \ldots, M - W + 1$

Section 2

1. (a) $\frac{7}{24}$ (b) $P(B_1 \mid A) = \frac{3}{7}$; $P(B_2 \mid A) = \frac{2}{7}$; $P(B_3 \mid A) = \frac{2}{7}$
3. 0.65
5. (a) $\dfrac{1}{n}\sum_{i=1}^{n}[\binom{n}{i}\binom{n}{i}/\binom{2n}{2i}]$ (b) $\dfrac{\binom{n}{k}\binom{n}{k}}{\binom{2n}{2k}}\Big/\sum_{i=1}^{n}\dfrac{\binom{n}{i}\binom{n}{i}}{\binom{2n}{2i}}$
7. $\frac{17}{32}$
9. $\displaystyle\sum_{k=0}^{5}\frac{\binom{9}{k}\binom{7}{5-k}}{\binom{16}{5}} \cdot \frac{6+k}{14}$

10. $\dfrac{(n^2-n+1)}{n^2(n-1)}$

11. (a) 0.52 (b) $\frac{12}{13}$

13. $\displaystyle\sum_{k=0}^{r} \frac{\binom{D}{k}\binom{M-D}{r-k}}{\binom{M}{r}} \cdot \frac{D-k}{M-r}$

14. $\displaystyle\sum_{k=0}^{3} \frac{\binom{10}{k}\binom{8}{3-k}}{\binom{18}{3}} \cdot \frac{\binom{10-k}{3}}{\binom{18}{3}}$

Section 3

1. 1
5. (a) 0.36 (b) 0.32 (c) 0.24
8. No; $P(AB \mid F) = P(A \mid BF) \cdot P(B \mid F)$
9. Pairwise independent, not independent
13. (a) $\frac{7}{15}$ (b) $\frac{2}{15}$ (c) $\frac{1}{30}$ (d) $\frac{11}{15}$
15. (a) 0.16 (b) 0.18 (c) 0.66
17. (a) 0.216 (b) 0.027 (c) 0.189 (d) 0.271 (e) 0.30
18. (a) $p^2(1+2p-3p^2+p^3)$ (b) $p^3(2-p)^2$
19. (a) 0.42 (b) 0.1218 (c) $(0.21)[0.3^{r-2}+0.7^{r-2}]$
21. $\dfrac{1}{2}\left(\dfrac{r_1}{r_1+s_1}+\dfrac{r_2}{r_2+s_2}\right)$
23. $\frac{1}{17}$
24. $\dfrac{(1-a)b^N}{1-ab^N}$
25. $\frac{4}{7}$

CHAPTER 4

Section 1

1. (a) 0, 1, 2, 3, 4, 5 (b) 1, 2, 3, 4, 5, 6 (c) 1, 2, 3, 4, 5, 6

3.

	$P^X((-\infty, 3])$	$P^X((1, 3))$	$P^X([2, \infty))$
(a)	$\frac{2}{7}$	$\frac{2}{21}$	$\frac{20}{21}$
(b)	$\frac{10}{21}$	$\frac{1}{7}$	$\frac{6}{7}$
(c)	$\frac{3}{7}$	0	$\frac{4}{7}$

5. (a) $\frac{1}{4}$ (b) $\frac{1}{8}$ (c) $\frac{7}{8}$ (d) $\frac{1}{4}$

Section 2

1.
$$F(x) = \begin{cases} 0, & x < -2 \\ \frac{1}{4}, & -2 \leqslant x < 3 \\ \frac{3}{4}, & 3 \leqslant x < 4 \\ 1, & x \geqslant 4 \end{cases}$$

3. (a) No (b) Yes (c) Yes (d) No (e) Yes
5. (a) $\frac{7}{12}$ (b) $\frac{1}{4}$ (c) $\frac{2}{3}$ (d) $\frac{1}{4}$ (e) $\frac{1}{3}$ (f) $\frac{7}{12}$ (g) $\frac{11}{12}$ (h) $\frac{2}{3}$
   (i) $\frac{1}{12}$ (j) $\frac{1}{3}$
9. (a) $m = 1$ (unique) (b) Any number in $[-1, 0)$ (c) $m = 4.5$ (unique)
10. (a) $e^{-1} - e^{-4} = 0.3496$ (b) $(e^{-2.25} - e^{-4})/(e^{-1} - e^{-4}) = 0.2491$

Section 3

1.
$$p_X(i) = \begin{cases} \frac{7}{50}, & i = 0, 3, 4, 5, 6 \\ \frac{3}{20}, & i = 1, 2 \end{cases}$$

3. (a) 0.3 (b) 0.3 (c) 0.5 (d) 0.6
5. $p_X(k) = \frac{1}{5}$, $k = 0, 1, 2, 3, 4$
7. (a) 0.1 (b) $\frac{1}{7}$ (c) e^{-2}
8. (a)
$$F(x) = \begin{cases} 0, & x < 5 \\ 1, & x \geqslant 5 \end{cases}$$
   (c)
$$F(x) = \begin{cases} 0, & x < 3 \\ \frac{1}{10}, & 3 \leqslant x < 4 \\ \frac{3}{10}, & 4 \leqslant x < 5 \\ \frac{6}{10}, & 5 \leqslant x < 6 \\ 1, & x \geqslant 6 \end{cases}$$
9. (a) $\frac{1}{420}$ (b) $\frac{1}{2}$ (c) $\frac{7}{15}$
10. $-\frac{1}{2} \leqslant x \leqslant \frac{1}{10}$
11. x takes values $-5, -1, 1.5, 3.5$ with respective probabilities $\frac{1}{12}, \frac{1}{12}, \frac{1}{6}, \frac{2}{3}$.
12. (a) 2 (b)
$$F(x) = \begin{cases} 0, & x < 1 \\ 1 - \dfrac{2}{[x] + 2}, & x \geqslant 1 \end{cases}$$
   (c) $2\left[\dfrac{1}{r+1} - \dfrac{1}{s+2}\right]$
13. $a = 2(3)^{-1/4}$, $r = \sqrt{2/\pi}$
15. (a) $\frac{3}{8}$ (b) $\frac{2}{5}$ (c) $\frac{3}{20}$
17. (a)
$$F(x) = \begin{cases} 0, & x < 0 \\ x^3/8, & 0 \leqslant x < 2 \\ 1, & x \geqslant 2 \end{cases}$$
   (b)
$$F(x) = \begin{cases} 0, & x < -1 \\ (1 - x^2)/5, & -1 \leqslant x < 0 \\ (1 + x^2)/5, & 0 \leqslant x < 2 \\ 1, & x \geqslant 2 \end{cases}$$
   (c)
$$F(x) = \begin{cases} 0, & x < 1 \\ 1 - \dfrac{1}{x}, & x \geqslant 1 \end{cases}$$
   (d)
$$F(x) = \begin{cases} 0, & x < -1 \\ (x+1)/4, & -1 \leqslant x < 1 \\ \frac{1}{2}, & 1 \leqslant x < 2 \\ (x+1)/6, & 2 \leqslant x < 5 \\ 1, & x \geqslant 5 \end{cases}$$
18. (a)
$$f(x) = \begin{cases} \frac{1}{2}, & 0 < x < 1 \\ 1/(2x^2), & x \geqslant 1 \\ 0, & \text{elsewhere} \end{cases}$$
   (c) $f(x) = 140x^3(1 - x)^3$, $0 < x < 1$; 0 elsewhere
   (e) $f(x) = -\ln x$, $0 < x < 1$; 0 elsewhere
19. $\dfrac{\sqrt{2} - 1}{\sqrt{2}}$
21. (a) $\frac{2}{9}$ (b) $\frac{2}{9}$ (c) $\frac{2}{3}$ (d) $\frac{4}{9}$ (e) 1 (f) $\frac{7}{9}$ (g) $\frac{12}{13}$ (h) $\frac{4}{7}$
23. $\frac{45}{128}$

CHAPTER 5

Section 1

1. $\binom{10}{6}(\frac{1}{3})^6(\frac{2}{3})^4$
3. (a) $1-(\frac{5}{6})^4$ (b) About 13 times
4. $p(0) = 0.336$; $p(1) = 0.452$; $p(2) = 0.188$; $p(3) = 0.024$
5. $p(k) = \binom{N}{k}\left(\frac{i}{N}\right)^k\left(1-\frac{i}{N}\right)^{N-k}, \quad k = 0, 1, \ldots, N$
11. $p(k) = \binom{n}{k}(0.88)^k(0.12)^{n-k}, \quad k = 0, 1, 2, \ldots, n$
13. 4; 0.2508
15. $p(k) = (0.1)(0.9)^k, \quad k = 0, 1, 2, \ldots$
17. $p(k) = \left(\frac{b}{b+w}\right)^{k-1}\left(\frac{w}{b+w}\right), \quad k = 1, 2, \ldots$
19. (a) $\binom{10}{6}(\frac{8}{27})^6(\frac{19}{27})^4$ (b) $(\frac{19}{27})^9(\frac{8}{27})$
20. (a) $\binom{n}{k}\left(\int_{-\infty}^{c} f(x)\,dx\right)^k\left(\int_{c}^{\infty} f(x)\,dx\right)^{n-k}, \quad k = 0, 1, \ldots, n$

    (b) $\left(\int_{c}^{\infty} f(x)\,dx\right)^{k-1}\left(\int_{-\infty}^{c} f(x)\,dx\right), \quad k = 1, 2, \ldots$
21. 0.3235
25. The most probable number of misprints is 4; the probability is 0.1917.

Section 2

1. (a) $f(x) = \frac{1}{8}$, $-3 \leqslant x \leqslant 5$; 0 elsewhere (b) $\frac{1}{4}$
3. (a) $F(x) = \begin{cases} 0, & x < 2 \\ (x-2)/7, & 2 \leqslant x < 9 \\ 1, & x \geqslant 9 \end{cases}$ (b) $\frac{4}{7}$ (c) 7.25
4. $b = 8$; $f(x) = \frac{1}{16}$, $-8 < x < 8$, 0 elsewhere
8. (a) -1.2 (b) 0.53 (c) 2.5 (d) -1.02
9. (a) 0.6915 (b) 0.5328 (c) 0.6915 (d) 0.1336 (e) 0.3753
11. 0.8664
13. (a) 4 (b) 9
15. 0.0606 (here $a = 40$, $b = 6$)
17. (a) 0.0228 (b) 0.7745
19. (a) 0.1587 (b) 0.0228 (c) 0.8413 (d) 0.9772
21. (a) $e^{-4.5} = 0.0111$ (b) $1-(1-e^{-0.9})^5 = 0.9264$

CHAPTER 6

1. Case 3, since $P(X=a) > 0 \Rightarrow P(Y=h(a)) \geqslant P(X=a) > 0$
5. (a) $p_Y(y) = \frac{(16+y)}{105}, \quad y \in \{-13, -7, -4, -1, 2, 5, 11\}$

    (b) $p_Z(1) = \frac{4}{35}$, $p_Z(2) = \frac{8}{35}$, $p_Z(5) = \frac{6}{35}$, $p_Z(10) = \frac{8}{35}$, $p_Z(26) = \frac{9}{35}$
10. $p_Y(i) = e^{-\lambda(i-1)}(1-e^{-\lambda})$, $i = 1, 2, \ldots$ (geometric distribution with parameter $1-e^{-\lambda}$)

11. $f_V(u) = \frac{3}{4\pi}$, $0 < u < \frac{4\pi}{3}$; 0 elsewhere
13. $f_Y(y) = 2ye^{-y^2}$, $y > 0$; 0 elsewhere
15. $f_Y(y) = \frac{1}{8\sqrt{y\pi}}$, $100\pi < y < 196\pi$; 0 elsewhere
19.
$$f_Y(y) = \begin{cases} \frac{1}{yb\sqrt{2\pi}} \exp\left\{-\frac{1}{2b^2}[(\ln y) - a]^2\right\}, & y > 0 \\ 0, & \text{elsewhere} \end{cases}$$
21. $f_Y(u) = \frac{1}{\sigma}\sqrt{2/\pi}\, e^{-u^2/(2\sigma^2)}$, $u \geqslant 0$; 0 elsewhere
23.
$$F_Y(y) = \begin{cases} 0, & y < 12 \\ 1 - e^{-(y-6)/6}, & y \geqslant 12 \end{cases}$$

CHAPTER 7

Section 1

1. (a) $\frac{14}{3}$ (b) $\frac{35}{3}$
3. (a) $E(X) = 1$; $\text{Var}(X) = 4.5$ (b) $E(X) = \frac{1}{7}$; $\text{Var}(X) = \frac{90}{49}$
(c) $E(X) = -\frac{4}{3}$; $\text{Var}(X) = \frac{11}{9}$ (d) $E(X) = 2$; $\text{Var}(X) = 2$
4. (a) $E(X) = \frac{3}{2}$; $\text{Var}(X) = \frac{3}{20}$ (c) $E(X) = 1$; $\text{Var}(X) = \frac{1}{6}$
(e) $E(X) = \frac{7}{4}$; $\text{Var}(X) = \frac{173}{48}$
5. $E(X) = 3.5$; $\text{Var}(X) = \frac{35}{12}$
7. $E(X) = \frac{N+1}{2}$; $\text{Var}(X) = \frac{N^2 - 1}{12}$
9. Expected number of matches is 1 for every n.
11. 3 13. 10.29 15. No 18. $\pi^2/6$
19. (a) 3 (b) 12 (c) 57 (d) 309
21. $E(Y) = 25$; $\text{Var}(Y) = 48$
23. $E(X^2) = 33 < [E(X)]^2$
25. $E(X^i)$ exists for $i = 1, 2, \ldots, n$; but $E(X^{n+1})$ does not exist.
27. $\frac{11}{7}$ 29. (a) 2.5 minutes (b) $\frac{19}{6}$ minutes

Section 2

1. 0.2019 3. (a) 200 (b) 20
4. $E(Y) = \frac{n(c_1 - 2c_2)}{3}$; $\text{Var}(Y) = \frac{2n(c_1 + c_2)^2}{9}$
7. (a) $n \int_{-\infty}^{c} f(x)\,dx$ (b) $\left[\int_{-\infty}^{c} f(x)\,dx\right]^{-1}$
12. (a) 0.9085 (b) There are two values, 3 and 4.
13. (a) $\frac{e^{-9.6}(9.6)^k}{k!}$, $k = 0, 1, 2, \ldots$ (b) 9.6
17. 0.5 19. (a) $\frac{1}{n+1}$ (b) $\frac{n^2}{(2n+1)(n+1)^2}$
21. $E(\ln X) = -1$; $\text{Var}(\ln X) = 1$
23. (a) 0.0228 (b) 0.3446 (c) 0.3087
25. (a) (i) $(0.9772)^{15}$ (ii) $1 - (0.9772)^{15}$ (iii) $\binom{15}{3}(0.0228)^3(0.9772)^{12}$
(b) $50(0.0228) = 1.14$
31. $E(X) = 0.8$; $\text{Var}(X) = 0.08$

CHAPTER 8

Section 1

1. $p(0,-3) = \frac{1}{8}, \quad p(1,-1) = \frac{3}{8}, \quad p(2,1) = \frac{3}{8}, \quad p(3,3) = \frac{1}{8}$
3. $f(x, y) = 2,\ 0 \leqslant y \leqslant x \leqslant 1$; 0 elsewhere. That is, the distribution is uniform over the triangle bounded by $x = y,\ y = 0,\ x = 1$.
5. $$p(x, y) = \frac{120!}{x!\,y!\,(120 - x - y)!}(0.6)^x(0.3)^y(0.1)^{120-x-y}$$
   where x, y are integers with $0 \leqslant x, y \leqslant 120$ and $x + y \leqslant 120$.
7. $p(X = i, Y = j) = \dfrac{1}{6i}, \quad 1 \leqslant j \leqslant i \leqslant 6,\ i, j$ integers
9. (a) $F(a_4, b_4) - F(a_4, b_1) - F(a_1, b_4) + F(a_1, b_1) - F(a_3, b_3) + F(a_3, b_2) + F(a_2, b_3) - F(a_2, b_2)$
   (b) $F(a_3, b_3) - F(a_1, b_3) - F(a_3, b_1) + F(a_1, b_1) - F(a_3, b_2) + F(a_2, b_2) + F(a_3, b_1) - F(a_2, b_1)$
13. (a) $F(x, y) = 0$ if $x < 0$ or $y < 0$; $F(x, y)$ for $x \geqslant 0, y \geqslant 0$ is given in the following table.

x \ y	$0 \leqslant y < 1$	$y \geqslant 1$
$0 \leqslant x < 1$	$\frac{1}{15}$	$\frac{1}{15}$
$1 \leqslant x < 2$	$\frac{4}{15}$	$\frac{6}{15}$
$x \geqslant 2$	$\frac{9}{15}$	1

   (b) $F(x, y) = 0$ if $x < 2$ or $y < -1$; $F(x, y)$ for $x \geqslant 2, y \geqslant -1$ is given in the following table.

x \ y	$-1 \leqslant y < 0$	$0 \leqslant y < 1$	$y \geqslant 1$
$2 \leqslant x < 3$	$\frac{2}{39}$	$\frac{6}{39}$	$\frac{12}{39}$
$x \geqslant 3$	$\frac{8}{39}$	$\frac{21}{39}$	1

   (c) $F(x, y) = 0$ if $x < 1$ or $y < 1$; $F(x, y)$ for $x \geqslant 1, y \geqslant 1$ is given in the following table.

x \ y	$1 \leqslant y < 2$	$y \geqslant 2$
$1 \leqslant x < 2$	$\frac{1}{12}$	$\frac{1}{6}$
$2 \leqslant x < 3$	$\frac{1}{4}$	$\frac{1}{2}$
$x \geqslant 3$	$\frac{1}{2}$	1

14. (a)
$$F(u, v) = \begin{cases} 0, & u < 0 \text{ or } v < 0 \\ u^2v^2, & 0 \leqslant u < 1,\ 0 \leqslant v < 1 \\ u^2, & 0 \leqslant u < 1,\ v \geqslant 1 \\ v^2, & u \geqslant 1,\ 0 \leqslant v < 1 \\ 1, & u \geqslant 1,\ v \geqslant 1 \end{cases}$$
   (c)
$$F(u, v) = \begin{cases} 0, & u < 0 \text{ or } v < 0 \\ u^3(4 - 3u), & 0 \leqslant u < 1,\ v \geqslant u \\ v^2(12u - 6u^2 - 8v + 3v^2), & 0 \leqslant u < 1,\ u \geqslant v \\ v^2(6 - 8v + 3v^2), & u \geqslant 1,\ 0 \leqslant v < 1 \\ 1, & u \geqslant 1,\ v \geqslant 1 \end{cases}$$

17. (a) $\frac{1}{14}$ (b) $\frac{1}{15}$ 18. (a) $\frac{1}{2}$ (c) $\frac{1}{2}$ (e) 2
19. (a) $\frac{5}{12}$ (b) $\frac{1}{4}$ (c) $\frac{2}{3}$ (d) $\frac{1}{2}$
21. (a) 0 (b) $\frac{3}{4}$ (c) $\frac{29}{144}$ (d) $\frac{29}{144}$ (e) $\frac{11}{16}$ (f) $\frac{11}{16}$
23.
$$P(X^2 + Y^2 \leqslant u) = \begin{cases} 0, & u < 0 \\ 1 - e^{-u/2}, & u \geqslant 0 \end{cases}$$
Z has the chi-square distribution with 2 degrees of freedom.

Section 2

1. (a) $p_X(1) = 0.2,\ p_X(2) = 0.8$ (b) $p_Y(-1) = 0.3,\ p_Y(4) = 0.7$
3.
$$F_X(t) = \begin{cases} 0, & t < 0 \\ \frac{1}{2}(1 - \cos t + \sin t), & 0 \leqslant t < \pi/2 \\ 1, & t \geqslant \pi/2 \end{cases}$$
$F_Y(t) = F_X(t)$
5. $p_X(i) = (1-p)^{i-1}p,\ i = 1, 2, \ldots$ $p_Y(j) = (1-q)^{j-1}q,\ j = 1, 2, \ldots$
7. $p(i, j) = \frac{1}{46},\ i = 0, 1, 2, 3,\ j = 0, 1, 2, \ldots, 9;\ i = 4,\ j = 0, 1, \ldots, 5$
$$p_X(i) = \begin{cases} \frac{5}{23}, & i = 0, 1, 2, 3 \\ \frac{3}{23}, & i = 4 \end{cases} \qquad p_Y(j) = \begin{cases} \frac{5}{46}, & j = 0, 1, \ldots, 5 \\ \frac{2}{23}, & j = 6, 7, 8, 9 \end{cases}$$
8. (a) $f_X(t) = 2t,\ 0 < t < 1;$ 0 elsewhere
$f_Y(t) = f_X(t)$
(c) $f_X(x) = 4x^3,\ 0 < x < 1;$ 0 elsewhere
$$f_Y(y) = \begin{cases} \frac{2}{3} - y + \frac{5}{3}y^3, & -1 < y \leqslant 0 \\ \frac{2}{3} - y + y^3/3, & 0 < y \leqslant 1 \\ 0, & \text{elsewhere} \end{cases}$$
(e) $f_X(x) = e^{-x},\ x > 0;$ 0 elsewhere
$f_Y(y) = 1/(1+y)^2,\ y > 0;$ 0 elsewhere
9. $f(x, y) = 2,\ 0 < y < x < 1;$ 0 elsewhere
$f_X(x) = 2x,\ 0 < x < 1;$ 0 elsewhere
$f_Y(y) = 2(1-y),\ 0 < y < 1;$ 0 elsewhere
11. (a) $\frac{3}{8}$ (b) $\frac{11}{72}$ (c) $\frac{7}{9}$
13. (c) An infinite family of joint pdf's which give the same marginal pdf's.

CHAPTER 9

Section 1

3. (a) $P(X = i \mid Y = j) = 1/(n-1),\ i = 1, \ldots, j-1, j+1, \ldots, n$
(b) $P(Y = j \mid X = i) = 1/(n-1),\ j = 1, \ldots, i-1, i+1, \ldots, n$
(c) $\frac{15}{19}$ (d) $\frac{4}{19}$ (e) $\frac{3}{19}$
4. (a) $f_{X|Y}(x \mid y) = 2x,\ 0 < x < 1;$ 0 elsewhere
$f_{Y|X}(y \mid x) = 2y,\ 0 < y < 1;$ 0 elsewhere
(c) $f_{X|Y}(x \mid y) = 2/(1-y),\ y/2 < x < \frac{1}{2};$ 0 elsewhere
$f_{Y|X}(y \mid x) = y/2x^2,\ 0 < y < 2x;$ 0 elsewhere
(e) $f_{X|Y}(x \mid y) = 2(1-x)/(1-y)^2,\ y \leqslant x \leqslant 1;$ 0 elsewhere
$f_{Y|X}(y \mid x) = 2y/x^2,\ 0 \leqslant y \leqslant x;$ 0 elsewhere

5. (a) $\frac{19}{24}$ (b) $\frac{109}{162}$ (c) $\frac{17}{135}$
7. (a) $f(x, y) = 120y(x-y)(1-x)$, $0 \leqslant y \leqslant x < 1$; 0 elsewhere
   (b) $\frac{13}{27}$ (c) $\frac{5}{32}$ (d) $\frac{57}{128}$
9. (a) $\frac{3}{4}$ (b) $\frac{3}{4}$ 11. $(3 \ln 2)/4$ 13. (a) $\frac{1}{2}$ (b) $\frac{1}{2}$

Section 2

3. $a = \frac{1}{3}$ and $b = \frac{1}{6}$; or $a = \frac{1}{6}$ and $b = \frac{1}{3}$
5. Either X is degenerate or Y is degenerate. *Generalization:* If X and Y have a joint probability function which is positive at an odd number of points, then one of the random variables has to be degenerate if the random variables are to be independent.
7. $F_X(x) \cdot F_Y(y)$ where

$$F_X(x) = \begin{cases} 0, & x < 0 \\ x/3, & 0 \leqslant x < 3 \\ 1, & x \geqslant 3 \end{cases} \quad \text{and} \quad F_Y(y) = \begin{cases} 0, & y < 0 \\ y^3/8, & 0 \leqslant y < 2 \\ 1, & y \geqslant 2 \end{cases}$$

9. $P(U=0, V=1) = \frac{1}{9}$, $P(U=0, V=3) = \frac{2}{9}$, $P(U=2, V=1) = \frac{2}{9}$, $P(U=2, V=3) = \frac{4}{9}$
11. Not independent in (a) and (c)
13. For example,

$$f_{UV}(u, v) = \begin{cases} \frac{1}{4}, & -1 < u < 1, \ -1 < v < 1 \\ 0, & \text{elsewhere} \end{cases}$$

15. (a) $\frac{1}{6}$ (b) $\frac{1}{3}$ 16. (a) 0.2839 (b) 0.4782 (c) 0.9563 17. $\frac{1}{18}$

CHAPTER 10

Section 1

1. With replacement: $\mu = 15$; $\sigma^2 = 12$
   Without replacement: $\mu = 15$; $\sigma^2 = 6$
2. (a)

z	0	1	2	3	4	5	7
$P(Z=z)$	$\frac{2}{42}$	$\frac{1}{42}$	$\frac{2}{42}$	$\frac{5}{42}$	$\frac{14}{42}$	$\frac{5}{42}$	$\frac{13}{42}$

(c)

v	1	3	4
$P(V=v)$	$\frac{5}{42}$	$\frac{10}{42}$	$\frac{27}{42}$

(e)

r	−4	−1	0	1	3	4	12
$P(R=r)$	$\frac{5}{42}$	$\frac{2}{42}$	$\frac{5}{42}$	$\frac{2}{42}$	$\frac{10}{42}$	$\frac{5}{42}$	$\frac{13}{42}$

7. 0.8488
9. (a) $$P(Z=i) = \begin{cases} (i+1)/(N+1)^2, & i = 0, 1, 2, \ldots, N \\ [2N-(i-1)]/(N+1)^2, & i = N+1, \ldots, 2N \end{cases}$$
   (b) $P(V=j) = (2j+1)/(N+1)^2$, $j = 0, 1, 2, \ldots, N$
   (c) $P(W=j) = [2(N-j)+1]/(N+1)^2$, $j = 0, 1, \ldots, N$

Section 2

1. (a)
$$F_{X+Y}(u) = \begin{cases} 0, & u<2 \\ \int_1^{u-1}\int_1^{u-x} e^{-(y-1)}\,dy\,dx, & 2\leqslant u<3 \\ \int_1^{2}\int_1^{u-x} e^{-(y-1)}\,dy\,dx, & u\geqslant 3 \end{cases}$$

(b)
$$F_{XY}(u) = \begin{cases} 0, & u<1 \\ \int_1^{u}\int_1^{u/x} e^{-(y-1)}\,dy\,dx, & 1\leqslant u<2 \\ \int_1^{2}\int_1^{u/x} e^{-(y-1)}\,dy\,dx, & u\geqslant 2 \end{cases}$$

(c)
$$F_{X/Y}(u) = \begin{cases} 0, & u<0 \\ \int_1^{2}\int_{x/u}^{\infty} e^{-(y-1)}\,dy\,dx, & 0\leqslant u<1 \\ 1-\int_u^{2}\int_1^{x/u} e^{-(y-1)}\,dy\,dx, & 1\leqslant u<2 \\ 1, & u\geqslant 2 \end{cases}$$

(d)
$$F_{\max(X,Y)}(u) = \begin{cases} 0, & u<1 \\ (u-1)(1-e^{-(u-1)}), & 1\leqslant u<2 \\ 1-e^{-(u-1)}, & u\geqslant 2 \end{cases}$$

3. (a)
$$F_{X+Y}(u) = \begin{cases} 0, & u<0 \\ 4\int_0^{u}\int_0^{u-x} x(1-y)\,dy\,dx, & 0\leqslant u<1 \\ 1-4\int_{u-1}^{1}\int_{u-x}^{1} x(1-y)\,dy\,dx, & 1\leqslant u<2 \\ 1, & u\geqslant 2 \end{cases}$$

(b)
$$F_{XY}(u) = \begin{cases} 0, & u<0 \\ 1-4\int_u^{1}\int_{u/x}^{1} x(1-y)\,dy\,dx, & 0\leqslant u<1 \\ 1, & u\geqslant 1 \end{cases}$$

(c)
$$F_{\max(X,Y)}(u) = \begin{cases} 0, & u<0 \\ u^3(2-u), & 0\leqslant u<1 \\ 1, & u\geqslant 1 \end{cases}$$

5. $k=2$
$$F(u) = \begin{cases} 0, & u<0 \\ 1-\dfrac{1}{1+u}-\dfrac{u}{(1+u)^2}, & u\geqslant 0 \end{cases}$$

7. $f_Z(u)=e^{-u},\ u>0;$ 0 elsewhere

Section 3

1. (a) $f_{X+Y}(u)=\lambda^2 u e^{-\lambda u},\ u>0;$ 0 elsewhere
   (b) $f_{X/Y}(u)=1/(1+u)^2,\ u>0;$ 0 elsewhere
   (c) $f_{\max(X,Y)}(u)=2\lambda e^{-\lambda u}(1-e^{-\lambda u}),\ u>0;$ 0 elsewhere
   (d) $f_{\min(X,Y)}(u)=2\lambda e^{-2\lambda u},\ u>0;$ 0 elsewhere

2. (a)

$$f_{X+Y}(u) = \begin{cases} \dfrac{b-a+u}{(b-a)^2}, & a-b<u<0 \\ \dfrac{b-a-u}{(b-a)^2}, & 0 \leqslant u < b-a \\ 0, & \text{elsewhere} \end{cases}$$

(b)

$$f_{XY}(u) = \begin{cases} \dfrac{2\ln b - \ln(-u)}{(b-a)^2}, & -b^2<u<-ab \\ \dfrac{\ln(-u) - 2\ln a}{(b-a)^2}, & -ab \leqslant u < -a^2 \\ 0, & \text{elsewhere} \end{cases}$$

3. $E(X+Y)=0;\ E(XY) = -\left(\dfrac{b+a}{2}\right)^2$
5. $f_E(u) = 3u(u - 2\ln u - 1),\ 0<u<1;$ 0 elsewhere
7.

$$f(t) = \begin{cases} t(t-4)/24, & 4 \leqslant t < 6 \\ t(8-t)/24, & 6 \leqslant t \leqslant 8 \\ 0, & \text{elsewhere} \end{cases}$$

9. $f_Z(u) = 4u(1-u^2),\ 0<u<1;$ 0 elsewhere
11. $f_{V,W}(s,t) = n(n-1)(s-t)^{n-2},\ 0 \leqslant s \leqslant t \leqslant 1;$ 0 elsewhere
13. At least 14

CHAPTER 11

Section 1

2. (a) 0 (b) 1.5 (c) 0.075 3. (a) 4 (b) 3 (c) $\frac{17}{30}$
5. $E(|X-Y|) = a/3;\ \mathrm{Var}(|X-Y|) = a^2/18$
7. (a) 0 (b) $-\left(\dfrac{b+a}{2}\right)^2$
8. $E(1/Y) = \dfrac{1}{b-a}\ln\dfrac{a}{b};\ E(X/Y) = \dfrac{(b+a)}{2(b-a)}\ln\dfrac{a}{b}$
13. $(3.5)^6$ 15. 70 17. $E(Z) = 11;\ \mathrm{Var}(Z) = 1254$
19. $E(\bar{X}) = \frac{5}{3};\ \mathrm{Var}(\bar{X}) = \frac{5}{1008}$ 21. $1/\sqrt{10}$ 23. $1/\sqrt{3}$
25. (a) $\sigma_X^2\sigma_Y^2 + \mu_X^2\sigma_Y^2 + \mu_Y^2\sigma_X^2$ (b) $\mu_Y\sigma_X^2$ (c) $\dfrac{\mu_Y\sigma_X}{\sqrt{\sigma_X^2\sigma_Y^2 + \mu_X^2\sigma_Y^2 + \mu_Y^2\sigma_X^2}}$
31. $M\left[1 - \left(1 - \dfrac{1}{M}\right)^n\right]$ 33. $\dfrac{-n}{36}$

Section 2

1. (a)

$$E(X \mid Y=i) = \begin{cases} 1.5, & i=2 \\ 2.5, & i=3 \\ 2, & i=5 \end{cases}$$

(b)

$$E(Y \mid X=i) = \begin{cases} \frac{13}{4}, & i=1 \\ \frac{18}{5}, & i=2 \\ \frac{29}{9}, & i=3 \end{cases}$$

3. (a) $-(n+1)/12$
   (b) $[n(n+1) - 2j]/2(n-1),\ j = 1, 2, \ldots, n$
   (c) Same as in (b)

5. (a) $E(X \mid Y=y) = (2y+1)/3,\ 0<y<1$
   (b) $E(Y \mid X=x) = 2x/3,\ 0 \leqslant x \leqslant 1$

7. (a) $3(\frac{3}{4})^n/(n+3)$
   (b) $\frac{27}{1280}$

9. $(1-y_0)^{n/2}/(n+1)$ if n is even; 0 if n is odd

13. 71.54

14. $P(Y=k) = 1/(n+1),\ k=0,1,\ldots,n;\ E(Y) = n/2$

CHAPTER 12

Section 1

1. $M_X(s) = e^{cs};\ E(X^r) = c^r,\ r=0,1,2,\ldots$
3. $M_X(s) = \frac{1}{6}e^{-2s} + \frac{1}{2}e^{3s} + \frac{1}{3}e^{4s};\ E(X) = \frac{5}{2};\ E(X^2) = \frac{21}{2};\ E(X^3) = \frac{67}{2}$
5. $E(X) = r/p;\ E(X^2) = [r^2 + r(1-p)]/p^2$
7. Uniform over $[-2, 4]$
9. (a) $E(X^r) = r!;\ E(Y^r) = r!\sum_{i=0}^{r}\frac{2^{r-i}}{i!}$
   (b) $M_Y(s) = M_{2X+1}(s)$; Y has the same distribution as $2X+1$.
11. (a) $M_X(s) = 1/(1-s^2),\ -1<s<1$
    (b) $E(X) = 0;\ E(Xe^{X/2}) = \frac{16}{9};\ \mathrm{Var}(X) = 2$
13. 0.3158
17. The triangular distribution
19. $M_Z(s) = \dfrac{e^{(\sigma^2 s^2)/2}}{1-s};\ E(Z) = 1;\ E(Z^2) = \sigma^2 + 2;\ E(Z^3) = 3\sigma^2 + 6$
    $E(Z^4) = 3\sigma^4 + 12\sigma^2 + 24$
21. 0.8185
23. $\dfrac{1}{30^5\Gamma(5)}\int_{140}^{\infty} x^4 e^{-x/30}\,dx$

Section 2

1. (a) $e^{\lambda(s-1)}$
   (b) $\dfrac{ps}{1-(1-p)s}$, if $(1-p)s<1$
   (c) $\left[\dfrac{ps}{1-(1-p)s}\right]^r$, if $(1-p)s<1$
   (d) $\dfrac{s^b - s^a}{(b-a)\ln s}$, if $s \neq 1$; 1 if $s=1$
   (e) $\frac{1}{5}s^{-5} + \frac{3}{10}s^{-3} + \frac{1}{10}s^{-2} + \frac{2}{5},\ s \neq 0$
7. $\lambda;\ \lambda^2;\ \lambda^3$; in general, $E(X(X-1)\ldots(X-n+1)) = \lambda^n$
9. $p(r) = \frac{3}{8}[(\frac{3}{4})^{r+1} - (\frac{1}{4})^{r+1}],\ r=0,1,2,\ldots$
11. $p(r) = 2(\frac{1}{3})^{r+1} - (\frac{1}{2})^{r-1} + 2(\frac{2}{3})^{r+1},\ r=1,2,\ldots$

CHAPTER 13

Section 1

6.

h	1.2	2	3
Chebychev's bounds	0.694	0.25	0.111
Exact probabilities	0.37	0.029	0

7.

h	1.2	2	3
Chebychev's bounds	0.694	0.25	0.111
Exact probabilities	0.111	0.050	0.018

9. Chebychev's lower bound is 0.75; the exact value is 0.9725.
11. $\sqrt{3}$

Section 2

1. X_n converges in distribution to X, where–

(a)
$$F_X(u) = \begin{cases} 0, & u < 0 \\ \frac{2}{3}, & 0 \leqslant u < 1 \\ 1, & u \geqslant 1 \end{cases}$$

(b)
$$F_X(u) = \begin{cases} 0, & u < 0 \\ 1, & u \geqslant 0 \end{cases}$$

3.

k	0	1	2	3	4
Poisson approximation	0.2231	0.3347	0.2510	0.1255	0.0471

5. (a) 0.16 (b) 0.575 (c) 0.692 (d) 0.271
7. 0.9544 9. 0.0526 11. 0.7422 13. 0.9525 15. 0.9544
17. (a) 0.6908 (b) 0.9713 19. 0.8351 21. 0.0228

Tables

Table A. Exponential Functions

x	e^x	e^{-x}	x	e^x	e^{-x}
0.00	1.0000	1.0000	2.5	12.182	0.0821
0.05	1.0513	0.9512	2.6	13.464	0.0743
0.10	1.1052	0.9048	2.7	14.880	0.0672
0.15	1.1618	0.8607	2.8	16.445	0.0608
0.20	1.2214	0.8187	2.9	18.174	0.0550
0.25	1.2840	0.7788	3.0	20.086	0.0498
0.30	1.3499	0.7408	3.1	22.198	0.0450
0.35	1.4191	0.7047	3.2	24.533	0.0408
0.40	1.4918	0.6703	3.3	27.113	0.0369
0.45	1.5683	0.6376	3.4	29.964	0.0334
0.50	1.6487	0.6065	3.5	33.115	0.0302
0.55	1.7333	0.5769	3.6	36.598	0.0273
0.60	1.8221	0.5488	3.7	40.447	0.0247
0.65	1.9155	0.5220	3.8	44.701	0.0224
0.70	2.0138	0.4966	3.9	49.402	0.0202
0.75	2.1170	0.4724	4.0	54.598	0.0183
0.80	2.2255	0.4493	4.1	60.340	0.0166
0.85	2.3396	0.4274	4.2	66.686	0.0150
0.90	2.4596	0.4066	4.3	73.700	0.0136
0.95	2.5857	0.3867	4.4	81.451	0.0123
1.0	2.7183	0.3679	4.5	90.017	0.0111
1.1	3.0042	0.3329	4.6	99.484	0.0101
1.2	3.3201	0.3012	4.7	109.95	0.0091
1.3	3.6693	0.2725	4.8	121.51	0.0082
1.4	4.0552	0.2466	4.9	134.29	0.0074
1.5	4.4817	0.2231	5	148.41	0.0067
1.6	4.9530	0.2019	6	403.43	0.0025
1.7	5.4739	0.1827	7	1096.6	0.0009
1.8	6.0496	0.1653	8	2981.0	0.0003
1.9	6.6859	0.1496	9	8103.1	0.0001
2.0	7.3891	0.1353	10	22026	0.00005
2.1	8.1662	0.1225			
2.2	9.0250	0.1108			
2.3	9.9742	0.1003			
2.4	11.023	0.0907			

Table B. Binomial Probabilities

A table of $\binom{n}{x} p^x (1-p)^{n-x}$ for $n = 2, \ldots, 10$

n	x	.01	.05	.10	.15	.20	.25	.30	$\frac{1}{3}$	.35	.40	.45	.49	.50
2	0	.9801	.9025	.8100	.7225	.6400	.5625	.4900	.4444	.4225	.3600	.3025	.2601	.2500
	1	.0198	.0950	.1800	.2550	.3200	.3750	.4200	.4444	.4550	.4800	.4950	.4998	.5000
	2	.0001	.0025	.0100	.0225	.0400	.0625	.0900	.1111	.1225	.1600	.2025	.2401	.2500
3	0	.9703	.8574	.7290	.6141	.5120	.4219	.3430	.2963	.2746	.2160	.1664	.1327	.1250
	1	.0294	.1354	.2430	.3251	.3840	.4219	.4410	.4444	.4436	.4320	.4084	.3823	.3750
	2	.0003	.0071	.0270	.0574	.0960	.1406	.1890	.2222	.2389	.2880	.3341	.3674	.3750
	3	.0000	.0001	.0010	.0034	.0080	.0156	.0270	.0370	.0429	.0640	.0911	.1176	.1250
4	0	.9606	.8145	.6561	.5220	.4096	.3164	.2401	.1975	.1785	.1296	.0915	.0677	.0625
	1	.0388	.1715	.2916	.3685	.4096	.4219	.4116	.3951	.3845	.3456	.2995	.2600	.2500
	2	.0006	.0135	.0486	.0975	.1536	.2109	.2646	.2963	.3105	.3456	.3675	.3747	.3750
	3	.0000	.0005	.0036	.0115	.0256	.0469	.0756	.0988	.1115	.1536	.2005	.2400	.2500
	4	.0000	.0000	.0001	.0005	.0016	.0039	.0081	.0123	.0150	.0256	.0410	.0576	.0625
5	0	.9510	.7738	.5905	.4437	.3277	.2373	.1681	.1317	.1160	.0778	.0503	.0345	.0312
	1	.0480	.2036	.3280	.3915	.4096	.3955	.3602	.3292	.3124	.2592	.2059	.1657	.1562
	2	.0010	.0214	.0729	.1382	.2048	.2637	.3087	.3292	.3364	.3456	.3369	.3185	.3125
	3	.0000	.0011	.0081	.0244	.0512	.0879	.1323	.1646	.1811	.2304	.2757	.3060	.3125
	4	.0000	.0000	.0004	.0022	.0064	.0146	.0284	.0412	.0488	.0768	.1128	.1470	.1562
	5	.0000	.0000	.0000	.0001	.0003	.0010	.0024	.0041	.0053	.0102	.0185	.0283	.0312
6	0	.9415	.7351	.5314	.3771	.2621	.1780	.1176	.0878	.0754	.0467	.0277	.0176	.0156
	1	.0571	.2321	.3543	.3993	.3932	.3560	.3025	.2634	.2437	.1866	.1359	.1014	.0938
	2	.0014	.0305	.0984	.1762	.2458	.2966	.3241	.3292	.3280	.3110	.2780	.2437	.2344
	3	.0000	.0021	.0146	.0415	.0819	.1318	.1852	.2195	.2355	.2765	.3032	.3121	.3125
	4	.0000	.0001	.0012	.0055	.0154	.0330	.0595	.0823	.0951	.1382	.1861	.2249	.2344
	5	.0000	.0000	.0001	.0004	.0015	.0044	.0102	.0165	.0205	.0369	.0609	.0864	.0938
	6	.0000	.0000	.0000	.0000	.0001	.0002	.0007	.0014	.0018	.0041	.0083	.0139	.0156

7	0	.9321	.6983	.4783	.3206	.2097	.1335	.0824	.0585	.0490	.0280	.0152	.0090	.0078
	1	.0659	.2573	.3720	.3960	.3670	.3115	.2471	.2048	.1848	.1306	.0872	.0603	.0547
	2	.0020	.0406	.1240	.2097	.2753	.3115	.3177	.3073	.2985	.2613	.2140	.1740	.1641
	3	.0000	.0036	.0230	.0617	.1147	.1730	.2269	.2561	.2679	.2903	.2918	.2786	.2734
	4	.0000	.0002	.0026	.0109	.0287	.0577	.0972	.1280	.1442	.1935	.2388	.2676	.2734
	5	.0000	.0000	.0002	.0012	.0043	.0115	.0250	.0384	.0466	.0774	.1172	.1543	.1641
	6	.0000	.0000	.0000	.0001	.0004	.0013	.0036	.0064	.0084	.0172	.0320	.0494	.0547
	7	.0000	.0000	.0000	.0000	.0000	.0001	.0002	.0005	.0006	.0016	.0037	.0068	.0078
8	0	.9227	.6634	.4305	.2725	.1678	.1001	.0576	.0390	.0319	.0168	.0084	.0046	.0039
	1	.0746	.2793	.3826	.3847	.3355	.2670	.1977	.1561	.1373	.0896	.0548	.0352	.0312
	2	.0026	.0515	.1488	.2376	.2936	.3115	.2965	.2731	.2587	.2090	.1569	.1183	.1094
	3	.0001	.0054	.0331	.0839	.1468	.2076	.2541	.2731	.2786	.2787	.2568	.2273	.2188
	4	.0000	.0004	.0046	.0185	.0459	.0865	.1361	.1707	.1875	.2322	.2627	.2730	.2734
	5	.0000	.0000	.0004	.0026	.0092	.0231	.0467	.0683	.0808	.1239	.1719	.2098	.2188
	6	.0000	.0000	.0000	.0002	.0011	.0038	.0100	.0171	.0217	.0413	.0703	.1008	.1094
	7	.0000	.0000	.0000	.0000	.0001	.0004	.0012	.0024	.0033	.0079	.0164	.0277	.0312
	8	.0000	.0000	.0000	.0000	.0000	.0000	.0001	.0002	.0002	.0007	.0017	.0033	.0039
9	0	.9135	.6302	.3874	.2316	.1342	.0751	.0404	.0260	.0207	.0101	.0046	.0023	.0020
	1	.0830	.2985	.3874	.3679	.3020	.2253	.1556	.1171	.1004	.0605	.0339	.0202	.0176
	2	.0034	.0629	.1722	.2597	.3020	.3003	.2668	.2341	.2162	.1612	.1110	.0776	.0703
	3	.0001	.0077	.0446	.1069	.1762	.2336	.2668	.2731	.2716	.2508	.2119	.1739	.1641
	4	.0000	.0006	.0074	.0283	.0661	.1168	.1715	.2048	.2194	.2508	.2600	.2506	.2461
	5	.0000	.0000	.0008	.0050	.0165	.0389	.0735	.1024	.1181	.1672	.2128	.2408	.2461
	6	.0000	.0000	.0001	.0006	.0028	.0087	.0210	.0341	.0424	.0743	.1160	.1542	.1641
	7	.0000	.0000	.0000	.0000	.0003	.0012	.0039	.0073	.0098	.0212	.0407	.0635	.0703
	8	.0000	.0000	.0000	.0000	.0000	.0001	.0004	.0009	.0013	.0035	.0083	.0153	.0176
	9	.0000	.0000	.0000	.0000	.0000	.0000	.0000	.0001	.0001	.0003	.0008	.0016	.0020
10	0	.9044	.5987	.3487	.1969	.1074	.0563	.0282	.0173	.0135	.0060	.0025	.0012	.0010
	1	.0914	.3151	.3874	.3474	.2684	.1877	.1211	.0867	.0725	.0404	.0207	.0114	.0098
	2	.0042	.0746	.1937	.2759	.3020	.2816	.2335	.1951	.1757	.1209	.0763	.0495	.0439
	3	.0001	.0105	.0574	.1298	.2013	.2503	.2668	.2601	.2522	.2150	.1665	.1267	.1172
	4	.0000	.0010	.0112	.0401	.0881	.1460	.2001	.2276	.2377	.2508	.2384	.2130	.2051
	5	.0000	.0001	.0015	.0085	.0264	.0584	.1029	.1366	.1536	.2007	.2340	.2456	.2461
	6	.0000	.0000	.0001	.0012	.0055	.0162	.0368	.0569	.0689	.1115	.1596	.1966	.2051
	7	.0000	.0000	.0000	.0001	.0008	.0031	.0090	.0163	.0212	.0425	.0746	.1080	.1172
	8	.0000	.0000	.0000	.0000	.0001	.0004	.0014	.0030	.0043	.0106	.0229	.0389	.0439
	9	.0000	.0000	.0000	.0000	.0000	.0000	.0001	.0003	.0005	.0016	.0042	.0083	.0098
	10	.0000	.0000	.0000	.0000	.0000	.0000	.0000	.0000	.0000	.0001	.0003	.0008	.0010

Table C. Cumulative Normal Distribution

$$\Phi(x) = \int_{-\infty}^{x} \frac{1}{\sqrt{2\pi}} e^{-t^2/2}\, dt$$

x	.00	.01	.02	.03	.04	.05	.06	.07	.08	.09
.0	.5000	.5040	.5080	.5120	.5160	.5199	.5239	.5279	.5319	.5359
.1	.5398	.5438	.5478	.5517	.5557	.5596	.5636	.5675	.5714	.5753
.2	.5793	.5832	.5871	.5910	.5948	.5987	.6026	.6064	.6103	.6141
.3	.6179	.6217	.6255	.6293	.6331	.6368	.6406	.6443	.6480	.6517
.4	.6554	.6591	.6628	.6664	.6700	.6736	.6772	.6808	.6844	.6879
.5	.6915	.6950	.6985	.7019	.7054	.7088	.7123	.7157	.7190	.7224
.6	.7257	.7291	.7324	.7357	.7389	.7422	.7454	.7486	.7517	.7549
.7	.7580	.7611	.7642	.7673	.7704	.7734	.7764	.7794	.7823	.7852
.8	.7881	.7910	.7939	.7967	.7995	.8023	.8051	.8078	.8106	.8133
.9	.8159	.8186	.8212	.8238	.8264	.8289	.8315	.8340	.8365	.8389
1.0	.8413	.8438	.8461	.8485	.8503	.8531	.8554	.8577	.8599	.8621
1.1	.8643	.8665	.8686	.8708	.8729	.8749	.8770	.8790	.8810	.8830
1.2	.8849	.8869	.8888	.8907	.8925	.8944	.8962	.8980	.8997	.9015
1.3	.9032	.9049	.9066	.9082	.9099	.9115	.9131	.9147	.9162	.9177
1.4	.9192	.9207	.9222	.9236	.9251	.9265	.9279	.9292	.9306	.9319
1.5	.9332	.9345	.9357	.9370	.9382	.9394	.9406	.9418	.9429	.9441
1.6	.9452	.9463	.9474	.9484	.9495	.9505	.9515	.9525	.9535	.9545
1.7	.9554	.9564	.9573	.9582	.9591	.9599	.9608	.9616	.9625	.9633
1.8	.9641	.9649	.9656	.9664	.9671	.9678	.9686	.9693	.9699	.9706
1.9	.9713	.9719	.9726	.9732	.9738	.9744	.9750	.9756	.9761	.9767
2.0	.9772	.9778	.9783	.9788	.9793	.9798	.9803	.9808	.9812	.9817
2.1	.9821	.9826	.9830	.9834	.9838	.9842	.9846	.9850	.9854	.9857
2.2	.9861	.9864	.9868	.9871	.9875	.9878	.9881	.9884	.9887	.9890
2.3	.9893	.9896	.9898	.9901	.9904	.9906	.9909	.9911	.9913	.9916
2.4	.9918	.9920	.9922	.9925	.9927	.9929	.9931	.9932	.9934	.9936
2.5	.9938	.9940	.9941	.9943	.9945	.9946	.9948	.9949	.9951	.9952
2.6	.9953	.9955	.9956	.9957	.9959	.9960	.9961	.9962	.9963	.9964
2.7	.9965	.9966	.9967	.9968	.9969	.9970	.9971	.9972	.9973	.9974
2.8	.9974	.9975	.9976	.9977	.9977	.9978	.9979	.9979	.9980	.9981
2.9	.9981	.9982	.9982	.9983	.9984	.9984	.9985	.9985	.9986	.9986
3.0	.9987	.9987	.9987	.9988	.9988	.9989	.9989	.9989	.9990	.9990
3.1	.9990	.9991	.9991	.9991	.9992	.9992	.9992	.9992	.9993	.9993
3.2	.9993	.9993	.9994	.9994	.9994	.9994	.9994	.9995	.9995	.9995
3.3	.9995	.9995	.9995	.9996	.9996	.9996	.9996	.9996	.9996	.9997
3.4	.9997	.9997	.9997	.9997	.9997	.9997	.9997	.9997	.9997	.9998

x	1.282	1.645	1.960	2.326	2.576	3.090	3.291	3.891	4.417
$\Phi(x)$	.90	.95	.975	.99	.995	.999	.9995	.99995	.999995

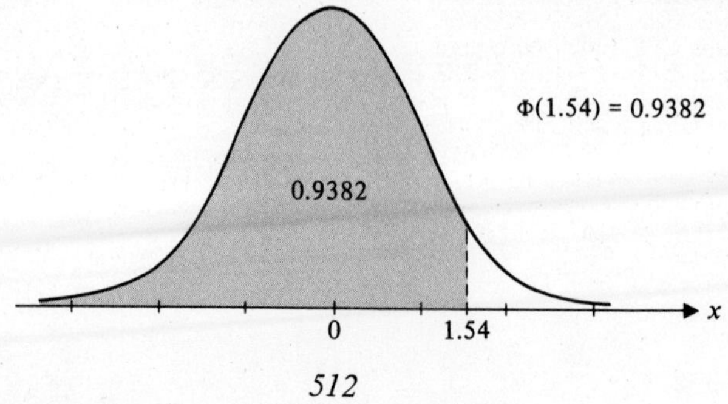

Index

A

Absolutely continuous distributions, 175-92
Absolutely continuous joint distributions, 280-8
Absolutely continuous random variables, 138-46
A posterior probabilities, 80-1
Arrangement, 46
Associativity, 7, 9
Asymptotically, 479, 481
At random, 430
Axiomatic theory, 4
Axiom of choice, 41

B

Bayes' rule, 80-3
Bernoulli distribution, 157-8; expectation of, 243; moment generating function for, 436, 445; properties of, 255; variance of, 243
Bernoulli law of large numbers, 468
Bernoulli random variable, 158
Bernoulli trial, 157
Binomial coefficients, 51
Binomial distribution, 158-64, 409-10; approximation by normal distribution, 483-4; approximation by Poisson distribution, 474-6, 484; expectation of, 243-4; moment generating function for, 437, 442, 445, 448; negative, 168-9, 255, 355, 407; properties of, 255; reproductive property of, 448; for several random variables, 353-4; variance of, 244
Binomial probabilities, 99
Binomial random variable, 158
Birthday problem, 63
Bivariate case, 261
Borel field, 25-7
Borel law of large numbers, 468-70
Borel-measurable functions, 331
Buffon's needle problem, 430-1

C

Cartesian product, 11-3
Cauchy distribution, 190-1, 381; point of symmetry of, 242
Central limit theorem, 460, 478-81
Characteristic function, 435
Chebyshev's inequality, 460-5
Chebyshev's theorem, 465
Chi-square distribution, 189-90; moment generating function for, 460, 446-7; reproductive property of, 449-50
Coefficient of linear correlation, 416
Combinations, 46, 50-1
Commutativity, 7, 9
Complement, 9
Composite experiment, 95
Composite function, 196
Conditional distribution: given an event of positive probability, 311-3; given a specific value, 313-25; for more than two random variables, 346
Conditional distribution function, 317
Conditional expectation, 236-9; of function of two random variables, 422-5
Conditional probability, 67-75

Conditional probability density function, 317-8
Conditional probability function, 314
Conditional variance, 424
Conditioning: expected value by, 425-8; probabilities by, 428-30
Continuity correction, 483-4
Continuity property, 38
Continuity theorem, 477-8
Convergence in distribution, 472-8
Convolution, 358
Correlation coefficient, 400; bounds on, 413-8
Countable additivity, 29
Counting principle, 47
Covariance, 399

D

Degenerate random variable, 113
Degrees of freedom, 383
DeMoivre-Laplace theorem, 483
DeMorgan's law, 21-2
Discrete joint distributions, 272-80
Discrete random variables, 130-8
Distribution function: conditional, 317; definition, 115-9; joint, 343-4; properties of, 119-28, 343-4
Distributions: absolutely continuous, 175-92; Bernoulli, 157-8, 243, 255, 436; binomial, 158-64, 243-4, 255, 353-4, 409, 437, 442, 448, 474-6, 483-4; Cauchy, 190-1, 381; Chi-square, 189-90, 440, 449-50; conditional, 311-25, 346; convergence in, 472-8; discrete, 157-72, 255; exponential, 188-9, 252-4, 256, 439; of extremes, 372; factorial moment generating functions for, 452-6; gamma, 185-8, 256, 378, 439; Gaussian, 179; geometric, 166-7, 245, 255, 355, 437; hypergeometric, 164-6, 255, 278, 410-1; joint, 261-88; Laplace, 192, 379; logistic, 155; marginal, 292-3, 298, 344; of maximum, 366-9; of minimum, 369-72; mixed, 146-9; moment generating functions for, 435-50; moments of, 232-3; negative binomial, 168-9, 255, 355, 437; normal, 179-85, 249-52, 256, 449, 483-5; Pascal, 168, 255; Poisson, 169-72, 246-8, 255, 354-5, 438, 442, 448, 474-6, 485; probability, 151; of product of random variable functions, 360-2; of quotient of random variable functions, 362-6; functions of random variables, 198-217; of range, 383-6; rectangular, 175; singular, 149-51; standard normal, 438-9; Student's *t*, 381-3; of sum of random variables, 357-60; symmetric, 242-3; triangular, 155, 375; trinomial, 279, 299, 315-6, 393-4, 401-2, 412; truncated, 236; uniform, 175-8, 248-9, 256, 438; uniform bivariate, 305. *See also* Probability
Distributive law, 10

E

Equiprobable, 43
Events, 17, 20; dependent, 87; impossible, 17; independent, 86-95; mutually exclusive, 88; mutually independent, 92; random, 24; simple, 17; sure, 17
Expectation: basic properties for functions of two random variables, 394-8; of Bernoulli distribution, 243; of binomial distribution, 243-4; conditional, 236-9, 422-5; by conditioning, 425-8; of exponential distribution, 252-3; of geometric distribution, 245; geometric interpretation of, 226-8; of linear combination, 395; of normal distribution, 249-50; physical interpretation of, 222; of Poisson distribution, 246; of product of random variable functions, 396-8; properties of, 231-3; of functions of random variables, 228-31; of random variables, 221-8; of several random variable functions, 389-418; of symmetric distributions, 242-3; of uniform distributions, 248-9
Expected value. *See* Expectation
Exponential distribution, 188-9: expectation of, 252-3; lack of memory property of, 254; moment generating function for, 439; properties of, 256; variance of, 252-3
Extremes, 372

F

Factorial, 49
Factorial moment, 232
Factorial moment generating function: definition of, 452-3; relationship to moment generating functions, 453
Functions: composite, 196; of random variables, 195-217.

G

Gamma distribution, 185-8, 378; moment generating function for, 439; properties of, 256
Gamma function, 185
Gaussian distribution, 179
General multiplication rule, 69
Generating function: factorial moment, 452-6; moment, 436-50; probability, 454-5
Geometric distribution, 166-7; expectation of, 245; moment generating function for, 437, 445; properties of, 255; for several random variables, 355; variance of, 245
Geometric random variable, 167

H

Hypergeometric distribution, 164-6, 410-1; generalization of, 278; properties of, 255
Hypergeometric probabilities, 55-63

I

Impossible event, 17, 30
Independent events, 86-95
Independent random variables, 311, 327-40; absolutely continuous case, 335-40; discrete case, 331-5
Independent trials, 95-7
Indicator random variables, 113, 408-12
Integer-valued, 454
Invariance of covariance under translation, 413

J

Joint distribution function for more than two random variables, 343-4

Joint distributions: absolutely continuous, 280-8; classification of, 272; definition of, 263-4; discrete, 272-80; properties of, 266-9; singular, 307
Joint probability density function, 280
Joint probability function, 272, 344
Joint probability mass function, 272

K

Khintchine's theorem, 466

L

Lack of memory property, 254
Laplace-DeMoivre theorem 483
Laplace distribution, 192, 379
Laws of large numbers: applied to Bernoulli sequence, 468; Borel law and, 468-70; Chebyshev's inequality and, 460-5; Chebyshev's theorem and, 465; Khintchine's theorem and, 466; Markov's theorem and, 465-6; strong, 467-8; weak, 465-7
Limits of monotone sequences, 13, 38-40
Linear combination: expectation of, 395; variance of, 405-7
Linear correlation, 416
Linear property, 395
Logistic distribution, 155

M

Maclaurin's series, 441
Marginal distributions, 344; absolutely continuous case, 300-7; definition of, 292-3, 298; discrete case, 296-9
Markov inequality, 470
Markov's theorem, 465-6
Matching problem, 62-3, 411-2
Maximum: distribution of, 366-9
Method of Jacobians, 351
Method of truncation, 466
Minimum: distribution of, 369-72
Mixed distributions, 146-9
Moment generating function: for absolutely continuous distributions, 438-40; definition of, 436; for discrete distributions, 436-8; factorial, 452-6; relationship to factorial moment generating functions, 453
Moments, 232-3
Moments of distribution, 440-3
Monotone property, 31
Monotone sequences, 13, 38-40
Monotonicity, 266-7, 344
Monte Carlo simulation, 487-8
Most probable number, 163
Multivariate case, 261
Mutual independence, 346
Mutually exclusive, 88
Mutually independent, 92

N

Negative binomial distribution, 168-9, 407; moment generating function for, 437, 445
Normal distribution, 179-85; approximation for binomial distribution, 483-4; approximation for Poisson distribution, 485; expectation of, 249-50; properties of, 256; reproductive property of, 449; standard bivariate, 303-4, 323, 403-4; variance of, 250

O

Ordered pair, 11
Outcome. *See* Events

P

Parameters, 158
Pascal distribution, 168; properties of, 255
Pascal's triangle, 63
pdf, 139
Permutations, 46, 48-9
Point of symmetry, 242-3
Poisson distribution, 169-72; approximation by normal distribution, 485; approximation to binomial distribution, 474-6, 484; expectation of, 246; moment generating function for, 438, 442, 445, 448; properties of, 255; reproductive property of, 448; for several random variables, 354-5; variance of, 246
Poisson random variable, 170
Power sets, 10-1
Probability: a posterior, 80-1; binomial, 99; compound, 69; conditional, 67-75; by conditioning, 428-30; definition of, 29; generating functions of, 454-5; hypergeometric, 55-63; total, 78-84. *See also* Distributions
Probability density function, 139; conditional, 317-8; joint, 280
Probability distribution, 151; center of gravity of, 222
Probability function, 131-8; conditional, 314; joint, 272, 344. *See also* Distribution function
Probability generating function, 454-5
Probability mass function, 131-8; joint, 272
Probability measure, 29, 38
Probability space, 29
Probability theory: application of, 1-2; axioms of, 29; classical, 2, 4; for finite sample spaces, 42-5; history of, 2, 4; sampling and, 45-63
Product: expectation of, 396-8

R

Random, 430
Random events, 2, 24
Random sample, 52
Random variables, 107-14; absolutely continuous, 138-46; Bernoulli, 158; binomial, 158; correlation coefficient of, 400; covariance of, 399; discrete, 130-8; expected value of, 221-8; geometric, 167; independent, 311, 327-40; indicator, 113, 408-12; integer-valued, 454-5; mean of, 221; mixed distributions, 146-9; Poisson, 170; several, 351-86; singular distributions, 149-51; standardized 413; uncorrelated, 400; variance of, 233-6. *See also* Random variables: functions of
Random variables: functions of: conditional expectation of, 422-5; continuous, 201-17; discrete, 198-201; distribution of, 198-217; expected value of, 228-31, 389-418; mathematical formulation of, 195-7; product of, 360-2; quotient of, 362-6; for several random variables, 351-86; sum of, 357-60. *See also* Random variables
Random vector, 261-2

Range, 383-6
Raw moments, 232
Rectangular distribution, 175
Regression, 422
Regression curve, 422
Relative frequency, 2
Replacement, 45
Reproductive property, 448-50

S

Sample mean, 396
Sample points, 15
Sample space, 15-22; finite, 17, 42-5; infinite, 17
Sampling techniques, 45-63; without replacement, 45-6, 48-51; with replacement, 45-6, 52-5
Sequences, 13; contracting, 13; expanding, 13; monotone, 13
Set function, 29
Sets: Borel, 26; Cartesian product of, 11-3; complement of, 9; definition of, 4-6; De Morgan's law for, 21-2; difference between, 9; empty, 5; equality of, 6; events as, 17, 20; intersection of, 6-7; monotone sequences of, 13; nonempty, 5; notation, 4-5, 20; operations on, 6-9; pairwise disjoint, 7; power, 10-1; sample space, 15; sigma field of, 24; theory of, 4-13; union of, 8-9; universal, 5-6; Venn diagrams for, 9-10; void, 5
Sigma fields: Borel field, 25-7; definition of, 24-5
Singular distributions, 139, 149-51; joint, 307
Standard bivariate normal distribution, 303-4, 323, 403-4
Standard deviation, 233
Standardized random variable, 413
Standard normal distribution, 180; moment generating function for, 438-9
Statistical estimation, 225
Stochastically independent, 87, 327
Strictly increasing, 215-6
Strong law of large numbers, 467-8
Student's t-distribution, 381-3
Subsets, 6; proper, 6
Sure event, 17
Symmetric distributions, 242-3

T

Theorem of compound probabilities, 69
Time interval, 169
Total probability theorem, 78-84
Tree diagram, 47-8
Trial, 95; Bernoulli, 157; independent, 95-7
Triangular distribution, 155, 375
Trinomial distribution, 279, 299, 315-6, 393-4, 401-2, 412
Truncated distribution, 236
Truncation, 466

U

Uniform bivariate distribution, 305
Uniform distribution, 175-8; bivariate, 305; expectation of, 248-9; moment generating function for, 438, 447; properties of, 256; variance of, 249
Uniqueness theorem, 445-7
Univariate case, 261

V

Variance: of Bernoulli distribution, 243; of binomial distribution, 244; conditional, 424, of exponential distribution, 252-3; of geometric distribution, 245; of linear combination, 405-7; of normal distribution, 250; of Poisson distribution, 246; of random variable, 233-6; of uniform distribution, 249
Vector-valued function, 262
Venn diagrams, 9-10

W

Weak law of large numbers, 465-7
Weibull density function, 142